Estates Gazette Law Reports

1978

Volume 1

Series edited by
Barry Denyer-Green LLM PhD FRICS
of the Middle Temple, barrister

Cases edited by
J Muir Watt OBE MA
of the Inner Temple, barrister

Audrey Boyle BA
Publisher's editor

Estates Gazette

Routledge
Taylor & Francis Group
LONDON AND NEW YORK

EG Books is an imprint of Routledge
2 Park Square, Milton Park, Abingdon, Oxon OX14 4RN
711 Third Avenue, New York, NY 10017, USA
Routledge is an imprint of the Taylor & Francis Group, an informa business

ISBN for complete set of 2 volumes: 0 7282 0224 7
ISBN for this volume 0 7282 0225 5
ISSN 0951-9289

For information on all EG Books publications
visit our website www.routledge.com

CONTENTS

TABLE OF CASES

Titles of cases shown in bold type
Names of parties reversed shown in ordinary type

INDEX OF SUBJECT-MATTER

COMPENSATION FOR ACQUISITION OF LAND — *continued*
Disturbance — *continued*

Off-licence premises
Tied tenancy — Claim for loss to brewer claimants based on average trade for period of years — Held, YP of 4.5 appropriate applied to adjusted annual income.
***Ind Coope (London) Ltd* v *Enfield London Borough (LT)*, 163**

Removal expenses
Claim for removal expenses, survey fees on abortive purchase and property purchased and travelling expenses — Held, tribunal allows claim in full.
***Succamore* v *London Borough of Newham (LT)*, 161**

Valuation
Land for access
Claimant contended subject land had value as access to adjoining development — Claim based on cost saving over alternative access route — Award based on 25% of cost saving of alternative access.
***JD Britton & Son Ltd* v *London Borough of Hillingdon (LT)*, 159**

Minus value
Listed building in advance state of decay acquired — Mathematically building did not have positive value — Held, hypothetical purchaser would be speculator who would make a small bid.
***Taylor* v *Cheltenham Borough Council (LT)*, 175**

DAMAGES
Measure of damages — See **Negligence** (Damages: measure and quantification of)
Nuisance — See **Nuisance** (Damages)
Wrongful eviction — See **Landlord and tenant** (Wrongful eviction)

DISTRESS — See **Civil procedure** (Distress)

ESTATE AGENTS
Commission
Introduction of purchaser
Estate agent in secret agreement with potential purchaser advised vendors to accept purchaser's bid — Advice not taken and property sold to another — Held, estate agent not entitled to commission — Had not caused sale to happen — By making of secret agreement disentitled to commission.
***Henry Smith & Son* v *Muskett (Sheraton Securities (International) Ltd, third party) (QB)*, 13**

Negotiations for purchase
Contract to find suitable leasehold property with fees to be paid on scale laid down by RICS — Not contract to negotiate terms with lessor — Estate agent's introduction to premises caused property to be leased — Held, RICS scale applicable only to contracts for seeking and negotiating — *Quantum meruit* damages — Task carried out not onerous — 2.5% of first year's rent.
***Reiff Diner & Co* v *Catalytic International Incorporated (QB)*, 16**

EVIDENCE
Expert
Meaning of word
Expert evidence on meaning of word "supermarket" in covenant in lease — Held, court to construe word as it would have been understood at date of lease — Meaning not established by date of appeal — Its use was somewhere intermediate between technical term and ordinary English word — Evidence of expert admissible.
***Calabar (Woolwich) Ltd* v *Tesco Stores Ltd (CA)*, 113**

Extrinsic
Construction of agreement
Application for retrial on basis of judge's interventions preventing admission of evidence of surrounding circumstances — Rule that such evidence admissible when established that written agreement not real statement of parties' intention — Held, such evidence had in fact been given and unlikely that could be established that case fell within rule — Application refused.
***Somma* v *Hazelhurst (CA)*, 69**

Planning condition
Question as to what evidence, other than planning permission, is admissible to establish reasons for planning condition — Held, where reasons given, under directory requirement of statute, no extrinsic evidence unless their *bona fides* in dispute or alleged that unfounded — Where no reasons are given, relevant extrinsic evidence is admissible, including oral evidence.
***RKT Investments Ltd* v *Hackney London Borough (QB)*, 146**

FAIR RENT — See **Landlord and tenant** (Rent Acts)

FEES — See **Estate agents** (Commission)

FORFEITURE — See **Landlord and tenant** (Forfeiture)

HOTEL — See **Rates and rating** (Valuation)

HOUSING
Repairs
"Reasonable expense"
Meaning of requirement, under s39(1) Housing Act 1957, that in determining whether house can be rendered fit for human habitation at reasonable expense "regard shall be had to ... the value which it is estimated the house will have when the works are completed" — Held, "value" means open market value — "Regard shall be had" implies that other factors can be taken into consideration on top of those expressly stated in s39(1).
***Inworth Property Co Ltd* v *London Borough of Southwark (CA)*, 24**

Notice served under s9 Housing Act 1957 — Appeal on ground that premises in question could not be rendered fit for human habitation at reasonable expense — Held, county court failed to treat, as parties had agreed, estimate of cost of repairs as a minimum — Failed to take proper account that interest on money landlord borrowed for repairs would exceed income from property, during and after repairs.
***Ellis Copp & Co* v *Richmond upon Thames London Borough (CA)*, 21**

INTEREST ON COMPENSATION — See **Compensation for acquisition of land** (Disturbance — Bank interest)

JUDICIAL REVIEW
Natural justice
Public inquiry
Refusal of inspector to allow objector to cross-examine witnesses at public inquiry — Held, breach of natural justice where person considering himself to be directly affected by proposal denied opportunity to cross-examine evidence upon which his case turns — Right to stop cross-examination by objectors if "fishing expedition", irrelevant, repetitious or when objection is not to proposed development but any development.
***Nicholson* v *Secretary of State for Energy (QB)*, 111**

LANDLORD AND TENANT
Agency
Undisclosed principal
Lease to two named tenants — Mother of one, occupying property, claims to be protected by Rent Act, as undisclosed principal of both and consequently tenant — Held, no principle that lessees can contract on behalf of undisclosed principal — Crucially important to landlord to know tenants given covenants and Rent Act.
***Hanstown Properties Ltd* v *Green (CA)*, 85**

Assignment
Unreasonable refusal to permit
Landlord not to unreasonably withhold consent to assignment to responsible and respectable person — Held, question of fact whether or not this was done — Judge entitled to hold that unreasonable to require personal guarantor of proposed assignee's obligations when corporate surety had been offered — Unreasonable to refuse on ground that did not know what use property to be put by proposed assignee — Real reason for refusing consent was that landlord wished to obtain possession for himself.
***Pakwood Transport Ltd* v *15 Beauchamp Place Ltd (CA)*, 27**

Business tenancies — See **Landlord and tenant** (Landlord and Tenant Act 1954 Part II)

Covenants
Restriction to use as supermarket
Whether freezer centre selling, *inter alia*, freezer cabinets constitutes business of supermarket and selling of those goods which are usually sold by and in supermarkets — Held, meaning of supermarket to be construed as it would have been at date of lease — Shop in which substantially the greater part of the articles on display are for self-service — On the facts a supermarket — No evidence to establish breach of second limb of covenant.
***Calabar (Woolwich) Ltd* v *Tesco Stores Ltd (CA)*, 113**

Forfeiture
Bankruptcy of surety
Lease containing proviso for right of re-entry upon bankruptcy of tenant's surety — Held, bankruptcy of surety a "breach of condition" under s146 Law of Property Act 1925 — Notice to be served in such cases as preliminary to forfeiture proceedings.
***Halliard Property Co Ltd* v *Jack Segal Ltd (Ch)*, 26**

Relief
Meaning of s146(2) Law of Property Act 1925 — Held, landlord who serves s146 notice is "proceeding to enforce a right of re-entry of forfeiture" — Court, therefore, has jurisdiction to grant relief once that notice has been served.
***Pakwood Transport Ltd* v *15 Beauchamp Place Ltd (CA)*, 27**

Landlord and Tenant Act 1954 Part II
Landlord's opposition to new tenancy
S30(1)(*g*) — Landlord claimed to intend to occupy premises for own business — Undertook to court that would use land for grazing — Held, neither the giving of such an undertaking nor ability to carry it out operated to create presumption of genuine intent — Judge entitled on facts to decide that landlord did not have *bona fide* intent.
***Lightcliffe & District Cricket & Lawn Tennis Club* v *Walton (CA)*, 35**

TOWN AND COUNTRY PLANNING — *continued*
Advertisements — *continued*
Discontinuance notice — continued

continued to display it — Held, information properly dismissed — No continuing offence.
***Kensington and Chelsea Royal Borough Council* v *Elmton Ltd (QB)*, 147**

Caravan site — See **Mobile homes** (Caravan site)

Development
Building operations
Stone-cladding on facade of house — Held, permitted development, under general development order — "Development within the curtilage of a dwellinghouse" includes exterior decoration – Restriction that development not to project "beyond the foremost part of any wall" defined such that "wall" includes moulded surrounds and sills.
***Bradford Metropolitan District Council* v *Secretary of State for the Environment (QB)*, 137**

Enforcement notice
Appeal to Secretary of State
Appellant's ground for statutory review was that having called high-level experts to express their opinions, the inspector, without any evidence to contradict them, was bound to accept them — Held, complete and total fallacy — Inspector entitled to rely on own experience, knowledge and common sense.
***Kentucky Fried Chicken (GB) Ltd* v *Secretary of State for the Environment (QB)*, 139**

Non-compliance
Information alleging offence under s89(1) Town and Country Planning Act 1971 dismissed by justices as out of time under s104 Magistrates' Court Act 1952 — New enforcement noticed issued with amended time for compliance — Held, s89(1) offence not continuing but rather once-and-for-all offence — Planning authority cannot use power, under s89(6), to extend time-limit in notice to avoid s104 nor to landowner's detriment.
***St Albans District Council* v *Norman Harper Autosales Ltd (QB)*, 140**

Inquiry procedure
Adjournment
Planning authority failed to make documents available to applicant until morning of inquiry — Request for long adjournment — Adjournment over lunch break granted — Held, applicant's counsel given insufficient time to reflect on evidence — Breach of natural justice — Risk of prejudice to applicant's case — Secretary of State's decision quashed.
***Performance Cars Ltd* v *Secretary of State for the Environment (CA)*, 143**

Planning conditions
Extrinsic evidence — See **Evidence** (Extrinsic)

Planning permission
Conditions
Planning condition — Office use restricted to specified public body — Statement at end of permission gives rise to allegation that imposed so that applicants would give up existing office-use rights elsewhere and develop for residential use — Held, condition in fact imposed to limit departure from development plan — Statement no more that encouragement to bring about planning gain.
***RKT Investments Ltd* v *Hackney London Borough Council (QB)*, 146**

Inspector's decision
Natural justice — Secretary of State's report crucially based on his view of an issue which the applicant's were not given an opportunity to present evidence on — Secretary of State confirmed the inspector's conclusions and recommendation — Held, by not affording the applicants an opportunity to present evidence addressing this issue, Secretary of State in breach of his duty *audi alteram partem*.
***H Sabey & Co Ltd* v *Secretary of State for the Environment (QB)*, 135**

Material consideration
Dismissal of appeal by Secretary of State against refusal to grant temporary planning permission for office use — Held, decision quashed — Secretary of State had purported to apply policy from development plan in making his decision — Misinterpreted that policy — A misdirection and had therefore acted in excess of his powers.
***Niarchos (London) Ltd* v *Secretary of State for the Environment (QB)*, 140**

Right of way
Appeal against refusal of planning permission for new means of access to land — Planning authority required proper visibility splays — Inspector found that appellant did not have control of land on which these were to be erected — Held, decision upheld — Appellant failed in argument that he had a right ancillary to grant of right of way to erect splays — Did not discharge burden of proof that entitled to land on which splays were to be erected.
***Hayns* v *Secretary of State for the Environment (QB)*, 134**

Use classes
Planning permission granted for use of site for warehousing and storage depot — Enforcement proceedings for removal of two caravans — Held, presence of caravans on site does not constitute use of that site for warehousing and storage depot, under use classes order — "Warehouse" implied covered storage.
***Hooper* v *Slater (QB)*, 137**

Secretary of State's decision
Natural justice
Appeal against refusal to grant listed building consent — Held, Secretary of State's decision to allow appeal based on findings of fact different from those made by inspector — Failed to notify planning authority of this — Decision quashed.
***Thanet District Council* v *Secretary of State for the Environment (QB)*, 142**

TREE ROOTS DAMAGE — See **Nuisance** (Damages)

VALUATION
Compensation — See **Compensation for acquisition of land**
Rates and rating — See **Rates and rating** (Valuation)
Rent review clause — See **Landlord and tenant** (Rent review clause)

VENDOR AND PURCHASER
Fiduciary relationship
Non-disclosure
Application for planning permission successfully made by agent of purchaser — Enhanced value of property — Made before exchange of contracts in name of vendors without their consent and knowledge — Had vendors known likely to have affected their decision whether to conclude contract — Fiduciary relationship between vendor and purchaser arose — Held, duty to disclose planning application before conclusion of contract — Non-disclosure gave rise to duty to account for profits made from planning application.
***English* v *Dedham Vale Properties Ltd (Ch)*, 118**

Warranty
Fit for habitation
House sold subject to warranty by vendor that built "so as to be fit for habitation" — House "doomed" — Built on hillside liable to landslips — Held, warranty not absolute, but breached — House unfit for human habitation by reason of defect which suitably qualified expert could have discovered.
***Batty* v *Metropolitan Property Realisations Ltd (CA)*, 148**

AGRICULTURE

A **COURT OF APPEAL**

February 23 1977

(Before Lord Justice CAIRNS, Lord Justice STEPHENSON and Lord Justice WALLER)

JAMES v LOCK

Estates Gazette April 29 1978

(1977) 246 EG 395

Agricultural Holdings Act 1948—Whether transaction a letting for 7 years, an agreement within the proviso to section 2(1) of the Act only for grazing or mowing during some B specified period of the year, or a licence to occupy the land for use as agricultural land so as to attract section 2(1) and result in a tenancy from year to year—Not a tenancy for 7 years because inter alia no evidence of fixed date for commencement of term—Not an agreement restricted to a specified period of the year—Held to be a licence within section 2(1) resulting in creation of a tenancy from year to year—Deputy circuit judge upheld but on different ground

In these proceedings the plaintiff, Ernest Rew James, in an action for possession tried by Deputy Circuit Judge Lowry in Portsmouth County Court appealed against the judge's C dismissal of his claim. The action concerned the recovery of pasture land at Barrett's House Farm, Halstock, lying partly in Dorset and partly in Somerset, used by the defendant, Ivan W Lock, for grazing his sheep in pursuance of an arrangement with the plaintiff the exact nature of which was the subject of dispute.

H Berry (instructed by Finnis, Foyer (Christopher) & Co, agents for Batten & Co, of Yeovil) appeared on behalf of the appellant; W D R Spens (instructed by Roche, Son & Neale, agents for Proctor & Horden, of Crewkerne) represented the respondent.

Giving the first judgment at the invitation of Cairns LJ, D STEPHENSON LJ said: This is an appeal by the plaintiff from a judgment of Deputy Circuit Judge Lowry given in the Portsmouth County Court on June 17 1976 dismissiing a claim for possession, damages for trespass agreed at £634.50, and an injunction. The case and the appeal concern 26¼ acres of pasturage at Barrett's House Farm, Halstock, in the counties of Dorset and Somerset. The land in question is bisected by a road which is the boundary between those two counties.

The plaintiff and the defendant are both farmers, and at all material times the defendant was also an agricultural contractor. It is common ground that the defendant grazed E his sheep on the land in 1972, 1973 and 1974 by the plaintiff's leave or permission under some sort of arrangement or agreement with the plaintiff. The question which the judge had to decide was what their legal relationship was, and had it come to an end. As the learned deputy circuit judge put it on the first page of his careful and comprehensive reserved judgment: "The real dispute between the parties is whether the defendant is, as the plaintiff says, a former owner of the grass keep of certain land of the plaintiff, or whether, as the defendant alleges, he was and still is the tenant of an agricultural holding in respect of the land in question." Whatever the legal relationship was, it was created by a conversation F (over a farm gate, we are told) in March 1972. What that conversation was is the subject of a conflict as acute as the dispute as to what are the correct inferences to be drawn from that conversation.

It is common ground that the plaintiff, who did not live as near to the land as the defendant, never visited the land during the three years after March 1972 or, at any rate, until January 1975, when he saw sheep grazing on the land and took objection to it. It is also common ground that the defendant paid £200 on December 1 1972, to the plaintiff, another £200 the next year on a different date, and £300 the following year, 1974, again on a different date, by in each case G deducting sums from invoiced amounts which were owed to him by the plaintiff for agricultural work which the defendant had done for the plaintiff. There was evidence—and evidence not only from the defendant himself—that he had manured the land as early as March 1972; that he had pastured his sheep and cattle, and the cattle of others, on the land at all seasons of the year—in winter as well as between April and October; and that he had cut hedges and maintained fences on the land. There was expert evidence that if what he was paying was a basic sum of £10 an acre that was about half way between the larger sum normally paid as the price for grazing rights over such land as this and the lesser sum H normally paid as rent for agricultural land in this area at this time.

We have no notes of evidence taken by the judge or by anybody else, but we have, as I have said, the benefit of the judge's full and careful reserved judgment. The case was tried in the Yeovil County Court, but the judgment was given later at Portsmouth. We therefore have to rely for the evidence of the plaintiff and of the defendant as to what took place, and what is alleged to constitute the necessary agreement under which the defendant did graze his sheep on these fields, on the learned judge's judgment. He said: J

> The two versions, given respectively by the plaintiff and the defendant, of the agreement between them are quite irreconcilable. The plaintiff says that early in 1972 he offered the defendant the land for grass keep at £200 from April to October inclusive "for two seasons or so," and that the defendant accepted. The price, he said, was arrived at by taking the rate at £10 per acre, ie £265, and deducting therefrom the odd £65 on the basis that the defendant would be responsible for trimming the hedges and for repairing the fences if the stock got out. The defendant says that the plaintiff told him he could not get rid of the land. He says that he (the defendant) said he would take it only if he could have it for at least seven years in order to "put body in" it, to which the plaintiff said, "Do what you like with the land." The K defendant says that he had in mind his own knowledge of the land, gained as a result of having cut it the previous year. He was aware of the distinction between grass keep and letting land. He knew that the plaintiff had offered the keep to other people—as I may say at this stage was plain from the evidence, because Mr Pring, who I have no doubt was telling the truth, said he was offered the keep by the plaintiff, who himself gave evidence to the like effect. The defendant denies that the sum of £200 was arrived at by the process described by the plaintiff. He said that no mention was made of the fences or hedges.

A The learned judge accepted the defendant's version, and accepted it in a manner which has opened wide the way to this appeal.

It is necessary that I should read in full what the judge said on this issue:

> The matter is made the more difficult by the fact that neither parties to this action struck me as obviously lying in the witness box. The conclusion to which I have come, however, is that the account which the defendant gave me of the conversation upon which he bases his claim to a tenancy is the right one. This is a conclusion which I have reached, of course, on the balance of probabilities, and this is not a case where the balance came down B very heavily on one side. Now the defendant bases his claim on his having told the plaintiff that he would not take the land unless he could have it for a minimum of seven years, to which the plaintiff had said, "Do what you like with it." That, said the defendant, he understood to relate to the length of time, not the particular use to which the land was to be put, but the defendant clearly did not accept, by that answer, that the arrangement was that he was to have the land merely for grazing for at least seven seasons. I am satisfied on balance that the defendant thought he was being offered a tenancy which was to last at least seven years. I am less confident that this is what the plaintiff meant to grant. I think it may well be that he merely meant to grant grazing rights for the next seven summer seasons. But I do not accept that he C made this clear to the defendant, and as I recall it, although he was in the end emphatic that he had actually spelt out the period—ie April to October—he at first simply referred to "the grazing season," or words to that effect, saying that the defendant knew what that meant. I do not think he said either. I do not think he remembered very clearly what had been said. I fear that this part of the plaintiff's evidence, together with his account of the defendant asking his permission to place cattle on the land in the winter and being paid two bales of straw for digging a ditch on this land probably represents an attempt retrospectively to clothe with the characteristics of a grazing agreement what the plaintiff had come to realise, even though probably he had not done so at the beginning, was an agreement for the letting of the land.

D There is then, I think, as I will indicate later, a gap in the learned judge's exposition or reasoning. He goes straight on:

> In any event, the onus is on the plaintiff to show that the agreement was within the proviso to section 2(1) of the Agricultural Holdings Act 1948: *Scene Estate Ltd v Amos* [1957] 2 QB 205 at p 212 per Parker LJ; *Lory v Brent London Borough Council* [1971] 1 WLR 823 at p 825.

Neither counsel has referred us to the second of those cases, a decision of Ungoed-Thomas J, but both counsel have referred us to the earlier case of *Scene Estate Ltd*.

I go on with the judge's judgment:

> E Further, the agreement in order to fall within the proviso, must have been made "in contemplation of the use of the land only for grazing or mowing during some specified period of the year." The "contemplation" must be mutual: see *Scene Estate Ltd v Amos* at pp 212-213. I have indicated a slight preference for the defendant's account of matters—a conclusion influenced by the very fact that he does not allege an entirely clearly spelt-out agreement, with its i's dotted and its t's crossed. But even if I were quite unresolved as to the facts, the case would have to be determined in the defendant's favour on the basis of the two passages from Parker LJ's judgment in the case to which I have just referred.

F That, as I read it, is a finding of a seven years' tenancy. The judge was accepting what had been pleaded in paragraph 4 of the defence:

> The plaintiff and the defendant entered into an oral agreement in 1972 that the plaintiff would let the said land to the defendant as an agricultural holding for a term of seven years from the 1st day of March 1972 at an annual rent of £200 and the defendant is in possession of such land pursuant to such agreement and has been so at all times material to this action.

That is supported by the judge's observations on the counterclaim. There was a counterclaim for a declaration to that effect, apparently, which was withdrawn and not persisted in. But the judge did say towards the end of his judgment G that he accepted

> that the defendant said that he could take the land only if he could have it for a minimum of seven years, and that he was then let on to the land, being told by the plaintiff, "Do what you like with it." I think it follows from that that the defendant holds the land for seven years from March 1 1972. However, the point was not argued on behalf of the defendant, and I doubt if it is strictly necessary for the determination of this dispute for me to do more than find whether or not the defendant has a tenancy.

It also follows from the passages of the judge's judgment which I have read that the judge was rejecting the plaintiff's H version of the way in which he had pleaded his case in paragraphs 1, 2, 5 and 6 of his reply. There he had alleged:

> 1. In 1972 by oral agreement the plaintiff let the grass keep of the land in question to the defendant for the period April 1 1972 to October 31 1972 at a price of £200. 2. In 1973 by oral agreement the plaintiff let the grass keep of the same land to the defendant for the period April 1 1973 to October 31 1973 at a price of £200. . . . 5. On a day early in 1974 the plaintiff spoke to the defendant about the grass keep for 1974 and told him that for the coming season the plaintiff proposed to charge £300. The defendant agreed to the increase. 6. The plaintiff denies that he made the agreement alleged in paragraph 4 of the defence. He has never let the land in question to the defendant as an agri- J cultural holding and the defendant is not in possession of the land under the alleged or any agreement.

Mr Berry, who has candidly confessed that he has a difficult task, but who has not been prevented by that from arguing the appeal, I venture to think, as well as it could be argued, challenges that finding of a tenancy. His attack on the judge's conclusions and his reasons for it are summed up in a single ground of his notice of appeal: "On the facts which he found the learned deputy judge misdirected himself in law in finding that an agreement had been reached between the parties and that such agreement constituted a tenancy." What he says first is that on the judge's own inconsistent K findings there was no agreement, because there was no consensus *ad idem*. The defendant thought that he was getting a tenancy for at least seven years. The plaintiff did not think he was giving him one; he thought he was giving him the grass keep. But the judge did not accept that he made that clear to the defendant in words, and in a passage which I have read, was only saying that it was possible that he meant it. "I think it may well be that he merely meant to grant grazing rights for the next seven summer seasons."

Mr Berry goes on to say that it is quite impossible to spell out of the defendant's version of the conversation, accepted by the judge, any tenancy or agreement for a L tenancy. "You cannot," he says, "have an agreement for a tenancy without the date for the commencement of the term being ascertained and agreed. True, it is alleged in the defence, and was apparently accepted by the judge, that this tenancy ran from March 1 1972, but there was no evidence of it. The evidence as to when the defendant went into occupation was only as precise as early in March. In those circumstances, it would be impossible, and a misdirection in law, to infer the creation of a tenancy from the 1st March." For that proposition of law Mr Berry relies on the case of *Harvey v Pratt* [1965] 1 WLR 1025 in this court, a case in which there was no actual possession. But the absence of a M commencement date would not be cured by possession having been taken in this case unless there was a precise date on which possession was taken, and there is no evidence of any such precise date on the evidence disclosed in the judge's judgment, and we ought not to assume that there was anything more precise. Further, there was no agreed date when the rent should be paid, and it would be quite impossible, and wrong, to invent one.

Another feature of a tenancy, of course, is exclusive occupation on the part of the tenant. Both in this connection and in the later part of his argument Mr Berry submits that

A it would be quite unsafe to infer here that the defendant was given exclusive occupation of these fields by the plaintiff. It is impossible to say what the position would have been if the plaintiff had wanted to pasture one of his cows on these fields for some period. Would he have had to ask the permission of the defendant, or had he a right to do it with or without the defendant's permission? Mr Berry relies, as I have said, on the inconsistency of the judge's findings about what the plaintiff may have thought he was doing and what the defendant assumed he was doing. If the plaintiff so acted as to induce in the defendant a belief that he was agreeing to give him a seven years' tenancy, the plaintiff's
B private thoughts on the matter would not have affected the agreement.

For my part, having considered the submissions of Mr Berry and the equally persuasive submissions of Mr Spens to the contrary on this part of the case, and bearing in mind that the grant of grazing rights would also confer exclusive occupation, as well as a tenancy would (see *Richards v Davies* [1921] 1 Ch 90) I am not satisfied that the defendant has established a tenancy of the land for seven years, and that was the basis on which the judge found against the plaintiff and dismissed his claim.

But that does not end the case. If this court is prepared to
C agree with the judge that there was a seven years' tenancy, there is, Mr Spens concedes, no need to resort to section 2 of the Agricultural Holdings Act 1948. But the defendant's case was put in the alternative. In his defence he alleged in paragraph 5: "Alternatively the defendant has been in possession of the said land as an agricultural holding since the 20th day of February 1972 as a tenant thereof on a yearly tenancy." Then paragraph 7: "The defendant claims the protection of the Agricultural Holdings Act 1948 as a tenant of an agricultural holding within the meaning of that Act."

It is agreed that that is a plea of section 2 of the Agricultural Holdings Act 1948, which is in these terms:

D
> (1) Subject to the provisions of this section, where under an agreement made on or after the first day of March, nineteen hundred and forty-eight, any land is let to a person for use as agricultural land for an interest less than a tenancy from year to year, or a person is granted a licence to occupy land for use as agricultural land, and the circumstances are such that if his interest were a tenancy from year to year he would in respect of that land be the tenant of an agricultural holding, then, unless the letting or grant was approved by the Minister before the agreement was entered into, the agreement shall take effect, with the necessary modifications, as if it were an agreement for the letting of the land for a tenancy from year to year: Provided that this subsection
> E shall not have effect in relation to an agreement for the letting of land, or the granting of a licence to occupy land, made (whether or not the agreement expressly so provides) in contemplation of the use of the land only for grazing or mowing during some specified period of the year. . . .

I need not read the rest of the subsection or subsection (2) of that section.

As this section, in so far as it deals with lettings, deals only with lettings to a person for use as agricultural land for less than a tenancy from year to year, it would have no application to a letting for seven years. But it also applies to the grant of a licence to occupy land for use as agricultural land,
F and it is submitted on behalf of the defendant that, whatever else may be the effect of the defendant's agreement with the plaintiff said to be made in March 1972, the plaintiff did thereby grant him a licence to occupy land for use as agricultural land. That was under an agreement made after March 1 1948, and the circumstances were such that if the tenant's interest was a tenancy from year to year he would in respect of that land be the tenant of an agricultural holding. No question of ministerial approval arises. It therefore follows, says Mr Spens on behalf of the defendant, that that agreement to grant a licence does take effect, with the necessary modifications, as if it were an agreement for the letting of the land on a tenancy from year to year, and the proviso does not G
apply, because, on the finding of the learned deputy judge that no specified period short of a year was mentioned, the agreement was not one for the grant of a licence to occupy the land made "in contemplation of the use of the land only for grazing or mowing" during some specified period of the year. It was, says Mr Spens, the grant of a licence for a period of at least seven years, or some variation of that, he might have to concede, such as the grant of a licence yearly, at £200 a year, to be increased or varied by agreement from year to year, and not to be determined before the end of seven years. For that construction of the section Mr Spens relies on the case of *Scene Estate Ltd v Amos*, to which the H
judge was also referred. He also relies on that case for the proposition that the onus of proving an agreement, which is *prima facie* within the subsection, is taken out of the subsection by being brought within the proviso rests on the owner, the landlord, or in this case the grantor of the licence.

Mr Berry submits that, again, whether you are considering a tenancy or a licence, there was no consensus. There was no agreement under which the licence was granted, because the minds of the two parties were not at one. He says that if he is wrong about that, if he is within the section, he comes out of it by means of the proviso, because when the proviso speaks of "in contemplation of the use of the land" it is J
referring to what the grantor contemplates and not what both parties contemplate. But I think he appreciated that this court is really constrained not to put that construction upon the section by what was said in this court in the case of *Scene Estate Ltd v Amos*, to which I now briefly refer.

The judgment of Parker LJ, with which Sellers LJ agreed, covers all the points on which Mr Berry relies in this part of the case. On p 212 of the report [1957] 2 QB 205 the learned Lord Justice said this:

> A landlord is, however, able to let or license the use of agricultural land without bringing himself within the Act if he can show that the conditions of the proviso are fulfilled. Quite clearly the K
> onus is upon him, the landlord, to satisfy the court that the agreement in question was "made (whether or not the agreement expressly so provides) in contemplation of the use of the land only for grazing or mowing during some specified period of the year." The words are not altogether easy to construe, but this at any rate, I think, is clear, first, that "specified period of the year" is equivalent to "specified part of the year"—that, I think, is the effect of the decision of this court in *Reid v Dawson* in 1955; secondly, that the "contemplation" must be not only that the use of the land shall be only for grazing or mowing but that the period of use shall be only a period which is less than a year. It seems to me that, as a matter of grammar and ordinary reading, the word "only" must apply not only to the use but to the period; L
> and indeed, if it were not so, it seems to me that the words "some specified period of the year" were quite unnecessary, because this is a proviso to a section which is only dealing with tenancies which are for part of a year. Thirdly, it seems to me that the "contemplation" must be mutual, both ono the part of the landlord and of the tenant.

Denning LJ gave the first judgment in that case and seems to have expressed agreement with what Parker LJ said about "contemplation" in what he said at p 211 of the report.

In my judgment, the plaintiff clearly conferred on the defendant a licence to occupy the land for use as agricultural land. It is true that such a licence has to confer the right to occupy M
exclusively, that is, to the exclusion of the grantor. That follows from the decision of the House of Lords in *Bahamas International Trust Co v Threadgold* [1974] 1 WLR 1514. But I find no difficulty in infering from the proved facts that the defendant had such a licence, had such exclusive occupation. No doubt the learned deputy judge did pay some attention to the subsequent conduct of both parties in considering what the original agreement was, and that is something which cannot usually be done in construing written agreements and may be objectionable even in considering an oral agreement of this kind. But he emphasised that he did not attach much

importance to any of the subsequent conduct of the parties. That conduct is consistent with exclusive occupation.

Where the judge's judgment does create a difficulty, as it seems to me, is in the omission from the passage which I have quoted from his judgment of any clear statement that he had considered the question whether section 2 of the Act was applicable to this agreement: because he goes straight, as I have indicated, from considering the question of tenancy to the proviso to the section. That ellipse (if that is the right word) does create in my mind a difficulty. But when he says, "In any event the onus is on the plaintiff," I cannot read his judgment as ignoring a consideration of the question whether the section applies at all. It was clearly pleaded and argued that it did; and, of course, the question of the application of the proviso only arises if the section has been considered. Although it would have been better if the learned deputy judge had given his reasons, in my judgment, he must have held that the section would have applied if he had not himself held that there was a letting, and a letting over a period of years which made the section inapplicable. When he said, "In any event the onus is on the plaintiff," I think he was intending to convey: "If I am wrong about the tenancy and this was a licence, it is a licence granted under an agreement within the section, and then I have to consider the submissions whether the proviso takes it out of the section, and I am clearly of opinion that it does not."

In my judgment, that is the right basis for the disposal of this case. The proviso is too limited in its terms for Mr Berry, on the facts found by the judge, to bring the plaintiff within it. I should perhaps say this: that it was not contended that the licence, if licence there were, was of such a character as to make it impossible to turn it by the necessary modifications into the tenancy from year to year which the statute requires. I am therefore of opinion that in occupying the land all the year round and not merely during the grazing season the defendant was not trespassing. His agreement with the plaintiff was not made in contemplation of such limited use as is defined by the proviso, interpreted as it was in the *Scene Estate Ltd* case, but under a licence which is now on the basis of the statute transformed into a tenancy from year to year.

I would therefore hold that the defendant holds the land not under a tenancy for seven years, but under a tenancy from year to year by operation of the statute. I would affirm the judgment of the learned deputy judge on that different ground and dismiss the appeal.

Agreeing that the appeal should be dismissed, WALLER LJ said: I agree not because I am satisfied that the deputy circuit judge came to the right conclusion. There are at least two pieces of evidence which I find difficulty in reconciling with his conclusions—and I refer to the receipt no 265 and to the payment of £300 in the third year when it had been £200 in the first and second years. But I agree that the appeal should be dismissed because there are findings of fact which the deputy circuit judge made which this court must accept. He was faced with the difficult task of choosing between two witnesses neither of whom, he said, was obviously lying in the witness box, but whose evidence was irreconcilable the one with the other. In those circumstances he made certain primary findings of fact on the balance of probabilities. He accepted the evidence of the defendant that he would take the land only if he could have it for seven years, that the plaintiff replied, "You can do what you like with it," and that he, the defendant, was let into possession. He rejected the plaintiff's evidence that he specified the period of the grazing season or said any such words which would lead to that result. He came to the conclusion—and this was in a way not a primary finding of fact—that the plaintiff probably did not intend to grant a tenancy, did not intend to grant an exclusive licence, but reluctantly came to the conclusion at some later date that he had done one or the other of those two things.

The question, therefore, is what inference is to be drawn from the findings of fact which the judge did make. Mr Berry argued, as my Lord, Stephenson LJ, has said, that the effect of the judge's findings is really that the parties were not *ad idem*, that they had misunderstood one another, and accordingly that the evidence was not sufficient to bring this case within the section of the Agricultural Holdings Act 1948 to which reference has already been made. Mr Spens, on the other hand, submitted not so. He submits that there was an agreement, and that it was in fact the grant of a tenancy for seven years. I do not propose to rehearse the details of the evidence, but it seems to me that on the findings of the judge it is difficult to come to a conclusion other than that there was an agreement for something, and the question really is: Was it a tenancy or a licence?

There are several factors which can be taken into account in considering that problem. I think the first is the finding which has already been read by my Lord, but which I will repeat, which the learned judge made in the course of his judgment: "I have indicated a slight preference for the defendant's account of matters—a conclusion influenced by the very fact that he does not allege an entirely clearly spelt-out agreement, with its i's dotted and its t's crossed." That perhaps is an understatement, if anything.

Secondly, the absence of a starting date. Although the learned deputy judge said later in his judgment by way of addition, "I think it follows from that that the defendant holds the land for seven years from 1st March 1972," it is difficult to see—true, we have no notes of the evidence—what grounds there were for saying that the tenancy started on March 1 1972. Although the pleading showed in contemplation that evidence would be forthcoming that the defendant had gone into possession on February 20 1972, there is nothing in the judgment to indicate that there was any movement into possession before some time in the month of March 1972. Therefore, it is difficult, in my view, to find any starting date. No such date for possession was found by the learned judge.

The third factor which makes it difficult to regard this as a tenancy is that payments were made on irregular dates: in 1972 it was in December, in 1973 in October, and in 1974 it was in either October or September. It is difficult to visualise such a wide variation of dates if there were a tenancy.

Fourthly, there is the variation in the amount paid, because in the first and second years, 1972 and 1973, £200 was paid, but in 1974 £300 was paid. Although the defendant's evidence was that he thought that the plaintiff should share in the good that he put into the land, it is inconceivable to me, at any rate, that £300 would be paid by the defendant at the plaintiff's request unless the defendant felt it was something which he was under an obligation to pay. That would not be consistent with a tenancy. In my view, therefore, it would be difficult to spell out the tenancy in this case.

On the other hand, there clearly was a licence of some sort, and I would accept that it was a licence to graze, having regard to the finding of the learned judge, for seven years. What amount was to be paid over the years it is difficult to decide. Happily, it is not a task for this court to decide that question. Section 2(1) of the Agricultural Holdings Act 1948, which has already been read, states: ". . . where . . . any land is let to a person for use as agricultural land for an interest less than a tenancy from year to year, or a person is granted a licence to occupy land for use as agricultural land, and the circumstances are such that if his interest were a tenancy from year to year he would in respect of that land be the tenant of an agricultural holding, then" certain consequences follow. It is not suggested in this case that the circumstances were such that if his interest were a tenancy from year to year he would not be the tenant of an agricultural holding. Therefore, having come to the conclusion that there was this licence to graze, in my view, it comes within section 2 of the Agricultural Holdings Act 1948 and is not caught by the proviso for the reasons which Stephenson LJ has stated, and with which I agree. Therefore, I come to the conclusion that this appeal must be dismissed.

Also agreeing, CAIRNS LJ said: This is an appeal on law only and, therefore, the learned deputy judge's findings of fact cannot be challenged where there was evidence to support them. He found, in accordance with evidence given by the defendant, that the defendant said that he would not take the land unless he could have it for a minimum of seven years, and that the plaintiff said, "Do what you like with it." Obviously that was not the whole of the conversation. As I understood him, Mr Berry, on behalf of the plaintiff, accepted that the conversation did result in an agreement between the parties whereby the defendant at least had granted to him a licence for the use of the land for grazing in consideration of an annual payment; though beyond that Mr Berry contended that nothing more was agreed, or, at least, that it was not shown that the parties were *ad idem*.

If I have misunderstood Mr Berry in thinking that he did accept that there was an agreement of some kind, I would conclude that on the admitted facts here there clearly must have been. It was common ground that as a result of the conversation between these two men the defendant did in following years use the land for grazing, and did make payments to the plaintiff, originally of £200 and later of £300 a year.

It may well be that no letting for seven years was created, as the judge thought it was, because there was no lease under seal, and in the absence of any evidence as to when the term was to commence, on the principle of *Harvey v Pratt* [1965] 1 WLR 1025 it would seem to follow that there was no agreement for a lease which in equity could take effect as a lease.

But if there was a licence for grazing, in my judgment, that has to be considered as a licence to occupy the land for use as agricultural land, and the circumstances were such that if the defendant's interest had been a tenancy from year to year he would be the tenant of an agricultural holding. It follows that under section 2(1) of the Agricultural Holdings Act 1948, and subject to the proviso to that subsection, the agreement did take effect, with the necessary modifications, as if it were an agreement for letting for a tenancy from year to year. It is not for us to consider what the necessary modifications would be. The proviso can have no effect, because even if the licence was for grazing only, as to which there may be some doubt having regard to the words, "Do what you like with it," the learned judge found that the agreement was not limited to a specified period or periods of the year.

The conclusion that I reach is that under the provisions of the Agricultural Holdings Act 1948 the defendant at the material time had a tenancy of the land and, therefore, that the plaintiff's claim was rightly dismissed. Accordingly, I agree with my brethren to the dismissal of the appeal.

The appeal was dismissed with costs.

COURT OF APPEAL

February 23 1978

(Before Lord SCARMAN, Lord Justice MEGAW and Lord Justice ROSKILL)

GLENDYNE (LORD) v RAPLEY

Estates Gazette May 13 1978

(1978) 246 EG 573

Rent (Agriculture) Act 1976—Whether gamekeeper employed in "agriculture"—Gamekeeper employed to rear pheasants for sport—Definition of "agriculture" for purpose of Act—"Livestock keeping and breeding" covers animals (including birds) kept for production of food, but not for sport—judge's finding that pheasants were kept for sport, although 80 per cent of those killed and retrieved were sold, conclusive against gamekeeper's claim to be protected by Act—Appeal dismissed

The appellant, John Rapley, appealed against a decision of Deputy Judge Massey, at Salisbury County Court, granting possession to Lord Glendyne of a cottage, known as Keeper's Cottage, at Hurdcott, Barford St Martin. The appellant had been employed as a gamekeeper by Lord Glendyne, the respondent to this appeal, and claimed that he was entitled to protection in his occupation of the cottage by virtue of the Rent (Agriculture) Act 1976.

M P Kolanko (instructed by Theodore Goddard & Co, agents for Batt, Broadbent & Beecroft, of Salisbury) appeared on behalf of the appellant; R C Pryor (instructed by Young, Jones, Golding, Patterson) represented the respondent.

Giving the judgment of the court, LORD SCARMAN said: On November 18 1977 Deputy Judge Massey, sitting in the Salisbury County Court, gave the plaintiff judgment against the defendant for possession of Keeper's Cottage, Hurdcott, Barford St Martin. The defendant, who now appeals against the judgment, contends that he is entitled, as an agricultural worker, to the protection of the Rent (Agriculture) Act 1976. The learned deputy judge held that he was not employed in agriculture and, therefore, not entitled to the protection of the Act. The only point raised in this appeal is whether the defendant was employed in agriculture. If he was, his appeal succeeds.

Mr Rapley, the appellant, was employed by Lord Glendyne, the respondent, from June 2 1972 until September 19 1977 as a gamekeeper. A term of his employment was that he should have a house, ie the Keeper's Cottage at Barford St Martin, rent and rates free. His rate of pay was to be under the terms of the agricultural wages structure which is itself derived from the Agricultural Wages Act. The learned judge found that his job was to keep and rear pheasants for sport; that, although paid for convenience by Lord Glendyne's farm manager, he was not employed on the farm; and that his wages did not go through the farm books. The defendant himself told the judge that his job was "to put birds in the air as targets for sportsmen." Upon this evidence the learned judge concluded that the defendant's job was to rear pheasants not for food but for sport. Though Lord Glendyne sold 80 per cent of the birds he and his guests shot, he would not have kept pheasants at all had it not been for the sport they provided. Accordingly, the judge held that the defendant was not employed in agriculture and that he was not entitled to the protection of the Act.

The appeal turns on whether keeping and breeding pheasants for sport is within the definition of agriculture to be found in section 1 of the Act. A number of cases in which similar definitions in other statutes have been considered by the courts were cited to us. They do not help, save to emphasise what might seem to be obvious, that not every rural pursuit is an agricultural activity.

The point turns on the definition itself. The section, so far as relevant, is in these terms: "(1) in this Act—(a) 'agriculture' includes—(i) . . . livestock keeping and breeding . . . ; . . . (2) For the purpose of the definition in subsection (1)(a) above—. . . 'livestock' includes any animal which is kept for the production of food, wool, skins, or fur, or for the purpose of its use in the carrying on of any agricultural activity, and for purposes of this definition 'animal' includes bird but does not include fish." Agriculture, therefore, includes "livestock keeping and breeding"; "livestock" includes "any animal which is kept for the production of food"; and "animal" includes birds but not fish.

The judge's finding that Lord Glendyne kept pheasants for sport, not food, clearly puts the defendant's employment outside the express terms of the definition. But it was urged before us that the definition was not intended to be exhaustive and that the statute should not be given a narrow meaning.

A Mr Kolanko, developing this submission for the appellant, made the following points: (1) the definition is introduced by the word " includes "; (2) this " inclusive " type of definition is used to enable the court to construe in a broad and flexible way the term " agriculture," so as to retain its essentially fluctuating nature; (3) keeping pheasants is part of the rural scene which includes agriculture; it is truly an agricultural activity; (4) the word " livestock " is similarly to be given a broad and flexible meaning; (5) Mr Rapley's contract was drafted so as to incorporate into it the rates of pay and holiday entitlement established under the agricultural wages legislation. Finally, Mr Kolanko reminded us that the
B statute has an important social purpose and should not be construed so as to introduce refined and difficult distinctions into rural and farming operations and country life.

C These arguments are entitled to respect. They fail because plainly not every rural or country activity is intended to be included in the definition of agriculture. Fishing, for example, is clearly excluded. The definition is really directed towards including all operations involved in farming land for commercial purposes of which the one relevant to this appeal is the production of food. The finding that these pheasants were kept for sport, though 80 per cent of those killed and retrieved were in fact sold, is, in our judgment, conclusive. Mr Rapley's employment was to promote not agriculture but a field sport. This is a country activity but not an agricultural one. The appeal is, therefore, dismissed.

D The appeal was dismissed. Legal aid taxation ordered of appellant's costs. Application by respondent for payment of costs out of Legal Aid Fund adjourned.

COURT OF APPEAL

February 16 1977

(Before Lord Justice CAIRNS, Lord Justice ROSKILL and Lord Justice BRIDGE)

UNION TRANSPORT FINANCE LTD v BRITISH CAR AUCTIONS LTD

Estates Gazette April 8 1978

(1977) 246 EG 131

Auctioneers' liability for conversion—Car subject to hire-purchase agreement put into car auction by dishonest hirer—Action against auctioneers by hire-purchase finance company—Auctioneers' liability in principle settled by House of Lords over 100 years ago—Whether hire-purchase company entitled to immediate possession so as to have right to sue for conversion—Bailor's right to immediate possession arises when bailee acts in a way wholly repugnant to contract—Auctioneers' appeal dismissed

This was an appeal by British Car Auctions Ltd against a decision by Lawson J holding them liable at the suit of Union Transport Finance Ltd for conversion as the result of the sale of a car at one of the appellants' car auctions, the car being the property of the respondents, Union Transport Finance Ltd. The car had been let on a hire-purchase agreement to a hirer, who after changing the number plates, entered it for auction with the appellants.

Gerald Owen QC and Thayne Forbes (instructed by Cripps, Harries, Willis & Carter, agents for Clive Fisher & Co, of Addlestone) appeared on behalf of the appellants; R Denyer (instructed by Meade-King & Co, of Bristol) represented the respondents.

Giving the first judgment at the invitation of Cairns LJ, ROSKILL LJ said: This is an appeal by the defendants from a judgment of Lawson J given by him at Bristol on May 10 of last year The appeal has, if I may say so, been admirably argued by Gerald Owen and his junior Thayne Forbes, but, for my part, I have no doubt that the learned judge's decision was right. The problem that we have to consider is the all too familiar one of which of two innocent parties has to suffer for the fraud of a third.

The plaintiffs in the action are Union Transport Finance Ltd, a well-known hire-purchase finance company, and, indeed, as their documents show, a subsidiary of the even better known Wagon Finance Corporation Ltd. The defendants are a well-known firm of car auctioneers, who, we are told, have passing through their hands throughout the length and breadth of this country many scores of thousands of cars every year. The trouble arose because on February 7 1973 the plaintiffs, in accordance with what to them was no doubt a perfectly ordinary hire-purchase transaction, bought an Audi car the number of which was CFH 138J from a firm of dealers for the purpose of letting it on hire-purchase to a gentleman who gave the not very unusual name of Harry Smith. Smith told them that he was a surveyor, and a contract was entered into on that date in the usual hire-purchase transaction form providing for a sum to be paid down and for some 36 monthly instalments to be paid, the first payment to be made later, namely, on March 7 1973.

The next day Smith (as he was thought to be) taxed the vehicle. On February 13 the plaintiffs, in accordance with the usual practice, informed Hire Purchase Information. On February 19 1973 Mr Smith—who by this time appears to have changed his name to Smithers—entered that car for auction with the defendants, not under its original number, but under number CFH 738J, presumably having done the necessary physical alterations to the number plates. He did not reveal that there were any outstanding hire-purchase instalments. On February 22 the defendants sold the car at an auction at Alexandra Palace for £1,045, from which they deducted their commission of just under £60. On February 28 1973 they sent Mr Smithers, as by then he was, a cheque for the difference.

Down to that time, of course, the plaintiffs knew nothing about Smithers' or Smith's dishonest activities. But come March 7, which was the day for the payment of the first instalment, strangely enough none was forthcoming. Nor was any instalment forthcoming on the next monthly date, April 7. The plaintiffs accordingly, as one would expect, on April 11 served a notice of default in respect of these two instalments. That letter reads thus: "We cannot trace a reply to previous reminders and unless the above amount"—which by this time was £90—"is paid in full within seven days from the serving of this notice by normal postal delivery the hiring of the above goods will be terminated forthwith and steps taken to recover our property together with your full liability under the terms of the hire-purchase agreement."

We were told by counsel that the vehicle had been bought by someone called Yates. He plays no part in this story. Whatever his part was, he was a completely innocent purchaser. We are told that he subsequently resold it, and the vehicle has now irretrievably disappeared. The sins of Smith, or Smithers, in due course were found out, and in yet another name of Day, in August 1973, he was sent to prison for obtaining this vehicle by dishonest representations. He might well, I should have thought, have been charged with the theft of the vehicle, though whether he was or not does not matter, nor is it clear from the papers.

That being the factual background of the matter, the action came before Lawson J at Bristol on agreed facts. There was no oral evidence adduced before the learned judge, and he was invited to decide on those facts whether or not the defendants were guilty of conversion. The defendants took two points in reply to the plaintiffs' claim. First—and I can deal with this quite shortly—they said that auctioneers who act as they acted on this occasion are not in law guilty of conversion; and second, they said: "Even if in law we as auctioneers can be guilty of conversion, you, the plaintiffs, on the date when we resold the car, February 22 1973, had no right to immediate possession, and, therefore, you cannot maintain this action for conversion against us." Those were the issues which Lawson J had to try.

I can, I hope without any discourtesy to Mr Owen, dispose of the first point speedily. The learned judge had no doubt—and he, too, dealt with this point very shortly—that before him it was not open to the defendants to assert that they were incapable as a matter of law of being guilty of conversion. The same position, as I think, obtains in this court. Mr Owen reserved his clients' right if this matter were to go to the House of Lords to reopen what has been long thought to be the law, ever since *Hollins v Fowler* (1875) LR 7 HL 757 about 100 years ago, and to put the matter to the test in the highest tribunal. I mention for the sake of completeness that the point that an auctioneer cannot in such circumstances as those be guilty of conversion, might have been, but was not, taken in this court in a case to which I shall refer in more detail in a moment, *North Central Wagon & Finance Co Ltd v Graham* [1950] 2 KB 7. If one looks at the statement of facts on pp 7 and 8 of the report one sees that the defendant, Graham, was an auctioneer against whom, as in the case of the present appellants, no allegation whatever of want of good faith was made. But counsel for Mr Graham did not take the point which it is sought to keep open in the present case.

I now turn to the second point that has been pressed upon us in this court. It is, I apprehend, an elementary branch of the law of tort, first, that an action for conversion depends upon injury to the right to possession and second, as Asquith LJ put it in the *North Central Finance Co* case: "When a bailor and lessor on hire-purchase wants to sue in conversion by reason of the wrongful sale by the bailee of the article bailed, it is essential for him to show that he is entitled at the time of the sale to immediate possession of the goods." The hire-purchase finance agreement between the plaintiffs and Smith contained a large number of clauses. It is not necessary for me to read more than a few in order to make plain the point that was argued before Lawson J and has been argued again in this court. If one looks at clause 1 one finds a number of covenants under the letters (a), (b) and (c) and through to (k) on the part of the hirer. Clause 1(c) provides: "To keep the goods in good repair and working condition . . . and not to remove or allow to be removed their present identification marks." Sub-clause (f) provides:

> Not without the previous consent of the owner to part with the possession or control of the goods . . . and not to sell or offer for sale nor assign pledge mortgage underlet give away or otherwise in any way legally or equitably deal or attempt to deal with either the goods or option to purchase herein contained or other interest of the hirer hereunder.

Pausing there, there can be no doubt that Smith's actions, first in altering the number plate by substituting the figure "7" for the figure "1," as I mentioned earlier, was a breach of clause 1(c), and then in handing over the car to the appellants as auctioneers for sale was a breach of clause 1(f).

Clause 3 provides:

> Unless and until the whole of the sums due under this agreement shall have been paid and purchase been made under clause 2(2) hereof the goods shall remain the absolute property of the owner and the hirer shall not have any right or interest in the same other than that of hirer under this agreement. (b) Should the hirer make default in any monthly rentals as agreed or commit any other breach of this agreement or be adjudicated bankrupt . . . the owner shall have the right at any time to declare the hiring to be terminated by forwarding a notice of default to the hirer's last known address and take action to retake possession of the goods and to recover against the hirer or his estate the amount stated in clause 3(f) hereof. . . .

I need not read sub-clauses (c), (d), (e) or (f), but I mention that (f) deals with what is the liability of the hirer.

Stated in a sentence the argument of the appellants is this. "True, Smith acted dishonestly on February 22 in handing this car over, but you the respondents, the legal owners, have no right to sue us, the appellants, in conversion, because as at the date when we sold this car and would be liable in conversion if an action for conversion lies against us at all, you had no right to immediate possession, and therefore your action is not maintainable against us." I hope that I do Mr Owen's argument no injustice if I say that in the ultimate analysis, however he sought to put it, in the end one comes back to the same point: on the true construction of the clauses to which I have referred in this agreement, is there anything to bar the plaintiffs from bringing this agreement to an end in any other way than as is there provided, and thus acquiring an immediate right to possession? That was the question before Lawson J, and it has been repeated in this court.

The learned judge well summarised the point when he said: "Mr Owen, on behalf of the defendants, submits this: That having regard to the terms of the hire-purchase agreement and, specifically, the terms of clause 3(b) of the hire-purchase agreement, unless and until the bailors, the plaintiffs in this action, have served a notice of termination, the right to immediate possession remains in the bailee, and that they, therefore, the bailors, the plaintiffs here, had no right to immediate possession because they had not in fact served a notice of termination under clause 3(b) as is the fact, of course; the position being that on February 22, as far as they, the owners of the car, were concerned the bailee was not in arrears of any instalment." There will have been observed, as I read that passage from the learned judge's judgment, the words "the right to immediate possession remains in the bailee." When Mr Owen read that passage all three members of this court intervened to ask whether that did represent his argument. He frankly said that it did, and that it was an essential part of his argument that at that point of time the bailee—Smith—alone had the right to immediate possession. I will come back to that point later.

Reliance was placed upon the decision in *North Central Wagon & Finance Co Ltd v Graham*, and Mr Owen laid much stress on the first of the judgments in that case delivered by Asquith LJ. The second judgment was given by Cohen LJ, and Singleton LJ merely said, "I agree." When one looks at Asquith LJ's judgment, it seems that he looked at the matter from the point of view of whether the particular hire-purchase agreement there before the court had the effect contended for by the defendants (a view which had been accepted by the trial judge, Lewis J) of preventing the plaintiff bailors from claiming to be entitled to sue for conversion as being the persons immediately entitled to possession. Asquith LJ went through the agreement in some detail at p 13, and he reached the conclusion on p 14 that that agreement on its true construction did not operate to prevent the plaintiffs from immediately terminating the agreement apart from its express provisions and thus claiming to be the persons entitled to immediate possession and thus to sue for conversion.

Cohen LJ, who began his judgment by saying that he agreed with Asquith LJ, and for the same reasons, then approached the matter somewhat differently, and, as I venture to think, on a rather wider basis. In the middle of p 15 the learned Lord Justice referred to a passage in the then current edition of *Halsbury's Laws of England* (2nd ed vol 1 p 736) which reads thus:

> "The act of the bailee in doing something inconsistent with the terms of the contract terminates the bailment, causing the possessory title to revert to the bailor and entitling him to maintain an action of trover." That view is stated in somewhat similar language in the following passage in *Pollock and Wright on Possession in the Common Law* (at p 132): "Any act or disposition which is wholly repugnant to or as it were an absolute disclaimer of the holding as bailee revests the bailor's right to possession, and therefore also his immediate right to maintain trover or detinue even where the bailment is for a term or is otherwise not revocable at will, and so *a fortiori* in a bailment determinable at will."

The learned Lord Justice then went on to say:

> Those passages, which seem to me, if I may respectfully say so, to be a correct statement of the law definitely support the view which my brother has expressed—namely, that the plain-

tiffs are in a position to maintain an action in trover. But Mr Elwes contends that they only apply where the contract is silent on the subject, and that here the contract provides what is to happen in the event of a breach of any stipulation and merely gives the right to terminate the contract. That, he says, is something distinct from a right to enter into possession, and the latter right is only exercisable after notice has been given to terminate the contract. As I have said, I do not so construe the contract, and I am glad to find that the construction which I place on the contract accords with the view of what the law would be in the absence of express provision and, if I may say so respectfully, with what is a very reasonable position in the circumstances of the case.

Cohen LJ was there presumably founding upon what he thought was the position at common law. He then went on to say, on the last page, that Mr Elwes, whose argument in that case must have resembled Mr Owen's argument in the present case, could not escape from the common law position by relying on the language of the contract there in question.

The learned judge in the present case drew attention to what Mr Owen had said, namely, that the current (4th) edition of *Halsbury's Laws of England* stated the law rather differently from the statement in the passage I have just read from the 2nd edition. However that may be, as my Lord pointed out during the argument, it seems plain that this court, in the persons of Cohen LJ and Singleton LJ, expressly approved that statement of the law in the 2nd edition. We therefore should follow that statement whatever may be stated in the 4th edition, to which we have not been referred beyond the reference to that passage in Lawson J's judgment.

It seems to me that there is no room for doubt that the position at common law is this: if the bailee acts in a way which, to use the phrase used in argument, destroys the basis of the contract of bailment, the bailor becomes entitled at once to bring that contract to an end, and thus at once acquires the right to immediate possession of the article bailed. That is consistent with what is stated in the 2nd edition of *Halsbury* and in their classic work on *Possession* by Sir Frederick Pollock and the first Mr Justice Wright.

In those circumstances, it seems to me that the only question that remains for consideration is whether the provisions of the present contract affect the basic common law position. Mr Owen strenuously argued that they do. He contends that because there is this express contractual right to bring this contract to an end only after notice of termination, there is no room for the survival as between the plaintiffs and Smith of the basic common law rule—and, indeed, the phrase he used in argument was that "clause 3 ousts the common law rule." I think, with respect, that this argument is misconceived for a number of reasons. When one looks at these clauses one can see why they are there. They give a modicum of protection to the hirer but they also give certain specific contractual rights to the bailor in the happening of certain events. They give him the right to bring the contract to an end and to retake possession in the event of certain things happening. But they do not expressly deprive the bailor of any other rights that he may have at common law; and still less, in my view, do these clauses, either expressly or impliedly, confer any right, possessory or otherwise, upon the bailee which would lead to a conclusion different from that which would follow at common law if the bailee deliberately, as happened here, tears up the contract of bailment by fraudulently selling the car through an auctioneer to an innocent third party. Therefore, it seems to me, following the reasoning in the *North Central Wagon Finance* case, that even if there be room in principle for the existence of a contract which may contract out of the basic common law rule, it would require very clear language to deprive the bailor of his common law rights in circumstances such as these; in the case of the present contract the language used is nothing like strong enough to achieve that result.

I mentioned a few moments ago a passage in the judgment of Lawson J in which, in setting out Mr Owen's argument, he referred to the fact that it had been contended that the right to immediate possession remained in the bailee—Smith. With respect to Mr Owen, that seems to me to be an impossible argument. I ventured to ask Mr Owen in whom he claimed the possessory title lay at the moment when the hammer fell and this car purported to be sold. As I understood him, he replied that it would then pass to the innocent purchaser, and would remain his as a possessory title unless and until it came to an end on the termination of the contract of hire-purchase by the giving of the appropriate notice. That argument involves that the innocent purchaser has a defeasible possessory title; which, being defeasible, would suddenly come to an end upon the true owner, the person in whom the legal title was vested, giving notice to determine the contract, to which the innocent purchaser would not be a party and of which he would never have heard. I venture, with respect, to think that that cannot possibly be right in principle.

During the argument my Lord referred to the well-known latin tag *nemo dat quod non habet*, which may be translated into English as: "You cannot give what you have not got." In the present case, it seems plain that Smith had nothing beyond a possessory title which he acquired when the agreement was entered into and he took delivery of the car, but which he lost the moment he destroyed the agreement by acting dishonestly, and that, therefore, he could not convey any title, possessory or otherwise, either to the auctioneer or to the innocent purchaser.

For those reasons, which I have given at some length in deference to the argument for the appellants, it seems to me clear that Lawson J gave the right answer, and I would dismiss the appeal.

BRIDGE LJ gave a judgment agreeing that the appeal should be dismissed.

CAIRNS LJ expressed agreement with both judgments.

The appeal was dismissed, with costs. Leave to appeal to the House of Lords was refused.

COURT OF APPEAL

January 13 1978

(Before Lord DENNING MR, Lord Justice ROSKILL and Lord Justice BROWNE)

R H WILLIS & SON v BRITISH CAR AUCTIONS LTD

Estates Gazette April 8 1978

(1978) 246 EG 134

Auctioneers' liability for conversion—Car subject to hire-purchase agreement entered by hirer at car auction—Reserve price not reached but car sold privately at special auction office to highest bidder under auctioneers' "provisional bid" procedure—Effect of this procedure—Auctioneers liable for conversion as if car sold under the hammer—No difference in principle—Hire-purchase firm entitled to immediate possession and so able to sue for conversion Auctioneers' appeal dismissed—Certain old cases doubted

This was an appeal by British Car Auctions Ltd from a decision by Judge Sheldon at Aldershot and Farnham County Court holding them liable in an action for conversion brought by a firm which provided finance for the purchase of cars, R H Willis & Son, of Southampton. The car which gave rise to the action was a secondhand Ford Mustang let under a hire-purchase agreement by R H Willis & Son to one Croucher, who put it into the appellants' car auction at Farnham, and shortly afterwards was made bankrupt.

D Friedman (instructed by Cripps, Harries, Willis & Carter, agents for Clive Fisher & Co, of Addlestone) appeared on behalf of the appellants; J Fulthorpe (instructed by Hepherd,

Winstanley & Pugh, of Southampton) represented the respondents.

Giving judgment, Lord Denning MR said: There is a firm of motor-car dealers in Southampton called R H Willis & Son. They sell secondhand cars. In May 1973 Mr Croucher, the licensee of a local public house, The City Arms at Nursling, wanted to buy one of their cars. It was a Ford Mustang eight years old. The cash price was £600. But he had not got that sum available. So the dealers allowed him to have it on hire-purchase terms at the hire-purchase price of £625. He paid £350 in cash or in part exchange and agreed to pay the balance of £275 by monthly instalments over the next six months. He signed the hire-purchase agreement in a familiar form. It contained this warning to Mr Croucher: "The goods will not become your property until you have made all the payments. You must not sell them before then."

Mr Croucher did not heed this warning. He was in financial difficulties with bankruptcy staring him in the face. He decided to raise some money by selling the car. He took it 50 miles away to Farnham in Surrey and put it into the car auction there. It was run by British Car Auctions Ltd. He put on it the reserve price of £450 net. He told the auctioneer that it was his own car and that it was not subject to any outstanding hire-purchase. He signed the entry form to that effect.

The auction was held subject to the auctioneers' usual "Conditions of Sale" which were plainly exhibited and also announced by the auctioneer. The car was entered as lot no 67 and put up for sale. It did not reach the reserve price of £450. The highest bid was £410. This was, however, not so very far short of the £450. So the auctioneers treated it as a "provisional bid" and followed their usual practice in regard to provisional bids. A clerk went down among the people and spoke to the man who had made the bid of £410. She said to him: "Will you stand on your bid?" He said "Yes." He was a Mr Worth, of Hillingdon near London. The auctioneers knew him because he had an account with them. The clerk then called over the loud-speaker system saying: "Will the vendor and highest bidder of lot no 67 please come to the provisional bid office." That was an office specially set up to deal with such provisional bids. Mr Worth went to the office. So did Mr Croucher; or if not, the auctioneers spoke to him on the telephone. The auctioneers told Mr Croucher that Mr Worth had bid at £410: "Will you accept it?" Mr Croucher said: "Yes, but only if you will reduce your commission." The auctioneers said: "Our usual commission is 5 per cent. That is £20.50 on £410. But we will reduce it to £15 if you will accept the £410." Mr Croucher said: "I am agreeable to sell the car on those terms." The auctioneers then treated the sale as concluded subject to the conditions of sale exhibited in the sale room. They filled in the entry form describing the car as sold for £410 and their commission as £15. They made out an invoice for the car to Mr Worth for £410 and in accordance with their conditions of sale added an "indemnity fee of £2 plus VAT at 10 per cent." They debited Mr Worth's account with that sum and gave him a purchase slip authorising him to take the car away. He did so.

Next day the auctioneers checked with Hire Purchase Information (HPI) so as to see if any of the finance companies had advanced any money on the car. HPI said they had no entry against it. So the auctioneers regarded Mr Croucher as having a good title to the car. They paid him the £410, less £15 commission and 50 pence service charge and £1.55 to check HPI, making £392.95.

A few days later Mr Croucher was made bankrupt. He had not paid the instalments due on the car. So R H Willis, the car dealers, sought to have recourse against someone or other. The purchaser had disappeared, so we are told, and so had the car. So they came down on the auctioneers. They said that the auctioneers were liable in conversion and they claimed from the auctioneers the £275 oustanding instalments on the car. R H Willis issued their plaint in the county court on November 5 1973. The case was adjourned for a long time awaiting the decision of the House of Lords in *Moorgate Mercantile Co Ltd v Twitchings* [1977] AC 890. It was then restored for hearing on this one point: Were the auctioneers liable in conversion? The county court judge held that they were. The auctioneers appeal to this court. They want to know how they stand in regard to "provisional bids," which are now an established feature of their trade.

The question that arises is the usual one: which of the two innocent persons is to suffer? Is the loss to fall on the motor-car dealers? They have been deprived of the £275 due to them on the car. Or on the auctioneers? They sold it believing that Mr Croucher was the true owner. In answering that question in cases such as this, the common law has always acted on the maxim *nemo dat quod non habet.* It has protected the property rights of the true owner. It has enforced them strictly as against anyone who deals with the goods inconsistently with the dominion of the true owner. Even though the true owner may have been very negligent and the defendant may have acted in complete innocence, nevertheless the common law held him liable in conversion. Both the "innocent acquirer" and the "innocent handler" have been hit hard. That state of the law has often been criticised. It has been proposed that the law should protect a person who buys goods or handles them in good faith without notice of any adverse title, at any rate where the claimant by his own negligence or otherwise has largely contributed to the outcome. Such proposals have, however, been effectively blocked by the decisions of the House of Lords in the last century of *Hollins v Fowler* (1875) LR 7 HL 757, and in this century of *Moorgate Mercantile Co Ltd v Twitchings* [1977] AC 890, to which I may add the decision of this court in *Central Newbury Car Auctions Ltd v Unity Finance Ltd* [1957] 1 QB 371.

In some instances the strictness of the law has been mitigated by statute, as for instance by the protection given to private purchasers by the Hire Purchase Acts. But in other cases the only way in which the innocent acquirers or handlers have been able to protect themselves is by insurance. They insure themselves against their potential liability. This is the usual method nowadays. When men of business or professional men find themselves hit by the law with new and increasing liabilities, they take steps to insure themselves, so that the loss may not fall on one alone, but be spread among many. It is a factor of which we must take account. See *Post Office v Norwich Union Fire Insurance Society Ltd* [1967] 2 QB 363 at p 375; *Morris v Ford Motor Co Ltd* [1973] 1 QB 792 at p 801.

Sales under the hammer

The position of auctioneers is typical. It is now, I think, well established that if an auctioneer sells goods by knocking down his hammer at an auction and thereafter delivers them to the purchaser—although he is only an agent—then if the vendor has no title to the goods, both the auctioneer and the purchaser are liable in conversion to the true owner, no matter how innocent the auctioneer may have been in handling the goods or the purchaser in acquiring them; see *Barker v Furlong* [1891] 2 Ch 172 at p 181 by Romer J and *Consolidated Co v Curtis & Son* [1892] 1 QB 495. This state of the law has been considered by the Law Reform Committee in its 12th Report (1966) Cmnd 2958 as to innocent acquirers; and in its 18th Report (1971) Cmnd 4774 as to innocent handlers. But Parliament has made no change in it: no doubt it would have done so in the Torts (Interference with Goods) Act 1977 if it had thought fit to do so.

Provisional Bids

Such is the position with sales "under the hammer." What about sales which follow a "provisional bid"? I see no difference in principle. In each case the auctioneer is an

A intermediary who brings the two parties together and gets them to agree upon the price. They are bound by the conditions of sale which he has prepared. He retains the goods in his custody. He delivers them to the purchaser on being paid the price. He pays it over to the vendor and deducts his commission. So, in principle I think that on a "provisional bid" an auctioneer is liable in conversion, just as when he sells under the hammer. There are two decisions, however, which suggest a difference. Each followed a dictum of Bramwell LJ in *Cochrane v Rymill* (1879) 40 LT 744 at p 746. One is the decision of the Court of Appeal in *National Mercantile Bank Ltd v Rymill* (1881) 44 LT 767. The other B is the decision of the Divisional Court in *Turner v Hockey* (1887) 56 LJQB 301. In those two cases it was held that the auctioneer was not liable in conversion, because he had not actually effected the sale. It had been made by the parties themselves. I doubt whether those decisions are correct. Although the auctioneer had not actually effected the sale, his intervention in each case was an efficient cause of the sale and he got his commission for what he did. To my mind those two decisions are a departure from the principles stated by Blackburn J in *Hollins v Fowler* (1875) LR 7 HL 757 at pp 766-7. That is the principle which should guide us, especially as it was inferentially accepted by the House C of Lords. I cannot help thinking that in those two cases the courts were anxious to protect the auctioneer, as an innocent handler, from the strictness of the law. In doing so they introduced fine distinctions which are difficult to apply. I do not think we should follow those two cases today, especially when regard is had to the insurance aspect to which I now turn. It is clear that the auctioneers insure against both kinds of sale equally. On every one of the sales, under the hammer or on provisional bids, the auctioneers charge an "indemnity fee" to the purchaser. He has to pay a premium of £2 on each vehicle purchased. In return for D it the auctioneers, British Car Auctions Ltd, through an associate company, the Omega Insurance Company Ltd, insure the purchaser against any loss he may suffer through any defect in title of the seller. So if the true owner comes along and retakes the goods from the purchaser or makes him pay damages for conversion, the auctioneers (through their associate company) indemnify the purchaser against the loss. The premium thus charged by the auctioneers (through their associate company) is calculated to cover the risk of the seller having no title or a defective title. That risk is the same no matter whether the true owner sues the auctioneer or the purchaser. The auctioneer collects £2 from E every purchaser to cover that risk. We are told it comes to £200,000 a year. Seeing that they receive these sums, they ought to meet the claims of the true owners out of it. This system is the commercial way of doing justice between the parties. It means that all concerned are protected. The true owner is protected by the strict law of conversion. He can recover against the innocent acquirer so that the loss is not borne by any single individual but is spread through the community at large. The insurance factor had a considerable influence on the Law Reform Committee. In view of it they did not recommend any change in the law: see the 18th Report, para 48 and note 2. So also it may properly have F an influence on the courts in deciding issues which come before them.

Conclusion

My conclusion is that, where goods are sold by the intervention of an auctioneer, under the hammer or as a result of a provisional bid, then if the seller has no title, the auctioneer is liable in conversion to the true owner. I would dismiss the appeal accordingly.

Agreeing that the appeal should be dismissed, ROSKILL LJ also expressed doubts about the correctness of the decisions in *National Commercial Bank Ltd v Rymill* and *Turner v Hockey*.

Also agreeing, BROWNE LJ said: Mr Friedman rightly G accepted on behalf of British Car Auctions Ltd (at any rate as far as this court is concerned): (a) that at the relevant time the plaintiffs were entitled to immediate possession of this car and were therefore entitled to sue in conversion (see the so far unreported decision of this court in *Union Transport Finance Ltd v British Car Auctions Ltd*, February 16 1977, Bar Library Transcript No 87); (b) that when an auctioneer sells under the hammer a chattel to which the purported seller has no title he is guilty of conversion—see *Cochrane v Rymill* (1879) 40 LT 744, 27 WR 776; *Barker v Furlong* [1891] 2 Ch 172; *Consolidated Co v Curtis & Son* [1892] 1 QB 495; *Union Transport Finance Ltd v British Car* H *Auctions Ltd*—above—(c) that for this purpose there is no distinction between a sale by an auctioneer under the hammer and a sale by an auctioneer by private treaty.

In this case there was no sale under the hammer. Mr Friedman submits that there was no sale by the defendants by private treaty, and that what the defendants did, something less than a sale by them, did not amount to conversion of the car by them; we had a long, able and interesting argument about the nature of the action of conversion, about the meaning and scope of such metaphorical (and I think unhelpful) phrases as "conduit pipe" (or "mere conduit J pipe") and "ministerial acts," about the meaning of "mere asportation" and about the authorities on these problems. But if on the facts of this case there was a sale of the car by the defendants none of these interesting problems arise.

Very naturally, by the time of the hearing no one from the defendants could remember anything about this particular transaction, and it can only be assumed that it was carried out in accordance with the defendants' usual practice as described by Mr Plumridge in his evidence. Mr Plumridge agreed that the intention of the defendants' fairly elaborate "provisional bid" system was for a sale to take place "because we are in commerce to make money and we are K delighted if a sale can be agreed so that we can have the negotiating fee." He was asked: "So would you accept my proposition that this provisional sale could be, as it maybe is, designed to help you sell cars and make money," and answered "Yes, I do." When a car is put up for auction and fails to reach the reserve, Mr Plumridge said (rightly I think) that the defendants were then without instructions as auctioneers. As he said ". . . it brings it into the realm of private treaty sale." But even before they get further instructions, the rostrum clerk goes to the highest bidder, taking with him the purchase slip. This is the same document as that which would be used if the car had been sold under L the hammer, and contains the following, to be signed by the purchaser: "I acknowledge purchased/provisionally purchased this lot at the price shown and in accordance with the conditions of sale as exhibited"; these are of course the conditions of sale to be incorporated in a contract between the defendants and the purchaser. It also contains a space for the price—and "+ £2 for indemnity policy with the Omega Insurance Co Ltd"; this refers to clause 20 of the conditions of sale, which requires the purchaser to take out a policy with Omega (a company associated with the defendants) "against any loss he may suffer . . . through any defect in title of the seller. . . ." When this "provisional M sale" was completed, the same entry form was used as if it had been sold under the hammer. The seller gets the same "guarantee" of his money from the defendants. The same insurance is effected. If the sale takes place on the same day, the defendants are entitled to the same commission from the seller (condition 9 of the conditions of entry), though in practice they are prepared to negotiate a reduction of their commission or "negotiating fee" as they did in this case. The invoice really seems to me conclusive in itself. It is headed "British Car Auctions Ltd"; it bears their VAT registration number and charges VAT: and it includes £2 for the insurance premium. I am clearly of opinion that this car was sold by the defendants to Mr Worth. Accordingly, they are in my judgment liable to the plaintiffs in conversion.

On my view of the facts the difficult questions raised in argument do not arise, and I think it would be wrong to express any detailed conclusions about them which would be *obiter*. I will only say—(a) whether or not *National Mercantile Bank v Rymill* (1881) 44 LT 767 was rightly decided, it is, in my view, clearly distinguishable from the present case on the facts; (b) I entirely agree with the criticisms of *Turner v Hockey* (1887) 56 LJQB 301 made by Collins J (as he then was) in *Consolidated Co v Curtis & Son* at pp 502-503.

The appellants have entirely failed to satisfy me that the judge's finding that they sold the car was wrong. I think it was right. I would therefore dismiss this appeal.

The appeal was dismissed with costs. Leave to appeal to the House of Lords was refused.

FEES AND COMMISSIONS

QUEEN'S BENCH DIVISION

June 22 1977

(Before Mr Justice MACKENNA)

HENRY SMITH & SON v MUSKETT (SHERATON SECURITIES (INTERNATIONAL) LTD, THIRD PARTY)

Estates Gazette May 20 1978

(1977) 246 EG 655

Estate agents' claim for commission on sale of houses—Any commission due to agents to be paid by purchasers—Claim for commission disputed on ground that an arrangement made by B (at that time a partner in the agents' firm) with one of the possible purchasers was not consistent with agents' duty to vendor—Arrangement concerned future letting agency for developed site comprising area of these and other houses—Agents held to be disentitled to commission on this ground and also because B's negotiations were not the effective cause of the sale

In this action Henry Smith & Son, estate agents, of Horsham, Sussex, claimed commission against the estate of the late Mrs Neville in respect of the sale of two houses, 44 and 46 Springfield Road, Horsham, to Sheraton Securities (International) Ltd, the third party in these proceedings. The defendant, Mr Muskett, was the personal representative of Mrs Neville. The facts are fully set out in the judgment of MacKenna J. It was stated at the hearing that Mr Baker, who was concerned in the negotiations described in the judgment, had ceased to be a partner in Henry Smith & Son some time before.

Alan H Ward (instructed by Lovell, Son & Pitfield, agents for Anderson, Longmore & Higham, of Petworth) appeared on behalf of the plaintiffs; Jeffrey Burke (instructed by L B Marks & Co, agents for G S Godfree & Co, of Brighton) represented the defendant and the third party.

Giving judgment MACKENNA J said: Henry Smith & Son, the plaintiffs, a firm of estate agents in Horsham, claim commission for services in connection with the sale of two houses once owned by Mrs Neville, now deceased, whose personal representative, Mr Muskett, is the defendant. The houses were sold to Sheraton Securities (International) Ltd, whom I shall call "Sheraton," who are the third parties, and it was one of the terms of this contract that Sheraton would on completion of the sale pay any commission owed by Mrs Neville to the estate agents. Sheraton do not dispute that, if Mrs Neville's estate is liable to pay the commission, they are bound to indemnify it. The only question in the action is whether Henry Smith & Son are entitled to commission from Mrs Neville and if so at what rate.

Mr K J Baker, formerly a partner in Henry Smith & Son, was the only witness called by the plaintiffs. I say at once that his evidence was not convincing and that I do not accept it, except where it is confirmed by the contemporary documents, or where it agrees with Mrs Neville's statement put in under the Evidence Act, or with the evidence of witnesses called by the other side.

The houses in question were 44 and 46 Springfield Road, Horsham, and they were sold to Sheraton on September 10 1973 at high prices, £23,050 for 44 and £34,575.60 for 46. There was a property boom at the time of the sale and Sheraton, who agreed to pay these prices, wished to develop for commercial purposes an area of Springfield Road which included 44 and 46 and other neighbouring properties on the even numbered side of that road. Another company, Mountjoy Investments Ltd, whom I shall call "Mountjoy," had similar schemes in mind and were competing with Sheratons for the purchase of the properties.

The case against the plaintiffs can be stated in a few sentences. It is that Mr Baker, unknown to Mrs Neville, had made an agreement with Mountjoy which gave his firm an interest in Mountjoy's acquiring Mrs Neville's houses, that in the negotiations for the sale, because of this interest, he favoured Mountjoy, and that Sheraton eventually became the owners not because of Mr Baker's efforts but in spite of them.

According to Mrs Neville's statement, which I accept, she went to see Mr Baker in April 1973. She wanted help in obtaining planning permission for the development of her properties, which, if it were obtained, would enable her to sell them at a better price. She was at this time acting along with the Dances who owned 40 and the Straughins who owned 42, and they too visited Mr Baker. He appears to have advised them not to apply for permission but to leave this matter to the property developers. On June 22 Clifford Dann & Partners, a firm of estate agents at Burgess Hill, wrote to Mr Neville. They said that they were acting for Mountjoy and asked whether he would consider selling them the freehold interest of 46. They did not know that the wife was the owner of 46 and that she also owned 44. Similar letters were written by Clifford Dann to the Dances and to the Straughins, and they, as well as Mrs Neville, referred the letters to Mr Baker.

On July 9 Mr Baker wrote to his friend Mr Squire of Clifford Dann's in these terms:

Dear Malcom,

40-46, Springfield Road, Horsham

Further to your letters to the owners of the above properties they have instructed me to act on their behalf. Perhaps we could therefore meet to discuss the matter further.

They met on July 17 or 18 in Mr Baker's office, along with Mr David Alcock, who was Mountjoy's director handling this business. The effect of their discussion is given in Mr Alcock's letter to Mr Baker of July 19, which I shall quote:

Dear Keith,

I confirm the meeting held at your office yesterday to discuss the possible development of 38-46 Springfield Road. I understand that you are acting for the owners of 40-46 and that we shall negotiate with you in the purchase of these properties. With regard to 38 and 38a it was agreed that you will act in the purchase of these properties jointly with Clifford Dann and Partners. Should we be successful in the purchase of all the properties and subsequently proceed with the redevelopment, I further confirm that you will be retained in the reletting on a joint basis with Clifford Dann and Partners, and paid fees based

A on the standard RICS scale of charges. I would be pleased if you would confirm that the above instructions are satisfactory to you.

Mr Baker told me that after the meeting he informed Mrs Neville that he had agreed to act for Mountjoy in the purchase of 38 and 38a and in the letting of the developed property, if that development took place. I do not accept Mr Baker's evidence on this point. I shall return to it later in this judgment.

On July 20, which was presumably the date when Mr Baker got Mr Alcock's letter, he wrote six letters, one to Mr Alcock confirming that he would be happy to deal with the matter as Mr Alcock had suggested, two to the owners of
B 38 and 38a, and three to Mrs Neville, Mrs Straughin and Mrs. Dance. The letters to each of the ladies was in common form. I shall quote that written to Mrs Neville:

44 and 46, Springfield Road

Further to our recent meeting I write to confirm that I shall be happy to act on your behalf, and I am at the present time negotiating with a number of developers who have expressed interest in your properties, but I feel that none of them will come up with a sensible offer until they can acquire more properties in addition to 44 and 46, and I have therefore taken the liberty of writing to the owners of 38 and 38a to find out whether or not they have been approached so that I can liaise with them to
C get the best price for you.

He wrote as follows to the owners of 38 and 38a:

We have been retained by Mrs Neville in connection with the possible sale of her properties, and I understand that your firm may have been similarly approached by developers asking whether you wish to dispose of your premises, and I therefore feel it would be a good idea if we could get together to discuss a mutual approach to these people.

Neither of the owners of 38 and 38a, nor any of the three ladies, would have understood from these letters that Mr Baker's firm were Mountjoy's agents to buy 38 and 38a, and
D it is clear to me that they were not meant to understand it. I should mention that nothing came of this approach to the owners of 38 and 38a.

A few days later, on July 25, Mr Baker wrote a number of letters in common form to property dealers in London, bringing to their attention three sites in North Street, Horsham, and two in Springfield Road, on opposite sides of that road, of which one I think must have been 40 to 46.

On August 7 Mr Rosen of Land and Town Development (London) Ltd wrote to Mr Baker:

Re: 38-46 Springfield Road

E Further to our conversation of yesterday I confirm that we are most interested in purchasing the above site, and would accordingly instruct you to forward the offer of £35,000 subject to contract for the freehold interest and vacant possession of nos 44 and 46 and £17,500 subject to contract each for nos 40 and 42 with vacant possession.

Next day Mr Baker wrote to Mrs Neville:

44 and 46 Springfield Road

I write to inform you that I am now beginning to receive offers for your property, and whilst I turned down some out of hand I would inform you that I have today received an offer of £35,000 subject to contract only for your two properties as above. I
F think it should be possible to get nearer to £40,000 and perhaps you would therefore kindly let me know what sort of figure you would be prepared to accept.

Mrs Neville's statement deals with this letter. She states:

Having got Henry Smith's letter of August 8 my husband went to see Mr Baker of Henry Smith and said that he thought the price should go to £40,000. Mr Baker told my husband that he thought £40,000 was a reasonable price and wanted to know whether my husband would be prepared to accept. My husband said that if Mr Baker thought that was reasonable, then fair enough he would. He (Mr Baker) told me eventually that this would have been shortly after August 8 that a firm called Mountjoy Investments were prepared to go to £40,000.

It is likely that Mr Baker had already been in touch with G
Mountjoy about Mr Rosen's offer before he wrote to Mrs Neville on August 8, and had got from them the figure of £40,000 which would outbid Mr Rosen's £35,000. It is certain that there was a meeting on August 10 between Mr Baker and Mr Alcock of Mountjoy to discuss the purchase of 40-46. Mr Alcock's letter to Mr Baker of August 11 deals with this discussion. It would seem from the second paragraph of the letter that Mountjoy had already offered £20,000 to the Straughins, who were negotiating for the purchase of another house, and that one of the matters discussed with Mr Baker was the completion of the sale of
42 at that price. The letter continues: H

I would further confirm that offers should be made to the owners of 40, 44 and 46 Springfield Road, in the sum of £20,000 each, subject to contract and subject to planning permission. I would be pleased if you would inform me as to the various vendors' reactions. Offers are also to be made for the freehold of nos 38 and 38a Springfield Road, in the sum of £50,000 each subject to contract and subject to planning permission. I will leave you to negotiate the best possible terms.

I would thank you for your co-operation and trust that we will achieve a successful event.

Mr Baker replied on August 15:

40-46 Springfield Road, Horsham J

Thank you for your letter of August 11 and I have instructed Mrs Straughin's solicitors accordingly. With regard to nos 40, 44 and 46, I have been able to ascertain that they would be prepared to accept £20,000 each, but subject to contract only and to your company paying legal and agents' fees. The owners would, however, agree to a delayed completion.

Unhappily for Mountjoy other estate agents were in touch with the Straughins. This was Mr Bruce Trent's firm who were acting for the vendors of a property which the Straughins hoped to buy with the proceeds of sale of 42, and who were also retained by Sheraton to find them proper-
ties suitable for development. Mr Trent heard from the K
Straughins that they were selling 42, and that Mrs Dance, who was, I think, Mrs Straughin's daughter, was selling 40 and he told Sheraton, who intervened before Mountjoy had completed the purchase of 40 and 42.

The rivals, Sheraton and Mountjoy, met on August 24 in the office of Tarran, Jones & Co, the Straughin solicitors. Mr Steadman was there representing Sheraton and Mr Alcock for Mountjoy. Each made a bid for 40 and 42. Mr Steadman's was the higher, and contracts for the sale of these properties were exchanged that afternoon.

Sheraton came to know that Mrs Neville was the owner
of 44 and 46 and was willing to sell. Mr Trent says that he L
was given the information by Mr Burstow of Messrs Tarran, Jones & Co on the afternoon of August 24 and that may be true.

Mr Baker says that he met Mr Steadman, whom he already knew, at least by sight, that same afternoon in the street in Horsham, and that Mr Steadman made a joking reference to his company having got nos 40 and 42. Mr Steadman says that he does not recollect this meeting.

There was a chance meeting that night in a public house in Horsham. Mr Steadman and Mr Makepiece, who was another Sheraton executive, were in the public house along
with their agent Mr Trent, when Mr Baker came in. He M
joined the others and they spent the rest of the evening together drinking. What they talked about is in dispute. Mr Baker says that the sale of 44 and 46 was mentioned and that £35,000 was offered by Mr Steadman and turned down by Mr Baker. Mr Steadman and Mr Trent say that there was no discussion of a sale, and that neither of them knew at this time that Mr Baker was Mrs Neville's agent. I do not know who is telling the truth about this, and I do not think it matters.

Four letters were written on August 28, two by Mr Baker and two by Mr Trent. Mr Baker's were to Godfree & Co,

A who were Mrs Neville's solicitors, and to Speechly, Mumford & Soames, who were Mountjoy's. Each of these letters told these solicitors that Mr Baker had that day arranged the sale of Mrs Neville's properties to Mountjoy for £40,000 subject to contract and to the payment of legal and estate agents' fees. Mr Trent's letters were to Mrs Neville and to her solicitors. The letter to Mrs Neville told her that clients of Mr Trent were keenly interested in buying 44 and 46 and that an immediate decision could be made and that contracts could be exchanged within seven days. The letter to the solicitors said much the same.

Mrs Neville passed this information on to Mr Baker, who
B spoke to Mr Trent on the telephone. A price was discussed between them. Less than £40,000 was offered at first, but on Friday August 31 Mr Trent increased his figure to £40,000. He was told by Mr Baker that he would take his client's instructions. Mrs Neville's statement does not mention any discussion of an offer of £40,000 by Sheraton. Mr Baker told me in his evidence-in-chief that he did discuss it with her, but I am doubtful whether that is true. Mr Trent confirmed his offer of £40,000 by letter of September 3 which he says he delivered by hand at Mr Baker's office. On the same day Mr Baker wrote to Mr Trent.

C *44 and 46 Springfield Road, Horsham*

Further to our letter of August 28 addressed to my client Mrs Neville and our subsequent telephone conversation, I write to inform you that I have submitted your revised offer to our client and she has decided to accept a higher offer from another source.

Mr Baker was questioned about this supposed higher offer. In his evidence-in-chief, he said that Mountjoy had by this time increased their offer from £40,000 to £46,000. That was certainly untrue, and in cross-examination he admitted the untruth. He tried to justify the use of the words, "a higher offer" in other ways, but in my opinion unsuccessfully. I believe that he untruly told Mr Trent that his client had a
D higher offer which she had decided to accept as a device to put Sheraton off, so that Mountjoy might get the properties.

Mr Steadman and Mr Trent were doubtful whether their offers were getting through to Mrs Neville, and Mr Steadman was rightly suspicious that there was some kind of tie-up between Mr Baker and Mountjoy. On September 4 Mr Steadman decided to approach Mrs Neville through her solicitors, Godfree & Co. He spoke on the telephone to Mr Muskett, a legal executive of that firm, who made a note of the conversation, which I shall quote:

4 September. Received telephone call from Mr Steadman of
E Sheraton Securities Ltd offering £40,000 for this property. I told him that I had instructions that I was not to negotiate and that Mrs Neville proposed to proceed with the sale that had been negotiated. I was asked whether an offer of £40,000 was likely to be entertained and I said "No." The offer was then increased to £46,000 and I was told that this was the maximum. I pointed out that I had instructions not to negotiate although I would report the offer to my client. This I did.

Mr Muskett must have told Mrs Neville, because she says in her statement, "I heard from my solicitors that £46,000 had been offered, and I went round to Mr Baker and told him of this. He told me that Mountjoys would go to this
F figure."

By September 5, which was a Wednesday, no answer had been received by Mr Steadman or Mr Trent to Sheraton's increased offer of £46,000. Mr Steadman telephoned Mr Baker, who told him that £46,000 was not enough. Mr Steadman raised the figure to £51,000. Mr Baker said that he would take instructions. I do not think that he did take Mrs Neville's instructions on this offer. Her statement does not mention any discussion of such an offer. Twenty-four hours went by without any communication from Mr Baker to Mr Steadman, who was by this time very suspicious. He rang Mr Baker on September 6 and asked him whether he had any decision. He said "Not at the moment." Thereupon, Mr Steadman withdrew the offer of £51,000 and G reinstated the lower figure of £46,000. I asked him why he did this. He answered, "I thought we were being given the runaround. I had a feeling the other company were working off our back. I thought our offers were being passed on to the other company." This was on the Thursday. Mr Steadman got no answer on that day or on the Friday morning.

He and Mr Trent made efforts to meet Mr Baker but did not succeed until lunchtime, when they met at a public house, the Wessex Oak. Mr Steadman, Mr Trent and a friend of theirs, Mr Williamson, local manager of the Abbey National H Building Society, were together in the bar when Mr Baker came in. There was a short conversation of which I have been given conflicting versions. I am satisfied of the following matters: (1) that Mr Steadman and Mr Trent asked Mr Baker whether Sheraton's offer of £46,000 was accepted or rejected; (2) that Mr Baker told them that he could not yet give them a decision; (3) that they asked him to let Mr Trent know by telephone that afternoon; (4) that Mr Baker said that he would; (5) that Mr Baker was not told on this occasion that the offer of £46,000 was Sheraton's last word; (6) that Mountjoy had not at this stage agreed to pay £51,000.

If, on this occasion, Sheraton's offer of £46,000 had been J unequivocally rejected by Mr Baker, and if Mr Steadman and Mr Trent had made it plain that this was Sheraton's last word, it would have been pointless to ask Mr Baker to telephone. There would have been nothing to talk about. But I have no doubt whatever that Mr Baker was asked to telephone and that he said he would. That was Mr Williamson's evidence, agreeing on this point with Mr Steadman and Mr Trent, and disagreeing with Mr Baker, and he was, I thought, a completely reliable witness.

I base the finding that Mountjoy had not at this stage agreed to pay £51,000 upon a note by Mr Muskett of a telephone conversation later that afternoon with Mr Radcliffe K of Speechly, Mumford & Soames, who were Mountjoy's solicitors. Mr Radcliffe told Mr Muskett that no price had been agreed, that Mrs Neville had been asked to say what she wanted, and that if Mountjoy thought this was a fair figure and they approved the terms of the contract they would sign it. An appointment was made to meet at Mr Muskett's office at ten o'clock on the Monday morning.

Mr Baker, who was about to go on his holidays, did not telephone Mr Trent but instead wrote him a letter which he could not receive until the Saturday morning after Mr Baker's departure. This is what he wrote: L

44 and 46 Springfield Road, Horsham

Further to your letters of September 3 and 5 and our various telephone conversations, I write to inform you that your clients' offer is still not acceptable to my client, Mrs Neville and she has therefore sold the property to the other party.

I am actually disappointed at this outcome, but my duty lay with Mrs Neville to obtain the best price.

The statement about his duty was true, but it was not true that the property had been sold to the other party. There was not even an agreement to sell it, and if there had been Mr Baker would not have been disappointed. Why had he not telephoned to Mr Trent? I suspect it was because he did not wish to negotiate with Sheraton but preferred to do a deal M with Mountjoy. Mr Trent got the letter on the morning of September 8 and told Mr Steadman. They were angry. They thought Mr Baker had let them down. They went straight to Mrs Neville at her home who put them in touch with Mr Muskett. They made him an offer which Mr Trent confirmed by telegram.

Re: 44 and 46, Springfield Road, Horsham. Our clients, Sheraton Securities, instruct us to confirm their offer of £55,500 subject to contract for the freehold interest of the above. . . will arrive at your office with my client by 9.30 am Monday to exchange and pay 10 per cent.

On the Monday morning, Mr Alcock of Mountjoy, Mr Steadman and Mr Trent met Mr Muskett in his office. At his request each side wrote down their offer on a piece of paper and gave it to Mr Muskett. Sheraton's offer for the two houses totalled £57,625.60 and Mountjoy's was £53,150. So Sheraton won. Mr Muskett accepted their offer and contracts were exchanged that afternoon. The contracts provided that Sheraton would pay fees and commission.

A few days later, on September 13, Sheraton's solicitors wrote to Godfree & Co. They referred to Mr Baker's letter of September 7 with its incorrect statement that the property had been sold to Mountjoy, and not unnaturally took the point that the writer of the letter "made no contribution to the eventual sale to our clients and on the contrary appears to have endeavoured to avoid such an agreement being entered into." Though Mr Steadman had for some time suspected there was a tie-up between Mountjoy and Mr Baker, he did not yet know of the agreement under which Mountjoy were to employ Henry Smith & Son as their agents to relet the developed site. Sheraton came to know of this agreement only when Mr Alcock's letter of July 19 was disclosed upon discovery.

I turn now to consider the issues in the action. I must first decide what agreement, if any, was made between Mrs Neville and Henry Smith & Son. I find that Mrs Neville employed this firm as her agents to negotiate a sale of her properties to Mountjoy or to any other buyer who might make a better offer for them, and that they were to be paid commission at the usual rate if they negotiated the sale. I infer the existence of this agreement from the correspondence, from Mrs Neville's statement, and from the probabilities of the case. The next question concerns the agreement of July 17 under which Mountjoy were to employ Henry Smith & Son as their agents in reletting the developed site if they acquired it. The making of this agreement was in my opinion inconsistent with the duty owed by Henry Smith & Son to Mrs Neville. It gave them an interest in Mountjoy's becoming the owner of these properties which might conflict with Mrs Neville's interest to sell her properties at the highest price, which might not have been obtainable from Mountjoy. I would be prepared to hold that the mere making of this agreement with Mountjoy disentitled the agents from obtaining commission under their agreement with Mrs Neville, even if it were clear that the agreement had in fact no influence on the agent's conduct of the negotiations. I think that Mr Baker did what he could to negotiate the sale to Mountjoy in preference to Sheraton and that in this matter he may have been influenced by the tie-up with Mountjoy. This is, I think, an additional reason for rejecting the claim for commission.

It was argued by Mr Ward that the sale of the two other properties, 40 and 42, to Sheraton on August 24 had frustrated the agreement of July 17 between the estate agents and Mountjoy, and that the agents' right to commission from Mrs Neville, if it had temporarily ceased to exist, thereupon revived. I do not accept this argument. Nor do I accept Mr Ward's other argument that because the agents in fact made no profit from their agreement with Mountjoy it cannot affect their right to commission.

Mr Burke on the other side argued that Mr Baker's negotiations were not the effective cause of the sale to Sheraton, and that this was a further reason for rejecting the claim to commission. I accept this argument, and for all these reasons reject the claim.

The plaintiffs' claim against the defendant was dismissed and the defendant's claim against the third party was dismissed. The plaintiffs were ordered to pay the defendant's costs, these costs to include the costs payable by the defendant to the third party. The defendant was ordered to pay the third party's costs.

QUEEN'S BENCH DIVISION

February 27 1978

(Before His Honour Judge **LAUGHTON-SCOTT QC**, sitting as a judge of the High Court)

REIFF DINER & CO v CATALYTIC INTERNATIONAL INCORPORATED

Estates Gazette May 27 1978

(1978) 246 EG 743

Disputed claim by estate agents for commission for acting on behalf of intending lessees of office premises—Agents not retained by lessors—Agreement by intending lessees to pay commission if agents found a property suitable in respect of locality, content and price and lease entered into in consequence—Agents not empowered to negotiate terms of lease—Agents' introduction held to be effective cause of transaction despite lessees' earlier knowledge of property—Question of appropriate fee—Quantum meruit as RICS Scale 24 did not apply—Critical observations on scales—One quarter of scale fee awarded

This was an action by Reiff Diner & Co, estate agents, of New Bond Street, London W1, claiming from Catalytic International Incorporated, construction engineers, commission alleged to be due to the plaintiffs in consequence of the leasing by the defendants of premises at Fanum House, Leicester Square, London WC2, the former headquarters of the Automobile Association. The events giving rise to the claim are fully set out in the judgment.

P Sheridan QC and S A Nathan (instructed by Culross, Lipkin & Co) appeared on behalf of the plaintiffs; N A Phillips and Robert Webb (instructed by Linklaters & Paines) represented the defendants.

Giving judgment, JUDGE LAUGHTON-SCOTT QC said: In this action, the plaintiffs, who are a firm of estate agents, claim from the defendants a very substantial sum as the commission asserted to be due to them on the entry by the defendants into a lease of premises at Fanum House, Leicester Square, London WC2. That lease was dated March 17 1977 and was for a period of 10 years from December 25 1975. The claim covers, additionally, commission in respect of options to take further parts of that property which were exercised by the defendants shortly after their entry into it.

The matter comes about in this way. The defendants are an international company of construction engineers specialising in oil-field equipment and industrial plants and are based in Philadelphia, Pennsylvania, in the United States of America. For some 22 years they have carried on a branch of their business in London. In 1974 they were lessees of premises in Kingsway, WC2, and two other nearby office premises into which their expanding business had required them to extend, but the lease of their main premises was due to expire at the end of 1975.

In the autumn of 1974 they began searches for new premises which could accommodate their London enterprise under one roof. The task of discovering a suitable place was delegated in turn to a number of their directors and employees under the general supervision of their local president and director, W E Hesler. He saw over a number of properties as early as November 1974, but was content to leave detailed questions of research to his subordinates, who acted successively in that capacity in the persons of Mr Lewis, an accountant, Mr Gore, their manager of procurement, Mr Gilmore, their vice-president and director, and finally, from the end of August 1975, Mr Parsons, a draftsman.

On November 20 1974 Mr Hesler inspected Fanum House, which was then at the beginning of a process of redevelopment after it had recently ceased to be the head office of the Automobile Association. That property was at that date to be let either as a whole, comprising some seven floors and 117,000 sq ft of office space, or as sub-divided units. Its letting had been placed in the hands of two joint sole agents, Healey & Baker and Michael Laurie & Partners, and at that time these agents were asking rentals of between £12 and £14 a sq ft per annum. Mr Hesler was probably first introduced to the property by the former of these two agents, and looked around it under their guidance. The defendants needed only some 30,000 to 40,000 sq ft or so and were at that point hopeful of finding an entire building with about that area of accommodation which could be called after the name of their company. The asking rent at Fanum House was certainly much higher than the defendants were willing to pay, and the building lay in an area more generally devoted to entertainment than business purposes. Their overtures in relation to this building then ceased until much later, although the letting agents continued hopefully to approach them from time to time. 1974 was a bad period for business generally and the property boom was passing, so the defendants, with time to spare, were content to look elsewhere.

In the months that followed, they made enquiries of a number of estate agents, and by the summer of 1975 had accumulated a large quantity of brochures and other literature about properties to let in central London. In May 1975 the defendants heard of a property which might suit them. This was a 52,000-sq ft building called Grosvenor House, Drury Lane, WC2, which is not, of course, to be confused with an hotel of that name in Park Lane. The letting agents were the plaintiffs, who were asking for that property a rental of only about £6 per sq ft per annum, which precisely coincided with the defendants' calculations. It had previously been the headquarters of a large firm of theatrical costumiers and was itself in the process of redevelopment. It was advertised as being suited to office, showroom and commercial space and stood in an area in which planning permission for entire office user, which was the defendants' requirement, was not easily to be obtained, but the building was likely to be ready for occupation by October 1975, and it is not surprising that the interest of the defendants in it was quickly aroused.

On June 12 1975 the plaintiffs by their partner, Mr Diner, sent a list of four other properties to the defendants which they thought might interest them. He thought the defendants would not get planning permission for their intended user of Grosvenor House. This letter was not replied to, but the defendants continued to show an increasing interest in Grosvenor House. On July 18 Mr Diner sent the plaintiffs a list of seven properties, adding this comment: "With regard to these properties, we are unable to be retained, and would therefore confirm that in the event of your company entering into a contract to take any of these buildings, we will be looking to you for the payment of a scale commission."

On July 28 1975, in a letter which has been lost, the plaintiffs wrote to the defendants offering them property at Greater London House, Hampstead, NW1, including, it is reasonable to assume, a similar formula. A reply dated August 1, which has survived, showed a disinterest in this property because of its location, but without making any reference to the question of commission. Meanwhile, the defendants had advanced, subject to contract, an offer to take Grosvenor House, and it is clear, in my judgment, that had planning permission been forthcoming they would in all likelihood have become the tenants. Shortly before August 1 the defendants held a meeting with the Greater London Council planning officer, at which they staged quite a costly and elaborate demonstration of their intended user of that property. On that very day, Mr Reiff, the other of the plaintiffs' partners, wrote a letter to the defendants in the person of Mr Gilmore, offering the plaintiffs' services to them as sole agents to obtain another property if the Grosvenor House proposal proved unacceptable to the planning authorities, on the basis of half fees related to the scales laid down by the Royal Institution of Chartered Surveyors, plus reimbursement of advertising costs. That letter also was not replied to.

On about August 25 1975 the planning authority informed the defendants that planning permission would not be granted for their proposed user of Grosvenor House. This put them into a position of difficulty. They had precisely four months, not only in which to acquire new premises, but also to move into them, with all the necessary adaptations required to be completed. They were galvanised into action and appointed their employee, Mr G Parsons, to work full-time in accordance with a written directive dated August 27 signed by Mr Gilmore, for the purpose of discovering suitable premises. His contacts with the plaintiffs are crucial to this dispute.

On Wednesday, September 3, Mr Parsons telephoned Mr Diner and what transpired then is not really in dispute. Mr Parsons announced that he had been appointed to find new headquarters for his company as the negotiations about Grosvenor House had proved abortive, and it was very urgent that he found premises for the company to move into by the end of the year, and asked Mr Diner to send him the details of anything that Mr Diner considered suitable. Mr Diner constructed a letter dated September 4 1975 setting out 17 properties, including Fanum House, which Mr Diner did not know, and could not have known, Mr Hesler had already seen some 10 months before. The letter ended with a repetition of the formula previously referred to and this letter was taken by Mr Diner to a meeting with Mr Parsons on the following day, September 5 1975.

Until taking up his assignment Mr Parsons had had no experience or training whatsoever in dealing with estate agents or with the complexities of business. He was a draftsman employed on a modest salary, who seems to have been selected for this duty simply because he had experience in office layout and design. His competence for the task assigned him is perhaps best to be gauged by the fact that he set about looking for vacant properties by walking about the streets of London to observe estate agents' boards, and by selecting estate agents to approach who happened to be advertising in the evening newspapers. Apart from the directive given him by Mr Gilmore, he was merely supplied with a filing cabinet full of estate agents' literature previously received by the company which, in his expressive phrase, was "all higgledy-piggledy."

At the meeting on September 5 it is common ground that Mr Diner went through his list of properties and then referred to the formula at the end, stating that if business resulted through this introduction commission would be payable. Mr Diner then asked Mr Parsons if he had read Mr Reiff's letter of August 1, and asked if the defendants were prepared to give his firm sole agency. Mr Diner gave evidence to the effect that Mr Parsons replied that he was not at that stage prepared to appoint them as sole agents, but he expected only to deal with the plaintiffs. Mr Diner also gave evidence that he told Mr Parsons that if business did result from any introduction of his, the defendants would have to pay commission, to which proposition Mr Parsons agreed. Mr Parsons' version of what happened differs from this. He agreed, at least at first, that he said that if the defendants did retain anyone, it would be the plaintiffs, but added that he did not understand Mr Diner to ask him for commission, even if he was appointed as sole agent. "I told him," he said, "we weren't going to pay them commission." This second version of that meeting compares unhappily with a letter written later by the defendants' lawyer D C Head who, on October 10 1975, replied to the plaintiffs' claim for agreed commission after an ample opportunity of seeing Mr Parsons, which makes no mention of Mr Parsons' refusal to pay commission, just as it does with Mr Diner's active efforts on behalf of the defendants, who on several occasions afterwards took Mr Parsons at his request around premises which might prove suitable to the defendants. Mr Diner had no prospect of obtaining any commission from lessors' agents, and it seems hardly likely that he would have exerted himself if his efforts were to go totally unrewarded. On that meeting I prefer the evidence of Mr Diner. Even if Mr Parsons did not specifically agree to pay commission, he certainly, in my judgment, did so by his conduct then and later.

Mr Gilmore assured the court that he had given the most explicit instructions to Mr Parsons to tell estate agents that they would get no commission from his company. He omitted to mention this fact in his directive of August 27 1975, when he could so easily have done so. I do not accept that this instruction had been made clear, at least by September 5. Mr Hesler in his evidence stated that he had given specific instructions that the defendants were not to employ estate agents for acquiring property, because, as he said, "We considered we had an in-house capability." But it transpired that this instruction was given as early as November 1974, when a very different situation confronted his company from that in September 1975. I do not consider that this instruction was ever made clear to Mr Parsons in early September 1975, who gave me the impression of being a person over-anxious to please everyone and was quite likely to have been amenable to Mr Diner's request, or at least to appear so.

Two other matters of significance emerged at that meeting. The first was that when Fanum House was mentioned by Mr Diner Mr Parsons did not say that he already knew of it. There was a good deal of evidence that the brochure on this building was for a long time lying in Mr Hesler's office. From this, the likelihood is to be concluded that very little, if any, documentation on that property was available to Mr Parsons at the outset of his enquiries. Moreover, I think that at the time Mr Parsons took over these duties, there were two lists of properties considered for possible acquisition by the defendants in existence and no others. These did not include Fanum House. This and many other questions relevant to the documentation available to the defendants at all material times is largely obscured by the unfortunate fact that Mr Parsons destroyed all papers pertinent to the defendants' searches for new premises when the defendants left Kingsway, and this happened despite the fact that the present dispute was then current.

The second important thing at that meeting was that Mr Diner's letter of September 4 included, in a reference to Fanum House, the fact that he had discovered through his enquiries that it might be possible to obtain a lease of the premises at a commencing rent of £8.50 per sq ft per annum, which very much more closely accorded with the defendants' preparedness to pay than did the lowest price hitherto quoted to them for the building at about £11 per sq ft.

On the following Monday. September 8, Mr Diner took Mr Parsons, at his request, round a number of properties, including Fanum House. By then, its internal reconstruction was nearing completion. Mr Parsons asserted in evidence that he had already heard of Fanum House through a number of estate agents, but the fact is that he chose Mr Diner to take him there. He had heard of the plaintiffs through Grosvenor House, and there was, he agreed, a special relationship between them and his company which was considerably more friendly than that with any other estate agents.

On September 8 and 9, Mr Diner sent Mr Parsons details of yet more properties in documentary form, again adding his conventional formula about commission, and at about this time Mr Parsons began drawing up a short list of properties which appeared to be the most suitable to him. These included Fanum House, but I find that that property was not added to that list until after September 18 1975. If it was, it seems likely that the defendant company informed the estate agents, Healey & Baker, inaccurately on that matter.

On September 18 Mr Diner took Mr Parsons to yet another property, called Enford House, Marylebone Road, NW1, and later to yet more properties, until finally, at Mr Parsons' request, he took him and Mr Hesler on October 1 1975 to see a property at 68 Newman Street, W1. During this visit Mr Diner heard Mr Hesler comparing that property with Fanum House. He realised that in his absence Mr Hesler had been there without informing him, and saw fit to raise the subject of his commission. Although there were three distinct versions of how that conversation continued, it is clear that Mr Hesler made out that he had seen the property long before, which would disentitle Mr Diner to any commission in respect of it.

Directly after that meeting Mr Diner wrote a letter to Mr Parsons pointing out that his firm expected commission on any letting of Fanum House to the defendants. The reply to that letter, which was written, after some delay, on October 10, by Mr Head, a letter which is already referred to earlier in this judgment, included this comment: "As it so happens, it seems probable that this company will not take a lease of any of the properties mentioned by you in correspondence. The question of a fee to you is therefore likely to be academic." This was perhaps unfortunate, since three days later a draft of an offer for that very property, or part of it, was prepared for Mr Hesler, who despatched it to Michael Laurie & Partners, the lessors' agents, on the day following. Two similar other properties, it is true, were advanced elsewhere at the time, but a bargain was struck over Fanum House, which by October 22 had been virtually concluded.

Before consideration is given to the issues in the case, a word is needed about the question of estate agents. Since the evidence about this was advanced only from the plaintiffs themselves and was left uncontradicted, the conclusions here reached can hardly be regarded as definitive. This seems to be the position: an estate agent may accept instructions from any intending vendor or lessor of property. He can do so either as sole agent, in which event he gets an entitlement on any sale or lease to scale commission, whoever is instrumental in effecting the contract; or as joint sole agent with another, when each gets three-quarters of the scale fees, whoever effects the contract; or merely act as one of a number of agents of whom only the party successful in bringing negotiations to fruition can recover commission. In each event he is, however, regarded as "retained" by the introducing vendor or lessor. An agent may also receive instructions from an intending purchaser or lessee. He may, if he can achieve it, arrange with the selling or letting agents to receive a share of their own prospective commission. In that event he is to be regarded as "retained" for the purposes of introductions to the property and he is not then allowed to look for any further commission from his own client. If, however, he is not successful in establishing any such retainer, he may then, and only then, look to his own client for commission. The expression frequently used by the plaintiffs in correspondence, "properties for which we are not retained," specifically meant that they could look for commission to the defendants if any introduction of theirs was effected for the defendants' purposes.

Mr Parsons, who made contact with a large number of agents, said that they all used this phrase. He had no idea what it meant and did not trouble to ask. He had a somewhat vague idea that if estate agents acted for his company they would get commission from the letting agents, but he failed to point this out to the plaintiffs. Nor, incidentally, did he tell the plaintiffs that he was in touch with a large number of other estate agents.

One question of fact is outstanding and should here be resolved. It was suggested to Mr Diner that he approached Michael Laurie & Partners for commission in respect of any lease to be granted for Fanum House. An implication in that suggestion may have appeared to have been that Mr Diner was trying to extract commission from both sides. He rejected it vehemently. Evidence was called that he or his firm did so act. What I consider happened was that the plaintiffs did inquire of this through the lessor's agents as a normal routine before advancing property to the defendants, which they did only after they had a clear understanding that no shared commission would be payable from Michael Laurie & Partners.

I now turn to the matters in dispute. First comes the question of whether Mr Parsons had any real or ostensible authority to make any contract with an estate agent involving payment of commission. On this point, the defendants rely, *inter alia*, upon the directive of Mr Gilmore dated August 27 1975. In this, Mr Parsons' instructions were placed under five heads, including one to make contacts with estate agents to seek available office premises. There was no restriction against employing estate agents, but merely a general statement at the end of the document

A saying, "Your responsibility will not include the commercial negotiations." In my judgment this referred, and was intended to refer, to the contractual negotiations preceding any agreement for a lease. Mr Gilmore may at some later stage have prohibited Mr Parsons from paying estate agents, but I do not consider that that can have occurred until well after September 8 1975. Mr Parsons had a wide brief of instruction and a good deal of technical knowledge, at least about the internal layout of office buildings, which would have been apparent to any estate agent, although his ignorance of business matters would not necessarily have been so obvious. In my judgment, he had at least an ostensible authority to make a contract with estate agents.

B Second comes the question of whether any contract was made between the parties. In view of the findings already set out in this judgment, my conclusion is that Mr Parsons made a contract with Mr Diner whereby Mr Diner undertook to act for the defendants as estate agents for reward. The matter can be expressed as shortly as that.

Third comes the question of what were the terms of that contract. My conclusion here is that the contract made was for the plaintiffs to discover or reveal and present a property suitable for the defendants in terms of locality, content and price for leasing for their business premises in consideration C for which, if they entered into a lease of such property, the plaintiffs would receive fees based on the scale laid down by their professional body, the Royal Institution of Chartered Surveyors. By this agreement, I do not find that the plaintiffs were employed to enter into any contractual negotiations as to the terms of any such lease. Such negotiations mean negotiations with the opposite party, that is, the intending lessor or his agents: see *Hoddell and Others v Smith* (1976) 240 Estates Gazette 295. There is no evidence to support such a provision, such as, for instance, any request of the plaintiffs to attempt to reduce any asking rent, and, on the contrary, there is much evidence to show that the defendants regarded themselves as fully equipped to do D their own negotiating.

Fourth comes the question of whether the plaintiffs earned any right to commission—a question complicated by the fact that the defendants, in the person of their president, Mr Hesler, had inspected the relevant property as early as November 1974 and were informed of its availability from time to time between then and July 1975. It is clear from the authorities that the plaintiffs cannot succeed at all unless they can prove that their introduction effectively caused the transaction.

The burden of the defendants' case was that from the very first they regarded Fanum House as a good proposition and had E it constantly in mind, so that, when they finally decided upon it, they were doing no more than opting for a property which always, in their opinion, would have proved an excellent solution to their problems. It was, said Mr Hesler, "a great property." The plaintiffs, on the other hand, aver that Fanum House was virtually rejected by the defendants when they first saw it and, were it not for the defendants' reintroduction to it by them, their intentions in regard to it would probably have remained dormant. The facts were that when Mr Hesler first saw Fanum House he had in mind a very much lower rent than £12 to £14 per sq ft. Also, as he agreed in evidence, the location was not entirely satisfactory. Undoubtedly, the defendants made no advances to F the letting agents at all until the autumn of 1975, and although Mr Hesler impressed upon me that he and others had had this property constantly in view, there is very little documentation to illustrate their interest. The letting agents, Healey & Baker, did send them a number of letters and details starting on April 30 1975, but such reactions from the defendants, as they themselves recall, were either lukewarm or negative, and certainly none of Healey & Baker's letters received any reply. In July 1975 these agents were told that the defendants had about £6 per sq ft in mind as the maximum rental they would pay, and at that date it seems probable that the landlords of Fanum House were unprepared to drop below about £11 per sq ft. Even a document dated as late as September 18 1975 records Mr Hesler as holding G the view that he did not like Leicester Square, although it was said that the defendants would include Fanum House on their short list. Mr Parsons, who did not come on the scene until about then, spoke of Mr Hesler saying that the area was "seedy."

In my judgment, although the defendants knew of the existence of Fanum House and its letting capacity, their attraction towards it was only stimulated by two factors. These were the emergency of their predicament in September 1975, and the fact that it had by then been brought to their notice by Mr Diner that a rental of £8.50 per sq ft per annum was for the first time being entertained by the landlords; a fact which also for the first time made the proposition feasible to the defendants. I consider that H any interest which the defendants displayed over this property did no more than glimmer initially and then, especially while the Grosvenor House negotiations were in progress, became virtually eliminated.

I find as a fact that when Mr Parsons was first introduced to the property on September 5 1975 any knowledge he had of Fanum House was at best sketchy, and probably non-existent, and even if he had any present knowledge of the property he would have disregarded it at that point as being outside his company's price range. His approach to Mr Diner, and Mr Diner's information, created an entirely fresh interest in that property which enabled negotiations to be opened at a stage within the defendant's time and price scale. J It follows that Mr Diner's introduction was, in my view, effective to bring about an agreement for the lease of Fanum House and effectively caused that transaction.

Lastly, I turn to the question of the remuneration which the plaintiffs are entitled to recover. The parties agree that this must either be the appropriate scale already indicated, or, if no such scale is appropriate, then on the basis of a *quantum meruit*. The relevant scales are to be found in the Royal Institution of Chartered Surveyors' Scales of Professional Charges, reprinted July 1975, clauses 19 and 24, where the pertinent passages read: K

> On letting unfurnished premises . . . (1) Where the letting is for a term certain of 12 months or more, 10% on one year's rent . . .
>
> For negotiating a tenancy or lease when acting on behalf of the tenant or lessee: (*a*) for seeking and negotiating the tenancy or lease of a property on the instructions of a client: the fee payable for a letting of the same property under the appropriate letting scale [ie 10 %] . . . (*b*) For negotiating the tenancy or lease of a particular property on the instructions of a client: three-quarters of the fee payable for a letting of the same property under the appropriate letting scale.

It is tempting upon the findings in this judgment pertaining to the terms of the contract to infer that the scale laid down provides L room, upon a proper construction, for a determination that an agent employed only to seek property should receive one-quarter of the scale fee. Both counsel reject this view, and in my judgment rightly, because seeking and negotiating, and negotiating for the lease of a particular property, are not quite one and the same thing. Counsel for the defendants argues that pure seeking is outside the scale altogether, and counsel for the plaintiffs argues that the whole 10 per cent is payable, because the plaintiffs' activities were bound to include an element of negotiation. If not, he asserts, then likewise the scale does not apply.

Because of my findings on the scope of the contract, I conclude that this scale does not apply and, therefore, in its place a *quantum* M *meruit* must be considered. What should this be? It should be something which fairly rewards the work which was done. In my view, the scales referred to give little assistance on this question. They seem, if I may say so, primarily designed for application to quite modest transactions rather than for commercial dealings on very expensive business properties. Whereas for purchasers of real property sliding scales are provided to take account of the purchase price, for tenancies and lettings they are not. Moreover, it is perhaps a little surprising to observe that the remuneration for letting, with all the costs of advertising and so forth which are involved, is the same as for seeking and

A negotiating a tenancy, and only 2.5 per cent more than a mere negotiating for a particular tenancy or lease of a particular property. It is also perhaps rather strange that these entitlements should so exactly or nearly approximate when in the first instance the interests of the principal and agent combine, and in the second they may conflict: the more the client has to pay, the more his agent receives.

Having said that, regard must be had not only to the great body of knowledge that the estate agent must amass in order to be enabled to serve his client at short notice, much of which is inevitably wasted, but also of the trust that he must create and retain with other members of his profession in order to gain an
B insight into the information available to them. It is clearly a hit-or-miss occupation, including many expensive misses.

Mr Phillips argues that the smallest appropriate sum which should be awarded if damages are to be recovered, is one-quarter of the scale fee. Although I think that the plaintiffs did act promptly and efficiently in this transaction, I do not consider C
that they had a very onerous task to perform, compared with the obligations arising in many similar transactions, before this contract was broken by the defendants, and that a remuneration equal to one-quarter of the scale fee is fair to both parties. Fanum House had been on the market for long previously and its availability was well known. The fact that this happens to coincide with the possible construction of the Royal Institution of Chartered Surveyors' scale fees, already referred to, is coincidental and irrelevant. The plaintiffs have abandoned their claim for certain declarations, and the defendants take no point on the plaintiffs' claim being equally referable to the options exercised by the defendants at Fanum House. D

It must follow that there will be judgment for the plaintiffs in the sum of £7,095.60, representing 2½ per cent of the first year's rent for those parts of Fanum House which the defendants entered as tenants in the autumn of 1975.

HOUSING ACTS

COURT OF APPEAL

December 7 1976

(Before Lord Justice CAIRNS, Lord Justice STEPHENSON and Lord Justice BROWNE)

ELLIS COPP & CO v RICHMOND UPON THAMES LONDON BOROUGH

Estates Gazette March 18 1978

(1976) 245 EG 931

Housing Act 1957—Appeal under section 11 against section 9 notice—Whether house capable of being rendered fit at reasonable expense—Test under section 39—Agreed cost of repair works subject to possibility of further works being required when house opened up—Widely different valuations by opposing surveyors of house when works completed—Judge failed to take into account the fact that the agreed cost was a minimum only and the high cost of borrowing money to carry out works—Only one conclusion possible on the evidence—House could not have been made fit at reasonable expense—Appeal allowed and notice quashed

This was an appeal by Ellis Copp & Co, estate agents, against a decision by Judge Figgis at Wandsworth County Court who had dismissed their appeal against a section 9 notice served on them by the London Borough of Richmond upon Thames in respect of a house at 11 Lodge Avenue, Mortlake, SW14. The notice had been served on them as estate agents under section 39(2) of the Housing Act 1957, which provides that the person receiving the rack rent, whether on his own account or as agent or trustee, shall be deemed to be the person having control of the house.

R H Bernstein QC and L J Belcourt (instructed by Bell & Co, of Cheam) appeared on behalf of the appellants; J C Harper (instructed by the Solicitor to the London Borough of Richmond) represented the respondent borough.

Giving the first judgment at the invitation of Cairns LJ, BROWNE LJ said: This is an appeal from a decision of His Honour Judge Figgis given on February 6 1976 in relation to a house at 11 Lodge Avenue, Mortlake, London SW14. The decision was given on an appeal to the county court against a notice served on the appellants by the respondent borough council under section 9 of the Housing Act 1957. I think it is convenient to refer straightaway to the relevant provisions of that Act.

Section 9 provides by subsection (1):

> Where a local authority, upon consideration of an official representation, or a report from any of their officers, or other information in their possession, are satisfied that any house is unfit for human habitation, they shall, unless they are satisfied that it is not capable at a reasonable expense of being rendered so fit, serve upon the person having control of the house a notice—(a) requiring him, within such reasonable time, not being less than twenty-one days, as may be specified in the notice, to execute the works specified in the notice, and (b) stating that, in the opinion of the authority, those works will render the house fit for human habitation.

A new subsection (1A) has later been added, but I do not think I need read that or the rest of the section.

The right of appeal to the county court is given by section 11, subsection (1) of which provides:

> Any person aggrieved by—(a) a notice under the foregoing provisions of this Part of this Act requiring the execution of works,

I can leave out paragraphs (b) and (c)

> may, within twenty-one days of the service of the notice, . . . appeal to the county court within the jurisdiction of which the premises to which the notice . . . relates are situate. . . .

Then subsection (3) provides:

> On an appeal to the county court under this section the judge may make such order either confirming or quashing or varying the notice . . . as he thinks fit and where the judge allows an appeal [under section 9(1) of this Act] requiring the execution of works to a house, he shall, if requested by the local authority so to do, include in his judgment a finding whether the house can or cannot be rendered fit for human habitation at a reasonable expense.

Section 39(1) provides:

> In determining for the purposes of this Part of this Act whether a house can be rendered fit for human habitation at a reasonable expense, regard shall be had to the estimated cost of the works necessary to render it so fit and the value which it is estimated that the house will have when the works are completed.

Those, I think, are the only sections which are directly relevant to this appeal, but Mr Harper, for the borough council, did draw our attention to the alternative powers of the council under sections 16 and 17 of the Act. Section 16 provides: "Where a local authority, on consideration of an official representation," and so forth, "are satisfied that any house—(a) is unfit for human habitation, and (b) is not capable at a reasonable expense of being rendered so fit," they are entitled to serve a notice requiring the owner of the house in effect to say what he is prepared to do about it.

The difference between the two procedures, of course, is that under section 9 the local authority takes the view that the house is capable of being rendered fit at reasonable expense, whereas under the section 16 procedure the local authority takes the view that it is not capable at reasonable expense of being rendered fit.

Then there are various provisions providing that the owner of the house can give undertakings or make offers, and by section 17, if no such undertaking is accepted by the local authority, or if an undertaking is not carried out, then

> the local authority shall forthwith make a demolition order. . . . Provided that if . . . the local authority consider it inexpedient to make a demolition order having regard to the effect of the demolition of that house upon any other house or building, they may make a closing order as respects that house instead of a demolition order.

Those are the two sets of powers available to a local authority when a house is in their judgment unfit for human habitation. They may also have power to buy it under section 12.

The notice under which this case arises was served by the local authority on April 17 1975. That notice was served on the present appellants, Ellis Copp & Co, who are estate agents, under the powers contained in section 39(2) of the Act, and is was accepted that the notice was properly served on them as being "the person who receives the rack rent of a house, whether on his own account or as agent or trustee for any other person." The present appellants appealed to the county court, and their notice of appeal to the county court stated that "The ground of our appeal is that the premises the subject of the notice cannot be rendered fit for human habitation at reasonable expense." That being the issue before the county court judge, he dismissed the appeal to him, and the appellants now appeal to this court.

The premises are described in a joint affidavit between the senior public health inspector of the borough council and a Mr G G Quilter, who was a surveyor acting for the appellants. If I may say so, it seems to be a very helpful and sensible way of putting the agreed facts before the county court and this court. That joint affidavit says this:

> Lodge Avenue Mortlake SW14 is in a residential area, about half a mile from Mortlake Station. The road is a narrow made-up public accessway maintained by the local authority. No 11 Lodge Avenue backs on to an enclosed yard and is a double-fronted 2-storey terraced Victorian cottage.

Then it describes the accommodation and the size of the property, and sets out the rateable value as £230 a year, £165 net. It then sets out the history. Para 8 says:

> A report had been made to the Housing Committee of the Richmond upon Thames London Borough Council on December 18 1974 stating that the house had been found to be unfit for human habitation; the report included an estimate of the cost of repairs at £3,498 and a valuation of the house after carrying out the repairs at £6,500.

That report having been made in December 1974, this affidavit was sworn in November 1975, and in para 10 it is said:

> It is now estimated that the cost of the works has increased from £3,498 to £4,034, both inclusive of VAT at 8 per cent.

Para 11 says:

> The weekly rent of house is now £3.21 inclusive of General Rates of £92.40 per year, and Water Rates of £6.10 per year.

It is common ground that the house is at present let on a controlled tenancy under the Rent Act at the rent there specified.

Before going any further I think it is convenient to go back for a moment to the notice which was served. That notice recites that the council:

> are satisfied that the above-mentioned house [that is, 11 Lodge Avenue] is unfit for human habitation. (2) The Council are not satisfied that it cannot be rendered fit at reasonable expense; and (3) in pursuance of section 9(1) of the Housing Act 1957 the Council require you within a period of one hundred and twenty days ending on the 26th day of August 1975 to execute the following works, which will in the opinion of the Council render the house fit for human habitation, namely, those works set out in the attached schedule.

The schedule sets out in considerable detail the various works, but, in so far as the cost is concerned, the schedule is not itemised as to the cost of any particular item. In view of one of the arguments put forward by Mr Bernstein, I think it is relevant to point out that the schedule includes, for example among external works:

> Cut out distorted and bulging brickwork to back and back addition walls, rebuild and properly tie in to sound brickwork; repair or renew broken and perished window cills.

And internally:

> *Passage.* Open up floor, remove all defective timber, properly secure joists and reconstruct floor using good sound timber. *Front Left Room.* Remove skirting, hack off all damp wallplaster to a sufficient height, treat walls for dampness, properly replaster and bring to a smooth absorbent finish; replace skirting. Open up floor, provide adequate sub-floor ventilation and underfloor site covering, cut out any defective timber, treat as necessary and reconstruct floor using good sound timber.

Mr Bernstein's point is that it is clear from that schedule that it might well be that as that work was carried out it would be discovered that further work not included in the schedule would be necessary. There are other items arising such as the removal of skirtings, taking down ceiling, and so on; open up floor in the front left room; open up floor in the front right room; and open up floor in the back addition room. So it would seem that all the floors had got to be opened up. At the hearing before the county court judge the agreed cost of the repairs at £4,034, referred to in the joint affidavit, stood, but it is clear that the agreement related only to the cost of the repairs specified in the schedule to the notice of April 17 1975, to which I have referred, and that the agreement was subject to the qualification that when the house was opened up to do those repairs it might be found that further work was necessary.

At the hearing two surveyors gave evidence for the appellants and one for the borough council. Subject to one or two points which I shall mention later, I think their evidence and the basis for their different valuations are sufficiently summarised in the judgment of the county court judge. That judgment was in these terms. The judge set out the history of the service of the notice and described the premises from the joint affidavit, and then went on:

> It is admitted that the works specified are required to render the premises fit for human habitation, and it is accepted that the premises are at present unfit. It is further agreed, subject to the qualification that when opened up further works may be required, that the works necessary to comply with the notice will cost £4,034 inclusive of VAT. The Housing Act 1957, section 9, provides:

and then he read the section

> It is agreed that the works specified will render the house fit. The sole ground of appeal is that the expense of carrying out the works is not reasonable.

The learned county court judge then read section 39(1), which I have also read, and continued:

> With these considerations in mind I turn to the evidence. Mr F J Roseberg [who was the first of the valuers for the appellants] valued the property after the works had been carried out at £3,650; this is less by £400 than the cost of the works. This figure was arrived at purely and simply by way of rental valuation. Spending £4,000 will increase the rent to £12.90. The gross rental, less the rates, management, etc, comes to a net rent of £475. He arrives at capital value by taking eight years' purchase. No account is taken of any possibility of vacant possession.

Pausing there, the present rent is, as I have said, £3.21 per week inclusive of rates. The figure of £12.90, as I understand it, includes the increase of 12½ per cent of the cost of repairs to which the landlord would be entitled under section 10 of the Housing (Rents and Subsidies) Act 1975.

The judge went on:

> Mr Quilter [that is the other valuer called by the appellants] quotes an even lower figure of £2,500 after the works had been completed. He made reference to certain comparables: three other houses with rents of from £3 to £5 per week and either controlled or regulated tenancies. The sale price of these properties was between £2,500 and £3,500 and all were occupied by people in their fifties and sixties. Mr Quilter regarded the possibility of vacant possession as an important factor and, therefore, the age of the tenants as important. He had been told that the occupants of the premises were in their fifties, although he referred to the old-age pension. His valuation takes no account of the increased rental which could properly be charged on completion of the works; he accepted that after the works had been completed either a higher rent would be recoverable or vacant possession would be obtained. Mr Quilter's valuation was based on a misconception, the tenants in fact each being 71 years of age, as was discovered during the lunch-break. No family are

living there, and the chances of vacant possession in the future are not all that far removed.

Mr D G F Tibble [the valuer called by the borough council] valued the property after the works had been done at £6,500 based on hopes of early possession and on certain comparables the usefulness of which was really limited to establishing a vacant possession price. He values the property at £12,000 with vacant possession as against Mr Quilter's valuation of £4,000. Mr Tibble puts a present valuation on the property of £4,000, not taking into account the present notice, and £2,500-plus with the notice, as against Mr Quilter's £500.

The differences in valuation show a wide bracket. A very material factor is when the owner might hope to get vacant possession and thereby a capital profit. There are many imponderables—age, health and job status of the tenant. Comparables are not altogether helpful as they do not necessarily include all relevant factors. On the evidence before me I am inclined to agree with Mr Tibble. I place the valuation on the property after the notice has been complied with at £6,000. In its present state Mr Tibble puts on it a value of £2,500-plus. I think that this is too high and value it at £2,000. The cost of works is agreed at £4,000. These works will have to be done in any event and will be useful whenever vacant possession is obtained. On these figures I am satisfied that the works can be done at reasonable expense having regard to all the matters which I have to. In these circumstances, under section 11 I shall make an order confirming the notice with a period of 120 days allowed for the purposes of section 10(1).

Mr Bernstein put before us a very helpful table showing a comparison of the valuations put forward by the various valuers in evidence and the judge's conclusion. I do not think I need go through them in detail. The effect of them is that in all cases the increase in the value of the property is less than the cost of the repairs, though only marginally less in the case of one of Mr Tibble's valuations, which has been called valuation "A," and the judge's valuation. Mr Tibble's valuation "A" was on the basis that he valued the house now subject to the notice which had been served. In the case of Mr Roseberg's and Mr Quilter's and Mr Tibble's valuation on the basis of the house as it is without a notice having been served, the difference is very substantial, varying between £884 in the case of Mr Roseberg and £2,034 in the case of Mr Quilter, Mr Tibble's valuation on that basis being £1,500 deficit.

It is now common ground that in this case the appeal to this court can only be on a question of law, because this case does not fall within any of the categories in section 109(2) of the County Courts Act 1959, which allows appeals on questions of fact. At an early stage in this case Mr Bernstein applied for leave to amend his notice of appeal to this court by adding certain additional grounds. Mr Harper objected, and we decided at that stage to hear the argument initially on the grounds as they originally stood, subject to one amendment to ground 1 to which I shall refer later, and then, if necessary, we said that we would consider the application for leave to amend again later. On the conclusion to which this court has come it is not necessary for us to consider further Mr Bernstein's application for leave to amend, apart from the one amendment to ground 1 which we allowed in the course of the argument.

Ground 1 reads as follows:

> That the learned judge was wrong in law in holding that the aforesaid house was capable at a reasonable expense of being rendered fit for human habitation having regard to his finding (a) that the value of the house when the required works have been completed was £6,000 (b) that the agreed cost of the works was £4,034 and (c) that the present value of the house while subject to the said notice was £2,000.

With the amendment which we allowed by the insertion of the word "minimum" between the word "agreed" and the word "cost" in sub-paragraph (b), that sub-paragraph now reads: "(b) that the agreed *minimum* cost of the works was £4,034."

Ground 2 of the grounds of appeal referred to the question of comparables and was not pursued by Mr Bernstein in this court.

Ground 3 is:

> That the learned judge was wrong in law when arriving at the aforesaid figure of £6,000 by failing to take into account that in order to spend the agreed sum of £4,034 on the property the owners or potential buyers of the property would have to borrow the said sum at the high commercial rate of interest currently charged for such loans, and that this by itself would result in a loss to the present owners or any potential buyer of the property.

Mr Bernstein's submission on ground 1 as he eventually presented it was this. He said that, although on the figures looked at by themselves as set out there was a deficit of £34 when you compared the value of the house after the repairs with the cost of repairs, he felt that that discrepancy was so small that he did not feel able to say that it was something that went beyond the discretion of the county court judge. But he said that the judge went wrong because he had treated the figure of £4,034 as being the cost of repairs, and had left out of account the agreed qualification relating to further work which might be found to be necessary when the other work was started and the house was opened up. He also said that the judge had been wrong to leave out of account the cost of decoration, which, although not covered by the notice, would be, as a matter of commonsense, essential for the landlord to do after he had done the repairs specified by the notice. So far as decorations are concerned, Mr Roseberg said in evidence at page 2: "The figure did not include decorations. Decoration would cost a minimum of £1,000."

The judge in the passage which I have already read on page 1 of his judgment did refer to the qualification of the agreement, but when he came to the end of his judgment he said, without qualification, "The cost of works is agreed at £4,000." Mr Harper says, I think, that the judge cannot have forgotten about the qualification between the beginning and the end of his judgment. Be that as it may, I agree with Mr Bernstein that that qualification is of great importance. The cost of £4,034 was in fact, as the ground of appeal now alleges, a minimum. It is plain that the probability was that the actual cost would be more than that. How much more nobody knows, and there was no maximum on the extra amount which might have to be spent on repairs; further, there was evidence that the decorations, which as a matter of commonsense would have to be done, would have cost a minimum of £1,000. It seems to me that the decision, even on the figures as they stood as originally set out in ground 1, whether or not the house could be made fit at reasonable expense was a very finely balanced one, and, in my judgment, if the judge had taken into account the additional expenses to which I have referred he could not have come to the conclusion to which he did come.

The matter does not stop there, because Mr Bernstein summarised the balance of reasonableness as affecting this house as follows. He said that if the landlord had spent only the £4,034 he would achieve two results: first, he would get an increase in capital value on the judge's findings of £4,000 if he sold the property; and secondly, he would get an increase in rent of 12½ per cent gross as against an expenditure in borrowing at a rate of 18 per cent. To summarise it, there would be an increase in capital value which would be marginally less than the expenditure on the figures. On the figures which the judge accepted he would get an increase of £4,000 on an expenditure of £4,034. That increase, of course, would only be realised if he sold the property and the buyer hoped eventually to get vacant possession. He would get the increase in rent of 12½ per cent of his expenditure. All that would only be on the footing that the £4,034 was the total expenditure, whereas in fact it is a minimum.

This point, it seems to me, is tied up with ground 3. The evidence of Mr Tibble before the judge was that the rate of interest which would have to be paid at that time on borrowing from a merchant bank would be 18 per cent—and that is where the figure of 18 per cent to which I have already referred comes from. Mr Bernstein's point on ground 3 raises two submissions. First of all, he said that the landlord would suffer loss because he would get no return at all on his expenditure on repairs while the house was empty and the repairs were being done; and secondly, that even after the house was occupied and he would get his rent he would still be losing the difference between £12.90 a week rent and the 18 per cent he was paying on the borrowed money. The £12.90 is, as I have said, 12½ per cent on the capital expenditure, but it is a gross figure and not a net figure.

These points on ground 3 are not mentioned in the judgment at all, and I am driven to the conclusion that the judge failed to take them into account. Even if he had them in mind, as Mr Harper suggested, he certainly does not mention them; and it seems to me that if he had taken them into account, and had taken into account the points which Mr Bernstein has made on ground 1 as to the minimum expenditure, he could not have come to the conclusion to which he did come. It seems to me, therefore, that the decision was wrong.

That leaves us with the problem of whether we should allow the appeal and quash the notice, or whether we should order a new hearing. We have before us all the evidence which the learned judge had. This is not a case which in any way turned on the assessment which the judge made of the various witnesses, nor, I think, would a new trial turn on the assessment which another judge might make of the witnesses. The points to which I have referred were, I think, all raised in the evidence before the county court judge, even though he did not refer to all of them in his judgment. For the reasons which I have given, it seems to me that on the evidence which the county court judge had before him there was really only one conclusion to which he could come, namely, that the house could not be made fit for human habitation at a reasonable expense. In those circumstances, it seems to me that we need not put the parties to the trouble and expense of a new hearing, and I would quash the notice. I would allow the appeal accordingly.

I should add this. Mr Harper appealed to us to give some guidance to local authorities about the principles on which they should act under section 9(1). I am afraid that I do not feel able to do that. This case does not seem to me to be a suitable case for trying to lay down any general guidance, and, speaking for myself, I very much doubt whether any hard-and-fast rules of guidance can be laid down in cases like this.

The other point raised by Mr Harper was that in considering the matter one should not look at it only from the point of view of the reasonableness of the expense from an investment point of view of the landlord. This may raise an interesting and difficult question which was not raised in the court below in this case. There was no cross notice, and this court cannot, I think, go into that question or express any opinion about it in this case.

I would, as I said, allow the appeal and quash the notice.

STEPHENSON and CAIRNS LJJ agreed.

The appeal was allowed and the notice under section 9 of the Housing Act 1957 ordered to be quashed. The appellants were awarded costs in the Court of Appeal and below, the costs below to be on scale 3.

COURT OF APPEAL

October 31 1977

(Before Lord Justice MEGAW, Lord Justice BRIDGE and Lord Justice WALLER)

INWORTH PROPERTY CO LTD v LONDON BOROUGH OF SOUTHWARK

Estates Gazette March 18 1978
(1977) 245 EG 935

Housing Act 1957—Appeal under section 11 against section 9 notice—Whether house capable of being rendered fit at reasonable expense—Test under section 39—Regard to be had to estimated cost and to value when works completed—" Value " means open market value—Meaning of " regard shall be had "—County court judge correct in finding that house could not be rendered fit at reasonable expense—Cost of repairs almost as much as open market value when repaired

This was an appeal by the London Borough of Southwark against a decision of Judge Figgis in Lambeth County Court that a house at 112 Cheltenham Road, London SE15, was not capable of being rendered fit for human habitation at a reasonable expense. The judge had, accordingly, quashed the notice under section 9 of the Housing Act 1957 which had been served on the owners, Inworth Property Co Ltd.

Stephen Sedley (instructed by J B Parker, deputy town clerk of the borough) appeared on behalf of the appellants; Richard Fawls (instructed by Stafford Clark & Co) represented the respondent company. Counsel for the respondent company was not called on by the Court of Appeal.

Giving judgment, MEGAW LJ said: This is an appeal from the judgment and order of Judge Figgis in the Lambeth County Court on May 26 of this year. The appellants in this court were the respondents below—the London Borough of Southwark. The respondents in this court were the appellants below—Inworth Property Co Ltd: I shall refer to them as " the owners," and to the appellants in this court as " the local authority."

The owners were, and presumably are, the owners of a house 112 Cheltenham Road, London SE15. By a notice dated January 13 1977 the local authority gave notice to the owners that the local authority were satisfied that the house was unfit for human habitation. That is a notice given by virtue of section 9(1) of the Housing Act 1957. In pursuance of the provisions of that Act, the council required the owners, within a period of 60 days ending on March 16 1977 to execute certain scheduled repairs.

Section 9(1) of the Housing Act 1957 provides, so far as is relevant for the purposes of this appeal, as follows:

> Where a local authority . . . are satisfied that any house is unfit for human habitation, they shall, unless they are satisfied that it is not capable at a reasonable expense of being rendered so fit, serve upon the person having control of the house a notice—(a) requiring him . . . to execute the works specified in the notice, and (b) stating that, in the opinion of the authority, those works will render the house fit for human habitation.

Then provision is made, by section 11, for a right of appeal by the owner where such a notice is served on him. It provides, in subsection (1), that a person aggrieved by a notice under the foregoing provisions requiring the execution of works may, within 21 days, appeal to the county court; and, by subsection (3): " On an appeal to the county court under this section the judge may make such order either confirming or quashing or varying the notice as he thinks fit. . . ." The remaining section of the Act to which it is necessary to refer for the purposes of this appeal is section 39(1). It provides:

> In determining for the purposes of this Part of this Act whether a house can be rendered fit for human habitation at a reasonable expense, regard shall be had to the estimated cost of the works

A necessary to render it so fit and the value which it is estimated that the house will have when the works are completed.

The owners appealed against the notice to execute the repairs. The first ground of their notice of appeal was: "The premises, being unfit for human habitation, are incapable of being rendered so fit at a reasonable expense." The notice contained other grounds, but the owners proceeded in the appeal only on that ground. The appeal was heard before Judge Figgis on May 26. He held that the house was not repairable at a reasonable expense. He allowed the appeal, and he quashed the notice. From that decision the local authority appeals to this court.

B Before the learned judge there was, as he records in his judgment, an agreement between the parties on certain figures. It was agreed that the present price of the house on the open market was £1,300. It was agreed, second, that the price of the house on the open market if repaired would be £1,900. It was agreed, third, that the cost of the required repairs, including value added tax, would be £1,800. So, taking those figures, in order to bring the house into a state where, on the open market, it would have a value of £1,900, the cost required so to do would be £1,800. That would produce a house the value of which, in the open market, was

C £1,900 compared with its present price, on the open market, of £1,300.

The principal argument put forward on behalf of the local authority by Mr Sedley in this court depends upon the meaning to be given to the word "value" in section 39(1). Clearly, if "value" in that subsection means "open market value," then the learned judge was right, on the agreed figures, to hold that, by reference to section 39(1), so far as he "had to have regard" to that, this house was not capable of being "rendered fit for human habitation at a reasonable expense." Therefore Mr Sedley's argument must be, and I understand that it is, that "value" in section 39(1) has some meaning

D different from "value on the open market." To my mind, there is no justification for that argument. It is certainly not to be found in either of the two authorities to which we were referred—*Ellis Copp & Co v London Borough of Richmond upon Thames* (see above) and *Bacon v Grimsby Corporation* [1950] 1 KB 272.

Mr Sedley further argues that section 39(1) provides only that for the purposes of section 9(1) "*regard shall be had* to the estimated cost of the works necessary to render it so E fit and the value which it is estimated that the house will have when the works are completed": therefore, other factors also fall to be taken into consideration for the purposes of deciding whether or not the local authority has arrived at the correct conclusion under section 9(1) when the judge on appeal has to consider that matter under section 11. That, in my judgment, does not arise in this case. I accept that "regard" having to be "had" to a matter indicates that it is not necessarily conclusive. In the present case, however, I have seen nothing to suggest that the learned judge was wrong in the view that he took, which was, that "It may be that there are factors in individual cases to be taken F into account which will affect the result"—that is, other than the section 39(1) calculations; but he found that in this case it was "unnecessary to take into account any of the other minor considerations which might be necessary in a borderline case, because none of these could, in my judgment, so adjust the figures as to show other than that these premises are not capable at a reasonable expense of being rendered fit for human habitation."

Mr Sedley complained that he, having asked the judge for leave to call evidence as to a different method of computation for the purposes of a section 39(1) calculation, was not G allowed to do so. There may well be different permissible methods of arriving at the "value"—both the present value of the house and the value when repaired—for the purposes of section 39(1). But, as I understand it, the evidence that was sought to be tendered here was for the purpose of showing such value being something different from the open market value. The open market value had been agreed between the parties and therefore evidence as to how it was arrived at would have been irrelevant. Accordingly, I do not see, once it is clear (as to my mind it is clear) that "value" means "open market value," that the learned judge could properly have admitted the evidence. He was right to H exclude it.

In my judgment, there is no substance in this appeal, and it must be dismissed.

BRIDGE and WALLER LJJ agreed.

Appeal dismissed with costs. Appellants given leave to apply, if so advised, for leave to appeal to the House of Lords.

LANDLORD AND TENANT
GENERAL

A **CHANCERY DIVISION**

October 19 1977

(Before Mr Justice GOULDING)

HALLIARD PROPERTY CO LTD v JACK SEGAL LTD AND OTHERS

Estates Gazette January 21 1978

(1977) 245 EG 230

**Law of Property Act 1925—Application for leave to amend statement of claim—Additional ground for forfeiture—Landlord claiming to re-enter on bankruptcy of surety—
B Whether statutory notice must be served under section 146 —Notice required—Leave to amend refused**

This was a summons by the plaintiffs, Halliard Property Co Ltd, seeking leave to amend a statement of claim by adding the bankruptcy of a surety as a further ground in an action to recover possession of the demised premises. The first defendant, Jack Segal Ltd, objected to the amendment because no notice was served before issue of the writ pursuant to section 146 of the Law of Property Act 1925.

R M Slowe (instructed by Tobin & Co) appeared for the plaintiff; J K Reynolds (instructed by Donne, Mileham & Haddock, of Brighton) represented the first defendant.

C Giving judgment, GOULDING J said that this summons was one of two in the same action and raised a short point in the law of landlord and tenant.

The first-named defendant, Jack Segal Ltd, was tenant under a lease dated June 24 1964, made between Isiah Segal, the landlord, the tenant, and two sureties, namely, the landlord Isiah Segal and Robert Emmanuel Segal. It was for a term of 21 years less three days from Lady Day 1964 and contained a proviso in a not unusual form for re-entry, in the event of the tenant's voluntary or compulsory liquidation, or the bankruptcy of the tenant or sureties. The present landlord, the plaintiff, had brought the action to recover
D possession of the demised premises on the ground of a forfeiture under the proviso by reason of a breach of covenant, namely, an unauthorised underletting of part of the premises to the second defendant.

The landlord now sought leave to amend its statement of claim by adding a further ground, namely the bankruptcy of one or both of the sureties. The tenant objected to the proposed amendment because, he said—and it was admitted as a fact—that no notice was served before issue of the writ pursuant to section 146 of the Law of Property Act 1925. In relation to the bankruptcy of sureties, his Lordship said, the application for leave to amend raised a short point whether a
E landlord claiming to re-enter under such a proviso must first serve the statutory notice as he must do in many other cases of a right to re-enter. He had been told that there was no direct authority on this particular question.

Mr Reynolds for the tenant had said quite plainly that where there was a condition for re-entry in case of the bankruptcy of a surety there was a breach of condition if the surety went bankrupt and the lessor must serve the statutory notice. The case was one of a right of re-entry for a breach of condition in the lease, and since the bankruptcy was not
that of the lessee, the case could not fall within any of the F
exceptions to section 146 defined in subsections (9) and (10). He said that clearly a right of re-entry arising on bankruptcy was *prima facie* within the section under the words of subsection (1); otherwise why should the original Act of 1881 have thought it necessary to except from the provisions of the statute a condition for forfeiture on the bankruptcy of the lesssee.

Mr Slowe for the landlord said that section 146 did not apply to all rights of re-entry under a lease but only to those rights of re-entry which arose under a proviso or stipulation in the lease for a breach of a covenant or condition
in the lease. Breach of a covenant or condition, Mr Slowe G
submitted, involved a voluntary or unjustifiable violation or transgression of some obligation. Where, therefore, the event on which the landlord under the condition had a right to re-enter was one such as the bankruptcy of a third party which the tenant had no power to control, it was quite incorrect to speak of a breach of condition and the section did not apply. Therefore no notice was required. Mr Slowe took a series of examples of cases in which under the provisions of a lease a landlord might have had a right to re-enter. The reason, Mr Slowe submitted, why the later legislation had required the notice in certain types of case
on re-entry for bankruptcy was because the policy of H
Parliament was to discourage collusive forfeitures on the tenant becoming insolvent.

He (his Lordship) did not propose to try to draw a theoretical line in the present judgment. It was quite apparent from the language of the section in its successive drafts enacted by Parliament that, subject to the statutory exceptions, a condition for forfeiture on the bankruptcy of the lessee was considered as a case of breach of condition. He was unpersuaded by Mr Slowe's attempted distinction between that case and the one of bankruptcy of a surety. A surety was a person put forward by the lessee in support
of his obligations under the lease, and the commonsense of J
the matter seemed to him to require the bankruptcy of the surety to be within the section if the bankruptcy of the lessee *prima facie* was included.

Breach of a condition was not necessarily a voluntary or avoidable act of the person bound by the condition. It was enough for him (his Lordship) to say that, applying his mind to the section as it stood, he was satisfied that it was a normal and reasonable interpretation of the words "breach of condition" in subsection (1) to hold that they applied to the bankruptcy of the surety in the present case.

The consequence was that leave to amend the statement K
of claim must be refused.

A **COURT OF APPEAL**

July 19 1977

(Before Lord Justice **BUCKLEY**, Lord Justice **ORR** and Lord Justice **GOFF**)

PAKWOOD TRANSPORT LTD v 15 BEAUCHAMP PLACE LTD

Estates Gazette January 28 1978

(1977) 245 EG 309

Rules governing relief against forfeiture of lease and unreasonable withholding of consent to assignment—Challenge by landlords of decisions favourable to tenants —Whether service of notice under section 146 of Law of
B Property Act 1925 a step in a "proceeding" to enforce a right of entry or forfeiture, so that relief can be claimed by a tenant—Whether grounds for refusal of consent to assignment of lease unreasonable—Landlords' real object to obtain possession and secure benefit of increased rental value, thus making good their previous failure to invoke effectively the rent review clause in the lease—Decisions in favour of tenants upheld on appeal

This was an appeal by landlords, a company known as 15 Beauchamp Place Ltd, against a decision by Foster J granting tenants relief against forfeiture and a declaration
C that the landlords had unreasonably withheld consent to a proposed assignment of an underlease of two rooms on the second floor of 15 Beauchamp Place, London SW3. The tenants in whom the underlease was vested were Pakwood Transport Ltd. The tenants were the plaintiffs in the action before Foster J and the respondents in the present appeal.

M F Harris (instructed by Rendell & Co) appeared on behalf of the appellants; B K Levy (instructed by Frere, Cholmeley & Co) represented the respondents.

Giving the first judgment at the invitation of Buckley LJ, ORR LJ said: This is an appeal by a company, 15 Beauchamp Place Ltd, who are the successors in title to the grantor of
D an underlease of premises consisting of two rooms on the second floor of 15 Beauchamp Place, London, against an order made on July 16 1976 by Foster J, whereby on an originating summons by Pakwood Transport Ltd, the assignee of the underlease, for relief against forfeiture and also for a declaration that the defendants had unreasonably withheld consent to a proposed assignment of the underlease, he granted the plaintiffs both the reliefs sought.

The underlease in question was dated December 4 1969 and was made between a Mr Pirkis as underlessor, Pakwood Transport Ltd (that being a different legal entity from the present respondents) as underlessee, and a Mr Tagart as
E guarantor of the underlessee's obligations. It was for a term from November 24 1969 to December 25 1979 at an annual rent of £600 and contained *inter alia* a provision, which has not in the event been operated, whereby the lessor could give notice requiring a review of the rent from the third year of the term; a covenant, contained in clause 2(k), against assignments without the previous consent in writing of the lessor, such consent not to be unreasonably withheld in the case of a responsible and respectable assignee; and also a right of re-entry (clause 4(a)) in the events *inter alia* of the rent being arrear or of the underlessee or its successors in title becoming bankrupt or, being a company, ". . . entering into liquidation
F whether compulsory or voluntary (except voluntary liquidation of a solvent company for the purpose of reconstruction or amalgamation)."

On August 9 1971 the appellants became successors in title to Mr Pirkis, the underlessee, and on December 1 1972 they granted to Shoreham Terminals Ltd, that being the new name of the original Pakwood Transport Ltd, a licence to assign to the new company, named Pakwood Transport Ltd (to which I shall refer as "Pakwood") Mr Tagart, the guarantor in the underlease, also joining in this licence as guarantor of the proposed assignee's obligations. On January 15 1976 Pakwood passed a resolution to wind up, having two days
earlier filed a declaration of insolvency, and a Mr Chilver was G
on the same day, January 15, appointed liquidator. But the fact of that resolution being passed did not come to the notice of the appellants until some time afterwards.

I turn at this point to the correspondence. On February 5 1976 Pakwood's solicitors wrote to the appellants' solicitors requesting consent under clause 2(k) of the underlease for Pakwood to assign the underlease to a company named County Properties (Sussex) Ltd and they enclosed references in respect of that company. On February 10 the appellants' solicitors replied in the following terms:

> Thank you for your letter of the 5th February enclosing refer- H
> ences. The references which you enclosed relate of course to County Properties (Sussex) Ltd, which is a limited company with a nominal capital of £1,000 and we would not be prepared to advise our clients to give licence to assign to a company without, of course, the supporting guarantees of the directors. If the directors are prepared to guarantee the rent and covenants, perhaps you would be good enough to supply references for them also, and our clients will consider the matter.
>
> We shall also require to know whether the assignees propose to occupy the premises themselves and, if so, for what purpose. Since there are restrictions on user in the lease, we would not be prepared to proceed with the application until this information
> is provided. J

To that the assignees' solicitors replied on February 11:

> We thank you for your letter of the 10th February. We have sent a copy of your letter to the proposed assignees' solicitors and immediately we have the information requested we will write to you again.

On February 24 the appellants' solicitors wrote as follows to the liquidator of Pakwood:

> We refer to the telephone conversation which we had with your office this morning, when we pointed out that the entry into liquidation by the company now known as Pakwood Transport
> Limited gives rise to a condition for the forfeiture of the lease K
> of two rooms on the second floor of 15 Beauchamp Place, London SW3, which had been vested originally in the company known as Pakwood Transport Limited who changed its name following the assignment to the company now known as Pakwood Transport Limited.
>
> We accordingly enclose notice under section 146 of the Law of Property Act 1925 in duplicate, and we should be obliged if you would kindly acknowledge receipt by signing the endorsement on the duplicate of this notice—

and they go on to say that they have also sent a copy of the notice to the solicitors for Pakwood.

On February 24 Pakwood's solicitors wrote as follows to L
the landlords' solicitors:

> With further reference to your letter of the 10th February, we understand that County Properties (Sussex) Ltd would prefer the rent and performance of the covenants to be guaranteed by J H & F W Green Ltd, a company that has an issued share capital of approximately £436,411. Would you please let us know whether this guarantee will be acceptable to your clients in principle as if so we will immediately arrange for references to be taken up.
>
> We are informed that County Properties will not use the premises other than for their own office premises and we look forward to hearing from you as quickly as possible.

I should mention at this stage that there had been earlier M
proceedings between the parties, the first of them being an application by Pakwood for relief against forfeiture for non-payment of rent—a sum of £300, half a year's rent. On that occasion the landlords had effected what they claimed was a peaceable re-entry and what the judge described as having been a clandestine re-entry, which he held to be ineffective, and he granted relief upon that application on November 22 1975. The other earlier proceeding was one in which the appellants claimed that they had effectively invoked the rent review clause. That was a matter of construction of the lease and, in a judgment delivered on January 22 1976, the

A judge held that the clause had not been invoked within the time allowed for that purpose.

On February 25 1976 the landlords' solicitors wrote as follows:

> We were somewhat surprised to learn on making a routine company search that Pakwood Transport Limited resolved to go into liquidation, and in the circumstances, as we have pointed out to the liquidator, our clients are entitled to forfeit the underlease pursuant to the proviso for re-entry contained in clause 4 thereof. We have, accordingly, today served notice under section 146 of the Law of Property Act 1925 and enclose a copy for your information.

B

They add, in a postscript to that letter:

> Since typing the above, we have received your letter of 24th February. Having regard to the notice which has now been served, the question of licence to assign will not fall for consideration.

On March 6 1976, they wrote again as follows: "With further reference to your letter . . . we are surprised to note the second paragraph of your letter, having regard to the enclosed particulars dated 5th March 1976. We can only assume that your clients are misleading you in those circum-
C stances." The particulars attached were estate agents' particulars of the premises at 15 Beauchamp Place, describing them as being on lease at a low rent with a moderate premium, and "available at rack rent without premium." That letter was acknowledged by Pakwood's solicitors on March 8, who said that they had written to the solicitors acting for County Properties (Sussex) Ltd requesting their views.

The summons came on for hearing on July 16 1976, there being before the judge three affidavits but no oral evidence being taken. The first affidavit was by the liquidator of Pakwood, Mr Chilver, who deposed that the underlease was of value and that he intended to assign if relief were granted.
D He referred to the earlier proceedings which I have mentioned and he claimed that the defendants had left no stone unturned in an endeavour to obtain the property, the market value of which, he deposed, was greatly in excess of the rent reserved. He claimed also that there was no prejudice if the guarantee which had been requested were given by a company. The second affidavit was by Mr Prior, a director of the defendants, who recited the relevant provisions of the underlease and claimed that the originating summons was premature, since proceedings for forfeiture had not yet been instituted, although he admitted that there was an intention to proceed. He denied that there had been unreasonable refusal of con-
E sent to the assignment and he claimed that the defendants required to be informed, before they could give their consent, as to two matters: first, the matter of the guarantee and the second, information as to the proposed occupation. For those reasons, and also the non-disclosure of the liquidation, and the advertisement, he invited the court to dismiss the application. The third affidavit was by Mr Green, who had earlier deposed that he was a person on whose instructions County Properties Ltd were accustomed to act and who now deposed that the acquisition of the plaintiffs' interest in the underlease was part of a larger transaction and that the proposed assignee was not necessarily intending to retain
F the premises; hence the advertisement by the house agents. On that material the learned judge granted to the plaintiffs both the reliefs sought by them.

On this appeal the court is in this difficulty, that there is not available any transcript of the judgment delivered by Foster J; nor was any note taken by either counsel or solicitors who were present on that occasion. The reason why no note was taken was that it was believed that the judgment was being tape-recorded, but it later turned out that that had not been done. In the result we have to deal with this appeal without the benefit of any transcript or any assistance by way of a note made by those present.

The first ground of appeal against the judgment is that the G
judge had no jurisdiction to grant relief against forfeiture, in that the defendants at the time of the issue of the originating summons were not proceeding by action or otherwise to enforce the right of re-entry conferred by the underlease. Section 146 subsection (2) of the Law of Property Act 1925 provides, so far as material to that ground of appeal, as follows:

> Where a lessor is proceeding, by action or otherwise, to enforce such a right of re-entry or forfeiture, the lessee may, in the lessor's action, if any, or in any action brought by himself, apply to the court for relief.

H

The material words for the present purpose in that subsection are the words "where a lessor is proceeding, by action or otherwise, to enforce such a right of re-entry or forfeiture." The appellants' argument has been that the giving of a section 146 notice does not form part of a "proceeding, by action or otherwise, to enforce a right of re-entry or forfeiture," but is merely a step preliminary to a proceeding. The respondents' argument is that the service of a section 146 notice is a step which the law requires to be taken by the landlord in order to enforce a right of re-entry, and that a landlord in serving such a notice may properly be said to be J
proceeding to enforce a right of re-entry or forfeiture.

In support of the latter argument reliance was placed on the decision in *Rogers v Rice* [1892] 2 Ch 170, a decision of this court which is authority for the proposition that under the provision, comparable to section 146(2) and in identical terms, previously contained in the Conveyancing and Law of Property Act 1881, section 14(2), a lessee could not apply for relief against re-entry or forfeiture after the landlord had obtained a judgment of the court entitling him to re-enter on a forfeiture; and it is claimed, and in my judgment rightly claimed, that the same principle must apply where the landlord has peaceably recovered possession. In other words, K
once he has either recovered possession or obtained an order for possession he can no longer be said to be "proceeding by action or otherwise to enforce a right of re-entry or forfeiture." On this basis the argument for the appellant appears to me to involve an absurdity, in that if the landlord has done no more than serve a section 146 notice it is too early for the tenant to apply for relief; but if the landlord's next step is peaceably to recover possession it is then too late for the tenant to apply. For my part, I am not prepared to accept an argument which leads to this absurdity, and I have no hesitation in holding that a landlord who serves a section 146 notice is at that stage "proceeding to enforce a right of L
re-entry or forfeiture" in that the service of such a notice is a step which the law requires him to take in order to re-enter or forfeit. I would therefore reject this first ground of appeal.

The second ground of appeal is that the judge misdirected himself in holding that the defendants had unreasonably refused a licence to assign. It was argued on behalf of the appellants that there were three grounds on which the appellants were entitled to withhold their consent to an assignment. It was claimed in the first place that the appellants were entitled in the circumstances to require as a condition of their granting a licence that there should be a personal M
guarantee in respect of the obligations of the proposed assignee, and it was argued in support of this claim that there had been, in the underlease itself, a personal guarantee by Mr Tagart, of the performance of the lessees' obligations, and that Mr Tagart had similarly joined in the licence of December 1 1972 as guarantor of the obligations of the proposed assignee.

On this issue the respondents relied on the decision of this court in the case of *Moat v Martin* [1950] 1 KB 175, where the tenant's covenant was not to assign without the consent in writing of the landlord, such consent not to be withheld in the case of a respectable and responsible person, and it was held that where the proposed assignee was respectable and responsible the landlord could not refuse consent even on some other reasonable ground. That case, however, differs from the present case in that the underlease here in question does not by its terms oblige the lessor not to withhold his consent when the proposed assignee is a respectable and responsible person, but only provides that consent to an assignment shall not be unreasonably withheld in the case of a responsible and respectable person. In my judgment the result of that is that it would have been open to the judge to conclude that in the particular circumstances it was a reasonable requirement that there should be a surety for the assignee. But this was essentially a question of fact for the judge, who was aware that a corporate surety had been offered in the form of a company having a large issued capital and had, so far as appears, no reason to doubt that that surety would be available, but provided for an interval of time before the order would be drawn up, so that, if any difficulty arose with regard to the terms of the order, it was open to the defendants to come back to the court; but they did not do so. In my judgment it is impossible to say that the judge came to a wrong decision in deciding that in all the circumstances the appellants were not entitled to insist on a personal guarantee.

The second ground on which the appellants allege that they could properly withhold their consent is that they were, so it is alleged, entitled to know to what use the premises were going to be put, and had cause for unease by reason of the estate agents' advertisement, and that that advertisement was not explained until Mr Green swore his affidavit on June 9 1976; but this was a month before the hearing, and in my judgment the appellants cannot claim that they were in that respect taken by surprise. Moreover, there were strong reasons why the judge could conclude that neither of these two considerations, that of the surety and that of the proposed user, was in the mind of the appellants as a ground of refusal, and that their real reason for refusing consent was that they wished to get back the premises because their rental value was far in excess of the rent; which would be an invalid reason for refusing consent to an assignment.

The third ground on which it is claimed that the appellants were entitled to withhold consent is the fact that they were seeking forfeiture of the lease, and this is the reason given in the postscript to their solicitors' letter of February 25 1976, which I have read. What is said in that letter is not, however, that the appellants wish to defer a decision, but that "the question of licence to assign will not fall for consideration." In other words, they refused to entertain the application and from that position they never resiled. In my judgment, whatever justification there might have been for deferring a decision, there can be none for declining outright even to reconsider the application.

In the result, in my judgment the judge was fully entitled to conclude, as he did, that there had been an unjustified refusal of consent. For these reasons I would dismiss this appeal.

Agreeing, GOFF LJ said: Mr Harris argued first that a breach of covenant or condition in a lease does not cause the forfeiture of the lease, but merely renders it liable to forfeiture; secondly, that there can be no forfeiture until the landlord thereafter takes some unequivocal step, in the majority of cases by the issuing and serving of a writ, but in some by obtaining peaceable possession of the premises, if he is able to do so; thirdly, that one cannot get relief against a liability to forfeiture, but only against an act of forfeiture; and finally, therefore, that a landlord cannot be said to be "proceeding by action or otherwise to enforce a right of re-entry or forfeiture" within the meaning of subsection (2) of section 146 of the Law of Property Act 1925 until he takes such an unequivocal step. I agree with my Lord, Orr LJ, that that argument leads to absurdity; first it means that in the case of peaceable re-entry the tenant can have no right to relief at all, because he cannot apply for it until there has been re-entry, by which time it is too late: see *Rogers v Rice* [1892] 2 Ch 170. Mr Harris sought to distinguish that case as being one, as indeed it was, in which the forfeiture was effected by action, but the principle of the case must be the same. Once the landlord has recovered possession, whether by action or by peaceable recovery of possession, the re-entry is complete. To avoid that difficulty Mr Harris suggested that where the landlord re-entered peaceably, he would still be proceeding to enforce the forfeiture for a reasonable time after he had recovered possession. I simply cannot accept that argument.

Then Mr Harris submitted that the court must have jurisdiction to relieve after the forfeiture has been completed, if it is peaceable, as it has in the case of forfeiture for non-payment of rent, and he said that there was no reason why the two cases should be different, but there is, because in the case of rent there is an old common law right to relief, whereas section 146 is a statutory right not previously enjoyed. In my judgment, however, there is indeed no difference in this respect between re-entry for breach of covenant or condition effected by order of the court and peaceable re-entry. I agree, therefore, that while under subsection (1) the service of a notice and, where the breach is capable of remedy, the lapse of a reasonable time thereafter are conditions precedent to the right to enforce the forfeiture, under subsection (2) the service of that notice, which is a prerequisite to completing the forfeiture, is a step by the landlord to that end, and properly within the words "where the lessor is proceeding to enforce his right." In my judgment, therefore, the application by the tenants for relief was not premature.

I turn now to the second point in the case, which is that the learned judge was, it is said, not right to hold that licence to assign had been unreasonably withheld. In my judgment he was perfectly correct in coming to that conclusion. It is clear that the reason why the landlord refused licence was because he saw what he had been seeking for some time, namely, an opportunity to destroy the lease to escape the consequences of his failure to operate the rent review clause, and that cannot be a good ground for refusing licence. I refer to the case of *Re Smith's Lease, Smith v Richards,* reported in [1951] 1 All ER 346 at p 349, where Roxburgh J said: "It has long since been settled that it is not a reasonable ground for refusing consent to an assignment that the landlord desires to obtain possession of the premises for himself." In my judgment that was his reason, and no other.

It is said, however, that the court can look at other reasons not advanced, or even entertained, at the time, but adduced before the court on the hearing of the application for relief. That may be so, but in the exercise of my discretion I do not think it right, on the facts of this case, that the landlords should now be allowed to turn round and rely on other grounds. But even if that be wrong, in my judgment there are no other grounds on which the landlords can rely. Mr Harris sought to rest himself on two points; first, the alleged failure of the tenants to provide a guarantor; and secondly, evidence which emerged from some estate agents' particulars that the tenants might be intending shortly to reassign the demised property.

With regard to the first there might, I think, be a question of construction whether the landlords could reasonably require a guarantor. As my Lord has said, the case of *Moat v Martin* [1950] 1 KB 175 is distinguishable; but I will assume that point in the landlords' favour. Even so, the tenants and their proposed assignees never refused to provide a guarantor; nor did they prevaricate. They made a perfectly proper inquiry whether the landlords would accept an apparently substantial company as guarantor instead of the directors. The landlords were of course entitled to require reasonable information about the company, and references, but the matter never got that far, because by the postscript to the letter of February 25 1976 the landlords cut the matter off and withheld consent, because they had one idea in mind, to seize on this winding-up—and, be it observed, a members' winding-up—as a means of ridding themselves of this burdensome lease. Moreover, they never resiled from that attitude, even in the affidavit in support of the summons. They never for one moment said that they were prepared to go on entertaining the question of an assignment, and I would refer to the affidavit of Mr Stephen John Prior, a director of the landlord company, in paragraph 9 of which he said:

> I respectfully submit that the relief claimed in paragraph 1 of the originating summons herein is premature as the defendants have not yet instituted proceedings for the forfeiture of the said underlease. I should like to make it clear, however, that the defendants intend to institute such proceedings on the grounds stated in the said notice served by the defendants under section 146 of the Law of Property Act 1925. I understand from the defendants' solicitors herein that counsel has been instructed to settle such proceedings.

The second ground, in my judgment, affords no reasonable cause for refusing licence, especially where, as in this case, the ground of forfeiture is a winding-up, and so the tenant must sell within one year or lose the right to apply for relief. It is unfortunate that, for the reasons which my Lord, Orr LJ, has explained, we have not had the advantage of a transcript of the learned judge's judgment, or even of any notes thereof; but in my view there were ample grounds on which the learned judge could reach the conclusion which he did, with which I entirely agree.

I agree with my Lord, Orr LJ, that this appeal should be dismissed and I agree with the reasons that he has given for proposing that order.

Also agreeing, BUCKLEY LJ said: I would only add one short observation upon the construction of section 146 (2) of the Law of Property Act 1925. The predecessor of that section, section 14 of the Conveyancing and Law of Property Act 1881, came before this court for consideration in the case of *Quilter v Mapleson* (1882) 9 QBD 672, in a case in which it is perfectly true, as Mr Harris in his admirably clear submissions has pointed out, the plaintiff was seeking to recover possession by an action and not by peaceable re-entry. But the language used by the three members of an exceptionally distinguished Court of Appeal, Sir George Jessell MR, Lindley LJ and Bowen LJ, was, it seems to me, simple, clear and in perfectly general terms. The Master of the Rolls put the matter thus: "The second subsection only applies where the landlord has not re-entered." Lindley LJ said: "So long as the tenant has not been turned out of possession, he is within the terms of the enactment, for the lessor is proceeding to enforce his right of re-entry;" and Bowen LJ said: "Under section 14 subsection (2) of the Act of 1881, the tenant may apply for relief until the landlord has re-entered."

As I say, the court was not there directly concerned with any question of peaceable re-entry, but the language which they used suggests that they thought that exactly the same principles would apply, and exactly the same construction would have to be placed upon the subsection, in respect of peaceable re-entry as of an action to recover possession; and if the subsection is applied to peaceable re-entry, the construction contended for by Mr Harris would, as my Lords have pointed out, result in a ridiculous situation. It cannot have been the intention of the legislature to produce that effect. In subsection (1) the words "by action or otherwise" qualify the word "enforceable;" in subsection (2), the words "by action or otherwise" qualify the words "proceeding to enforce," and the two things are not, I think, the same. When a landlord has served his notice under subsection (1) he has taken the first step towards enforcement of his right to re-enter and he is, from then on, "proceeding to enforce." It is perfectly true that he may not have made an irrevocable decision to take the necessary steps eventually to recover possession; he has not committed himself in any way to issuing a writ or to entering into peaceable possession of the property, but he is in my judgment proceeding to enforce his right of re-entry as soon as he has given his notice.

For the reasons that have been elaborated by my Lords in the judgments which have been delivered, I agree that this appeal fails.

The appeal was dismissed with costs.

COURT OF APPEAL

November 11 1977

(Before Lord DENNING MR, Lord Justice LAWTON and Lord Justice GOFF)

DRANE v EVANGELOU

Estates Gazette April 8 1978

(1977) 246 EG 137

Wrongful eviction of tenant—Eviction accompanied by gross provocation and persistent defiance of the law—Breach of covenant for quiet enjoyment and trespass—County court judge's award of exemplary damages against landlord upheld—Appellant's "monstrous behaviour"—Claim for exemplary damages need not be specifically pleaded in county court action—£1,000 not too much—"Not to be weighed in nice scales"

This was an appeal from a decision of Judge Lonsdale at Barnet County Court awarding the tenant, Anthony Malcolm Drane, £1,000 exemplary damages for actions which the judge held to constitute trespass as well as a breach of the covenant for quiet enjoyment. There were three defendants in the county court action. The first defendant, George Evangelou, appealed. The dwelling in question was a maisonette in a house owned by the appellant at 172a Bowes Road, New Southgate. The tenant, respondent to this appeal, had occupied it with his common law wife. The sequence of facts is stated in Lord Denning's judgment.

E Cousins (instructed by Griffinhoofe & Co) appeared on behalf of the appellant; T Eaton (instructed by Pegden & Co) represented the respondent.

Giving judgment, LORD DENNING MR said: "Monstrous behaviour"—that is how the judge described it. He said that it called for exemplary damages. He awarded £1,000. The defendant appeals. These are the facts.

Mr George Evangelou is the owner of a leasehold house, 172a Bowes Road, New Southgate. He let a maisonette in it to a young man Anthony Malcolm Drane, who lived there with a woman, Ann Watts—not his wife—but who lived with him as if she were his wife. The maisonette was let to them furnished at a rent of £25 a week inclusive of rates from August 31 1974.

On July 11 1975 Mr Drane applied to the rent officer for a revision of the rent. This annoyed the landlord Mr Evangelou greatly. So on August 21 1975 the landlord gave Mr Drane notice to quit. That was not effective because the tenant was protected by statute from eviction. On October 8 1975 the rent officer fixed the rent and adjudged it to be £16 a week exclusive. Note the date—October 8 1975. Six days later on October 14 1975 the landlord behaved atrociously. He waited until the young couple were out—when Mr Drane had taken Ann Watts to college in the morning—and then got three men to invade the maisonette. The judge described what Mr Drane found on his return: "When he came back a little later, I think at 9.30, there he found that a large Greek Cypriot was barring the entrance; all his belongings had been put outside in the back yard; the lock had been hammered in of the door; the door was bolted on the inside, about four to five people were inside his premises and two women among them . . . some of their belongings were

A broken and books were damaged." Mr Drane called the police. They told Mrs Evangelou—that is the landlord's wife —that she was committing an offence and that it would be reported to the Town Hall. Nevertheless, she did not let the tenants back into their maisonette. They had to go and stay with friends. They stored some of their belongings in their friends' garage, and slept on the living-room floor of their friends' house.

Mr Drane went to the county court and asked for an injunction so that he and Ann Watts could be restored to their premises. The judge on October 31 1975 granted an injunction against the landlord. But the landlord did not
B obey it. He had moved his wife's father and mother into the maisonette. The landlord or his in-laws appealed to this court. This court heard the appeal on November 27 1975 and rejected it. Lawton LJ, giving the judgment of the court, said: "I am surprised that this appeal has been made to this court. The defendant's behaviour was reprehensible . . . it is right and just that the plaintiff should be put back where he is entitled to . . . this court should take every step it can to see that landlords who behave like the defendant in this case has behaved should get no benefit whatsoever from what they have done." So the appeal was dismissed on November 27 1975. Still the landlord did not go out.
C Mr Drane had to apply again for an injunction to the county court on December 19 1975. The landlord and his in-laws were ordered to leave by 6 pm on Saturday December 21. They gave Mr Drane a key, but it did not fit. So he could not get in. On December 23 Mr Drane applied to commit them for contempt. That at least brought results. The in-laws left. It was only then, on December 23, that the in-laws and the defendant went out. The plaintiff had been kept out for 10 weeks. Mr Drane eventually moved in on January 1. He found everything dirty and damaged: and went on with his action for damages.

D Now there is an appeal to this court. The first point taken on behalf of the defendants was a pleading point. The particulars of claim alleged that the landlord "had interfered with the right of the plaintiff and his *de facto* wife Ann Watts to quiet enjoyment of the said premises by unlawfully evicting them from the said premises on Tuesday October 14 1975." Counsel for the defendant submitted that that claim was for breach of a covenant for quiet enjoyment. He cited a passage from *Woodfall* 27th ed (1968), paragraph 1338: "Since the claim is in contract, punitive or exemplary damages cannot be awarded." The judge at once said: "What about trespass? Does the claim not lie in trespass?"
E Counsel for the defendant urged that trespass was not pleaded. The judge then said: "The facts are alleged sufficiently, so it does not matter what label you put on it." The judge was right. The plaintiff in the particulars of claim gave details saying that three men broke the door, removed the plaintiff's belongings, bolted the door from the inside: and so forth. Those facts were clearly sufficient to warrant a claim for trespass. As we said in *Re Vandervells Trusts* [1974] Ch 269 at p 321: "It is sufficient for the pleader to state material facts. He need not state the legal result. If, for convenience, he does so, he is not bound by, or limited to, what he has stated. He can present, in argument, any
F legal consequence of which the facts permit."

Another pleading point was not taken before the county court but was taken before us. It was said that the particulars of claim did not expressly claim exemplary damages and therefore they could not be awarded. In the case of *Broome v Cassell & Co Ltd* [1971] 2 QB 354 at p 378 H we said that exemplary damages need not be pleaded expressly. In the House of Lords Lord Hailsham in [1972] AC 1027 at p 1038 F said that he was content to accept the view of the Court of Appeal but he thought the practice should be altered. As a result of his words, the High Court Rules have been amended. Order 18, rule 8, subrule (3) says: "A claim for exemplary damages must be specifically G pleaded together with the facts on which the party pleading relies." Does that amendment apply to the county court? In my opinion it does not. The County Court Rules have not been amended. So it is not necessary in the county court to plead exemplary damages. The old practice still applies there, where exemplary damages can be awarded, even though not pleaded. In any case this point was not raised in the county court. So it cannot be raised here. If it had been raised in the county court, I am quite sure it would have been met immediately by an amendment to claim exemplary damages.

The next point was one of substance. Mr Cousins, for H the defendant, has taken us through Lord Devlin's judgment in *Rookes v Barnard* [1964] AC 1129, referring us especially to the passages from pp 1226 to 1230. He said that the general principle nowadays is that in a civil action damages are awarded by way of compensation for damage actually done or for any aggravation by way of injured feelings of the plaintiff; but the court cannot in the ordinary way award punitive damages over and above that which is compensation because punishment is the prerogative of the criminal courts and should have no place in the civil courts.

That exclusion of exemplary damages has not found favour J in the other common law countries, such as Canada, Australia, New Zealand and the United States of America. But since *Broome v Cassell & Co Ltd* [1972] AC 1027 it must be accepted in England. Lord Devlin, however, acknowledged that there are some categories of tort in which exemplary damages may still be awarded. This case seems to me to come within the second category. Lord Devlin said in *Rookes v Barnard (supra)* at p 1227:

> This category is not confined to moneymaking in the strict sense. It extends to cases in which the defendant is seeking to gain at the expense of the plaintiff some object—perhaps some property which he covets—which either he could not obtain at K all or not obtain except at a price greater than he wants to put down. Exemplary damages can properly be awarded whenever it is necessary to teach a wrongdoer that tort does not pay.

To my mind this category includes cases of unlawful eviction of a tenant. The landlord seeks to gain possession at the expense of the tenant—so as to keep or get a rent higher than that awarded by the rent tribunal—or to get possession from a tenant who is protected by the Rent Acts. So he resorts to harassing tactics. Such conduct can be punished now by the criminal law. But it can also be punished by the civil law by an award of exemplary damages. In the recent case of *McCall v Abelesz* [1976] 1 L QB 585 it was held that the provisions of the Act of 1965 against harassment only created a criminal offence: but I said at p 594: "I see no need to give any new civil remedy for harassment. As I understand it, the law already gives a perfectly good civil action for damages." So in a case of this kind damages can be awarded not only by way of compensation but also by way of exemplary damages.

It was said that, in any event, the sum of £1,000 was far too high even as exemplary damages. It was suggested that this sum is out of scale altogether. In my opinion a sum awarded by the way of exemplary damages is not to be weighed in nice scales. It is a question for the judge, having M heard all the evidence, to award such sum as he thinks proper. As this case unfolded before us—the circumstances in which this young couple were forced out of the house, the in-laws being pushed in, the defendant or the in-laws not complying with the injunction and not leaving until there was an application to commit—it did seem to me a case in which, in Lord Devlin's words, it was necessary "to teach the defendant a lesson." The judge thought £1,000 was appropriate. No doubt he felt strongly about it. So would any jury or any judge. I do not think we should interfere with the sum of £1,000.

A There was a point about the costs. The judge said that it was not a case where he could bind the registrar to the higher scale, and he would give him a discretion to award a higher sum under order 21, rule 2. This is not a matter on which it would be right to alter the judgment of the county court judge.

This appeal should be dismissed.

Agreeing, LAWTON LJ said in the course of his judgment: Mr Cousins on behalf of the defendant at times seemed to be suggesting that this was a comparatively minor dispute between a landlord and a tenant. I emphatically dissociate myself from that. To deprive a man of a roof over his head
B in my judgment is one of the worst torts which can be committed. It causes stress, worry and anxiety. It brings the law into disrepute if people like the defendant can act with impunity in the way he did. Indeed I would add that C
when the police arrived they were going to report the matter to the Town Hall. If they did, it is surprising that this defendant was not prosecuted for an offence under section 30 of the Rent Act 1965. Parliament has said that this kind of conduct is sufficiently serious to be made a criminal offence. It follows, in my view, that the court is entitled to approach cases of this kind on the basis that a very grave wrong indeed has been done to the tenant who is evicted in the sort of way that this plaintiff was evicted.

GOFF LJ also delivered a judgment agreeing that the appeal should be dismissed.

The appeal was dismissed with costs, respondent's costs to D
be taxed in accordance with legal aid taxation. Interest was awarded at 9 per cent from the date of the judgment.

LANDLORD AND TENANT

BUSINESS TENANCIES

A **COURT OF APPEAL**

January 21 1977

(Before Lord Justice CAIRNS, Lord Justice ROSKILL and Lord Justice BRIDGE)

TEGERDINE v BROOKS

Estates Gazette January 7 1978

(1977) 245 EG 51

Landlord and Tenant Act 1954—Validity of section 25 notice—Omission of notes prescribed in the statutory form—Whether notice "substantially to the like effect" as the statutory form—Authorities reviewed—Omitted notes
B **irrelevant in the particular case—Notice held valid—County court judge's decision reversed**

This was an appeal by William Fordham Tegerdine, the plaintiff in an action under the Landlord and Tenant Act 1954 before Judge Head in the Lowestoft County Court. Judge Head decided in favour of the defendant, Cyril Brooks, that a notice served by the plaintiff as landlord under section 25 of the 1954 Act was invalid because it did not include certain notes prescribed by the Landlord and Tenant (Notices) Regulations 1957 as amended. The subject-matter of the tenancy was a fish-and-chip shop at 65 Pakefield Street,
C Lowestoft, let by the plaintiff to the defendant on a weekly tenancy.

J Ross Martin (instructed by Hobbs & Durrant, of Lowestoft) appeared on behalf of the appellant, but was not called on; R J Mitchell (instructed by Bailey, England & Co, of Lowestoft) represented the respondent.

Giving judgment, CAIRNS LJ said: This is an appeal from a decision of His Honour Judge Head in the Lowestoft County Court in a case under the Landlord and Tenant Act 1954. The question was whether a landlord's notice determining a tenancy of business premises under section 25 of that Act was valid under the Act and the regulations made
D under it. The premises concerned were a fish-and-chip shop at 65 Pakefield Street, Lowestoft. They were let by the plaintiff landlord to the defendant tenant on a weekly tenancy. The landlord desired to terminate the tenancy but was willing for the tenant to have a new tenancy. He gave his notice accordingly. The defendant tenant relied simply on the allegation that the notice was not a valid one. The learned judge held that it was not a valid notice and accordingly gave judgment for the defendant. The plaintiff now appeals.

Section 25 of the Act of 1954 provides in subsection (1):

> The landlord may terminate a tenancy to which this Part of
> E this Act applies

that is to say, a business tenancy

> by notice given to the tenant in the prescribed form specifying the date at which the tenancy is to come to an end. . . .

It is provided under subsection (5) of that section that:

> A notice under this section shall not have effect unless it requires the tenant, within two months after the giving of the notice, to notify the landlord in writing whether or not, at the date of termination, the tenant will be willing to give up possession of the property comprised in the tenancy.

Subsection (6) provides: F

> A notice under this section shall not have effect unless it states whether the landlord would oppose an application to the court under this Part of this Act for the grant of a new tenancy and, if so, also states on which of the grounds mentioned in section thirty of this Act he would do so.

Section 66(1) provides:

> Any form of notice required by this Act to be prescribed shall be prescribed by regulations made by the Lord Chancellor by statutory instrument.

Regulations were made accordingly and have been amended from time to time, and the regulations current at the G time in question were the Landlord and Tenant (Notices) Regulations as amended by the Landlord and Tenant (Notices) Regulations 1973. Those regulations provide by paragraph 4:

> The form in the appendix to these regulations, or forms substantially to the like effect shall be used for the following purposes; that is to say

and then subparagraph (7):

> A notice under the provisions of Section 25 of the Act, being a notice terminating the tenancy to which Part II of the Act applies, shall . . . be in Form 7.

H

Form 7 as prescribed by the amending regulations sets out the form of the body of the notice, together with certain notes numbered from 1 to 7, and note no 4 begins with the words:

> The court has no power to make an order for the grant of a new tenancy if the landlord, having stated in his notice that he will oppose an application to the court on one of the grounds specified in the Act, establishes that ground to the satisfaction of the court

and then proceeds to set out what those grounds are. Note 5 provides:

> J If the only grounds for opposing an application for the grant of a new tenancy stated in paragraph 3 of this notice are grounds set out in (a), (e), (f) and (g) above,

and so on, with further provisions as to the position where grounds are given for opposing an application for a new tenancy. Note 7 is:

> The term 'landlord' in this notice does not necessarily mean the landlord to whom the rent is paid: it means the person who is the landlord for the purposes of Part II of the Act. The term 'business' includes a trade, profession or employment and any activity carried on by a body of persons whether corporate or unincorporate. K

Now the notice given in this case was in all respects in accordance with the regulations, except that note no 4 and note no 5 were omitted, the note numbered in the regulations 6 was included as note 4, and note no 7 was omitted. The reason why notes 4 and 5 were omitted was that these notes refer only to the case where the landlord is opposing the grant of a new tenancy, and the notice given in this case expressly stated that the landlord would not oppose the grant of a new tenancy. Note no 7 was omitted because in this case there could be no possibility of any uncertainty as to who was the landlord, and no uncertainty as to whether a business was carried on at the premises. It is not a case where any other parties were concerned in any way in relation to the premises as head landlord, mesne landlord or anything of that kind, but simply the landlord who was the plaintiff in these proceedings and the defendant who was the tenant. Quite obviously, with premises let as a fish-and-chip shop there could be no doubt that they were let for business purposes. The learned judge, however, placing a strict construction upon the language of the regulations, held that the notice was not a valid one.

I think that what this court has to consider is as to whether in the circumstances of this case it could properly be said that the form was substantially to the like effect as the form set out by the regulations. In my view, it was, because there could be no purpose in setting out in that notice notes which were irrelevant.

We have been referred to certain authorities. The first was a decision of Barry J in *Barclays Bank Ltd and Another v Ascott* [1961] 1 All ER 782, where the landlords stated in their notice of termination that they would not oppose the grant of a tenancy if the tenant could find a guarantor, but did not make it clear whether they were opposing it or not, and, anyhow, did not indicate, as they were required to do if they were opposing it, any grounds under section 30 of the Act. The learned judge held, and I have no doubt at all in those circumstances that he rightly held, that the notice was not a valid one. But he did certainly by *obiter dictum* express at p 786 a view which seems to me to be a right one. He said:

> It appears to me that the real gist of these decisions

he had referred to some earlier decisions

> was summarised by Hodson LJ in *Bolton's (House Furnishers) Ltd v Oppenheim*. As I understand Hodson LJ's judgment, the question which the court really has to consider is whether the notice given by the landlord has given such information to the tenant as will enable the tenant to deal, in a proper way, with the situation (whatever it may be) referred to in the notice. It is clear, I think, from the authorities which have been cited to me that this notice should be construed liberally, and provided that it does give the real substance of the information required, then the mere omission of certain details, or the failure to embody in the notice the full provisions of the section of the Act referred to, will not in fact invalidate the notice.

I adopt that statement as being a correct one. *Bolton's (House Furnishers) Ltd v Oppenheim* [1959] 1 WLR 913 was in fact not a decision in relation to compliance with the regulations as to the form of the notice, but was a corresponding question as to setting out the grounds under section 30 upon which the landlord was relying.

Two other cases were cited in the court below. One is the case of *Re Bleacher's Association Ltd's Leases, Weinbergs Weatherproofs Ltd v Radcliffe Paper Mill Co Ltd* [1957] 3 All ER 663. That was a judgment of Harman LJ, which has not been relied on in this court. It was a case where the notice quite plainly on the face of it did not begin to be a notice in accordance with the provisions of the Act, and, indeed, it was conceded that it was not. The other case referred to below was *Lewis v MTC (Cars) Ltd* [1974] 1 WLR 1499, a decision of Templeman J, where, so far as concerns this case, the relevant question was whether the failure to strike out certain words in a printed form of notice rendered the notice invalid. Templeman J at the foot of p 1501 of the report said:

> Paragraph 3 of the prescribed form which was used in this case contains two sentences enclosed in brackets, and intended to be alternatives. The first sentence runs as follows: "[I would not oppose an application to the court under Part II of the Act for the grant of new tenancy . . .]" The second sentence is in these terms "[I would oppose an application to the court . . . under Part II of the Act for the grant of a new tenancy on the ground that. . . .]" Then there is left a space in which the grounds should be specified. The form of the notice is so designed that the landlord may strike out the sentence which does not accord with his intention, and if he intends to oppose, he must set out the statutory grounds for his opposition in the space provided. In the notice now in dispute neither sentence was struck out, but there was added in the space which follows the second sentence the following

and then there are set out the grounds of opposition. It goes on:

> No one reading that notice can be in the slightest doubt; the landlord accidentally failed to strike out the first sentence and intended to oppose an application to the court for the reasons set forth in the notice.

Therefore, so far as it goes, that decision was a clear decision by the learned judge that the omission to strike out something which in order to comply with complete strictness with the regulations should have been struck out did not invalidate the notice.

We were referred to two further authorities, one a decision of Shaw J (as he then was), *Bond v Graham* (1975) reported in 236 ESTATES GAZETTE 563, where there was not a strict compliance with the form of a notice inasmuch as the reference to rateable value was wrong, because under the earlier regulations the rateable value that was relevant was £500, but that figure had been changed to £2,000, and then after that to £5,000. In the notice the out-of-date figure of £2,000 was given, but it was immaterial because the rateable value in the circumstances of that particular case was below £2,000. Shaw J held that the notice was a valid one, and in my judgment rightly so held, although it could be said that the notice was not strictly in compliance with the form in the regulations.

The case principally relied upon by Mr Mitchell on behalf of the tenant here was *Sun Alliance and London Assurance Co Ltd v Hayman* [1975] 1 WLR 177. That was a decision of this court, concerned with whether or not a notice under this section was valid having regard to the fact that it referred to the "receipt" of the notice, whereas under regulations, as amended, it required to refer to the "giving" of a notice. It is perfectly true that in that case the court said that it was immaterial whether or not the tenant had been misled by the form of the notice. I, of course, accept that proposition. It is immaterial whether the tenant has been misled. But it is, in my opinion, relevant whether the departure from the form prescribed is such as to be immaterial to the whole of the facts of the case, and if it is, in my view, such departure is not one which will render the notice a notice which does not substantially conform with the regulations.

For those reasons, I would allow the appeal.

Agreeing, ROSKILL LJ said: I would only add a few words because we are differing from the learned judge and out of respect for Mr Mitchell's clear argument. He began his submissions, as it were, by apologising for taking a technical point. A good point does not become a bad point merely because it is technical. The point here is technical, but that does not of itself prevent it from being good. The learned judge thought that it was a good point. He thought it was a good point because the landlord's surveyors, in no doubt a desire for economy of language and time, omitted from the notice which they gave on behalf of their client three of what are called the statutory notes, notes 4, 5 and 7. Those omitted notes, as my Lord has said, were wholly irrelevant to the matter to which the notice related, because the landlord was not at that stage going to oppose the grant of a new tenancy. But because that irrelevant matter was omitted it is said that the notice was bad. I confess that I should be reluctant to give a decision which made it necessary to hold that the law required that wholly irrelevant matter had to be included in a form in order to make a notice good. I do not think we are compelled as a matter of construction of the regulations, or as a matter of authority, to reach that lamentable result. If,

A as Bridge LJ said during the argument, a notice is incomplete or inaccurate in a relevant respect, then it does not matter that the tenant has not been misled by that inaccuracy. But if the omission or inaccuracy is wholly irrelevant, as was the omission of notes 4, 5 and 7, then *a fortiori* there can be no question of the tenant being misled and I see no reason why we should hold here that this was a bad notice.

The learned judge was referred to a number of authorities which his very clear note lists. He was not, however, directly referred to what I regard as the most important of the authorities, the decision of this court in *Bolton's (House Furnishers) Ltd v Oppenheim* [1959] 1 WLR 913. The judg-
B ments of Hodson and Harman LJJ show the approach which is to be adopted in these cases. That case was referred to in the later decision of Barry J, to which my Lord referred, *Barclays Bank Ltd and Anr v Ascott,* which was drawn to the attention of the learned judge. Like my Lord, I would respectfully adopt as a correct statement of the law the passage in the middle of p 786:

> It is clear, I think, from the authorities which have been cited to me that this notice should be construed liberally, and provided that it does give the real substance of the information required, then the mere omission of certain details, or the failure to embody in the notice the full provisions of the section of the Act referred
> C to, will not in fact invalidate the notice.

That seems to me to be an admirably compendious and correct statement of the law. Applying that, it seems to me plain that the plaintiffs were entitled to the order sought.

I too would allow the appeal and enter judgment for the plaintiff and order possession.

Also agreeing, Bridge LJ said: The question is a short question of construction. Was the notice served by the plaintiff under section 25 of the Act a notice substantially to the like effect as the notice prescribed in Form 7 of the relevant regulations? What is the meaning of that phrase? I think its
D meaning is: "Was the notice to the like effect as to its substance?" What then is the substance of such a notice? That question is to be answered in the light of section 66(2) of the Act, which provides:

> Where the form of a notice to be served on persons of any description is to be prescribed for any of the purposes of this Act, the form to be prescribed shall include such an explanation of the relevant provisions of this Act as appears to the Lord Chancellor requisite for informing persons of that description of their rights and obligations under those provisions.

The substance, then, of a notice given for any particular purpose and in any particular circumstances under the Act is
E that which informs the recipient of his rights and obligations under the relevant provisions of the Act. One can test the argument for the respondent in this case by asking the question: If the omitted notes had been included, would they have formed any part of the substance of the notice? The answer to that question is clearly "No." To have solemnly set out, as note 4 requires, all the grounds upon which a landlord can oppose the grant of a new tenancy in a notice in which the landlord has already said he will not oppose such grant in any event cannot be a matter of substance; it would be wholly irrelevant. Like my Lords, in reaching that conclusion I would endorse as correct and accurate the
F succinct statement of principle cited by Cairns LJ from the judgment of Barry J in *Barclays Bank Ltd and Anr v Ascott.*

I would also refer to a couple of sentences in the judgment of Harman LJ in the case of *Bolton's (House Furnishers) Ltd v Oppenheim* [1959] 1 WLR 913, at p 920, where he said:

> The reason why notices of this sort are so elaborate is again tenderness for the tenant. The tenant is supposed to be a person who is either almost illiterate or so poor that he cannot get advice, and, therefore, he is to be shown very carefully everything which may help him to resist the landlord's demands to get his own property back.

Could inclusion of any of these omitted notes here have G
helped this tenant, however poor or illiterate he may have been, to resist the landlord's demands to get his property back? As regards notes 4 and 5, which are concerned with grounds of opposition to an application for a new tenancy, the answer is clearly "No." The answer in the case of note 7 is perhaps not quite so crystal clear, but, as I understand it, the only conceivable purpose of telling the tenant that the landlord does not necessarily mean the person to whom he pays rent is because under the definition of "landlord" in section 44 of the Act the landlord for the purpose of the Act need not necessarily be the immediate landlord; but if the landlord, as here, is in fact the immediate landlord, that piece of H
information is not going to be of any interest or assistance to the tenant.

I confess that I do not really understand what is the purpose of the second part of note 7, which sets out the full statutory definition of "business." The only earlier parts of the statutory form which refer to "business" are the heading, "Landlord's Notice to Terminate Business Tenancy," and notes 3 and 6, in which, for certain purposes, it may be material to inquire, where premises are only partly occupied for business purposes, what parts are so occupied. In relation to the definition of "business," this tenant is not "a body of persons corporate or unincorporate"; he is an J
individual, Mr Cyril Brooks. He is not carrying on a profession, but he is carrying on a trade. Could it conceivably be said, when he occupies, as is common ground, the whole premises let to him as a fish-and-chip shop, that he was in any way denied relevant information by not being told that "business" includes a trade? I cannot see that it could.

I also would allow the appeal.

The appeal was allowed with costs in the Court of Appeal and below on Scale 4. The judgment given below for the defendant was set aside and judgment was given for the plaintiff for possession within two months from January 21 K
1977 with mesne profits.

COURT OF APPEAL

March 31 1977

(Before Lord Justice STEPHENSON, Lord Justice SHAW and Sir DAVID CAIRNS)

LIGHTCLIFFE AND DISTRICT CRICKET AND LAWN TENNIS CLUB v WALTON

Estates Gazette February 4 1978
(1977) 245 EG 393 L

Landlord and Tenant Act 1954—Application by club for a new tenancy—Application opposed by landlord on ground of intention to occupy holding for purposes of business—Landlord's undertaking to use land only for his business as a farmer—Evidence of landlord's intention—Whether intention fixed and genuine—Existing land occupied by landlord not fully farmed—County court judge entitled to conclude intention not fixed and genuine—Landlord's appeal dismissed

This was an appeal by the landlord, Harold Walton, from a decision of Judge Suddards granting an application by the M
Lightcliffe and District Cricket and Lawn Tennis Club for a new tenancy of land held by the club from Mr Walton. The application had been opposed by the landlord under section 30(1)(g) of the Landlord and Tenant Act 1954.

I McCullough (instructed by Newstead & Walker, of Otley) appeared on behalf of the appellant; J Meredith (instructed by F G & H E Smith, of Bradford) represented the respondent club.

Giving the first judgment at the invitation of Stephenson LJ, Shaw LJ said: This is an appeal from a decision of His Honour Judge Suddards given on October 7 1976, when he granted an application made by the present respondents, the Lightcliffe and District Lawn Tennis Club, for a new tenancy of their ground which they held from Mr Walton, who was the landlord. The application was opposed on the ground set out in section 30(1)(g) of the Landlord and Tenant Act 1954, namely, that the landlord intended to occupy the holding for the purpose of a business to be carried out by him.

The main ground of appeal is that in arriving at his conclusion the learned judge misdirected himself in that he failed to give due weight to an undertaking which was offered by the landlord in the course of his evidence, repeated by his advocate in his final address and reiterated in somewhat broader terms by counsel in this court. The undertaking was that when the land comprised in the club's lease fell into possession it would be used for and only for the purposes of the appellant's business as a farmer. There is a cross-notice by the respondent club which arises from the fact that by agreement the parties asked for a determination only of the question whether or not a new tenancy should be granted so that the conditions of the new tenancy, if granted, could then be dealt with between themselves. The learned judge overlooked that arrangement and fixed a term of five years. It is conceded by Mr McCullough on behalf of the landlord that that part of the order cannot stand in any circumstances.

The history of the matter is a long one. The cricket club has been established on this particular ground, where there are also some tennis courts, for more than a century. The appellant came to the land in 1939 when he took a lease of some 20 acres. The part which is occupied by the respondent club is some 3½ acres and forms an island in the middle of the whole of that land. The remainder of it, some 16 acres, together with a further 9 acres, of which the landlord acquired a lease also is used by him as land for grazing cattle. That has been the position for a substantial time. When the lease which the club held at the time the appellant acquired the land was due to expire, they applied for a new tenancy. They were granted one for a term of 10 years. On the expiration of that tenancy they sought a further lease in the present proceedings.

The opposition by Mr Walton was based on the fact that farming was the family business and he intended it to be continued by his son and other members of his family. Accordingly, he wished to extend the use of his land and to farm it all in its entirety. This had been, so he asserted, his fixed intention for a long time past. When he gave his evidence it became apparent that he had not used the land already available to him to its fullest advantage. His own estimate was that it was used to 50 per cent of its capacity. This evidence was both cogent and material and the learned judge rightly attached considerable importance to it. The question which he had to decide in the face of that evidence was whether the undertaking offered by the landlord represented a bona fide intention, so that it could be accepted with confidence. The appellant put it in these terms: "I am prepared to give an undertaking that if the tenancy is refused I shall use this land for grazing." He estimated he could put something like nine extra beasts out to graze if he had the extra land occupied by the club. There was evidence that the land was of poor quality. Furthermore, that part which had been occupied by the club for so very long a time would require considerable treatment before it could become suitable for grazing. That would involve a substantial expenditure by the landlord. All these factors and others were before the learned judge. It was incumbent upon him to examine the whole of the situation in deciding the answer to the question propounded: "Was the intention of the landlord genuine and bona fide?"

When he came to give his judgment, which was in admirably concise terms, he said:

> If I had to decide if there was a case of greater hardship, the matter would be easy, but that is irrelevant. The fact that Mr Walton is land-hungry, as many farmers are, is irrelevant. The question whether he needs the land or wants the land is irrelevant. The fact that he could help to do a great public disservice is irrelevant. The fact that he would probably make himself one of the most unpopular men in the district is irrelevant.

What the learned judge was seeking to do was to reject and exclude those factors which did not impinge upon the essential problem. His last comment referred to the fact that this long-established cricket club was one of the great amenities of the district and was looked upon with pride by people who dwelt in and about. The club itself has achieved some distinction in the cricket world. Quite obviously, Mr Walton, whatever his ideas about his need for this land, would make himself somewhat unpopular in many quarters, but the learned judge wished to make it quite clear that he would not allow such irrelevant considerations to affect his decision of the problem before him. He said: "What I have to decide is whether he at this moment has the genuine intention of using the land for business." He said "at this moment" because he had in mind the decision in the case of *Betty's Cafés Ltd v Phillips Furnishing Stores Ltd* [1957] 2 All ER 223, where Danckwerts J, as he then was, first came to the conclusion that the material time at which the landlord had to exhibit an intention to use premises for the purposes of his business was when he gave notice of opposition. Accordingly, if he had not exhibited that intention until the matter came before the court for decision, it was too late and could not adversely affect the tenant's claim. On the landlord's appeal the Court of Appeal came to the opposite conclusion. They held that the material time was when the matter came before the court. Romer LJ said: "I would . . . remit the case . . . for . . . determination whether the intention at which the landlord company arrived on April 23 1956 possessed the necessary qualities of fixity and genuineness." The fact that the Court of Appeal thought fit to remit that question for decision demonstrates that an undertaking is not *ipso facto* decisive of the matter. It must be looked at by the court which has to decide whether it possesses the necessary qualities of fixity and genuineness. It so happened in that case that the landlord was a registered company and the undertaking followed from a resolution of its board. It may well be that for practical purposes somewhat different considerations arise in deciding whether an undertaking which is given by a substantial company of good reputation is more dependable than an undertaking given by an individual who may for personal reasons find it necessary or expedient to change his mind.

The learned judge went on to say: "He has undertaken to graze cows on it and it is suggested that that has great value. I say this is interesting but in this case it does not help me to decide whether or not his intention is genuine." Mr McCullough, who put forward every possible argument which might support the proposition that the learned judge misdirected himself in law, argued that the brushing aside of the landlord's undertaking shows that the learned judge did not really apply his mind to the problem. Counsel submitted that he ought at least to have asked himself whether there was any good reason for rejecting that undertaking as lacking bona fides or not showing fixity of intention. What the learned judge was saying however was: "That is relevant but in this case it does not really take my mind to the conclusion that the intention which is behind it is a genuine one." He goes on to explain why that is so. He says:

> If he needed the land it would give a hint to me as to whether his intention is genuine because if his land is already stuffed with cattle then obviously there is a greater need, and obviously there is a more genuine intention to use every scrap of land one can get. But here on his own evidence he is using his own land to 50 per cent of its capacity.

That consideration alone may not be conclusive one way or the other but the judge was entitled to look at it as a factor when he came to decide whether the undertaking represented a genuine intention. He added the comment: "I think I am entitled to say that if I don't think he was telling me entirely the truth then I am entitled to say 'Has he been telling me the truth as to his genuine intention?'"

The learned judge does not say in which particular aspect of the landlord's evidence he found it impossible to accept what he was saying, but he makes it clear that there were aspects of that evidence which cast doubt in the judge's mind upon the landlord's veracity and upon the bona fides of the intention which he had expressed in the undertaking which was offered. The judge added: "Since he knew that this matter would be brought to court he has taken steps to get estimates of buildings which will allow him to increase his stock. That I do not regard as evidence in his favour of a genuine intention. The whole question on the balance of probabilities is: 'Do I think he has a genuine intention of using this land in his business?' Not: 'Does he need it?' not 'Is it reasonably required?' but 'Has he that intention?'"

Mr McCullough cited a number of other cases including *Expresso Coffee Machine Co Ltd v Guardian Assurance Co Ltd* [1959] 1 All ER 458. That was described by Harman J at first instance as a very peculiar case. The judgment was affirmed in the Court of Appeal in favour of the respondent landlords who had given an undertaking which the court accepted as "honest, present and real." This is the critical issue which has to be decided. It is by no means the case that if an undertaking is given there is a presumption that it represents a genuine intention. It must be examined in the light of all the circumstances of the case. The only avenue really open to Mr McCullough upon this ground of appeal is one that leads to the conclusion that the learned judge had no evidence before him upon which he could reasonably have doubted the honesty of the undertaking. But there was much in the evidence which could warrant such a view. This did not necessarily involve the fact that the landlord was a dishonest man—of course not—but it meant that there were factors which told against the reality and bona fides of the intention which he stated he had. It is not as though there is nothing in the history since the landlord acquired the property in 1949 which does not tell against the genuineness of his intention in relation to the land. When the last application was made for a fresh tenancy, and the tenancy of 10 years was granted, the landlord put forward a different projected user for the land. That might be inconsiderable if it stood alone.

None the less, it is something which is part of the whole picture, and in coming to his final conclusion, the learned judge said this: "Is the real reason 'I want that land, it's mine,' and I think that is all he wants." I do not think he ever thought what he intended to use it for and that is another way of saying "I do not think the intention that he expresses is a bona fide one." The judge goes on: "I don't know what his intention is. I doubt if he does. In my conclusion he has no genuine intention and I am prepared to grant a new lease to the club. I am not going to decide rent; that is unnecessary. The lease is for five years."

There, in what are unequivocal terms, the judge is saying "Having considered the matter by and large, and looking at all the evidence before me, I find myself unable to accept the undertaking as being genuine."

The matters went a little further, because when the learned judge came to sign the notes of his evidence, he added this: "I did not believe the respondent"—ie the landlord—"had a genuine intention to use this land as a farm as in evidence he is not fully farming the land he already has." In the mind of the learned judge that was a significant circumstance as it could be. In all the cases which have been cited to us, the decisions went on their special facts. It is true that it is not necessary for a landlord to prove that he requires or needs the particular land. All he has to show is that he has an intention to use it for the purposes of his trade or business. None the less, absence of need, the lack of requirement, are incidental matters which either go to support that intention or to show that it is not to be accepted as being that fixity and genuineness adverted to by Danckwerts J in *Betty's Cafés Ltd v Phillips Furnishing Stores Ltd* [1957] 2 All ER 223 at p 225.

By and large, the landlord has failed to show that the learned judge misdirected himself in law or in fact. On the contrary, it seems to us that he apprised himself of the true nature of the problem, and that he approached the question of whether or not the undertaking was "honest, present and real" in the right way. Accordingly, there is no ground, since the question is very much a question of fact as well as of law, which would justify this court in coming to a different conclusion from that which the learned judge formed.

There is a further incidental matter in relation to the question of costs. The learned judge, having made his order in favour of the tenant club, said when he did so "There is no reason why the landlord should not pay the costs since he was able to turn down an offer of £10,000." Perhaps that might have been more happily expressed; but Mr McCullough did not pretend that that would not represent a proper exercise of the judge's discretion as to the incidence of costs. All in all, there is no ground for departing in any respect from the decision of the learned judge.

I would dismiss this appeal.

Sir David Cairns agreed and had nothing to add.

Also agreeing Stephenson LJ said: The question in this case for this court is that which was stated by Lord Evershed MR in the *Expresso Coffee v Guardian Assurance* case [1959] 1 All ER 459 at p 463. The question was "Was the judge entitled to find as a fact, as he did, that at the date of the hearing before him . . . the landlords intended to occupy [*this site*] for the purposes of their own business?"

In my judgment, the judge asked himself the right question and he was entitled to arrive at the answer at which he did. There was evidence before him which supported an honest or genuine or settled or fixed intention. There was the undertaking, and there was undoubtedly the ability of the landlord to carry it out. But there was also evidence to the contrary. There was evidence that he had no need whatever for this field. There was evidence that he had taken no preparatory steps to carry it out, such as the steps which were taken by the resolutions of the companies in the cases to which Mr McCullough has referred us. Mr McCullough has failed to satisfy me that the judge discarded the undertaking because of the language which he uses about it to which my Lord Shaw LJ, has referred, or that the learned judge failed to consider the ability of the landlord to carry out his intention simply because the learned judge did not mention it; nor has he satisfied me that the judge regarded the absence of need as conclusive simply because it was the reason which he gave in that manuscript note at the end of his judgment.

Finally, he has not satisfied me that the learned judge regarded the intention as genuine in the sense of being bona fide and fixed simply because he was kind enough not to say everything which he thought about it. It would, in my judgment, be quite wrong for this court to hold that giving an undertaking, whoever gives it and in whatever circumstances, creates any legal presumption in favour of a landlord. Equally, it would be wrong to hold that uncontradicted evidence of a landlord's ability to carry out his intention creates such a presumption. Nor are we bound by any authority which has been cited to us so to decide against commonsense and to treat as a matter of law what is, it seems to me, quite clearly a question of fact.

Support for the view that an undertaking is not conclusive cannot possibly be derived, as Mr McCullough sought to derive it, from the observations of Romer LJ in the case of *Expresso Coffee*. It is quite clear from what that learned

A Lord Justice said at p 463 of the report that when he is talking about an undertaking being perfectly decisive of fixity of intention, he is referring to the kind of undertaking which had been given by a responsible body like the landlords in the case of *Expresso Coffee* and in the earlier case decided by Danckwerts J of *Betty's Cafés Ltd v Phillips Furnishing Stores Ltd*.

Equally, I find it impossible to draw the conclusion which Mr McCullough asked us to draw, that the ability to carry out an intention proves its existence, from the decision or observations of the Lords Justices in the case of *Reohorn v Barry Corporation*, reported in [1956] 1 WLR 845 to which B he also referred. In that case, Denning LJ said at p 849 that "'intention' connotes the ability to carry it into effect." In that case, there was no ability to carry the intention into effect and so the intention was negatived. That does not in my judgment help us to decide the converse that where there is ability to carry out intention the judge ought in the absence of any other circumstances to find that the intention is genuine and settled.

I can detect no error of law in the judgment of the learned judge, and I would, therefore, agree that this appeal should be dismissed. I have nothing to add on the question of costs.

C The appeal was dismissed with costs and the respondents' cross-appeal allowed.

CHANCERY DIVISION

February 2 1977

(Before Mr Justice OLIVER)

DERBY & CO LTD v ITC PENSION TRUST LTD

Estates Gazette February 18 1978

(1977) 245 EG 569

D **Landlord and Tenant Act 1954—Tenants' application for new lease—Landlords' claim that rent and other terms had been agreed and set out in engrossed but unexecuted underlease—Negotiations subject to contract broken off before sealing of underlease—Held that there had been no agreement for purpose of an application within the meaning of sections 32 to 35—Argument by landlords based on estoppel also rejected**

Derby & Co Ltd, tenants of office premises at Moor House, London Wall, were plaintiffs in proceedings by originating summons against their landlords, the ITC Pension Trust Ltd, claiming under the Landlord and Tenant Act 1954 an order E for the grant of a new tenancy. The defendant landlords by a summons in these proceedings contended that the rent and terms of the proposed new tenancy had been agreed between the parties following negotiations and that, in effect, the court was bound to make an order in the terms agreed. This matter was adjourned by the master to be decided by the judge.

J Harper (instructed by Coward Chance) appeared on behalf of the plaintiffs in the action; D Gidley Scott (instructed by Trower, Still & Keeling) represented the defendants.

Giving judgment, Oliver J said: This is a summons by the defendants in a proceeding commenced by the plaintiffs under the Landlord and Tenant Act 1954, the particular proceeding F being a summons for the grant of a new lease pursuant to the provisions of that Act. In this summons the defendants, who are the landlords, claim first of all, and this is the primary relief claimed:

> An order that the duration of the rent reserved by and the other terms of the tenancy to be granted by the defendants to the plaintiff pursuant to the plaintiff's application herein shall be as set out in the engrossed but unexecuted underlease which is intended to be exhibited to the affidavit to be filed in support of this application.

The background facts relating to this matter can be fairly shortly stated. The plaintiffs are tenants by assignment of some office premises at Moor House, London Wall, and their lease, which was dated July 26 1971, expired contractually G on March 23 1975. It was a lease at a rent of £11,500 pa and the defendants are the present reversioners. Notice to determine the lease was given on July 26 1974 pursuant to the provisions of the Landlord and Tenant Act. An intimation was made that the landlords would not oppose the grant of a new tenancy and the usual counternotice was given. An application for a new tenancy was duly made. There is no dispute about the plaintiffs' right to have a new tenancy under the Act; the only dispute now before me relates to the terms.

The originating summons under the Landlord and Tenant Act was issued on November 15 1974, and, as so very often happens, negotiations then ensued between the parties to see H whether it was possible to dispose of the matter out of court. What happened was that on November 26 1974 Coward Chance, the solicitors for the plaintiffs, wrote acknowledging receipt of the originating summons endorsed with acceptance of service on behalf of the defendants, and saying:

> We note that you have written to your clients' managing agents asking them to contact our clients with a view to negotiating the terms of the new tenancy and that you are prepared to consent to the proceedings being adjourned generally pending negotiations.

They enclosed a consent to adjourn and asked that it be signed. That was done, the proceedings were duly adjourned, J and the matter then got into the hands of the parties' respective agents; the agents for the landlords being Richard Ellis & Co and the agents for the tenants being Kemsley Whiteley & Ferris. On May 19 1975 it appears that the agents had arrived at a broad outline, at any rate, of the sort of terms that they were prepared to recommend to their respective clients; and on that day Kemsley Whiteley & Ferris wrote to Richard Ellis saying:

> We refer to our recent discussions with your Mr Walshe and confirm the terms that have been agreed on behalf of the tenants Derby & Co without prejudice to their rights under the Landlord and Tenant Act procedures. K

They then set out various terms as to the premises to be included, the rent (there specified as £23,000 pa), the length of the term (14 years from March 25 1975), certain proposals as to service charges, and so on; and they concluded:

> We look forward to receiving your written confirmation to the above terms as soon as possible.

The letter was then marked as a matter of precaution, no doubt, by them, "Subject to contract". Those are hallowed words which have received attention in numerous cases and their meaning is not in doubt.

Then the matter got into the hands of the solicitors for the L respective parties and a draft lease was prepared, and it was submitted to Coward Chance by the solicitor acting for the defendants. The draft went to and fro with amendments being inserted, objected to, taken out, reinserted, and so on; until finally, on October 2 1975 the draft appeared to have reached a final stage. Coward Chance wrote a letter in which they agreed to the deletion of certain clauses to which the landlords had objected. They also agreed to the inclusion of a clause which the landlords wanted, and they said:

> We now believe that the underlease is agreed and we look forward to receiving from you the counterpart underlease for execution by our clients. M

That letter was responded to on October 6 1975 in a letter from the solicitor acting for the landlords in which he said:

> It appears that we are now agreed on the form of the draft lease and I am putting in hand the preparation of the counterpart underlease for execution by your clients.

That underlease was duly engrossed. It was sent for sealing by the plaintiffs, but there appear to have been some typing errors and it was returned. It was then re-engrossed and resubmitted in December 1975 and was sent by Coward Chance to their clients for sealing. But then on January 16 of last year there occurred an event which put an end to the negotiations, because the plaintiffs wrote a letter direct to Richard Ellis, saying:

> Since negotiations for the renewal of the lease on the above premises commenced last year there have been considerable changes in the state of the property market. Accordingly we would wish to discuss with you a number of the items contained in the proposed lease, including the level of rent and the rent review provisions. We look forward to hearing from you as soon as possible.

The parties being unable to agree, on March 4 of last year Coward Chance wrote saying:

> . . . we can now let you know that on instructions from our clients we have seen counsel on our clients' position as against your clients and counsel and we consider that there is no agreement between our clients and your company so that section 28 of the Landlord and Tenant Act 1954 cannot apply since such an agreement has to be a concluded agreement and in this case the agreement has not been concluded.

Then there was some discussion about agreement of a present market rent.

As a result of that, this summons was issued by the defendants because, the attempts to dispose of the matter out of court now having failed owing to the refusal of the plaintiffs to execute the underlease in the agreed terms, it appears inevitable that the matter must now be resolved by the court. Mr Scott, who appears for the defendants on this application, points to sections 32 to 35 of the Landlord and Tenant Act 1954 and says that there has been an agreement of the terms; and that the court really has very little to do now, except to direct that a new lease be granted on the terms which have been agreed in the correspondence to which I have referred and are now embodied in the engrossed underlease.

Sections 32 to 35 of the 1954 Act provide for a number of matters in relation to the grant by the court of a new lease. The sections deal with different terms of the tenancy, but broadly they are all on the same lines. Section 32 deals with the property to be comprised in the tenancy, and provides that:

> Subject to the next following provisions of this section an order under section 29 of this Act for the grant of a new tenancy shall be an order for the grant of a new tenancy of the holding, and in the absence of agreement between the landlord and the tenant as to the property which constitutes the holding the court shall in the order designate that property by reference to the circumstances existing at the date of the order.

Section 33 deals with the duration of the new tenancy. The new tenancy:

> shall be such tenancy as may be agreed between the landlord and the tenant, or, in default of such an agreement, shall be such a tenancy as may be determined by the court to be reasonable.

Section 34 deals with rent, and, again, the rent is to be:

> such as may be agreed between the landlord and the tenant, or as in default of such agreement, may be determined by the court.

Finally, section 35 deals with other terms of the new tenancy and provides, simply, that the terms of the tenancy granted by the order:

> (other than terms as to the duration thereof and as to the rent payable thereunder) shall be such as may be agreed between the landlord or the tenant or as, in default of such agreement, may be determined by the court.

Mr Scott's primary submission is that here you have got the rent, the duration and, indeed, all the terms of a tenancy agreed between the landlord and the tenant; agreed, it is admitted, in a negotiation which was subject to contract; agreed, it is admitted, in a document whcih has no contractual force at this stage; but nevertheless, Mr Scott says, agreed between the landlord and the tenant, and therefore agreed for the purposes of the Act.

The provisions to which I have referred, Mr Scott submits, are simply provisions relating to the ascertainment of particular terms to be included in a tenancy to be granted by the court. Where therefore, he says, the sections refer to agreements between the landlord and the tenant, or to something being agreed between them or, in default of agreement, fixed by the court, they are not referring necessarily to a contractual agreement between them; indeed, he says, there would not be a contractual agreement *ex hypothesi* because up until the stage where the court makes an order there would be no subsisting contract between the parties for the new tenancy.

I have been referred to section 69 of the Act. Section 69(2) provides that:

> References in this Act to an agreement between the landlord and the tenant

except in certain sections which one need not bother about for present purposes

> shall be construed as references to an agreement in writing between them.

And, says Mr Scott, you have here an agreement in writing, because it has been reduced to the terms of an engrossed lease, and if one looks at the correspondence it is clear that the terms of the underlease were agreed between the solicitors of the parties. That, he says, is all that is required here, and the court now does not need to go outside the terms of the agreements—indeed, on the wording of the sections, it cannot go outside the terms of those agreements, because all it can do is to make an order for the grant of a tenancy either in terms agreed or, in default of agreement, on such terms as it may think right. There being no default of agreement on Mr Scott's definitions, the court's hands are, he says, bound.

I find myself unable to accept that contention. I have very considerable sympathy with the defendants, who were undoubtedly led to believe that they were going to get a final underlease at the end of the day by the correspondence between the solicitors, which seemed to suggest that there was nothing more to be done—and, indeed, of course, apart from execution and exchange, there was not at that stage if the plaintiffs had not backed out of the negotiations. At the same time I think that the agreement which is referred to in the sections to which I have made reference must be, as a matter of construction, an agreement for the purposes of an application before the court. Indeed, one finds that in section 33 it is provided that "where *on* an application under this part of this Act the court makes an order for the grant of a new tenancy, the new tenancy shall be such a tenancy as may be agreed between the landlord and the tenant." As I read that, it must be agreed on the application. I am not impressed with the submission which Mr Harper for the plaintiffs makes, that the only agreement which can be referred to here is an agreement in the face of the court at the hearing. But I think that at least when the sections refer to terms being agreed, they must be terms which were agreed for the purposes of the application by the tenant under the section.

On the facts of this case it does not seem to me that one has that here. What one has is an attempt by the parties to avoid the necessity of proceeding with the application. One has that introduced by a letter which is expressly and in terms said to be without prejudice to their rights under the Landlord and Tenant Act procedures. So, as I see it, the negotiations are leaving the Landlord and Tenant Act procedures out of the question for the moment. The attempt being made is to see whether, apart from those procedures, agreement can be reached. The negotiations are expressed to be "subject to contract". Now whatever else may be meant by the word "agreed" in the sections to which I have referred, it must, I think, refer to an unconditional and final agreement. I do not mean necessarily an agreement in the contractual sense, but an agreement which is final and not subject to any suspensive or other condition.

The meaning of the words "subject to contract" is tolerably clear now and has been the subject matter of a number of

A recent decisions. I have been referred by Mr Harper to *D'Silva v Lester House Development Ltd* [1971] Ch 17 and *Leveson v Parfum Marcel Rochas (England) Ltd* (1966) 200 EG 407. Those are merely, I think, examples of the same sort of situation as that in this case. But Mr Scott himself does not contend that the negotiations gave rise to any contractual liability on either side.

In my judgment it is not simply a question of contract. Where negotiations are carried out subject to contract, that means—and I think this is clear from the authorities—that they are conditional upon the final engrossment, execution and exchange of the formal lease, where the case is one of nego-
B tiation for the grant of a lease or tenancy. In the instant case there never was such an exchange. That condition was never fulfilled, and the bargain remained conditional. It does not seem to me, therefore, that one has, until that condition has been fulfilled, a final agreement for the purposes of the sections of the Act to which I have referred. Mr Scott says—and I see the force of this—that whether one arrives at a concluded agreement in negotiations or whether one goes to the full length of making application to the court, the ultimate result is the engrossment and exchange of a lease. He says therefore that to treat the condition of exchange as being, as it were, sus-
C pensive of the operation of the agreement in so far as the Act is concerned is illogical, because one is looking at the matter from the outset on the basis that nothing is going to be binding upon anybody until the leases are exchanged. I do not believe that that is the right analysis. I think what one has to look for is to see whether an agreement has been entered into between landlord and tenant, not for the purposes necessarily of the grant of a new lease, but for the purposes of the application before the court. It does not seem to me that an agreement subject to contract, in an endeavour to avoid the necessity of proceeding with the court application, can be held to be binding for the purposes of the application if those negotiations are
D unsuccessful. I therefore rather reluctantly feel that I must reject that contention of Mr Scott's, and I turn to the alternative way in which he puts his case.

He says that this is a case where he is entitled, on the basis of such cases as *Crabb v Arun District Council* [1976] Ch 179 to rely upon estoppel, because, he says, he had a right under section 24A of the Act to make an interim application for payment of an increased rent. The course of the negotiations was such, he says, that he was led to believe that there was going to be a lease granted at the end of October 1975, and in reliance upon that hope—which he says was encouraged
E by the defendants—he did not make such an application and therefore has changed his position, so that the defendants are not now, or should not now be, permitted to take the point against him that there has been no agreement. Again I find myself unable to accept that submission. It seems to me again that where parties negotiate on a basis "subject to contract" everybody knows there is a risk that, at the end of the day, either side may back out of the negotiations, up to the point where leases are exchanged. I do not think that a party who relies upon the other side not to back out can be said to have estopped that party from backing out simply because he has not done something which he might have done in the inter-
F vening period. I notice in fact that although notice was given in January that the plaintiffs were backing out, the present summons—which asks, in its second claim for relief, for an order pursuant to section 24A determining the rent—was not issued until the end of April of this year. Certainly the defendants do not seem to have been particularly active about getting on with the matter even when they knew that the plaintiffs had backed out. In any event I do not think that there is any evidence before me on which I could say that the defendants had been encouraged not to make an application. It was a thing which was always open to them and they could have done it at any time. The negotiations stretched out over several
G months. There was nothing whatever to stop them making an application if they wanted to. There is no evidence that they were requested by the plaintiffs not to do so. In those circumstances I think I must also reject that claim.

In the event, therefore, I cannot make the order which Mr Scott seeks under paragraph 1 of his summons.

It was ordered that the costs of the adjournment of this matter from the master to the judge should be paid by the defendants in any event.

H

COURT OF APPEAL

June 1 1977

(Before Lord Justice ORR and Lord Justice GEOFFREY LANE)

LEWIS AND ANOTHER v WOLKING PROPERTIES LTD

Estates Gazette February 25 1978

(1977) 245 EG 655

Landlord and Tenant Act 1954—Tenants' application for new tenancy under Part II—Application made in time but service of application on landlords out of time—Question whether under County Court Rules the time for service
J **could be extended—County court judge's decision to grant an extension of time upheld**

This was an interlocutory appeal from a decision of Judge Ruttle at Westminster County Court. The proceedings concerned an application by Philip Lewis and Roy Charles Weksler for a new tenancy of garage premises at Haunch of Venison Yard, Brook Street, London W1. The county court judge had dismissed the landlords' application to strike out the tenants' originating application for a new tenancy and had granted the tenants' application for an extension of time for the service of their originating application on the landlords. The landlords appealed against this order.

K Michael Rich (instructed by Underwood & Co) appeared on behalf of the appellants (landlords); Norman Primost (instructed by Miller, Clayton & Co) represented the respondents (tenants).

Giving judgment, ORR LJ said: This is an appeal by the landlords (the title having been assigned since the trial and new landlords substituted as appellants) of premises known as Garage No 1, Haunch of Venison Yard, Brook Street, London, against an order made by Judge Ruttle on February 2 of this year in the Westminster County Court, by which he dismissed an application by the landlords to strike out an
L originating application by the tenants for a new tenancy of those premises, and on the tenants' application to extend their time for service of the originating application he granted an extension of three days from February 21 1977.

The facts of the case are, briefly, these. On August 23 of last year the tenants made application for a new tenancy of the premises under Part II of the Landlord and Tenant Act 1954, and that application was made in good time, the last date for making it being September 3 1976. But the County Court Rules, by Order 40 rule 8 subrule (1A) and Order 8 Rule 35 require service of such an application to be effected within a month. On that basis the last date for service was
M September 23 1976, but in fact service was not effected within that period and it appears from a letter from the Westminster County Court dated October 26 1976 that this failure was due, in part at least, to the application having been attempted to be served by post at the appellants' registered office on August 25 and the letter having been returned by the Post Office marked "Gone Away," the landlords having by then transferred their registered office to an address in Hounslow.

On September 9 1976 the solicitors for the respondents wrote to the solicitors for the landlords informing them that they had been told by the county court that the summons

A had not been served; and that letter ends with this paragraph: "We would be pleased, accordingly, if you would let us know the correct address of Wolking Properties Ltd so that the summons can be served." To that they received a reply dated September 17 from the appellants' solicitors: "Thank you for your letter of September 9, the contents of which are noted. On the assumption that the summons was issued within the required time, from service of the section 25 notice, we will agree to accept service on behalf of our clients." It is unfortunate that in that letter the appellants' solicitors did not provide the answer for which the respondents had asked, namely, the correct address of the landlord B company.

Before the judge the tenants claimed that the court had jurisdiction to extend time for service, both under the County Court Rules and under the inherent jurisdiction, and in the further alternative they alleged waiver by the appellants of any objection they might have had to the form of service, but all three of those claims were challenged by the appellants. In his judgment the judge found that there was jurisdiction under the rules; and the first issue in this appeal is whether he was right in so holding. The second and third issues, which only arise if the judge was wrong in holding that there was jurisdiction under the rules, are whether he C had inherent jurisdiction to make the orders he did and whether he was right in rejecting, as he did, the respondents' plea of waiver.

The County Court Rules directly relevant to the first issue are as follows: Order 40 Rule 8(1) and (1A), and Order 8 Rule 35(1), (2) and (4).

Order 40 Rule 8:

(1) An application for a new tenancy under section 24 of the Act of 1954 shall be made by originating application in Form 335. (1A) Order 8, Rule 35, shall apply in relation to an originating application under this Rule as it applies in relation to a default D summons but with the substitution for the references to 12 months of references to one month.

Order 8 Rule 35:

(1) The time within which a default summons may be served shall, unless extended under the next succeeding paragraph, be limited to a period of 12 months from the issue of the summons. (2) Where reasonable efforts have been made to serve the summons within the said period and service has not been effected, the registrar may, on application, order that the time be extended for a further period not exceeding 12 months or for successive periods not exceeding 12 months each: Provided that the time shall not be extended for any period unless the application is made within the currency of the last preceding period. . . . (4) E Where the summons has not been served within the time allowed for service by this Rule, the action shall be struck out.

Mr Rich in this court, as before the learned judge, placed much reliance on the very positive character of the words in subrule (4), but, as the learned judge pointed out, there are other rules to which regard is to be had, in particular Order 37 Rule 4 and Order 13 Rule 5.

Order 37 Rule 4:

(1) Where there has been a failure to comply with any requirement of these Rules, the failure shall be treated as an irregularity and shall not nullify the proceedings, but the court may set aside the proceedings wholly or in part or exercise its powers under F these Rules to allow such amendments, if any, and to give such directions, if any, as it thinks fit. (2) No application to set aside any proceedings for irregularity shall be granted unless made within a reasonable time, nor if the party applying has taken any step in the proceedings after knowledge of the irregularity. (3) Where any such application is made, the objections intended to be relied upon shall be stated in the notice. (4) The expression "proceedings" in paragraph (1), and where it first occurs in paragraph (2), includes any step taken in the proceedings and any document, judgment or order therein.

Order 13 Rule 5:

(1) Subject to the provisions of these Rules, any of the times fixed by these Rules or by any judgment, order or direction for doing any act may be enlarged or abridged by consent of all parties G or by the court on the application of any party. (2) An order enlarging time may be made although the application therefor is not made until after the expiration of the time allowed or appointed.

On the basis of those provisions and on a consideration of the rules generally, the judge came to the conclusion that there was power under the rules to extend time for service of the tenants' application. On this appeal Mr Rich has sought to persuade us that that conclusion was wrong. He argued that there are to be found in the rules three different kinds of time provision, the first of them being where the rule does not provide what is to happen if the time limit is H extended (and he gave us as examples of that Order 8 Rule 4 and Order 9 Rule 4); the second, where there is such a provision but the consequences may, under the rules, be avoided (and he gave as examples Order 7 Rule 9 and Order 9 Rule 1); and finally, the third class (a variant of the second), where the consequences cannot be avoided (and he gave as an example, and I think as the sole example, of that, Order 8 Rule 35). I for my part have not been satisfied that that provision falls outside Mr Rich's second class; and he was, through no fault of his, unable to support his argument by any cogent authority. Mr Primost was able to rely on a considerable number of authorities, to which I do not find J it necessary to refer in detail, but they include the speeches in the House of Lords in *Kammins Ballrooms Co Ltd v Zenith Investments (Torquay) Ltd* [1970] 2 All ER 871; they include, in addition, *Westminster City Council v Chapman* [1975] 2 All ER 1103, *Harkness v Bell's Asbestos & Engineering Ltd* [1967] 2 QB 729, a decision of this court; and *R v Bloomsbury and Marylebone County Court, ex parte Villerwest Ltd* [1976] 1 All ER 897. I have borne in mind, in reference to an additional authority on which Mr Primost relied, *Sheldon v Brown Bayley's Steel Works Ltd* [1953] 2 QB 393, that there was a difference in wording involved in that case. But I have derived, for the purposes of the decision K of this case, considerable help from the passages to which we were referred in the speeches in the House of Lords in the *Kammins Ballrooms Co* case. In addition, Mr Primost was in my judgment entitled to claim that the construction which he invited us to put on the provisions in question is in complete harmony with the provisions contained in Order 2 rule 1 of the Rules of the Supreme Court and the note which follows that rule in the *Annual Practice*. I would add that it is a necessary consequence of Mr Rich's argument that, if it is right, no relief at all could be obtained by an applicant for a new lease who, for reasons quite beyond his control, has been unable to serve notice of his application on the L landlord within the permitted time; and for this reason, quite apart from authority, I should for my part have been reluctant to adopt the construction which Mr Rich advances unless compelled to do so by the language of the rules. I do not, for the reasons I have indicated, find myself so compelled.

For these reasons (and I intend no disrespect to the arguments on both sides in dealing with the matter shortly) I am left in no doubt that the judge's decision on the main issue in this case was right for the reasons given by him; and I find it unnecessary to consider the other issues, which on that basis do not arise. M

I would dismiss this appeal.

LANE LJ gave a judgment agreeing that the appeal should be dismissed.

The appeal was dismissed. Leave to appeal to the House of Lords was refused.

CHANCERY DIVISION

December 21 1977

(Before Mr Justice BRIGHTMAN)

RE: PITSEA ACCESS ROAD, BASILDON, ESSEX: LAND RECLAMATION CO LTD v BASILDON DISTRICT COUNCIL

Estates Gazette April 29 1978

(1977) 246 EG 401

Landlord and Tenant Act 1954, Part II—Preliminary issue on point of law—Whether grant of a right-of-way for a term of years is a tenancy to which Part II applies—Whether a legal easement can be "occupied" for purposes of a company's business—Right-of-way can be "used" but not "occupied"—Mere right-of-way, as distinct from the actual roadway, cannot be "occupied" for purpose of business or any other purpose

This was a preliminary issue in proceedings transferred by consent from the county court to the High Court concerning the application by the Land Reclamation Co Ltd for a new tenancy from Basildon District Council. The new tenancy claimed, following a previous lease for seven years, was of a right-of-way over a roadway vested in the council to land belonging to the company and used for the company's business of waste disposal. The council took the point that the lease, being the grant of a term of years in an incorporeal hereditament, did not qualify for the security of tenure given by Part II of the Landlord and Tenant Act 1954.

A F B Scrivener QC and N M Lowe (instructed by Ellison & Co, of Colchester) appeared for the plaintiffs; R H Bernstein QC and C B Priday (instructed by J L Knight, of Basildon District Council) for the defendants.

Giving judgment, BRIGHTMAN J said: This is the trial of a preliminary issue arising on an application under the Landlord and Tenant Act 1954. The issue is whether the grant of a right-of-way for a term of years is a tenancy to which Part II of the Act applies.

The District Council of Basildon is the owner of a private roadway called Marsh Road, which runs from the neighbourhood of Pitsea southwards towards the Thames estuary. On each side of the roadway are areas of marshland, derelict farmland and mudflats. The plaintiff, Land Reclamation Co Ltd, is the owner of a large part of such land. The company uses the land for the disposal of waste and says that this is the largest waste disposal site in the United Kingdom. I was told that the site was particularly well suited to the disposal of waste, including toxic waste, because the subsoil consists of a saucer of impermeable clay which acts as a container for waste matter without polluting the surrounding land or the estuary. I was told, but there was no sworn evidence to this effect, that the site was of great importance to the country's industry, because there are few places so suitable for toxic waste disposal. The Basildon District Council, however, is conscious of local opposition and does not wish to be a voluntary party to the continued use of the site for this purpose.

The roadway was formerly vested in the Crown, and the company at that time used the roadway, which is the only existing access to the waste disposal site, under a licence granted by the Secretary of State for Defence. The local authority which was the predecessor of the present district council became the owner of the roadway in 1969. On August 14 1970 this authority, which I will call the urban district council, granted under seal the lease which is the subject-matter of the application now before the court.

By clause I of the lease the urban district council, in consideration of a yearly rent of £1,000, demised to the company:

> Full right and liberty for them and their assigns, their agents, servants or licensees, to pass and repass, with or without vehicles, along the private road in common with the lessor, its servants, agents, tenants and licensees, and others having a like right and liberty, between the hours of 6 am to 8 pm Mondays to Fridays, and 6 am to 2.30 pm Saturdays (hereinafter called the permitted hours) for all purposes in connection with the use of the company's land for the dumping and disposal of waste materials; subject nevertheless to, and reserving unto the lessor, the right at all times hereafter, or at any times or time, to erect, renew and maintain a gate or gates across the said roadway at the points marked 'G' on the said plan, with all necessary fittings and fixtures, but so that the same shall not be locked or be so erected or maintained as to impede or obstruct the free use and enjoyment of the right and liberty of way hereby granted in accordance with the tenor hereof.

By clause II the company entered into a number of covenants with the urban district council dealing with the maintenance of the roadway. For this purpose, the roadway was divided into four stretches, lettered from the south "A," "B," "C" and "D." The company covenanted to make up at its own expense the stretch from "A" to "B." It covenanted to pay to the urban district council the cost of putting "B" to "C" into repair; it covenanted to pay a fair proportion, according to user, of the cost of maintaining the stretch "C" to "D," its proportion being fixed for the first year at three-quarters. Clause III of the lease is a covenant for quiet enjoyment. Clause IV contains power for the urban district council to determine the lease in the event of breach of covenant by the company.

The term of the lease was due to end on March 25 1977. On April 20 1976 the company served notice on the Basildon District Council requesting the grant of a new tenancy, to which the district council responded with a counternotice. The company then applied to the Brentwood County Court. The Basildon District Council took the point that the lease, being a grant of a term of years in an incorporeal hereditament, was outside the protection afforded by Part II of the Landlord and Tenant Act 1954. By consent, an order was made removing the application to the High Court. The only question before me is the preliminary issue.

The case depends principally on section 23(1) of the Act. This provides as follows:

> Subject to the provisions of this Act, this Part of this Act applies to any tenancy where the property comprised in the tenancy is or includes premises which are occupied by the tenant and are so occupied for the purposes of a business carried on by him or for those and other purposes.

Subsection (3):

> In the following provisions of this Part of this Act the expression "the holding" in relation to a tenancy to which this Part of this Act applies, means the property comprised in the tenancy, there being excluded any part thereof which is occupied neither by the tenant nor by a person employed by the tenant and so employed for the purposes of a business by reason of which the tenancy is one to which this Part of this Act applies.

"Tenancy" is defined by section 69(1), but not in a manner which indicates whether Part II of the Act was intended to apply to an easement held for an interest equivalent to a term of years.

To come within Part II there are, shortly stated, two conditions which must be satisfied by the applicant. (1) The applicant must have a tenancy and not a mere licence or contractual permission which falls short of an interest in the land. (2) The applicant must establish that the property comprised in the tenancy is or includes premises which are occupied by him for the purposes of a business carried on by him.

The Basildon District Council first submitted that the legal relationship between it and the company was that of licensor and licensee, and not that of landlord and tenant. In support of this submission counsel, while recognising that the wording of the document was wholly appropriate to a lease, relied upon two factors. First, the absence of a specific description of the company's land to which the right-of-way could be appurtenant. Secondly, the fact that the grant was only for a period of seven years, although both grantor and grantee were fee simple owners. There is no substance in either of these points.

The Law of Property Act 1925, section 1, recognises that a legal easement may be created for an interest equivalent to a term of years absolute, and it does not matter whether the term is seven years or 700 years. It is also well established that a right-of-way can exist as a legal interest notwithstanding that the dominant tenement is not precisely identified in the deed of grant. The general reference to the company's land in the lease is sufficient for this purpose, extrinsic evidence being admissible. Neither of these propositions of law was disputed by the Basildon District Council; the factors mentioned were only relied upon as a pointer to the true construction of the deed. However, it seems to me clear that the urban district council granted to the company a legal easement for a term of years and not a mere licence.

This does not conclude the matter, because it is necessary for the company to establish that it is an occupier for business purposes of the property comprised in the lease. The property comprised in the lease is the right-of-way and not the roadway itself. There is no doubt that the right-of-way is exercised by the company for the purposes of its business. The company's lorries pass along the roadway at the rate of about one every four minutes. It is equally true that the roadway is used for the purposes of the company's business. But is it possible to say that the property comprised in the lease, ie the right-of-way, is, or includes, premises which are occupied by the company? In ordinary speech, a right-of-way is not a possible subject-matter of occupation as distinct from user, nor is it easy in normal circumstances to say that a roadway is occupied by a person who has a mere right-of-way thereover.

The first case to which the company referred was *Whitley v Stumbles* [1930] 1 KB 393. This arose under the Landlord and Tenant Act 1927. The plaintiff was the assignee of a lease which included an hotel and the exclusive right of fishing certain adjacent waters. The county court judge held that the fishing rights were not premises within the meaning of section 17(1) of the 1927 Act, and that accordingly the plaintiff, when applying for a new lease of the hotel, could not require the fishing rights to be included therein. This decision was reversed by the Divisional Court and the Divisional Court was upheld by the Court of Appeal [and House of Lords—see below].

Section 17(1) of the 1927 Act was worded as follows:

> The holdings to which this Part of this Act applies are any premises held under a lease, other than a mining lease, made whether before or after the commencement of this Act, and used wholly or partly for carrying on thereat any trade or business and not being agricultural holdings within the meaning of the Agricultural Holdings Act 1923.

There is a significant difference between that Act and the 1954 Act. In the 1927 Act the formula is "premises used for carrying on thereat any trade or business." In the 1954 Act the formula is "premises occupied for the purposes of a business carried on by him."

Greer LJ expressed strong doubts whether the word "premises" would cover a separate and distinct lease of an incorporeal hereditament. Slessor LJ perhaps leaned the other way. The doubt did not need to be resolved, because the fishing rights were demised along with corporeal hereditaments. The subject-matter of the lease was the hotel with certain rights attached thereto, and the Act applied to the entirety. This view was accepted by the House of Lords at [1930] AC 544, leaving unresolved the doubts expressed by Greer LJ.

The company also sought at one time to rely on *Bracey v Read* [1963] Cha 88, but more fully reported in [1962] 3 All ER 472. In that case a lease had been granted of the right to train and exercise racehorses on certain gallops on the Berkshire Downs. The 1954 Act was held to apply to that lease. On a closer examination of the case, and particularly of the facts as set out in the report in the *All England* series, it became clear that the lease was construed as a demise of the actual strips of land and not merely as a grant of a right to train and exercise horses. The decision accordingly is not an authority that a grant of a term of years in an incorporeal hereditament is by itself within the Act.

Counsel also relied on *Lee-Verhulst (Investments) Ltd v Harwood Trust* [1972] 3 All ER 619, where Sachs LJ compared the wording of the 1927 Act with the wording of the 1954 Act, and observed at p 624:

> On being asked what was the relevant distinction between "used" and "occupied," neither counsel—both very experienced in this branch of the law—could after full and helpful consideration suggest such a distinction; on the other hand, we were referred by them to statutes in which the word "used" seemed to have the same meaning as "occupied."

This observation has to be read in the context of the case then under consideration, namely a tenancy of premises consisting of 20 separate apartments in full occupancy. As occupation is a kind of user, it is difficult to envisage an occupation of land or buildings which is not also a user. The reverse does not apply; not every use is an occupation, and obviously many things capable of being used are incapable of being occupied. The words are not fully interchangeable, only interchangeable in some contexts.

I am not justified in reading section 23(1) of the 1954 Act as if it said "premises which are used by the tenant and are so used for the purposes of a business." So rewritten, a right-of-way for a term might well be within the Act, for I see no reason why a right-of-way should not be premises. Subsection (3), which defines "the holding" as the property comprised in the tenancy "there being excluded any part thereof which is occupied neither by the tenant nor by a person employed by the tenant," also suggests to my mind that occupation is not employed in the Act in the sense of user. There are, in fact, a number of sections of the Act which are to some extent inappropriate if occupation is equated with use, and if the Act is interpreted as including within its scope an incorporeal hereditament, such as an easement of way. It would be tedious to refer to them all. Notable examples can be found in section 30, and it will be noticed that subsections (1) and (7) of section 57, and section 58, contain the expression "the use or occupation of the property."

It was also submitted on behalf of the Basildon District Council that parts of the 1954 Act are geared to the rateable value of the holding: see section 63 which allocates jurisdiction to the High Court or the county court according to the rateable value of the holding, and section 37 which measures compensation for disturbance by reference to the rateable value. A right-of-way is not *per se* rateable; it would only be rateable if on the facts of the case it involved the exclusive occupation of the roadway, or something approaching exclusive occupation. The Act does not seem to envisage as within its compass a property right which is not rateable.

In *Jones v Christie* decided on April 30 1963 (Bar Library Transcript No 119) the Court of Appeal were concerned with the application of the 1954 Act to the lease of a house and fishing rights. The tenant was not able to claim that he occupied the house for business purposes, but he did claim to carry on the business of letting the fishing rights. The claim failed for a reason which is not material to be explained, but the learned Master of the Rolls doubted whether the Act could apply to a mere incorporeal hereditament on its own. He said, "If the letting were solely of an incorporeal hereditament, such as the right to fish, I doubt whether the Act would apply to it. When the Act speaks of 'premises' being

A ' occupied,' it may well refer to corporeal hereditaments and not to incorporeal hereditaments."

Counsel for the company sought to say that, although the property comprised in the tenancy consisted only of the right-of-way, the exercise of such right necessarily involved the occupation, although not the exclusive occupation, of the roadway; therefore, to that extent, the property comprised in the lease, namely the easement, included premises, namely the roadway, which were in the occupation of the company for the purposes of its business; in other words, occupation of the roadway resulted from the easement. I do not feel able to accept that submission. In my judgment, the
B company uses the roadway for the purposes of its business, but it does not occupy either the easement or the roadway. I take the view that Part II of the 1954 Act has no application to a mere right-of-way standing by itself, because such a right is not property or premises capable of being occupied for the purposes of a business, or indeed for any other purpose. I will make a declaration accordingly.

CHANCERY DIVISION

March 4 1977

C (Before His Honour Judge JOHN FINLAY QC, sitting as a deputy judge of the division)

LOVELY & ORCHARD SERVICES LTD v DAEJAN INVESTMENTS (GROVE HALL) LTD

Estates Gazette May 20 1978

(1977) 246 EG 651

Landlord and Tenant Act 1954—Application for new lease—Preliminary points—Whether tenants' proposals in originating summons constituted an offer accepted by landlords—Held that, even if proposals constituted offer capable of acceptance, landlords had not accepted but had made counter-offer—Principles for determination under section
D **34 of rent for new tenancy—Rent to be determined as at date of hearing at figure at which premises might reasonably be expected to be let in open market for a term commencing when new tenancy is to commence**

These proceedings were concerned with preliminary points arising in an application by tenants Lovely & Orchard Services Ltd, for a new lease of part of the ground floor of Africa House, Kingsway, London WC2. These points, which are fully explained in the judgment, were ordered to be dealt with as preliminary issues before the trial of the action.

R C Pryor (instructed by Waltons & Morse) appeared on
E behalf of the plaintiffs; R M Slowe (instructed by Tobin & Co) represented the defendants.

Giving judgment JUDGE JOHN FINLAY said: This is an application under the Landlord and Tenant Act 1954, made by the tenants by originating summons dated August 7 1975, whereby they applied for an order for the grant of a new lease upon the terms proposed in the summons, or, alternatively, upon such other terms as to period of rent and otherwise as the court may determine. Two preliminary points have arisen in the application the first of which is whether the tenants' originating summons contains an offer as to the terms of the new tenancy which the tenants propose which is capable of
F being accepted and whether in the events which have happened the offer has been accepted.

The summons is in the form provided for by Order 97 and contains the particulars required by the rules of the order. These particulars after describing the premises, which are part of the ground floor of Africa House, Kingsway, London, specified the current tenancy as being a term of years commencing on October 29 1971 and expiring on September 26 1975, granted by a lease dated April 11 1972, and continued in these terms:

> The defendants have served on the plaintiffs a notice of termination under section 25 of the said Act dated 9th April 1975, specifying as the date at which the tenancy is to come to an end G the 14th October 1975, and stating that an application to the court for a new tenancy would not be opposed. The plaintiffs have by a counter notice dated 25th April 1975 notified the defendants that the plaintiffs will not be willing at that date of termination to give up possession of the property comprised in the tenancy.

There is a new paragraph:

> The plaintiffs' proposals as to the new tenancy are: 1. Period: 15 years from the date of termination of the current tenancy; 2. Rent: £27,500 pa with rent reviews every five years. 3. Other Terms: As in the above-mentioned lease dated 11th April 1972.

The landlords submit by Mr Slowe, who appears on their H behalf, that that originating summons containing these particulars, and, in particular, the paragraph beginning "The plaintiffs' proposals as to the new tenancy" contains an offer which was capable of being accepted and has, in the event, been accepted by the landlords.

The subsequent history of the proceedings in brief was as follows: The landlords having served the section 25 notice and apparently not knowing, because the tenants' summons was not immediately served, of the issue of the tenants' summons, themselves issued a summons under the Act for an order determining the interim rent, that application being made under section 24A of the 1954 Act. That summons J has, by order for consolidation, been ordered to come on for hearing as consolidated with the tenants' summons. Furthermore, in May 1976 affidavit evidence was filed on behalf of the tenants proposing a rent for the new tenancy of £21,000 per annum and a further affidavit on behalf of the tenants served on or about February 25 1977 proposed a rent of £15,300. It will be observed that the successive proposals made in the tenants' summons and in the two affidavits which I have mentioned all proposed at the various dates rents of decreasing amounts.

By a letter dated February 23 1977 the landlords by their K solicitors wrote to the tenants' solicitors in these terms, after referring to the property:

> We are instructed by our clients, Daejan Investments (Grove Hall) Limited, to accept the offer made by your clients, Lovely & Orchard Services Limited, which same was made on the 7th August 1975 in their application to the High Court in the action, the short title of which is . . .

and then it is stated:

> The offer was set out in the Particulars thereto, namely, to take a lease of all that Ground Floor 'B' of the premises known as Africa House, Kingsway, London WC2 for a period of 15 years from the 14th October 1975 at a rent of £27,500 L per annum with rent reviews every five years and otherwise on the terms as in the Lease of the 11th April 1972 set out. It is this offer that on behalf of our clients we do hereby accept. By this acceptance, the application by your clients is compromised and the new rent will become payable three months from the date of delivery of this letter, which will be the date first above written.

I am asked to determine whether the proposals put forward by the tenants' summons are capable of constituting an offer. It is, I think, clear that in an application for a new tenancy under the 1954 Act it would be possible for a tenant to make proposals in such terms as would constitute an offer capable of being accepted and giving rise to a contract. The question M is whether that has been done in this case. I will assume for the moment that the proposals put forward by the tenant were capable of constituting an offer. The question then arises: what upon the true construction of the summons the offer was? That raises a question of construction because of the reference in the first term of the offer to the period of the tenancy as being 15 years from the date of termination of the current tenancy. The summons, as I have indicated, contained a reference to the section 25 notice as specifying the date at which the tenancy was to come to an end on October 14 1975 and referred to the counter notice by the tenants that

A the tenants would not be willing to give up possession at that date of termination.

In my judgment, the reference, in the proposal that the term should be 15 years from the date of termination of the current tenancy, to the date of termination of the current tenancy is not a reference to that date of termination which was specified in the section 25 notice. On the construction of the originating summons, my view is that the reference to a 15-year term commencing on the date of termination of the current tenancy is a reference to the date of termination which will result from the application of the provisions of B section 64 of the Landlord and Tenant Act 1954. This provides, in brief, that in a case where notice has been given under section 25 and an application has been made to the court and, apart from the section, the effect of the notice would be to terminate the tenancy before the expiration of the three months beginning with the date at which the application was finally disposed of (which would be the case here) the effect of the notice is to terminate the tenancy at the expiration of the said period of three months and not at any other time. Accordingly, I construe the offer, assuming that it be such, as an offer to take a tenancy for 15 years from the date falling three months after the date when C the application is finally disposed of.

That in my judgment gives rise to an initial difficulty in regarding the proposals, as put forward in the summons, as constituting an offer, because it leaves uncertain the date when the tenancy is to commence, making that date ascertainable only by reference to the determination of the proceedings under the provisions of section 64 of the 1954 Act. That uncertainty is, of course, capable of being resolved in that in due course in one way or another the proceedings will be finally disposed of and, therefore, the date of termination as ascertained under section 64 will be determined.

D This circumstance, however, that the ascertainment of the date of commencement of the term depends upon the date of termination of the proceedings, suggests to me that the proposals as put forward in the summons were not intended to be put forward as an offer. However, assuming that they were so put forward, it is submitted by Mr Pryor, on behalf of the tenants, that the offer made by the originating summons was an alternative offer, that is, an offer either to have the terms of the tenancy determined by the court, or, alternatively, to have the terms of the tenancy be such as were set out in the summons. If, as he suggests, it was an alternative offer, then the landlords by filing evidence in the E proceedings, as they have done, have elected to pursue the matter in one way, ie to have the terms of the tenancy determined by the court, and it is not now open to them to accept the alternative offer, which was an offer to have the terms determined by offer and acceptance by the parties. I do not think that it is necessary for me for the purpose of disposing of this preliminary point to decide whether that submission is or is not well founded, because, in my view, the only acceptance of the suggested offer which is relied upon is that contained in the letter of February 23 1977 and that letter, in my judgment, does not constitute an acceptance of the offer.

F The letter falls into several paragraphs, and in the first the letter states: "We are instructed by our clients . . . to accept the offer made by your clients," that is by the tenants, "which same was made on August 7 1975 in their application to the High Court." Assuming that the summons contained an offer, that paragraph would constitute an acceptance of it, if it were not for the point about the alternative offers being made by the summons. But giving that paragraph the construction that it is applicable only to the proposals as to a new tenancy, not to the alternative that the matter should be determined by the court, which I think is a fair construction to put upon the first paragraph of the letter, G there then follows a second paragraph which states that: "The offer was set out in the Particulars thereto, namely, to take a lease of all that the ground floor," which is more fully described, "for a period of 15 years from October 14 1975 at a rent of £27,500 per annum with rent reviews every five years and otherwise on the terms as in the lease of April 11 1972 set out." The letter then says: "It is this offer that on behalf of our clients we do hereby accept." These latter words in particular, as well as the whole of the preceding paragraph, make it impossible, in my judgment, to construe the first paragraph of the letter as containing an acceptance of the offer simpliciter. H They accepted an offer of a tenancy of a lease for 15 years from October 14 1975. I have already held that on the true construction of the summons that is not the offer, assuming there be an offer, which the summons puts forward. The purported acceptance therefore of a lease for a term of 15 years from October 14 1975 is not a true acceptance but in the nature of a counter offer.

Furthermore, the letter continues: "By this acceptance, the application by your clients is compromised and the new rent will become payable three months from the date of delivery of this letter, which will be the date first above written." J That paragraph proceeds on the basis that the acceptance of the offer will terminate the proceedings and thus under section 64 of the Act determine the date upon which the new tenancy is to begin, namely, the expiration of three months from the date of the acceptance of the offer. I cannot construe section 64 as having that effect. Section 64, in my judgment, results in the date of termination of the current tenancy falling three months after the date on which the application is finally disposed of, and the application is finally disposed of, in accordance with section 64(2), on ". . . the earliest date by which the proceedings on the application (including any proceedings on or in consequence of an appeal) have been determined and any time for appeal- K ing or further appealing had expired," save when the application is withdrawn or any appeal is abandoned in which case the reference to the determination or to the proceedings having been finally disposed of is construed as a reference to the date of the withdrawal or abandonment. The date upon which the parties come to terms is not, in my judgment, a date which starts to run the period of three months which is mentioned in section 64. Accordingly, this paragraph of the letter which suggests that the new rent would become payable "three months from the date of the delivery of this letter" is itself a new term proposed and differing from L the terms suggested in the summons, and, therefore, for this reason also, the letter would fall to be construed as a counter offer and, therefore, not an acceptance of the terms put forward by the tenant.

For these reasons I have come to the conclusion that there has been no determination of these proceedings in the sense of an acceptance by the landlords of any offer which may have been made by the tenants in the originating summons which they issued on August 7 1975.

That being so, there falls for determination the second preliminary point, which is as from what date the rent falls to be determined for the purposes of section 34 of the 1954 M Act. That section provides that:

> The rent payable under a tenancy granted by order of the court under
>
> Part II of the Act
>
> shall be such as may be agreed between the landlord and the tenant or as, in default of such agreement, may be determined by the court to be that at which, having regard to the terms of the tenancy (other than those relating to rent), the holding might reasonably be expected to be let in the open market by a willing lessor, there being disregarded

various matters which I do not think for the present purpose I need read.

It is to be noted that the duty imposed upon the court by the section is to determine the rent at which, "the holding might reasonably be expected to be let in the open market by a willing lessor." That wording itself suggests to my mind that the determination must be of the rent which would be appropriate either at the date when the determination is made or, at any rate, at some later date and that it is not very likely that having regard to the wording the statute contemplates the determination of some rent payable as from some earlier date. It is submitted, however, on behalf of the landlords, that for a variety of reasons the proper date for the determination of the rent at which the premises might reasonably be expected to be let should be the same date as that from which an interim rent would be payable if it fell to be determined under section 24A, that is to say, the date when apart from the application under the Act the tenancy would have come to an end by reason of the section 25 notice and the date of termination therein specified, namely, October 14 1975.

There are three possibilities as to the date on which the reasonable rent should be ascertained. The first is the date I have mentioned, October 14 1975, the second is the date of the hearing, and the third is the date as from which the new lease will commence, that is to say, the date when the present tenancy comes to an end having regard to the joint effect of the provisions of section 25 of the Act and section 64 thereof, that last mentioned date being in the event of the matter having to be determined at the hearing of the summons, as is the case, the date falling three months after the hearing or three months after the hearing plus the time allowed for appealing or such later date as may be determined if there are any further proceedings by way of appeal in accordance with section 64 (2).

Mr Slowe, who appears for the landlords, submits that there are numerous reasons, all I think based upon reasons of convenience, why the date for consideration of what the appropriate rent is should be October 14 1975. First, he submits, that although there is nothing in section 34 which gives any direct guidance upon it, in order to give effect to section 34 it would be highly convenient if the rent falling to be determined under that section had to be determined as at the same date as an interim rent to be determined under section 24A because the latter is usually taken as a percentage reduction of the former. Furthermore, he submits that there would be commercial difficulties in determining the rent at any other date than the date of termination specified in the section 25 notice and that if the rent fell to be fixed at some date in the future by reference to the operation of section 64 there would be very great uncertainty in knowing at what date the market should be regarded, since the date to be determined by section 64 would, of necessity, remain unascertained until it had actually happened.

These submissions have, I think, a certain amount of attraction about them, but I am not persuaded that the proper date for determination of the rent is the date upon which the tenancy comes to an end as specified in section 25. Section 34 requires the rent to be ascertained as being that at which the holding might reasonably be expected to be let in the open market by a willing lessor, there being disregarded various matters set out in section 34(1). The tenancy in respect of which that rent falls to be determined is one which under the provisions of the Act is not to commence until after the date of the hearing. In these circumstances, it appears to me to be very likely, upon the reading of the words in the section, that the notice thereto is contemplating that the market will be regarded as at the date of the hearing. But there is this consideration to be borne in mind that, since the tenancy is not to commence until a date after the date of the hearing, it would be likely that the provisions should contemplate that the court would take into account changes which might be reasonably expected to take place between the date of the hearing and the date when the term commences. It seems to me to be *prima facie* unlikely that the section requires the court to determine the rent as at the date which may be long past if there be the possibility that in the meantime circumstances may have radically changed so that what was a reasonable rent as at the date specified at the termination of the tenancy in the section 25 notice may no longer be a reasonable rent at the date of the hearing.

There is no direct authority on it, but in the case heard by Megarry J, as he then was, in 1972, ***English Exporters (London) Ltd v Eldonwall Ltd*** [1973] 1 Ch 415, there fell to be considered certain questions as to the principles to be applied in determining the interim rent on an application under section 24A. In the course of dealing with that matter, Megarry J also determined the substantive rent to be paid in respect of the new tenancy, and in dealing with that it is evident from the report that his determination proceeded on the footing that the witnesses, whose evidence was heard as to the duration of the lease, were giving evidence, as at the date of the hearing, relevant to a tenancy which was to begin at a future date, the parties there being agreed as to the date at which the tenancy should commence. Having referred to those difficulties, as he did, at p 419, Megarry J says at p 431:

> A determination under section 34 alone is necessarily prospective, fixing a new rent as from the future commencement of the new tenancy, whereas a determination under section 24A will usually, if not always, be in some degree retrospective, applying in part to a period that has already run.

and then he goes on to consider the section 24A point. He then says:

> The facts that section 34 is prospective, and that under section 24A(3) there is a hypothetical yearly tenancy, seem to me to provide sufficient grounds for holding that the values to be applied should be those existing when the interim period begins to run.

There appears to have been no argument in that case as to the date at which the section 34 determination should be considered or, to put it another way, as to the appropriate date for the purposes of the section 34 determination, but the clear statement made by Megarry J in the passage which I have read that a determination under section 34 alone is necessarily prospective appears to me to indicate clearly the view, which accords with my own, that under section 34 what falls to be done is to ascertain what would be the rent at which the premises might reasonably be expected to be let in the open market as at the date when the tenancy is going to commence. If that be so, then it is submitted on behalf of the landlords that the result of that is that because of section 64 and the uncertainty introduced by that section as to when the tenancy will commence (having regard to the possibility that the proceedings may not be determined at the hearing at first instance but may then go to appeal) that uncertainty introduces a complication which renders this mode of approach to section 34 highly impracticable.

I do not accept that that result follows. The duty of the court in determining the rent under section 34 appears to me to be to determine the rent under that section as at the date of the hearing, but having regard to such evidence as there may be that indicates that any changes may have to be taken into account, being changes that the evidence shows are likely to occur between the date of the hearing and the date when, assuming that the court's order stands as it is made by the court, the new tenancy will commence. That will be the period of three months from the date of the hearing, plus any time that falls to be allowed under section 64(2) for an appeal being instituted. The court cannot, it seems to me, reasonably determine the rent for the purposes of section 34 on the footing that proceedings are going to be prolonged for an indefinite and unascertained period. The duty of the court is to determine the rent as at the date of the hearing, but it must do so having regard to what is the likely date of the commencement of the tenancy. Accordingly, I must reject the submission made on behalf of the landlords that the proper date to have regard to for the purpose of section 34 is October 14 1975 and I hold that the date is that which I indicated.

Mr Slowe relied upon the decision of the Court of Appeal in *Lawrence Frederick Ltd v Freeman, Hardy & Willis Ltd* [1959] 3 All ER 77, where the question arose as to whether the period of five years referred to in section 30(2) of the Act had fallen to be ascertained by reference to the date of termination specified in section 25, or the date of termination of the tenancy, as ascertained in accordance with section 64, and he submitted that the practical difficulties that were there elaborated in the judgment of Romer LJ of applying section 64 to the provisions of section 30(2) were practical difficulties which were equally applicable if the provisions of section 64 were regarded for the purpose of determining the date at which the rent fell to be ascertained under section 34 of the Act. I do not think that the conclusion follows. The period of five years referred to in section 30(2) is a period ending with the termination of the current tenancy. The practical difficulties that arise from construing that reference to the termination of the current tenancy by the application of section 64 are not, by any means, the same practical difficulties as arise by the process of having regard to section 64 and its effect upon the date of the commencement of the new tenancy in relation to the question of when the rent shall be determined under section 34 of the Act.

Accordingly, in my judgment, the proper date for the purposes of determination under section 34 is the date of the hearing but having regard to matters which can reasonably be expected to happen between that date and the date when the term is to be expected to commence, if the order made at the hearing takes effect in the usual course, in so far as these events subsequent to the date of the hearing have a bearing upon determination of the question: At what rent the premises might reasonably be expected to be let in the open market by a willing lessor for a term commencing at that date of commencement.

The costs of these preliminary issues were reserved to be dealt with at the end of the hearing of the action.

CHANCERY DIVISION

January 19 1978

(Before Mr Justice GOULDING)

CHARLES CLEMENTS (LONDON) LTD v RANK CITY WALL LTD

Estates Gazette May 27 1978

(1978) 246 EG 739

Landlord and Tenant Act 1954, Part II—Application for new tenancy of premises used for business of a retail cutler—Proposal by landlords to modify covenant restricting user of demised premises—Existing covenant prevented user for purposes other than business of retail cutler without landlords' consent in writing—Proposed addition of words "such consent not to be unreasonably withheld"—Proposal, which would relax restriction, but increase rent by about £1,750 per annum, resisted by tenants—Objection upheld—Tenants not to be pressed to accept concession in return for higher rent

In this case the plaintiff tenants' originating summons for a new tenancy of premises at 125 Regent Street, London W1, used by them for the business of a cutler, was adjourned into court for the judge to decide whether the new tenancy should include the particular modification of the user covenant proposed by the defendant landlords and described in the judgment.

R C Pryor (instructed by Laytons) appeared on behalf of the plaintiffs; Aron Owen (instructed by Richards, Butler & Co) represented the defendants.

Giving judgment, GOULDING J said: I have before me an originating summons which has been adjourned into court. The summons was taken out by the plaintiffs, Charles Clements (London) Ltd against the defendants, now named Rank City Wall Ltd, pursuant to the Landlord and Tenant Act 1954, for the purpose of obtaining a new tenancy, under the terms of the Act, of business premises at 125 Regent Street in Westminster. I shall try to refer to the plaintiffs as "the tenant" and to the defendants as "the landlord."

The matters that have been argued before me and on which I am invited to express an opinion before the final disposal of the originating summons involve the application in particular of sections 33, 34 and 35 of the Act of 1954. I will read the material parts of those sections before stating the facts. Section 33 deals with the duration of any new tenancy granted pursuant to the Act as follows:

> Where on an application under this Part of this Act the court makes an order for the grant of a new tenancy, the new tenancy shall be such tenancy as may be agreed between the landlord and the tenant, or, in default of such an agreement, shall be such a tenancy as may be determined by the court to be reasonable in all the circumstances, being, if it is a tenancy for a term of years certain, a tenancy for a term not exceeding 14 years, and shall begin on the coming to an end of the current tenancy.

Section 34 deals in particular with rent, and under subsection (1):

> The rent payable under a tenancy granted by order of the court under this Part of this Act shall be such as may be agreed between the landlord and the tenant or as, in default of such agreement, may be determined by the court to be that at which, having regard to the terms of the tenancy (other than those relating to rent), the holding might reasonably be expected to be let in the open market by a willing lessor

and then follows a direction to disregard four classes of fact in ascertaining what is reasonable. Subsection (2) of the section is ancillary to one of those provisions for disregarding special facts and subsection (3) provides that "Where the rent is determined by the court the court may, if it thinks fit, further determine that the terms of the tenancy shall include such provision for varying the rent as may be specified in the determination." I ought, in connection with section 34(1), to remember that under section 23(3) it is provided that "the holding" in relation to a tenancy means "the property comprised in the tenancy," with an exclusion which is immaterial for present purposes. Finally section 35 deals with terms of a new tenancy other than duration and rent in this manner:

> The terms of a tenancy granted by order of the court under this Part of this Act (other than terms as to the duration thereof and as to the rent payable thereunder) shall be such as may be agreed between the landlord and the tenant or as, in default of such agreement, may be determined by the court; and in determining those terms the court shall have regard to the terms of the current tenancy and to all relevant circumstances.

Having stated the statutory provisions I return to the originating summons in the present case which, as required by the rules of the court, sets out the material facts. The premises are at 125 Regent Street and are a lock-up shop forming the ground floor and basement at that address. They were let by the landlord to the tenant by a lease dated September 15 1972 for five and a half years from Midsummer 1971 at a rent of £4,000 per annum. The rateable value is £7,388. The tenant occupies the premises for the purpose of the tenant's business as a cutler. The proper notice and counternotice have been served under the Act.

By the originating summons the tenant makes the following proposals for a new tenancy: period, 14 years from the termination of the current tenancy with reviews at 5 and 10 years; rent, £9,000 per annum; other terms as in the existing lease. The evidence on both sides is formal and does not include any expert or other evidence as to values. The landlord, in an affidavit by its secretary, makes a counter-proposal as to the new tenancy. The landlord agrees to a term of 14 years from the termination of the tenant's current tenancy with rent reviews at the beginning of the 6th and 11th years (which I understand so far to be the same as the tenant's proposal) and then departs in two ways from what is suggested by the tenant. First of all, as to rent, the landlord asks for £15,000 per annum as the rent for the first 5 years; secondly, while in general adopting the proposal that the terms of the new tenancy shall be those of the old, the landlord asks for one alteration, and that is in clause 2(L) paragraph 2 of the existing lease. The paragraph follows on a preceding paragraph of the same subclause which prohibits a great number of specified trades and other acts on the premises. Then comes the material sub-paragraph as follows:

> At all times during the said term to keep open use and occupy the premises for the purpose of the permitted trade or business as defined in the 4th schedule

that being, put shortly, the retail business of cutlers

> and not at any time during the said term to use or permit to be used the premises or any part thereof for any other purpose whatsoever without the landlord's previous consent in writing but nothing herein contained shall prevent the landlord demising and letting other shops belonging to it and adjoining the said New Gallery Cinema for use as shops for the sale of fancy goods in connection with any other trade or business.

That is part of a clause containing covenants on the part of the tenant. The landlord desires in the new tenancy to modify it by inserting after the words "without the landlord's previous consent in writing" the further words "such consent not to be unreasonably withheld."

No evidence has been adduced as to the reasons for and effect of the proposed change. It is, however, agreed between counsel that if and when the rent of the premises for the purposes of a new tenancy is determined in accordance with section 34 of the Act—which it will be remembered requires the court to have regard to the terms of the tenancy other than those relating to rent—the rent will be higher if the landlord's additional words are included than if they are omitted, leaving the form of the new lease similar to that of its predecessor. It is said in argument, though there is no agreement on the point, that the difference would be, very approximately, £1,750 per annum. So the landlord wants to relax the restraint on the tenant's liberty of trading in the premises to a certain extent in order that the open market rent, having regard to the terms of the tenancy, may be some £1,750 higher. The tenant, which is content with the existing restrictions and does not want them relaxed, objects to paying that increased rent.

The master, at the request of the parties, has adjourned the originating summons into court in order that I may express my view upon that particular proposal of the landlord regarding the modification of the tenant's covenants. I have at various points during the hearing entertained doubt whether I ought to do so in isolation. Although the sections I have read provide a scheme of approach to the determination of the different elements in a new tenancy, it is apparent that duration, rent, restrictive covenants, and indeed other matters if there were any falling to be determined, must interact on one another and in the end the court can only approve or impose a complete set of provisions. I was anxious that I might be deciding a hypothetical question in considering this particular point, in the absence of any evidence or determination as to the new rent and the duration of the tenancy. However, counsel assure me that there is a firm agreement, whatever I may decide on this point, that the duration shall be that proposed by both parties, namely 14 years with two rent reviews, and also that they or their expert advisers will have little difficulty in agreeing the open market rent according to the formula of the Act (section 34) when it is known whether the covenant is to be modified or not.

I am also encouraged by an observation of Widgery LJ (as he then was) in the case of *Cardshops v Davies* [1971] 1 WLR 591 at p 596, where he said this:

> Section 35 of the Landlord and Tenant Act 1954 provides that the terms of the tenancy have to be fixed by the court, in the absence of agreement; and section 34, dealing with the fixing of rent, requires that the rent itself shall be fixed according to the other terms. It is, therefore, I think, quite obvious that the proper sequence is for the court to fix the other terms first and then to consider, in the light, among other things, of those terms, what the rent should be.

Accordingly, though not entirely without misgiving, I think I ought to express a view on this point in isolation as the parties desire. Let me say at once that, although I shall have to deal with the arguments that have been addressed to me, and discuss the terms and application of the Act, it is not my intention to express any general propositions as to its interpretation. There is already a certain amount of authority and it is an Act of wide and everyday application. One thing that is clear is that the court in fixing the provisions of a new tenancy has a very wide discretion. Nothing I say can or should affect the width and generality of that discretion. If any guide lines are to be laid down it should be, in my view, by an appellate court and anything I say is to be read strictly with reference to the facts of this case and not treated as of general application.

As I have said, no evidence has been adduced in favour of the proposal except the informal evidence by way of agreement between counsel that the landlord will under section 34 be able to obtain a higher rent if the tenant's obligations are relaxed in the way proposed. Mr Owen, appearing for the landlord, starts with this general approach to the Act, that it is one that interferes with the landlord's enjoyment of his legal property and his freedom to contract regarding his property as he thinks fit. Therefore in interpreting and applying the statute one should adopt a cautious approach, not interfering with the common law rights of the landlord more than necessary, and, while the authorities show, as Mr Owen agrees, that the court will not by imposing terms on a new tenancy restrict the scope of the tenant's business, or, at the wish of the landlord, impose some novel burden upon the tenant, none the less there is no reason why the court should prevent the landlord, if he wishes, from conferring a valuable additional asset or privilege on the tenant. On the other side of the picture Mr Owen submits that by resisting the change the tenant here is trying in part to sterilise the landlord's valuable shop in Regent Street which forms the holding with which the court is concerned, and which, it is submitted and I will assume, though I have no evidence, is capable of use for many different retail trades or other activities within the existing planning permission.

More particularly, Mr Owen summarised the effect of the authorities, and the further contentions that he says the court should accept, in five propositions which I will read. First, he says that under section 35 of the Act there is a very wide discretion in the court as to the terms of the new tenancy; secondly, it is *prima facie* wrong to introduce a restriction on user which deprives the tenant of protection in respect of his existing business; thirdly, it is equally wrong to introduce a new and burdensome covenant imposed upon the tenant; fourthly, if the landlord wishes to restrict and cut down the current user of the premises he must justify the proposed restriction; similarly, if he wishes to introduce a new or burdensome covenant upon the tenant he must show special reasons based on sufficient evidence; fifthly, on the other hand, where the landlord, far from imposing a restriction or introducing a burdensome covenant, wishes, without in any way altering the tenant's present user of the premises, to remove a restriction it is quite proper for him to do so.

The proposal in the present case, submits Mr Owen, will not only provide a higher rent for the landlord, it will have advantages to the tenant, too. First of all the existing lease (in the positive portion of the subparagraph that I have read) obliges the tenant to continue to carry on the existing type of business throughout the term, and it prevents the tenant, either by itself or through an underlessee or assignee, from carrying on a different business. That limitation would in effect be removed by the proposed change. Again, although the landlord cannot under the existing provisions unreasonably refuse consent to an assignment or underlease of the whole of the premises, the restriction as to user would prevent any such assignment or underlease being made on reasonable terms if the trade of a retail cutler were to become unprofitable. There again the relaxation of the restriction, it is said, will confer a valuable benefit on the tenant. All that will be given to the tenant, in return for the advantage of a higher rent to the landlord, if the landlord's proposal is accepted. Conversely, it is submitted, if the landlord's proposal is rejected the court will be preventing the landlord from getting what it was meant to get under the Act as the condition of a renewed tenancy, namely a truly open market rent by current values for the landlord's property.

In my judgment those contentions are ill-founded. I have been referred to a number of authorities and I rely in particular on *Gold* v *Brighton Corporation* [1956] 1 WLR 1291, *Aldwych Club* v *Copthall Property Company* (1963) 185 ESTATES GAZETTE 219, and *Cardshops* v *Davies* [1971] 1 WLR 591. Of those the first and last are decisions of the Court of Appeal, the *Aldwych Club* case is a decision of Pennycuick J. Approaching the Act with the guidance derived from those authorities I find it clear that the protection intended to be given by Part II of the Act is to the tenant in his character of an occupying business tenant carrying on his trade on the premises. The advantages promised to the tenant as delineated by Mr Owen are advantages that would benefit the tenant if and when it wishes to discontinue the present business. The obligation to continue the business, the restriction on carrying on some different business, is no burden to the tenant while it continues to do what it was the policy of the Act to protect it in continuing to do. Thus those advantages are to be regarded, I think, for the purpose of exercising the court's discretion as essentially collateral. On the other hand the correlative burden on the tenant of paying a higher rent to the landlord is something that immediately affects the tenant in the carrying on of its business because there is a further outgoing, perhaps £1,750 per annum (it matters not what is the exact figure) that has to be met before any profit can be arrived at. Thus the argument that the landlord should not be prevented from conferring a benefit on the tenant by modifying the restriction is one that seems to me unpersuasive.

Similarly I think there is an error of reasoning in the argument that, in the circumstances of this case, the landlord, if its proposal is not accepted, will be prevented from getting the open market rent intended by the Act. Section 34 of the Act says that the rent under the new tenancy, in default of agreement, is to be that at which, having regard to the terms of the tenancy, the holding might reasonably be expected to be let in the open market by a willing lessor. That, as Widgery LJ pointed out, presupposes that the terms are known, and under section 35 the terms of a tenancy granted by order of the court are, in default of agreement, to be determined by the court having regard "to the terms of the current tenancy and to all relevant circumstances." That does not mean that the terms of the current tenancy are necessarily decisive, because there may be other circumstances that outweigh them, but it is a guide to the court that Parliament wanted the terms of the current tenancy to be considered in the forefront in every case, along with any relevant circumstances of the particular case. Here I have no evidence of any special facts affecting the property that are not common to all retail premises in high-class business areas. If the parties are to be at liberty to insist on changes in the terms of the existing tenancy simply because they consider them beneficial to themselves, a field would be opened which I think the court would find it bewildering to traverse. Every tenant could volunteer to accept increased liabilities as to this or that matter for the purpose of getting, on terms that he thought might ultimately profit him, a lower rent; every landlord might press this or that concession on the tenant because it would enable him to say the open market rent was a higher rent. It therefore appears to me, for the purposes of the present case, that the right course is to take the terms of the existing lease as a sufficient guide in this respect, and not to make the alteration desired by the landlord since no special reason is shown for it and the tenant objects to it.

I think it may well be that if I now adjourn the originating summons back to chambers no order need be drawn up, but should it be necessary to do so I suggest that the order should be on the following lines but without prejudice, of course, to the discretion of the registrar to settle it in such form as may be in accordance with existing practice or the views of the registrar's department. For myself, I think that, reciting the documents that I have read, and that I have heard counsel on both sides, the order should recite "The judge being of opinion that any new lease ought not to incorporate the modification of the tenant's covenants in the existing lease proposed by paragraph 3 of the affidavit of John Beech Rutherford"—and then comes the substantive part of the order—"adjourn the originating summons back to chambers." So if an order has to be drawn up the substantive part, save for any part dealing with costs, would simply be an adjournment back to chambers, prefaced by an indication of the view I have just expressed.

Summons adjourned back to chambers. Plaintiffs (tenants) to have costs of adjournment into court in any event. Leave to appeal granted.

COURT OF APPEAL

June 18 1976

(Before Lord Justice CAIRNS, Lord Justice STEPHENSON and Lord Justice BRIDGE)

MICHAEL KRAMER & CO. v AIRWAYS PENSION FUND TRUSTEES LTD

Estates Gazette June 10 1978

(1976) 246 EG 911

Landlord and Tenant Act 1954—Discontinuance by tenants of application for a new tenancy—Effect of discontinuance on landlords' application under section 24A for determination of an interim rent—Whether landlords' application lapsed with discontinuance of proceedings by tenants—Landlords' application not "parasitic"—Analogy of counterclaim—High Court and county court rules of procedure reviewed—Principles for exercise of court's discretion under section 24A discussed—Object is to promote justice—Tenants' appeal dismissed

A In their appeal the tenants, Michael Kramer & Co, a firm of solicitors, sought to have set aside decisions of Judge Stockdale at Westminster County Court under the Landlord and Tenant Act 1954 in relation to part of premises at 137-141 Regent Street, London W1, of which the appellants at the material time had a subtenancy. The appellants had applied for a new tenancy and the respondents for the determination of an interim rent. Subsequently the appellants gave notice of discontinuance of their application. Judge Stockdale held that the respondents' application for an interim rent was not defeated by this discontinuance and he determined an interim rent substantially higher than the rent payable under the subtenancy.

B B Green (instructed by Michael Kramer & Co) appeared on behalf of the appellants; W Marshall (instructed by Richards, Butler & Co) represented the respondents, but was not called on.

Giving the first judgment at the invitation of Cairns LJ, STEPHENSON LJ said: This is a tenants' appeal against two judgments of His Honour Judge Stockdale given in the Westminster County Court, the first on a preliminary point on September 24 1975, and the second on the substance of the matter on October 30 1975.

The matter arises under a lease granted to the tenants, a firm of solicitors, Michael Kramer & Co, by the respondents, Airways
C Pension Fund Trustees Ltd, on February 16 1968. What was then granted was a business tenancy or subtenancy of premises 137-141 Regent Street, Piccadilly, in this city. It was a tenancy of the fourth floor and part of the fifth floor of that building for seven years from September 29 1967. It was at an annual rent of £2,800.

On March 29 1974 the landlords served a notice under section 25 of the Landlord and Tenant Act 1954 to determine that tenancy on October 5. On April 4 the tenants wrote a letter, to be taken as a counter-notice that they were not willing to give up the premises, which is required by subsection (2) of section 29 of the Act. On July 25 the tenants applied under section 24 of
D the Act for a new tenancy for 14 years from October 5 1974 at an annual rent of £4,200, and they made that application to the High Court. The landlords informed the tenants that they did not object to the proposed rent, but that they did object to the new tenancy. On November 28 the matter was transferred from the High Court to the Westminster County Court under section 63 of the Act. As a result of a decontrol notice dated February 12 1975 the freeze upon business rents which had been imposed by the counter-inflation legislation was lifted by virtue of the Counter-Inflation (Business Premises) (Decontrol) Order 1975.

E On April 30 the landlords applied for a determination of an interim rent under section 24A of the 1954 Act, and, I must add, because it is a matter of cardinal importance to Mr Green's submissions on behalf of the tenants, under Order 13, Rule 1 of the County Court Rules. That application was adjourned until the hearing of the tenants' application; and before the hearing of either could come on, the tenants on July 17 gave notice of discontinuance in these terms: "Take notice that the plaintiffs wholly discontinue this action."

On July 31 the learned county court judge heard the tenants' application for a new tenancy, coupled with the counter-application, if I may call it that, by the landlords for an interim rent.
F He heard argument on the question whether the landlords' application for the determination of interim rent had, as the tenants contended, lapsed by reason of the tenants' discontinuing their action or application for a new tenancy. On that matter he reserved his judgment, and on September 24 he gave judgment rejecting the tenants' submissions. He said that the landlords' application in his opinion "instituted proceedings under section 24A which are quite distinct from the proceedings instituted by the tenants' application for a new tenancy, though the former was made in the course of the latter. The landlords' application cannot therefore be brought to an end by the tenants' unilateral discontinuance of their new tenancy application." He went on to hear evidence. On September 30 he made his second order G determining the interim rent at a figure of £9,000, a substantial increase on the £2,800 contracted for by the lease.

In this appeal Mr Green has, with learning and considerable courage in the face of fairly early disclosed opposition on the part, I think, of each member of this court, contended, first of all, that the learned judge was wrong to dismiss his preliminary point and ought to have held that the landlords' application lapsed with the tenants'. Secondly, he has contended, with, I think he would admit, less confidence and much more briefly, that although the quantum of the interim rent which the learned judge fixed is not unreasonable, or, at any rate, is not appealable, the learned judge was wrong in exercising his discretion under section 24A of the H Act to order any interim rent at all. In other words, the tenants should have been entitled to retain the continuation of their tenancy provided for them by the statute at the old unreasonably low rent of £2,800 for a period of 5½ months which is in question, namely, the period which it is agreed runs from the date of the landlords' application for interim rent on April 30 1975 to the expiry of the continued tenancy on October 17 1975, which means a loss to the landlords of £2,841.

I have omitted in my summary of the history of this matter one event, and that is that on July 29 1975 the landlords made a second application for a determination of the interim rent, this J time under a different county court rule, namely, Order 6, Rule 4. They made that application *ex abundanti cautela* in case the learned judge accepted the tenants' submission that their first application was one which he had no jurisdiction to hear because it had lapsed with the discontinuance of the tenants' application. That second application is not before us, and on the view which I take of the preliminary point taken by Mr Green on behalf of the tenants it does not arise for consideration in this appeal.

What does arise for consideration is the rules on which Mr Green relies in support of his interesting but, he concedes, unmeritorious submission. The claim for interim rent is brought under section 24A of the Landlord and Tenant Act 1954, a K section introduced into the Act by section 3 of the Law of Property Act 1969 to meet, to a considerable extent at any rate, the conduct of tenants who drag out proceedings under Part II of the 1954 Act so as to remain in possession of business premises at unreasonably low rents. That new section provides:

> (1) The landlord of a tenancy to which this Part of this Act applies may—(*a*) if he has given notice under section 25 of this Act to terminate the tenancy; or (*b*) if the tenant has made a request for a new tenancy in accordance with section 26 of this Act; apply to the court to determine a rent which it would be reasonable for the tenant to pay while the tenancy continues by virtue of section 24 of this Act, and the court may determine a rent accordingly. (2) A rent determined in proceedings under this section shall be deemed to be the rent payable under the L tenancy from the date on which the proceedings were commenced or the date specified in the landlord's notice or the tenant's request, whichever is the later. (3) In determining a rent under this section the court shall have regard to the rent payable under the terms of the tenancy . . .

and then there is a provision which I need not read.

If this matter had remained in the High Court it is unlikely that there would have been any preliminary point available to Mr Green. The Rules of the Supreme Court contain a provision in Order 97, rule 9A (1):

> An application under section 24A of the Act of 1954 to determine an interim rent shall—(*a*) if the tenant has begun proceedings for a M new tenancy under section 24 of the Act, be made by summons in those proceedings, and (*b*) in any other case, be made by originating summons. (2) The application may be heard and determined in chambers.

That means that in the instant case, the tenant having begun proceedings for a new tenancy under section 24, the rules in the High Court provide that the landlords' application would have been made by summons in the tenants' proceedings for a new tenancy. Furthermore, it is provided by the Rules of the Supreme Court that discontinuance of any proceedings usually requires the leave of the court. By Order 21, rule 3, it is provided:

> Except as provided by Rule 2

A which does not apply to this case

> a party may not discontinue an action. . .or withdraw any particular claim made by him therein, without the leave of the Court, and the Court hearing an application for the grant of such leave may order the action. . .to be discontinued, or any particular claim made therein to be struck out, as against any or all of the parties against whom it is brought or made on such terms as to costs, the bringing of a subsequent action or otherwise, as it thinks just.

In the county court there are two differences. First, there is no rule regulating the making of applications under section 24A. There is at p 1737 of the current edition of the *County Court Practice* this note on procedure:

> B If the tenant has already applied to a county court for a new tenancy an application under this section may be made in the course of those proceedings in accordance with Order 13, Rule 1, or in the landlord's answer. If the tenant has not already applied for a new tenancy an application to the county court under this section should be made by originating application in accordance with Order 6, Rule 4.

It was no doubt a consideration of that note which led the landlords' solicitors to make the application of April 30 under Order 13, Rule 1, in the tenants' proceedings, and also to make the application of July 29 under Order 6, Rule 4, by originating application.

Next, there is no provision in the County Court Rules requiring the court's leave before a plaintiff can discontinue. By C Order 18, Rule 1, it is provided:

> If a plaintiff desires to discontinue wholly or in part any proceedings against all or any of the parties thereto, he shall give notice thereof in Form 58 to the registrar, and to every party against whom he desires to discontinue.

Order 13, Rule 1, deals with applications in the course of proceedings, and it provides:

> Where by any Act or Rule any application in the course of an action or matter before or after judgment is expressly or by implication authorised to be made to the court or to the judge or to the registrar, then subject to the provisions of the particular Act or Rule applicable thereto, and so far as not inconsistent therewith, the following D provisions shall apply.

Then there are detailed provisions requiring the application to be in writing, where made on notice. Order 6, Rule 4, on the other hand, deals with originating applications in matters. It provides:

> (1) Any proceedings authorised to be commenced in a county court and not required by any Act or Rule to be commenced otherwise, may be commenced by originating application.

Then follow detailed provisions requiring the application to be in writing and referring to Form 23.

The landlords' two applications were made in conformity with those rules. The first of them, on April 30, was made on E notice in Plaint no 7503765. It appears (we have no information about it) that when the matter of the tenants' application was transferred from the High Court, where it had a number, 1974 M4404, it was given a plaint no 7503765. By that notice, headed, "In the Westminster County Court Plaint No 7503765," between the tenants as plaintiffs and the landlords as defendants, it is said: "Take notice that the defendants intend to apply to the Judge . . .that: (1) He determine a rent which it would be reasonable for the plaintiffs to pay whilst their tenancy of the premises at the Fourth and part Fifth floors of 137-141 Regent Street, London W1, continues by virtue of section 24 Landlord F and Tenant Act 1954. (2) The costs of this application be costs in the cause." Their later application of July 29 is a longer document. It receives a new number, "Number of Matter 7509148," and it is (I do not think I need read it) in the form prescribed for an originating summons, Form 23.

Mr Green contends that there is nothing wrong in a landlord applying in the tenants' proceedings under Order 13, Rule 1, but if he does so he takes the risk of having his application discontinued by the discontinuance of the tenants' proceedings. If he wants to avoid that risk he ought to make his application by an originating application under Order 6, Rule 4, or by counterclaim so designated. In this case he has made it by an interlocutory application in the proceedings under Order 13, Rule 1, G and therefore it has gone with the discontinuance of the tenants' application. It is plainly laid down both in the County Court Rules and in the Rules of the Supreme Court that a counterclaim does not fall with the discontinuance of the action in which it is made. Order 23, Rule 6 of the County Court Rules provides: "Where the defendant sets up a counterclaim and the claim of the plaintiff is discontinued, struck out, stayed or dismissed, the counterclaim may be proceeded with and the defendant, on proof thereof may have judgment." Order 15, Rule 2 of the Rules of the Supreme Court is to the same effect.

Mr Green adopts as part of his argument the opinions to be found in two well-known works. The first is *Atkin's Court* H *Forms*, 2nd ed, annual supplement 1974, at p 106, where, dealing with a tenant's application for a new tenancy, the author says:

> If the tenant decides to discontinue the action he can only do so with leave, and the court can fix the interim rent when giving leave. This can be dealt with in chambers. In the county court the claim can be made by application in the tenant's action or by originating application. If there appears to be any possibility of the tenant discontinuing his claim for a new lease the latter course would be preferable as the originating application will survive, whereas an application in the action would not, as no leave is required to discontinue in the county court. In practice it will usually be convenient to have the application determined at the same time as the main action.

That, says Mr Green, is right. J

The second is "the White Book." In the Second Supplement to the *Supreme Court Practice* 1976, there is this addition of a new subparagraph to the notes on Order 97, rule 9A:

> It is better to ask the Master, on the first hearing, for a direction under Order 28 r 7, that the summons stands as a counterclaim, it will not then automatically fall if the Originating Summons is dismissed or discontinued.

Mr Green argues that that statement too is right: a counterclaim would not automatically fall with the discontinuance of the tenant's application, but an interlocutory application made in the proceedings must.

Order 28, rule 7 (1) of the Rules of the Supreme Court provides: K

> A defendant to an action begun by originating summons who has entered an appearance to the summons and who alleges that he has any claim or is entitled to any relief or remedy against the plaintiff in respect of any matter (whenever and however arising) may make a counterclaim in the action in respect of that matter instead of bringing a separate action.

Rule 7 (2) gives the court power to direct how such claim shall be made. So there is in the High Court a provision for making a claim into a counterclaim in proceedings started by originating summons.

We have not been referred to any similar provision in the County Court Rules. We have been referred to Order 9, rule 9, L dealing with counterclaims in ordinary actions; and there is also Order 10, rule 3, dealing with counterclaims in default actions. But there would not appear to be any rule dealing with counterclaims in matters started by originating applications in the county court. Nevertheless, Mr Green contends that unless the application is in form a counterclaim or in form originating it perishes with the discontinued action.

It is quite true that the first application which the landlords have made is an application under Order 13, rule 1, and is in a sense interlocutory. Indeed, such applications are so described, I think, in Order 13, rule 1 (1) (g), which refers to their costs, (The learned judge referred to the landlords' application of April M 30 as an "interim application," by which he may have meant only that it was an application to determine interim rent.) But interlocutory though it may be, in the sense that it is an application made in the proceedings, none the less it is a separate application. That, indeed, was recognised by the order of May 7 which was made by His Honour Judge Ruttle ordering that that application for interim rent be adjourned to the date of the hearing, that is, the date for hearing the tenants' proceedings. It was not an application made, as it could have been made according to the note in the *County Court Practice*, in any answer to the tenants' application. It is perhaps to be noted that, unlike many a counterclaim, although it relates to the same premises as the application, it really relates to a separate matter. The tenants are applying for a new tenancy; the landlords are counter-applying for an increase of the rent payable under the continued old tenancy.

I would have thought, unless compelled by authority to take the contrary view, that it would be unjust and unreasonable that such an application should be treated as so parasitic, to use the learned judge's phrase, as to fall with the tenants' application. So far from there being any authority compelling me to take the view which Mr Green puts forward, there is the authority of this court, on which the learned judge relied, in the case of *McGowan v Middleton* (1883) 11 QBD 464. That was a case in which an action had been discontinued after a counterclaim had been delivered, and the court decided that a plaintiff cannot put an end to such an action so as to prevent the defendant from enforcing against him causes of action contained in the counterclaim. There had been a difference of judicial opinion in the court below based on a difference of judicial opinion in earlier cases, and that was resolved in favour of a counterclaimant. But the learned judge relied, and rightly relied, on statements which went beyond a case of a counterclaim and went to the reason for holding a counterclaim to survive the discontinuance of the action in which it had been set up. Sir Balliol Brett MR said at p 469: "The defendant has set up by counterclaim a wholly distinct cause of action from that of the plaintiffs." He went on to say: "For the plaintiffs, it is contended that within a limited time they may from caprice or any other motive put an end to the action; but surely it is more reasonable that the defendant should be at liberty to go on with his claim against the plaintiffs." Then at p 470 he made the statement upon which the learned judge relied: "The plaintiffs' action being discontinued, that which is only a defence to it drops with it; but anything beyond a defence, anything in the nature of a claim against the plaintiffs, must be treated separately and cannot be discontinued." That is why the Master of the Rolls came to the conclusion that a counterclaim had to be treated separately, because it was in the nature of a claim against the plaintiffs. Bowen LJ's judgment was to the same effect. He said at p 473: "In the present case the defendant sets up a claim for independent relief." And he, like the Master of the Rolls, took the view that the plaintiffs could only discontinue their own case, and not thereby bring down the defendant's, in so far as it made, as he put it, "a claim for independent relief."

It seems to me that those observations of the Master of the Rolls and Bowen LJ exactly cover this case. It was following that decision that the rule was introduced into the Rules of the Supreme Court which saved a counterclaim from dropping with the discontinuance of the action. But I do not read the rule as intending to cut down—if, indeed, it could—the principle which was enunciated, in accordance, it seems to me, with commonsense, reason and justice, in *McGowan's* case.

The landlords' application of April 30 is, in my judgment, not parasitic to the tenants' application of July 25 1974, but a wholly distinct claim for independent relief. It is a claim distinct from the tenants' proceedings for relief independent of those proceedings. It does not drop or fall with the discontinuance by the tenants of their own application, but it survives. It must be treated separately, and it cannot be discontinued simply by the discontinuance of the tenants' proceedings. However described, and whatever its form, the landlords' application is in substance an originating application. However labelled, it originates, or starts, or initiates a claim, in some sense countering the tenants' application though affording no defence to it, and has its own distinct and separate life just as much as if it had been initiated by what was in form an originating summons or application or had been ordered to stand as a counterclaim, which in substance it was. By making the application in the form in which it was made in the course of the tenants' proceedings for discontinuance, instead of by a differently headed and worded application alongside those proceedings, the landlords have not turned it into a perpetual parasite. It can be detached and stand alone in its present form without any alteration of its heading or its wording. It would be unreasonable and unjust if by abandoning your attack you could compel your opponent to call off his counterclaim, particularly if it is mounted to attain an objective of its own. I do not find myself compelled by any arguments, rules or authority put before us to treat this counter-attack, if it is correctly so described, as called off by the tenants' abandoning their attack. To do so would be, I think, to disregard the realities of these two applications. They are quite distinct in the relief which they claim. The one is, in the form in which it has been made, and understandably made, independent enough of the other to survive its discontinuance. For these reasons, I regard the judgment of the learned judge as entirely right on the preliminary point, and I proceed to consider the second part of Mr Green's argument for the tenants.

The learned judge, after disposing of other issues which are not now the subject of dispute, reached the conclusion, as he said, without any hesitation, "that this is a case where it would be just to determine interim rent and unjust not to do so. While it is true that the period during which the interim rent will bite is a relatively short one, namely, between April 30 1975 and October 17 1975, if Mr Green's submission with regard to the effect of section 64 (2) is correct, as I think it is, I do not regard such a period as trivial." (That period is the correct period as decided by Sir John Pennycuick V-C in the case of *Stream Properties Ltd v Davis* [1972] 1 WLR 645. I see no reason to suppose that that case is not rightly decided, and, indeed, it has not been challenged by Mr Green.) After saying that he did not regard the period as trivial, the learned judge went on: "If one takes it for ease of calculation as half a year and recalls that the difference between £9,000 and £2,800 is £6,200—the non-determination of an interim rent in this case would mean a loss of the order of £3,000"—Mr Green has given us the more accurate figure of £2,841—"to the landlords who happen to be Pension Fund Trustees who have already had their existing rent frozen between October 5 1974 and April 30 1975 as a result of the counter-inflation legislation—another six months." Then he refers to a submission by Mr Green, repeated in this court, "that the landlords had so far taken no active steps to find a new tenant at the current market rent and that this was a reason why the court should not determine an interim rent." He rejected those arguments and held that this was a case in which he ought to exercise his discretion to determine an interim rent, and that it was just for him to do so: and he determined it at the now undisputed figure of £9,000 per annum.

Mr Green has submitted that, although that sum of nearly £3,000 is not trivial, the period of 5½ months from April 30 to October 17 over which it is payable is, and that is a matter which the court is entitled to take into account in considering whether to exercise the statutory discretion undoubtedly conferred upon it by section 24A. The triviality of the period is a matter which was referred to by Megarry J in the case of *English Exporters (London) Ltd v Eldonwall Ltd* [1973] Ch 415 at p 425. We have been referred to that case for what it says generally about the exercise of this statutory discretion. Emphasis has been laid on the fact that it is a discretion, and the learned judge is entitled to refuse to exercise it in a proper case. But it is to be noted that Megarry J in that case said at p 432: "Without laying down any formal rule that the onus lies on the tenant to show why the discretion should not be exercised, I would say that in most normal cases the court's discretion under section 24A ought to be exercised, in that to do so will usually promote justice." That was a passage which the learned judge quoted in his judgment.

Mr Green argues that this was not a normal case, first of all, because, as I have said, the period was trivial. Secondly, he repeats the submission that the landlords have taken no active step to make prospective arrangements to relet these premises. They could not make arrangements to let them before the

A period of the continued tenancy had actually expired, but Mr Green says that they ought to have shown an earlier interest in reletting them immediately after the expiration of that period. I cannot for the life of me see how such a consideration is in any way relevant to the exercise of the judge's discretion under section 24A. There is no evidence, or relevant evidence, about the reasons for that inactivity, certainly no evidence to which Mr Green has called our attention and no evidence of improper conduct on the part of the landlords; and, with all respect to him, I am still in the same position as the learned judge in being unable to give any weight to that factor, and least of all to allow it to prevent the exercise of the discretion which, I agree with
B Megarry J, ought to be exercised in the landlords' favour in most normal cases.

Finally, Mr Green submits that the learned judge in taking into account the fact that the landlords happened to be a trustee pension fund, as their name clearly shows, and in taking into account the fact that they had already been compelled by the counter-inflation legislation to be paid an uneconomically low rent for a considerable period, was taking into account matters which were irrelevant to the exercise of his discretion. For my part, I regard them as relevant matters which the judge was right to take into account, and right to take into account in favour not
C of refusing to exercise, but of exercising his discretion to determine an interim rent.

I am, therefore, of opinion that the learned judge was right on this point, as on the other, and that this appeal should be dismissed.

CAIRNS and BRIDGE LJJ agreed.

The appeal was dismissed with costs.

COURT OF APPEAL

December 9 1977

D (Before Lord Justice BUCKLEY, Lord Justice BRIDGE and Sir DAVID CAIRNS)

COVELL MATTHEWS & PARTNERS v FRENCH WOOLS LTD

Estates Gazette June 17 1978

(1977) 246 EG 1007

Landlord and Tenant Act 1954—Tenants' originating summons claiming a new lease of business premises—Subsequent application by tenants for leave to discontinue proceedings —Whether order for discontinuance can operate from a date earlier than order—Date of discontinuance is date of
E **order under Order 21, rule 3—No power to make order retrospective—Observations on contrast between High Court and county court rules**

This was an appeal by the plaintiff tenants, Covell Matthews & Partners, from an order of Graham J giving the plaintiffs leave to discontinue their proceedings claiming a new lease. The plaintiffs claimed that the judge had a discretion to make the order operate retrospectively and that he should have exercised that discretion. Graham J rejected this submission (see 241 EG 915). The plaintiffs' interest in contending for a retrospective operation was that it was likely to have a favourable effect on the rent which they remained
F liable to pay.

Paul Batterbury (instructed by May, May & Merrimans) appeared on behalf of the appellants (plaintiffs); Michael Barnes (instructed by Theodore Goddard & Co) represented the respondents (defendants).

Giving the first judgment at the invitation of Buckley LJ, BRIDGE LJ said: This appeal raises a novel and interesting point under the Landlord and Tenant Act 1954 and under the Rules of the Supreme Court as they apply to certain proceedings under that Act. I should like at the outset to express my indebtedness to learned counsel on both sides for the help that they have given us in our task of resolving the point.

The plaintiffs (the tenants) appeal from an order of Graham G J made on November 1 1976. The form of the order, after reciting certain undertakings given by counsel on behalf of the plaintiffs, was that the plaintiffs be at liberty to withdraw an originating summons in which they had claimed a new lease of certain business premises, "such withdrawal to take effect as from July 5 1976 (being the date agreed for such purpose between the plaintiffs and the defendants should this court accede to the said application) and that the same be withdrawn accordingly." The language of the rule of the Supreme Court under which the application had been made to the court contemplates, in fact, not the withdrawal of the proceedings but their discontinuance; but in my judgment H nothing turns on this distinction of language, and the order should be understood as if it had been an order giving the plaintiffs leave to discontinue their proceedings begun by the originating summons and that the said proceedings be discontinued accordingly. The reason why the order is to take effect, not from the date when it was made but from an earlier date, July 5 1976, was because the parties, as the order recites, had made an agreement to that effect.

The points which are raised in the appeal are, first, whether independently of any agreement between the parties the court had power, as the tenants (as I shall call them on this appeal) J contend, to make an order for discontinuance operating retrospectively from some earlier date than the date upon which the order was made. Secondly, if the court had such a power the question arises whether, in the exercise of its discretion in the circumstances of this case, it was proper that the power should have been exercised, and if so, to what earlier date the order for discontinuance should effectively have been backdated.

The history of the matter begins in 1960, when the predecessors in title of the defendants in these proceedings, whom I shall call the landlords, granted to the predecessors in title of the present tenants a lease for 14 years of premises 7-11 K Lexington Street, London W1, at rentals rising during the period of the lease to a maximum, which was the rent payable at any material time with which we are concerned, of £5,600 a year. The contractual term granted by that lease was due to expire on June 24 1973. Apart from any notice given pursuant to Part II of the Landlord and Tenent Act 1954, the lease would of course have continued under the provisions of that statute. But on December 21 1972 the landlords served on the tenants, pursuant to section 25 of the Act of 1954 (which is very familiar and to the provisions of which I need not refer), a notice to terminate the tenancy on the contractual L terminal date, namely, June 24 1973; and in that notice the landlords indicated that they would not oppose an application by the tenants for the grant of a new lease. In accordance with the statute, on February 5 1973 the tenants duly served a counternotice intimating their intention to apply for a new lease, and on March 13 1973 they duly made their application in the appropriate form. The rateable value of the premises being marginally in excess of the county court limit, which is £5,000, they commenced their proceedings, claiming a new lease, in the Chancery Division in the High Court and their proposal, as pleaded, was that they should be granted a new lease for the term of seven years at the currently payable M rent of £5,600 a year.

As soon as that application was made, the provisions of section 64 of the Landlord and Tenant Act 1954 came into operation; those provisions, as far as material for present purposes, read as follows:

> (1) In any case where—(a) a notice to terminate a tenancy has been given under . . . Part II of this Act . . . and (b) an application to the court has been made under the . . . said Part II . . . and (c) apart from this section the effect of the notice or request would be to terminate the tenancy before the expiration of the period of three months beginning with the date on which the

> application is finally disposed of, the effect of the notice or request shall be to terminate the tenancy at the expiration of the said period of three months and not at any other time.
>
> (2) The reference in paragraph (c) of subsection (1) of this section to the date on which an application is finally disposed of shall be construed as a reference to the earliest date by which the proceedings on the application (including any proceedings on or in consequence of an appeal) have been determined and any time for appealing or further appealing has expired, except that if the application is withdrawn or any appeal is abandoned the reference shall be construed as a reference to the date of the withdrawal or abandonment.

Following the issue of the tenants' originating summons there were negotiations between the parties which continued over the next two years. What we are told is that by 1975 the parties had come to terms, at all events in principle, as to the new lease which should be granted to the tenants. The landlord's contention is that there was a concluded and binding contractual agreement reached between the parties for the grant of such a new lease. The position presently taken up by the tenants, however, is that any agreement reached in the course of those negotiations was an agreement subject to contract, and accordingly one which never became binding upon the tenants. The negotiations having come to an end on this issue—or perhaps " broken down " is a more accurate expression—what happened in the event was that in or about December 1975 the tenants vacated the premises, leaving subtenants in occupation of certain parts of them. We are not concerned with the precise detail of that, although it may be important to note that there were subtenants left when the tenants vacated the premises.

Some time before the end of 1975, the tenants gave to the landlords a notice to terminate the tenancy pursuant to section 27(2) of the statute. That subsection reads as follows:

> A tenancy granted for a term of years certain which is continuing by virtue of section 24 of the Act may be brought to an end on any quarter day by not less than three months' notice in writing given by the tenant to the immediate landlord, whether the notice is given after the date on which apart from this Act the tenancy would have come to an end or before that date, but not before the tenant has been in occupation in right of the tenancy for one month.

The question whether or not the notice given by the tenants pursuant to that subsection was an effective notice is not before us for decision. Both parties invite us to assume for the purposes of the present argument (and I proceed upon this assumption without making any pronouncement on the question of whether it is right, because it may be a difficult question) that the section 27 notice given, as it was, long, long after the original terminal date specified in the landlords' section 25 notice had passed, was ineffective to determine the tenancy pursuant to the provisions of section 27(2).

Reverting to the further history, it was on January 23 1976 that the tenants made application pursuant to Order 21, rule 3 of the Rules of the Supreme Court for leave to discontinue the proceedings begun by the originating summons. As we shall see in a moment, when I turn to the provisions of the rules, there is no doubt that in the High Court proceedings such as these, claiming a new lease by way of originating summons in the Chancery Division, cannot be discontinued by the applicant otherwise than pursuant to the leave of the court.

On February 12 1976 the landlords made an application to the court to determine an interim rent pursuant to section 24A of the Act of 1954, to the provisions of which I need not make detailed reference. Both applications, the tenants' application for leave to withdraw their proceedings and the landlords' application for an order determining interim rent, came before Master Cholmondeley-Clarke on March 26 1976. On the interim rent application he merely gave appropriate directions and nothing turns upon that. On the application for leave to withdraw, the master expressed himself as minded to grant that leave, but in accordance with the well-established practice of the Chancery Division, when the landlords intimated that they were not content to accept the order which the master was minded to make, the master adjourned the matter to the judge.

The original hearing before the judge was fixed for July 5 1976; in the event that date had to be vacated primarily, as I understand it, for the benefit of the landlords, and it was because there was a vacation of that date that the agreement was made between the parties that if in the event the judge, on hearing the application, were to grant the tenants' application for leave to discontinue their proceedings, any order made that the proceedings be discontinued should be treated by agreement between the parties as if it had been made on the original hearing date of July 5. In the event the matter came before Graham J on October 20 1976 and he delivered his reserved judgment on the matter on November 1 1976.

Both before the master and before the judge the landlords were contending that no leave to discontinue the proceedings should be granted—in other words that the tenants' application should be refused—and the basis of that contention was, so it was argued, that the issue as to whether there was a binding agreement between the parties for the grant of a new lease ought to be decided as a preliminary point in the proceedings commenced by the originating summons, notwithstanding that the landlords also intimated that it was their intention, failing such decision, that the alleged binding agreement between themselves and the tenants for the grant of a new lease should be the subject of proceedings claiming an order of specific performance. The main argument before Graham J turned upon the issue so raised; on the issue so raised the learned judge came to a conclusion favourable to the present appellants, the tenants, but as that issue does not arise in this court, I need do no more than go straight to the learned judge's conclusion in the matter, in which he decided that leave to discontinue should be granted, but should be granted on terms.

What he said, having rehearsed the arguments between the parties, was this:

> In my judgment, therefore, I think that in principle, the master was correct in his grant of leave to the tenant

I think that means " in his intended grant "

> to withdraw their application of February 1973

That is an erroneous date; it should be March 23 1973

> for a new tenancy and I hold that such leave should be given. At the same time, for the landlords' protection, the order being for leave to discontinue and not for dismissal, the tenants should be put on terms: (1) Not to make any fresh application for a new tenancy. This may in fact be academic in view of the time which has elapsed since the first application, but such a term positively safeguards the position. (2) The grant of leave should, as requested by Mr Harper,

Mr Harper being counsel for the landlords before the learned judge

> be expressed to be given without prejudice to the landlords' contention that a binding agreement has been reached between the parties, and without prejudice to the effect of such an agreement if it is eventually decided that it is binding. (3) An appropriate order for costs against the tenants should be made.

I pause at that point to interject that, as Mr Barnes very properly concedes in the argument before us, if at any earlier stage the landlords had intimated to the tenants that they would be willing to agree to discontinuance of the tenants' proceedings on the terms which the learned judge eventually imposed, it is an almost irresistible inference that the tenants would readily have concurred in the voluntary imposition of those terms.

On the question which arises for our decision relating to the back-dating of the discontinuance order, the matter was argued before the learned judge by both parties on the footing that the judge had a discretion under the relevant rule of court to back-date his order if in all the circumstances he thought it just to do so. Indeed, not only was this view conceded by counsel then appearing on behalf of the landlords, but that counsel positively advanced this proposition in support of the contention that this provided the court with all the power it needed to meet an argument which had been advanced on behalf of the tenants, that otherwise the tenants might suffer hardship through the proceedings being unreasonably protracted by the behaviour of the landlords resisting a tenants' application for leave to withdraw. It was on this footing, on the basis that he had an admitted discretion as to whether or not the order should be made retrospective, and if so to what extent, that the learned judge expressed his conclusion in these terms: " In my judgment I consider that the normal rule in the absence of avoidable delay or unusual circumstances should be that the date of the judgment giving leave to withdraw is the date of withdrawal within the section." Again I interpose to say that if one were to follow the language of the relevant rule, to which I shall refer shortly, the words " discontinue " and " discontinuance " would be more apt than the words " withdraw " and " withdrawal."

In this court the concession made and the arguments advanced on behalf of the landlords to the effect that this is a matter within the jurisdiction of the court, namely to make an order of discontinuance with retrospective effect, have been withdrawn by a counternotice served by the landlords, and it is in these circumstances that the primary question which we have to determine is whether under the relevant rule there is indeed any discretion in the court to make an order for discontinuance but to make it with retrospective effect. The importance of the issue from the tenants' point of view, and indeed from the landlords' point of view, is obvious. On Mr Barnes' contention the date of withdrawal within the meaning of section 64 of the Landlord and Tenant Act 1954 in a case where the party seeking to discontinue has to obtain the leave of the court to do so can only mean the date when that leave was given. On the footing that that is right, unless the order for discontinuance in this case can be given retrospective effect to some date earlier than July 5 1976, then until a date three months after that date, which would be October 5 1976, the tenants remained liable to pay rent, not only at the original contractual rate but at any rate which might be determined pursuant to the landlords' application for an interim rent under section 24A of the statute.

Before I turn to the terms of the relevant rule of court, I should explain the fundamental basis of Mr Batterbury's argument, which he relies on both in support of what he says should be the court's inclination on the point of construction, and also what he says should govern the court's exercise of any discretion which it may have; and that is this: in the county court, under the corresponding rules which govern applications for new tenancies, which will be applications in relation to all business premises having rateable values not exceeding £5,000, no leave of the court is required by an applicant who desires to withdraw or discontinue his proceedings; he can do so as of right pursuant to the provisions of Order 18 of the county court rules. So, according to Mr Batterbury, the argument runs thus: Parliament must, when it enacted section 64 of the Landlord and Tenant Act 1954, have contemplated that any tenant, having instituted proceedings for a new lease, could discontinue those proceedings at any time voluntarily and thereby limit his liability to continue paying rent for the relevant premises to a period of three months after the date when he does so. That is illustrated by the County Court Rules; it is a provision comparable with, and parallel to, the provision to which I made reference earlier, in section 27, which certainly gives a tenant a right to terminate his statutory tenancy under the Act at any time before an application for the grant of a new tenancy is made. I see the force of those contentions, and I approach the matter of construction bearing them in mind. But what we have to construe are the provisions of Order 21 of the Rules of the Supreme Court, to which I now turn.

So far as it is relevant for the purposes of this case, rule 2 provides in subrule (1):

> The plaintiff in an action begun by writ may, without the leave of the court, discontinue the action, or withdraw any particular claim made by him therein, as against any or all of the defendants at any time not later than 14 days after the service of the defence on him or, if there are two or more defendants, of the defence last served, by serving a notice to that effect on the defendant concerned.

I pause to interpose that of course that provision has no application at any time to proceedings begun by originating summons, as these proceedings were. I need not read subrules (2) or (3); subrule (4) reads:

> If all the parties to an action

which I interpose to say would include proceedings commenced by originating summons

> consent, the action may be withdrawn without the leave of the court at any time before trial by producing

to the appropriate person, who is then spelled out in the rule

> a written consent to the action being withdrawn signed by all the parties.

Then we come to the crucial provision, which is rule 3:

> (1) Except as provided by rule 2, a party may not discontinue an action (whether begun by writ or otherwise) or counterclaim, or withdraw any particular claim made by him therein, without the leave of the court, and the court hearing an application for the grant of such leave may order the action or counterclaim to be discontinued, or any particular claim made therein to be struck out, as against any or all of the parties against whom it is brought or made on such terms as to costs, the bringing of a subsequent action or otherwise as it thinks just. (2) An application for the grant of leave under this rule may be made by summons or motion or by notice under Order 25, rule 7.

Then for completion I had better read, although in my judgment nothing turns on it, the provisions of rule 4:

> Subject to any terms imposed by the court in granting leave under rule 3, the fact that a party has discontinued an action or counterclaim or withdrawn a particular claim made by him therein shall not be a defence to a subsequent action for the same, or substantially the same, cause of action.

Counsel on both sides have developed interesting arguments directed at illuminating the problem of construing this rule, based upon a consideration of other provisions under the Rules of Court, or under a variety of statutes in which the court may or may not, as the case may be, make orders with retrospective effect. Mr Batterbury draws our attention to a number of instances where the court's orders may operate retroactively; Mr Barnes, on the other hand, draws analogies with cases where no retroactive operation is permissible. I hope neither counsel will think it discourteous on my part if I do not examine those suggested analogies in detail; in the end it seems to me that this appeal turns on what is really a very short point on the construction of the particular language of Order 21, rule 3, in the context in which it is found, and more particularly as applied to the situation which arises when an applicant for a new lease who has commenced proceedings by originating summons in the Chancery Division is seeking leave to discontinue those proceedings because he wishes, for the purposes of avoiding further liability to pay rent, to withdraw his proceedings within the meaning of section 64 of the Act at the earliest possible date.

The first thing I would say—and I doubt whether this was really in controversy at the end of the argument—is that it seems to me clear beyond doubt that the date of withdrawal, as that phrase is used in section 64 of the Act, can only mean in relation to the withdrawal of proceedings by originating summons in the High Court, the date when an order is made that those proceedings be discontinued, subject always to any power of the court to make such an order with retrospective effect. I reach that conclusion because until an order has been made giving leave to the applicant to discontinue his proceedings it must remain uncertain whether such leave will or will not be granted. It is perfectly clear from authorities to which we have been referred that the granting of leave to discontinue is no mere formality; that an applicant for leave to discontinue may be put in a position, by the terms which the court is minded to impose upon him as a condition of ordering discontinuance of his proceedings, which will be so onerous that he would not wish to accept them. Without going into the details, an illustration of that is to be found in the recent decision of Whitford J, to which we were referred, in *Young Austen & Young Ltd v British Medical Association* reported in [1977] 1 WLR 881. Accordingly one must wait until the date when the order was made before knowing whether the tenant's attempt to withdraw his proceedings will be successful or unsuccessful; it is quite impossible, subject to the court's power, if it has one, to make an order with retrospective effect, to say that any other date than the date of the order is the effective date of withdrawal. So we come to the central question of whether the order can be back-dated and, as I say, it is in relation to that question that the numerous analogies have been drawn to our attention.

If the order is to be back-dated, the language of the rule which authorises the introduction of such a provision into the order must be found, if anywhere, in the words "on such terms as to costs, the bringing of a subsequent action or otherwise as it thinks just." The first conclusion that I have reached is that it is quite impossible to construe these words as giving power to the court to order discontinuance with retrospective effect, for this reason: the terms which are there contemplated are terms to be imposed on the applicant—terms adverse to the applicant as a condition of granting him the relief which he seeks. It is the applicant who is asking the court to grant him leave to discontinue his proceedings; it is against the applicant that any terms will operate if the court is minded to say: "Yes, we will grant you leave to discontinue your proceedings, but only if you comply with the following terms." To my mind it would be really quite an astonishing interpretation of this language to say that a power to impose terms upon an applicant as a condition of granting him certain relief for which he is asking should extend to a power to impose onerous terms upon the respondent, who is resisting the application.

But if that is the wrong approach to the matter, I should still reach the same conclusion by another route. If one pauses to consider what is the position which will have been reached by the time the parties get before the master or, as it may be, the judge, who is to make, if he is persuaded it is proper to do so, the relevant order of discontinuance having granted the leave sought, it is this. Up to that time the tenancy will have been continuing; up to that time whether or not he has exercised the right, the tenant will have had the right to occupy the premises; up to that time the tenant, and the tenant alone, in such circumstances as are applicable here, will have been entitled to receive, and will perhaps have received, the rent payable by the subtenants; up to that time, other than by agreement with the tenants, the landlord will have had no right to resume possession of the premises and will not have been in a position to let them to anyone else; nor indeed will he have known until the time comes when the court decides whether leave is to be granted or not, whether the tenancy is to come to an end on the date when the order is made or is to continue thereafter indefinitely. He will have been entitled to contemplate not only that the tenancy will continue as it will have done up to the date when the judge makes his order, but that it will continue in the ordinary way for three months thereafter. In this situation, to my mind it would require the strongest and clearest statutory language—and by "statutory" I include, of course, the language of the rules—to give the court a power retrospectively to deprive the landlord of the rights which by then will have accrued to him.

For these two reasons, if for no others, I have reached the conclusion that there is no power conferred on the court by Order 21, rule 3 to make an order for the discontinuance of proceedings such as these, or, so far as I can tell, of any other proceedings (but it may not be of any importance in another context), which will have operation retrospectively from a date earlier than the date on which the order is made.

Having said that, I remain conscious of the startling anomaly between the situation under the High Court rules and the situation under the county court rules, to which Mr Batterbury has drawn our attention. We have heard an interesting argument as to how, if amendment is called for, the rules ought to be amended. Mr Batterbury suggests that the High Court rules should be brought into conformity with the county court rules; Mr Barnes, for the landlords, perhaps understandably, suggests precisely the contrary. It does seem to me that this may be a situation which has escaped the attention of the rule-makers who are responsible for both sets of rules, and it may be that, now this case has thrown the point up, the relevant rule committees, either severally or in consultation with each other, will wish to consider whether any, and what, amendment to the rules might be appropriate to remove the anomaly. But I certainly do not regard it as any part of the task of this court to suggest what, if any, amendment might be appropriate.

For the reasons I have sought to explain, I would dismiss this appeal.

Agreeing that the appeal should be dismissed for the reasons given by Bridge LJ, SIR DAVID CAIRNS said: In addition to contending that the words "on such terms as to costs, the bringing of a subsequent action or otherwise" give jurisdiction to backdate the order, Mr Batterbury on behalf of the tenants argued that there is inherent jurisdiction in the High Court to backdate its orders. In my view there is no such general jurisdiction. In some forms of order—for example, as to periodical payments which have accrued due before the making of the order, or as to the remission of such payments—there is a jurisdiction to backdate: see the decision of this court in *McDonald v McDonald* [1964] P 1. Other forms of order clearly could not be backdated. Two instances which occur to me are: a money judgment, which clearly could not be backdated to an earlier date so as to carry interest under the Judgments Act from that earlier date; or a decree of divorce, which plainly could not be made to operate from a date earlier than the order granting the decree.

The type of order with which Order 21, rule 3 is concerned, seems to me to be obviously a rule of such a character that it would be something of an absurdity to say that it could be backdated, the effect of backdating being to say that an action which had in fact been in existence up to the date of the order had ceased to be in existence at some earlier date.

For those reasons I agree to dismissing the appeal. Agreeing with both judgments BUCKLEY LJ said: What cannot be done without leave clearly cannot be done unless and until leave has been obtained. Consequently, it seems to me impossible that a tenant, by making an application for leave to withdraw or discontinue his application for a new tenancy under the Landlord and Tenant Act 1954 could, even in some contingent manner, discontinue or withdraw his application merely by issuing a summons for leave to do so.

The court can of course kill an action by ordering its discontinuance, but our attention has been drawn to no power conferred upon the court to direct not merely that the action

A should be killed but that it should be deemed to have died at some date earlier than the date of the order. I am fully satisfied that the court has no inherent jurisdiction to make any such order, and I do not think that any such power is to be found in Order 21, rule 3. That rule does empower a court to whom application is made to discontinue an action upon terms; but those terms must, I agree, be terms which are imposed upon the applicant; it cannot refer to terms imposed upon the respondent to the application.

For the reasons which have been given by my brethren I agree that this appeal fails. It seems to me that there are substantial grounds for thinking that consideration should be B given to the question whether in some way the relevant High Court and county court rules should not be revised so as to make them accord with one another in relation to withdrawal of an application for a new lease under the Act of 1954. But that, as Lord Justice Bridge has said, is not a matter upon which we could properly express any definite views, either as to whether amendment of the rules should be made, or what form any amendment might take.

In those circumstances the appeal will be dismissed.

The appeal was dismissed with costs.

LANDLORD AND TENANT

LEASEHOLD REFORM

A **COURT OF APPEAL**

January 28 1978

(Before Lord Justice STEPHENSON, Lord Justice ORR and Lord Justice GOFF)

JONES v WROTHAM PARK SETTLED ESTATES AND ANOTHER

Estates Gazette April 15 1978

(1978) 246 EG 223

Leasehold Reform Act 1967—Appeal from Lands Tribunal —Important question of principle affecting many cases— Effect on price payable by tenant for enfranchisement of B interposition of a concurrent lease between tenant and freeholder—Device to increase substantially value of freehold —Whether valuation on assumption that concurrent lessee had actually granted a subtenancy to tenant—Concession by tenant's counsel questioned—Assumption not decided to be correct—Landlords' concurrent lease device held to be invalid under section 23 as "purporting" to modify statutory right of enfranchisement—Order of Lands Tribunal set aside

This was an appeal by Mrs Lena Jones, widow, the occupying tenant of a house at 45 Wellesley Crescent, Potters Bar, from a decision of the Lands Tribunal, under section 21 of C the Leasehold Reform Act 1967 (reported at (1976) 240 EG 804) that the price payable by her for the exercise of her right of enfranchisement under the Act was £4,000 in respect of the freehold and nil in respect of an intermediate (concurrent) lease.

N T Hague (instructed by Andrew W Rowntree, of Potters Bar) appeared on behalf of the appellant (tenant); T M E B Etherton (instructed by Farrer & Co) appeared for the first respondents, the concurrent lessee (Wrotham Park Settled Estates) but took no part in the argument except as to costs; E G Nugee (instructed by Boodle, Hatfield & Co) represented the second respondents, the freeholders (Wentworth Securi- D ties Co Ltd).

Giving the first judgment at the invitation of Stephenson LJ, ORR LJ said: This is an appeal by the lessee of a house, 45 Wellesley Crescent, Potters Bar, against a decision of the Lands Tribunal (V G Wellings) on October 15 1976 whereby it was determined, on a reference to the tribunal under section 21 of the Leasehold Reform Act 1967 (which I will call "the Act"), that the price payable by the lessee on the exercise of her rights under the Act to acquire the ownership of the premises is, in respect of the freehold estate in the premises, £4,000, but that in respect of a superior lease of the premises, which I shall call "the concurrent lease," no sum is E payable.

The facts of the case can be shortly stated. On October 2 1962 a company named Dugdale Hill Estates Co Ltd leased the premises to the appellant, Mrs Jones, and her husband who has since died, at a yearly rent of £15 for a term of 87 years from December 25 1961 (and therefore expiring in December 2048), and by a leaseholder's notice under the Act, dated October 5 1973, but not served until October 8, notice was given in the names of Mr and Mrs Jones to the second respondents, Wentworth Securities Company Ltd, who had acquired the freehold title from Dugdale, of their desire to F acquire the freehold. It is common ground that had the matter rested on these facts the sum payable for the freehold would have been £300, but on October 6 1973, between the date of the tenant's notice and the date of its service, the landlords, Wentworth, interposed between their freehold and the lease to Mr and Mrs Jones the concurrent lease to which I have earlier referred, being a lease of the premises to Wrotham Park Estate Co, since renamed Wrotham Park Settled Estates and the first respondents in this appeal (and whom I shall call "Wrotham"), for a term of 300 years from December 25 1970, subject to and with the benefit of the lease to Mr and Mrs Jones. We have been told, and it is not in dispute, G that a large number of similar leases were on or about the same date granted by Wentworth to Wrotham in respect of some 100 other houses in the same locality; that the tenants were informed of what was being done; and that Wentworth and Wrotham are connected companies having at least one common director. All these leases granted to Wrotham a term of 300 years from December 25 1970, subject to and with the benefit of the pre-existing leases, the rents payable under the concurrent leases being, by paragraph 1 of the 4th schedule to each lease, a peppercorn (if demanded) until December 25 2048 (the date of expiry of the Jones's and the other original leases), and thereafter, by paragraphs 2 and H 3 of the same schedule, rack rents to be determined by an arbitrator for the periods from December 25 2048 to December 25 2098, and from the latter date until the end of the term; such rack rents to be based on values at the commencement of each period. Paragraph 5, which is the crucial provision for the purpose of the present appeal, and paragraph 6 of the same schedule provided as follows:

(5) If and so often as a sublease (not being a tenancy from year to year or any lesser interest) is granted by the Tenant of the whole of the Premises the annual rent payable hereunder shall for the period beginning with the commencement of the term J granted by the sublease and ending with the termination thereof (however the same determines and if the term is extended by virtue of any enactment then ending with the termination of the term so extended) be (in lieu of the foregoing rents) an amount equal to the best yearly rent for which the Premises could be let in the open market at the commencement of such terms with vacant possession free from all incumbrances (including the existing lease) for a term equal to the term granted by the sublease such amount to be ascertained (in default of agreement) by arbitration as aforesaid.

(6) Provided always that if (whether by virtue of the provisions of paragraph (5) of this Schedule or otherwise) the yearly rent which would (apart from this paragraph (6)) be for the time K being payable hereunder shall be more than the yearly rent reserved by any sublease granted by the Tenant (whether in pursuance of the provisions of the Leasehold Reform Act 1967 or any attachment or re-enactment thereof or otherwise howsoever) then the yearly rent payable hereunder under paragraphs (2) and (3) of this Schedule shall be reduced to a peppercorn during a period (hereafter called 'a rent free period') from the actual termination of the term of such sublease (howsoever the same shall determine and if the term shall be extended by Statute then from the termination of the term as so extended) on

December 25 2048 or the expiration of any rent free period referable to any earlier sublease (whichever shall be the later date) until the expiration of the term hereby granted (whichever shall last occur).

Before turning to the proceedings before the Lands Tribunal it will be convenient to refer to the relevant provisions of the Leasehold Reform Act 1967. [His Lordship then summarised or quoted the provisions of sections 1, 8(1), 9(1)(a), 14(1), 15(2), 23(1) and Schedule 1, paragraphs 1, 2(a), 4(1) and (2), 5(1)(2) and (3), 7(1)(a) and 10(1). He also mentioned that it was common ground that Wrotham was the reversioner for the purposes of Schedule 1 and that Wentworth's *locus standi* was derived from paragraph 5(1) of Schedule 1 and that they had, by virtue of a notice to that effect, obtained a right to deal directly with the claimant.]

In consequence of the leaseholder's notice, and in default of agreement between the parties, an application was made to the Lands Tribunal under section 21 of the Act to determine the price payable for the house and premises, 45 Wellesley Crescent, under section 9 of the Act, and on the hearing the applicant, Wrotham and Wentworth were all represented by counsel, but counsel for Wrotham took no part in the argument. There were before the tribunal alternative agreed valuations, the first of which, relied on for the applicant, was prepared on the basis that the valuers were required to take account of the concurrent lease dated October 6 1973 but to ignore paragraph 5 of the 4th Schedule thereto, and on that basis produced agreed figures of £50 payable to Wentworth in respect of the freehold estate and £250 payable to Wrotham in respect of the concurrent lease, making a total of £300, and this basis of valuation the tribunal referred to as Assumption 1. The other valuation was prepared on the basis, which the tribunal referred to as Assumption 2, that all the provisions of the concurrent lease, including paragraph 5 of the 4th Schedule, were to be taken into account, and produced an agreed figure of £4,000 payable to Wentworth in respect of the freehold estate but nothing payable to Wrotham in respect of the concurrent lease.

On behalf of Wentworth it was argued before the tribunal that the proper assumption to be applied in the valuation was assumption 2, because (1) section 9(1) of the Act required the price payable for the freehold to be assessed on the double basis that the freehold was subject to the 1962 lease and that such lease was to be extended for 50 years; (2) that for this reason, taken in conjunction with paragraph 10(1) of the 1st Schedule to the Act, a new lease under section 14 must be deemed to be granted by Wrotham; and (3) that in the circumstances the deemed new lease under section 14 was a sublease of the whole of the premises granted by Wrotham within the meaning of paragraph 5 of the 4th Schedule to the concurrent lease, with the result that, for the purpose of assessing the compensation payable to Wentworth in respect of the latter's freehold estate in the premises the rent must be taken to be the best yearly rent for which the premises could be let in the open market as defined in that paragraph.

On behalf of the applicant it was conceded that the concurrent lease was a valid lease and not a sham and that the submissions made by counsel for Wentworth were correct subject only to arguments which counsel for the applicants proceeded to develop, based on section 23(1) of the Act to the effect, first, that paragraph 5 of the 4th Schedule to the concurrent lease was an agreement relating to the applicant's tenancy within that subsection and that it purported to modify the tenant's right to acquire the freehold by providing for a greatly increased purchase price compared with the £300 which, but for the execution of the concurrent lease would have been payable under the Act, and also contravened the same subsection by providing for the imposition of a penalty or disability on the tenant in the event of her claiming the right to acquire the freehold.

In its decision the Lands Tribunal observed that there could be no doubt that the interposing of the concurrent lease between the freehold and the occupying tenant's lease was intended as a device to increase the value of the freehold under section 9 of the Act, and that the question at issue was whether this device had succeeded. Having said this, and having referred to the concessions made by counsel for the applicant, the tribunal proceeded to reject his submissions based on section 23, holding, first, that the reference in that subsection to "any agreement relating to a tenancy" applies only to an agreement to which the tenant is a party and could not therefore apply to the concurrent lease to which the tenant was not a party, and secondly that the concurrent lease did not purport to modify the tenant's rights, which remained intact and unqualified, nor did it provide for the imposition of any penalty or disability on the tenant in the event of her acquiring or claiming any right under the Act. Accordingly the tribunal held that assumption 2, whereby regard was to be had to all the terms of the concurrent lease, including paragraph 5 of the 4th Schedule, was the proper basis of valuation and determined that the price payable for the freehold estate was £4,000 and that no sum was payable in respect of the concurrent lease.

Against this determination the applicant now appeals and (Wrotham not being represented in this court) we have heard, on behalf of the appellant and of Wentworth, argument limited to the issues arising in relation to section 23 of the Act. In the course of the argument, however, questions were put by Goff LJ to counsel for the appellant as to the concession made by him before the Lands Tribunal that, for the purpose of assessing the compensation payable to Wentworth, it was to be assumed that a sublease had actually been granted by Wrotham to the applicant within the meaning of paragraph 5 of the 4th Schedule to the concurrent lease; one of the points put being that the apparent intention of section 9 is to secure to the tenant, when assessing the sum payable by him for the freehold, the benefit of his right, although not exercised, to have an extended lease at a ground rent and without paying a premium, and that all that is required for this purpose is to treat the tenancy as extended without postulating the grant of any actual subtenancy; and another point being that the assumption required to be made by section 9(1)(a) of the Act is that the tenancy "was to be extended" which, as Mr Nugee conceded, appears to recognise that there was not in fact going to be any new subtenancy. For these reasons Mr Hague for the appellant was asked whether he wished to reconsider his concession, but he decided not to do so and in the result we heard no argument upon it. In these circumstances I express no view on this matter save to say that it appears to me to be a very arguable one, and I turn to the issues in relation to section 23, and first to the broad arguments advanced on each side, which may be summarised as follows. Mr Hague argued that it was the intention of Parliament that a qualified tenant should have the right given him by the statute; that the object of what was done by the landlords in the present case was to defeat that right and that if in any doubt the court should resolve the matter in favour of the applicant.

Mr Nugee for the respondents submitted two broad arguments, the first that it is a fallacy to suppose that there exists, independent of the provisions of the Act, any set of fair terms on which Parliament intended that the tenant should be entitled to obtain the freehold; the words "on fair terms" in section 1 being merely an introductory reference to the terms defined in subsequent sections of the Act, and it being common ground that the sum produced by those sections in the circumstances of this case is £4,000. In the second place Mr Nugee relied on the distinction drawn in decided cases in relation to the Rent Acts between, on the one hand, attempted exclusion of those Acts, and on the other the parties' bona fide entering into a situation where those Acts have no application; a distinction which is clearly stated in

A the following passages in *Megarry, Rent Acts,* 10th ed at pp 18-19:

> (a) *Genuine Transactions.* There is nothing, however, to prevent the parties from so arranging matters that there is nothing to which the Acts can apply, provided the transaction in question is a genuine transaction and not a mere sham, such as a tenancy disguised as a contract for sale; "real and lawful intentions cannot be dismissed as shams merely because they are disliked." The difference is between on the one hand a provision attempting to exclude the Acts from a transaction to which they apply, and on the other hand entering into a bona fide transaction to which the Acts have no application.
>
> B (b) *Evasion and avoidance.* "There is every difference between evasion and avoidance." "You do not evade an Act by doing something which is not forbidden by the Act, but you do evade the Act by doing something which is prohibited under the guise of doing something else."

In relation to these passages Mr Nugee relied strongly on the admission made by Mr Hague before the Lands Tribunal that the concurrent lease was not a sham, but I do not pause upon these passages because it is in my judgment clear that they cannot assist the respondents' case if the crucial provisions of the concurrent lease are rendered void by section 23, nor in my judgment can they assist the appellant if her C case fails under that section.

The first issue under section 23 is whether the concurrent lease is "an agreement relating to the tenancy." It is claimed by Mr Nugee that it is not, on the ground that the tenant is a necessary party to such an agreement, and he sought to support this argument by a passage from the judgment of Diplock LJ (as he then was) in *Joseph v Joseph* [1967] Ch 78, a case involving section 38(1) of the Landlord and Tenant Act 1954, much of the language of which has been adopted in drafting section 23 of the Leasehold Reform Act 1967, and which reads as follows:

> D Any agreement relating to a tenancy to which this Part of this Act applies (whether contained in the instrument creating the tenancy or not) shall be void in so far as it purports to preclude the tenant from making an application or request under this Part of this Act or provides for the termination or the surrender of the tenancy in the event of his making such an application or request or for the imposition of any penalty or disability on the tenant in that event.

The *Joseph* case is authority for the proposition that in section 38 the words "in so far as it purports" denote "in so far as it has the effect of" and on that authority I have no doubt that the same meaning should be given to the same E words in section 23 of the Leasehold Reform Act. In addition, however, Diplock LJ in his judgment (at p 90) said "In my judgment this subsection does render void any provision of an agreement between landlord and tenant whereby the tenant undertakes to do in the future any act which will have the effect of disqualifying him from applying for a new lease under sections 24(1) and 29." But the agreement in issue in that case was between landlords and tenants and it is clear in my judgment that Diplock LJ, who had posed (on p 89) as the question which he answered in the passage above quoted "Can he (the tenant) . . . agree with the landlord?" was not, in the passage quoted, putting any general construc- F tion on the words "agreement relating to a tenancy." For my part I can see no good reason in the context of section 23 for adopting the narrow construction that the words "any agreement relating to a tenancy," which are in my judgment very wide words, are restricted to agreements between landlord and tenant, and I would find it very surprising if they did not apply to an agreement between, for example, the landlord and a relative of the tenant that on the tenant obtaining enfranchisement the relative would pay to the landlord the amount by which the enfranchisement price fell short of a specified sum, and I would find it equally surprising if such an agreement fell outside the section because the promise was made, not to the landlord, but to a company in which G he was interested. It was not necessary for Mr Nugee in the present case to argue that the landlord must be a party to "an agreement relating to a tenancy" under section 23 since the landlord in the present case was a party to the concurrent lease, and he was content to argue no more than that the landlord was probably a necessary party, but if the words "agreement relating to a tenancy" involve that the tenant must, but the landlord need not, be a party I would have expected the section to say so in terms. For these reasons, with great respect, I consider that the Lands Tribunal was wrong in the only reason which it gave for holding that the agreement in question was not "an agreement relating to H the tenancy." It remains, however, to consider whether the agreement in question was an agreement "relating to" the tenancy. In my judgment it was, because paragraph 5 in the 4th Schedule to the concurrent lease was capable of applying to the statutory right of enfranchisement and was intended by both parties to do so, and such right, although not a term of the tenancy, was a right annexed to the tenancy by statute and therefore, in my judgment, an incident of the tenancy.

The next question is whether the agreement purports to "modify" any right to acquire the freehold or an extended lease or right to compensation under Part I of the Act. J The Lands Tribunal held that it does not, on the ground that, while the effect of paragraph 5 is that the appellant has to pay £4,000 instead of £300 for enfranchisement, her right to obtain the freehold "remains absolute and unqualified"; in other words, although the value of the right has been drastically reduced, the right itself has remained unaffected. In my judgment, however, again with great respect to the Lands Tribunal, this reasoning ignores the fact that until the act of enfranchisement £300 remained the proper price for the freehold, calculated in accordance with the Act, and apart from paragraph 5 of the 4th Schedule the exercise of the K right of enfranchisement could not increase the price, but paragraph 5, coupled with the concession, makes it do so, and in that respect, in my judgment, modifies the right.

Having reached these conclusions I do not need to decide whether paragraph 5 of the 4th Schedule "provides for . . . the imposition of any penalty or disability" on the tenant in the event of her acquiring or claiming any right under Part I of the Act, but in my judgment this part of section 23 has no application in the present case since it is prefaced by the words "provides for" and paragraph 5 does not provide for the imposition of any penalty or disability.

For these reasons I would allow this appeal and in lieu L of the Lands Tribunal's determination that the price payable for the house and premises is £4,000 would substitute, subject to any observations that counsel may wish to make, a determination that the price is £300, being £50 payable to Wentworth and £250 payable to Wrotham.

Agreeing that the appeal should be allowed, GOFF LJ said: The facts which give rise to the question we have to determine are fully stated in the judgment of Orr LJ, and I need not take up time by repeating them.

I desire to stress, however, that I am not at all satisfied that the assumption which section 9(1)(a) of the Leasehold Reform Act 1967 requires to be made for the purpose of M ascertaining the price payable for the premises requires the two elements of the reversion, namely the concurrent lease to Wrotham and the ultimate estate in fee simple remaining in Wentworth, to be valued on the hypothesis that Wrotham had actually granted a subtenacy to Mrs Jones. It is true that where the tenant claims an extended lease, and not the freehold, the machinery provided by section 14 of the Act is to bind the reversioner to grant a new lease, and the existing lease is merged or surrendered by operation of law, but where the freehold is claimed, although the valuation has pursuant to section 9 to be made "if the tenancy has not

A been extended under this Part of this Act, on the assumption that . . . it was to be so extended," I think that must be construed in the light of the context and clear intention of the section. That intention was to secure to the tenant, when assessing the price he has to pay for the freehold, the benefit of his right, albeit not in fact exercised, to have an extended lease at a ground rent only and without paying a premium. All that is required for that purpose is to treat his tenancy as if it were what it would be if it had been extended. It is not I think necessary to postulate the grant of a new tenancy. The words " was to be " are in my view significant, because as Mr Nugee conceded they recognise that there is not going
B to be any actual new tenancy.

It was common ground, however, that section 9 upon its true construction does require the valuation to be made upon the footing that Wrotham are to be deemed to have granted a subtenancy within the meaning of, and so as to bring into operation the provisions of, clause 5 of the 4th Schedule to the lease of the October 6 1973, which I will call the concurrent lease, subject only to the effect, if any, of section 23 of the Act, and Mr Hague, although offered the opportunity to do so, refused to resile from that position. We must, therefore, decide this case on that assumption, but without deciding that it is correct, and should any case be brought upon any
C of the other leases to which my Lord, Orr LJ, has referred, the point as to the true effect of section 9 will be open, and as I think well arguable. Having said this, I now turn to the argument on section 23 subsection (1) on which Mr Hague rested his case.

That subsection is qualified by opening words " except as provided by this section," but there is nothing relevant for present purposes in that qualification. The subsection continues as follows

> . . . any agreement relating to a tenancy (whether contained in the instrument creating the tenancy or not and whether made
> D before the creation of the tenancy or not) shall be void in so far as it purports to exclude or modify any right to acquire the freehold or an extended lease or right to compensation under this Part of this Act, or provides for the termination or surrender of the tenancy in the event of a tenant acquiring or claiming any such right or for the imposition of any penalty or disability on the tenant in that event.

Mr Hague submits that clause 5 of the 4th Schedule to the concurrent lease is an agreement relating to Mrs Jones' tenancy which purports to exclude or modify her right to acquire the freehold, or provides for the imposition of a penalty or disability on her in that event. This raises several
E questions of which the first is whether the agreement embodied in clause 5 relates to the tenancy, and in my judgment it does. The Lands Tribunal held that it does not, because in their view every such agreement must be one to which the tenant is a party. It is unnecessary to consider the position with regard to the landlord, because, of course, it was in fact a party to the concurrent lease, but in my judgment it is not essential that the tenant should be a party. In common with Orr LJ I cannot doubt that an agreement by a close relative of the tenant, or anyone else for that matter, that if the tenant should exercise his right to enfranchisement and if the price payable calculated in accord-
F ance with the Acts should be less than £X he would pay the landlord the difference would be an agreement relating to the tenancy within the meaning of section 23(1) notwithstanding that the tenant was not a party to the agreement. So in my judgment the agreement contained in clause 5 may fall within section 23(1) of the Act notwithstanding Mrs Jones was not a party, but the question remains whether it did relate to her tenancy and I agree with my brother that it did.

On the construction of section 9 on which this case proceeds one of the events on which the operation of clause 5 is made to depend is an assumed grant of a subtenancy by
G virtue of that section, and that is not even a casual or fortuitous result. It was intended that it should be so, and the respondents' whole case is that it is. Therefore, as it seems to me, the agreement in clause 5 clearly relates to the statutory right of enfranchisement. Now that right, although not a term of the tenancy, is an incident of it and so the agreement in clause 5 relates also to the tenancy itself. It is true that clause 5 would also have come into operation if Mrs Jones or her successors in title had forfeited her tenancy or allowed it to expire, and Wrotham, thus being entitled to grant a subtenancy, had actually done so, or if indeed she or they had in fact taken a sublease from
Wrotham, but that seems to me to be an irrelevant H
consideration.

It was argued that the concurrent lease including, of course, clause 5 of the 4th Schedule is a fair bargain between Wrotham and Wentworth and that if that clause were invalidated by section 23 of the Act it would become unfair, but in my judgment that is not so, because section 23 subsection (1), if it applies, invalidates clause 5 only in so far as it purports to exclude or modify the statutory right, and therefore, in any event leaves its operation untouched in the supposed case of an actual, as distinct from an assumed, subtenancy.

But then does clause 5 purport to exclude or modify the J
right of enfranchisement? Again in my view it does. In this context, of course, " purport to " cannot bear the meaning it often has, which may very well be its primary meaning, that is to say " is expressed to but not effectively." It means " has the effect of "—see per Lord Denning MR in *Joseph v Joseph* [1967] Ch 78 at p 87, and per Diplock LJ at p 90. I ask myself, therefore, " does it have that effect?" And it seems to me that there can be only one answer " Yes, it does." Immediately before the concurrent lease the price for the freehold calculated in accordance with the Act was £300. Immediately after the execution of that lease right down at
least until Mrs Jones had actually exercised the right of K
enfranchisement that was still the proper price, because the freeholder Wentworth could not get more than a nominal rent from any concurrent lessee during the subsistence of the tenancy. Apart from clause 5 the exercise of the right of enfranchisement obviously could not increase that price. However, on the construction of clause 9 adopted in this case that exercise does have that result, and it does so solely because of the impact of clause 5. Therefore, that clause does " purport to " that is " has the effect of " modifying the statutory right.

In my judgment the respondents' argument to the contrary L
on the ground that the increase in price is simply the result of making the valuation on the statutory hypothesis is unsound, since it assumes that clause 5 of the 4th Schedule to the concurrent lease stands unaffected by section 23 of the Act, and, therefore, begs the question. The alternative way in which this argument was presented, namely that the valuation pursuant to and as directed by the Act is £4,000, and there is nothing on which section 23 can bite, is in my judgment equally untenable for the same reason.

For these reasons I would allow this appeal and it is unnecessary to consider Mr Hague's further argument that
clause 5 is invalidated in its application to the right of M
enfranchisement as being an agreement which

> . . . provides for the termination or surrender of the tenancy in the event of a tenant acquiring or claiming any such right or for the imposition of any penalty or disability on the tenant in that event.

but were it necessary to decide this I would, I think, reject the submission both because the verb governing this part of the section is " provides for " and not " purports to " and because it does not seem to me that clause 5, or the effect of its coming into operation, can properly be described as a penalty or disability.

Also agreeing, STEPHENSON LJ said: The question referred to the Lands Tribunal was

> the prices payable on the acquisition of the intermediate leasehold and the freehold of the property described above under section 9 of the Act, . . .

"There can be no doubt," said the member giving the decision of the tribunal,

> that the purpose of the interposition between the freehold and the occupying tenant's lease in each case was intended as a device to increase the value of the freehold under section 9 of the Act of 1967. The question in the present case is whether it has succeeded.

He decided that it had succeeded on the grounds that section 23 of the Act had no application. The question upon which our decision is desired is stated in the case to be

> whether the tribunal was correct in holding that section 23 of the Leasehold Reform Act 1967 had no application.

The notice of appeal asks us to set aside or vary that decision on the ground that the tribunal misdirected itself in holding that on its true construction section 23 had no application and did not render paragraph 5 of the 4th Schedule to the intermediate or concurrent lease void. Mr Hague for the appellant conceded that, for the purpose of the assumption in section 9(1)(a) that the appellant's tenancy was to be extended under Part I of the Act, it was a sublease granted by the (intermediate) tenant within the paragraph. That led Goff LJ to question whether the concession was rightly made.

If the appellant had given notice of her claim to acquire an extended lease, not of her claim to acquire the freehold, the concurrent tenancy and paragraph 5 would be relevant in deciding the terms of the new tenancy under section 15. But I cannot see why in making an assumption required by section 9 to be made to determine the amount of the purchase price of the freehold the tenancy should be treated as notionally extended in the actual circumstances of a concurrent tenancy. I would not feel compelled without further argument to make the further assumption that the notional new tenancy is a sublease granted by the intermediate tenant. I share the doubts expressed by Goff LJ upon the question whether this concession was rightly made, but I must accept it as right when I consider the question whether section 23 applies to invalidate the concurrent lease or paragraph 5 of the 4th Schedule to it.

The purpose of section 23 appears from its terms to be the preservation of the statutory rights to acquire the freehold or an extended lease and to compensation for being taken away, or restricted, or made more difficult for the tenant to acquire or claim, by agreement. If a tenant qualifies under section 1 of the Act by the rent and the term of his tenancy, by the rateable value of the leasehold house and premises and by his occupation of the house as his residence, he is not to contract or be contracted out of the right to enfranchisement or extension conferred on him by that section, either completely or partly, or out of the right to compensation conferred on him by section 17(2) and section 18(4). He is not to be deterred from acquiring or claiming any of those rights by being required to terminate or surrender the tenancy, or by being penalised or disqualified, in the event of his acquiring or claiming the right.

The obvious method of contracting him out of that right or deterring him from acquiring or claiming it is by agreement between him and his landlord from whom he must acquire and claim the right to enfranchisement or extension, or by whom he must be paid compensation. So there can be little doubt that it is to agreements between landlord and tenant that section 23 is primarily addressed. The agreements excepted from subsection (1) by subsections (2) and (3) appear to be agreements between landlord and tenant. An agreement which "provides for the termination or surrender of the tenancy" within subsection (1) would seem also to be such an agreement. So it is not to be wondered at that Mr Hague in his book on *Leasehold Enfranchisement* (1967) p 199 considers subsection (1) as a subsection "which prevents the parties to a lease contracting out of Part I of the Act" and that Diplock LJ considered the identical terms of section 38(1) of the Landlord and Tenant Act 1954 as concerned with agreements between landlord and tenant: *Joseph v Joseph* 1967 Ch 78 at p 90. But in neither Act does the subsection say "any agreement between landlord and tenant"; nor does it say "any agreement" or "any agreement whatsoever." It says "Any agreement relating to the tenancy." Those are wide words, and the succeeding parenthesis indicates that they are wide and does not necessarily limit them to "any agreement between landlord and tenant," because, as was pointed out in argument, a guarantee by a third party might be "contained in the instrument creating the tenancy." I see no reason to exclude from the subsection an agreement made between one party to the lease and a third party if it relates to the tenancy and either purports to exclude or modify the statutory rights or provides for any of the matters specified in the subsection. I do not regard Mr Hague the advocate as hoist with the petard of Mr Hague the textbook writer. As the tenancy with which the whole Act, including this section, is concerned is a long tenancy at a low rent from which a tenant with the other qualifications required by section 1 derives the rights conferred on him by the Act, any agreement which affects or bears upon any of those statutory rights of a tenant does in my judgment relate to his tenancy.

I am therefore of the opinion, respectfully dissenting from the decision of the Lands Tribunal and concurring with my brethren on this point, that an agreement between the landlord and a third party which grants that party a lease of the leasehold house of which the qualified tenant has the necessary long tenancy at a low rent from the landlord, and indeed turns that tenant into a subtenant of the third party and his lease into a sublease, is an agreement relating to the tenancy. Is it also an agreement which purports to do or provides for the things which section 23(1) declares void?

I agree with the Lands Tribunal, for the reasons given by my brethren, that the concurrent lease does not provide for the imposition of any penalty or disability on the tenant in the event of his acquiring or claiming his right to acquire the freehold. It would be straining the language of the subsection to hold that paragraph 5 provides for a penalty of £3,700. But I am of the opinion, reached with some hesitation after the excellent arguments addressed to us by both counsel, but supported by both the judgments which have been delivered, that the concurrent lease by paragraph 5 of the 4th Schedule "purports to modify" the tenant's right to acquire the freehold. It is, of course, true that section 1 confers on a qualified tenant a right to acquire on fair terms the freehold or an extended lease of the house and premises and that those terms include the purchase price payable for the freehold of the house and premises on the statutory assumptions, including the assumption under section 9(1)(b) that the tenancy was to be extended under Part I of the Act, and the terms of the extended tenancy set out in section 15. There is therefore some force in Mr Nugee's argument for the landlord that the tenant's right to acquire the freehold is a right to acquire it for the purchase price payable under section 9, and *that* right is not modified by the interposition of the concurrent lease and the consequent purchase price of £4,000 because £4,000 is the purchase price payable by the tenant for the freehold at the time when he gives the landlord notice.

The opening words of sections 8(1) and 14(1), "Where a tenant of the house *has* under this Part of this Act *a right . . . and gives* to the landlord written *notice* of his desire," may suggest that the right accrues before notice is given. The reference in section 5(1) to "the rights and obligations of

A the landlord and tenant arising from the notice" perhaps indicates the contrary. But Mr Nugee's argument appears to me to be met by the point which has been so clearly stated in both judgments just delivered that notwithstanding the execution of the concurrent lease the price payable in accordance with section 9 was £300 at least until the tenant exercised her right of enfranchisement.

We are bound by the decision of this court in *Joseph v Joseph* to construe "purports to exclude or modify" as "has the effect of excluding or modifying." There can be no doubt that paragraph 5 has the effect of increasing the price B payable for the freehold by the tenant from £300 to £4,000. Mr Nugee indeed concedes that that is the purpose of the clause. When I suggested to him that it was a barefaced attempt to increase the price, he admitted the attempt but objected to the epithet. The attempt can only succeed if the terms on which the freehold is acquired are not modified by increasing the purchase price from £300 to £4,000 because those terms and the amount payable as the purchase price are to be considered at or after the time when the right to acquire is exercised, and if the actual exercise of the right has the effect of introducing the price payable in accordance with the Act without regard to the price payable, or likely C to be payable, until that time.

I cannot believe that the legislature intended to allow a landlord—I must not say an unscrupulous landlord, because he may be a trustee acting in the best interests of widowed or orphaned beneficiaries in need of every penny he can get for them—to reduce the apparent value of the tenant's right by ingenious devices of this kind. It was to defeat such devices or bargains that Parliament enacted section 23 in language wide enough to hit them. The landlords' device does, in my opinion, purport to modify the right to acquire the freehold conferred on the tenant by section 1 by having the effect of increasing the purchase price. D

I agree, therefore, that we should allow the appeal, set aside the order of the Lands Tribunal and declare that section 23 of the Leasehold Reform Act 1967 does apply to paragraph 5 of the 4th Schedule to the concurrent lease and that the prices payable under section 9 of the Act are £50 and £250. E That is subject to what counsel may have to say on the form of our order.

Appeal allowed. In lieu of Lands Tribunal's determination that the price payable for the house and premises was £4,000, determination substituted that the price was £300, being £50 payable to Wentworth and £250 payable to Wrotham. The respondents ordered jointly to pay the appellant the costs of the appeal to the Court of Appeal. The respondents also jointly to pay the appellant half the costs of the proceedings before the Lands Tribunal. Leave to appeal to House of Lords given conditionally on terms that the order for costs F in Court of Appeal and Lands Tribunal should not be disturbed and that appellants to House of Lords should not ask for costs in the House.

LANDLORD AND TENANT

RENT ACTS

QUEEN'S BENCH DIVISIONAL COURT

November 30 1977

(Before Lord WIDGERY CJ, Mr Justice PARK and Mr Justice MAY)

R v KENSINGTON AND CHELSEA RENT TRIBUNAL EX PARTE BARRETT

Estates Gazette February 4 1978

(1977) 245 EG 397

Jurisdiction of rent tribunal—Whether applicant to tribunal had the status of a tenant or contractual licensee at date of application—Applicant the deserted wife of a person whose holiday tenancy had expired before date of application—No evidence that applicant had sufficient status to give tribunal jurisdiction—Tribunal's decision quashed

In these proceedings William Barrett, of 29 Arundel Gardens, London W11, applied for an order of certiorari to quash a decision of the Kensington and Chelsea Rent Tribunal given on May 18 1977 whereby, on an application by Jennifer Ridley under Part VI of the Rent Act 1968, they fixed a rent of a room occupied by her at 29 Arundel Gardens at £56.80 per four weeks, inclusive of rates, and, in the absence of a direction for the substitution of a shorter period, afforded her six months' security of tenure.

K A D Hornby (instructed by Myers, Ebner & Deaner) appeared on behalf of the applicant; Harry Woolf (instructed by the Treasury Solicitor) represented the tribunal.

PARK J, giving the first judgment at the invitation of Lord Widgery CJ, said that by an agreement dated December 10 1976 the applicant let to one, Mohamed Atabani, a furnished room at 29 Arundel Gardens. The agreement showed it to be a holiday letting for a period of 17 weeks from December 15 1976 to April 13 1977 at a rental of £64 for every four weeks. At the time of that agreement Mr Atabani was married to Jennifer Ridley, but Mr Atabani left the premises after a short time and his wife, using the name of Ridley, continued in occupation and paid rent until towards the end of March 1977. She was still in occupation when the letting to Mr Atabani expired on April 13. On April 18 she applied to the Kensington and Chelsea Rent Tribunal for consideration of rent and security of tenure. She described herself as a "lessee." On the face of the application there appeared to be a Part VI tenancy. There is evidence that after the expiry of the letting to Mr Atabani Mrs Ridley had asked the applicant, Mr Barrett, to let the room to her and that he had refused. Mr Barrett, however, said that he agreed not to press for her to leave until June 22 1977. Mrs Ridley signed a letter written by Mr Barrett purporting to withdraw her application to the tribunal. A few days later she wrote to the tribunal stating that she did not wish to withdraw her application.

The tribunal hearing took place on May 18 1977. Both parties appeared in person. Mr Barrett took the point that the letting was a holiday letting and that in consequence the tribunal had no jurisdiction. As shown in their "reasons for decision" the tribunal were not satisfied on the evidence before them (and in the absence of the production of any written agreement) that there had been a holiday letting. There was no reference in these reasons to consideration having been given to the precise status of Mrs Ridley at the date of the application.

Mr Hornby had argued that the vital point was that at the date of the application Mrs Ridley did not have any status at all. There was no evidence that she had become a tenant or a contractual licensee. Mr Woolf had argued that there were two pieces of evidence which supported her status. The first was that she had paid rent to Mr Barrett. She had not, however, paid rent after a date towards the end of March. The earlier payments of rent were not sufficient to enable the tribunal to decide that she had acquired a tenancy. The affidavit on behalf of the tribunal does not go on to say that there were grounds for an inference to be drawn that she had become a tenant or a licensee. The second piece of evidence mentioned by Mr Woolf was the letter dated April 26 written to the tribunal by Mr Barrett and signed also by Mrs Ridley stating that they had agreed to apply to the county court for a possession order to take effect in eight weeks' time and for an order for £128 mesne profits. The objection to this was, however, that this arrangement was made after the date of application. Events taking place after the application could not be looked at as evidence of the status of Mrs Ridley at the date of the application. There was in fact no evidence before the tribunal on which they could properly find that Mrs Ridley was a tenant or contractual licensee. Consequently the tribunal had no jurisdiction. It followed that the order of certiorari should go.

MAY J agreed.

LORD WIDGERY CJ, also agreeing, said that the question of a holiday letting had turned out to be irrelevant to the decision of the rent tribunal. The applicant was awarded not confer on her a status before the tribunal.

It was ordered that certiorari would go to quash the decision of the rent tribunal. The applicant was awarded costs.

QUEEN'S BENCH DIVISIONAL COURT

December 7 1977

(Before Lord WIDGERY CJ, Lord Justice CUMMING-BRUCE and Mr Justice PARK)

R v LAMBETH RENT OFFICER, EX PARTE FOX

Estates Gazette February 18 1978

(1977) 245 EG 569

Motion for mandamus directed to rent officer—Tenant's application for determination of fair rent—Tenant evicted before consultation—Tenant's failure to notify new address—Whether rent officer in error in treating application as withdrawn or considering that he was deprived of jurisdiction—Rent officer should have proceeded to determine rent—Order of mandamus issued

In these proceedings counsel moved on behalf of Patrick Fox, a pensioner, for an order of mandamus directed to the

A Lambeth rent officer requiring him to hear and determine according to law an application by Mr Fox for him to fix a fair rent for ground-floor premises formerly occupied by Mr Fox at 121 Strathleven Road, Brixton, London. The rent officer had purported to dismiss the application after Mr Fox had left the premises pursuant to an order of the county court granting possession to the landlords.

William Bowring (instructed by James Woolley, Lambeth Community Law Centre) appeared for the applicant; Harry Woolf (instructed by the Treasury Solicitor) represented the respondent rent officer.

Giving the first judgment at the invitation of Lord Widgery B CJ, CUMMING-BRUCE LJ said that the applicant's case was that the rent officer had wrongly declined to hear his application to have a fair rent fixed. The applicant applied in March 1976 for a fair rent to be fixed but vacated the premises in July before the application was heard. It was in these circumstances that the rent officer treated the application as withdrawn and dismissed it on the ground that he had no jurisdiction. In July 1976, the rent officer wrote to the applicant and landlords inviting them to a consultation. The letter to the applicant was sent to the only address known to the rent officer, the premises which were the subject of the application. There was no reply to that letter. The applicant's only C interest now was in recovering any difference in rent which might be due to him from the date of the application to the date when he left the premises. That sum might amount to about £30.

It was clear from the decision of the Court of Appeal in *Hanson v Church Commissioners for England* [1977] 2 WLR 848 that the rent officer was wrong in the view he formed that, without any intimation from a party, he could safely regard the application to fix a fair rent as withdrawn. The rent officer's decision that he had no jurisdiction was also wrong. So far as the question of withdrawal was concerned, D Lord Denning MR had pointed out in *Hanson* (p 854) that when the dispute is one in which there is a public interest involved it may not be permissible for one of the parties to withdraw on his own without the assent of the other, and even if they both agree, he may not be able to withdraw unless the tribunal consents. It was, of course, a question relating to the withdrawal of an objection in that case. A rent officer should be extremely cautious in the public interest in dealing with withdrawals. It was easy to see how the rent officer in the present case came to think that he had solid reasons for inferring that the applicant had lost interest in the application. The applicant had not replied to letters. He had E failed to notify the rent officer of his change of address.

The rent officer had based his decision on an alternative ground. He had considered that he had no jurisdiction because the applicant had vacated the dwelling before the rent officer could proceed to registration. That view was wrong. The relevant date for the purpose of jurisdiction was the date of the application. The rent officer's jurisdiction persisted despite the applicant's subsequent departure: see *R v West London Rent Tribunal, ex parte Napper* [1967] 1 QB 169.

The next question was whether, in the exercise of its discretion, the court should allow mandamus to go. There F were arguments for and against. On the one hand the court should not put itself in the position of being a superior but expensive debt-collecting agency, the applicant had been slack in not notifying his change of address, the amount involved was small and there might be difficulties in recovering it. On the other hand, the rent officer had been in error, there was a public interest in the correct administration of the statute, and the determination of rent would be *in rem* and would be capable of affecting comparable properties. In all the circumstances mandamus should go.

LORD WIDGERY CJ and PARK J agreed.

The application for mandamus was granted with costs.

COURT OF APPEAL G

December 20 1977

(Before Lord Justice MEGAW, Lord Justice LAWTON and Lord Justice SHAW)

STUBBS v ASSOPARDI AND ANOTHER

Estates Gazette March 18 1978

(1977) 245 EG 937

Rent Acts—Effect of amendments made by Rent Act 1974 (now consolidated in Rent Act 1977)—Two successive fixed terms, one granted before and one after the commencement of 1974 Act—Whether second fixed term resulted in tenancy becoming protected—Held not, because exclusion of protected status was H by virtue of transitional provisions in Schedule 3 to 1974 Act, not by virtue of section 5A of Rent Act 1968 (inserted by Rent Act 1974, section 2 (3) and Schedule 2)—Court of Appeal settles point in doubt since 1974.

In this appeal Jeffrey Adams, one of two joint tenants of a flat, appealed against a decision of Judge Curtis-Raleigh at Bloomsbury and Marylebone County Court in favour of the landlord, Mrs Anna Maria Josepha Stubbs, who claimed possession of the flat. Mrs Stubbs was a resident landlord and the question was whether the grant of a second fixed-term tenancy after the commencement of the Rent Act 1974 (August 14 1974) following an earlier fixed term granted before such J commencement resulted in the tenancy becoming a fully protected tenancy. The point turned upon the true construction of section 5A (5) (*b*) of the Rent Act 1968 (inserted by the Rent Act 1974) and the transitional provisions in Schedule 3, paragraph 1, to the Rent Act 1974. The point had been the subject of some uncertainty and discussion since 1974.

Derek Wood (instructed by Douglas-Mann & Co) appeared on behalf of the appellant (second defendant); the respondent, Mrs A M J Stubbs (plaintiff), appeared in person.

Giving the judgment of the court, MEGAW LJ said: The plaintiff, Mrs Anna Maria Josepha Stubbs, claimed against the defendants, Miss Lydia Assopardi and Mr Jeffrey Adams, K possession of the ground-floor flat at 13 Norland Square, London W11. The particulars of claim as formulated by the plaintiff in the Bloomsbury and Marylebone County Court averred that by an agreement in writing dated July 1 1975 the flat had been let to the defendants, furnished, at a rent of £60.68 per month, for 11 months, which fixed-term tenancy expired on June 1 1976.

In this court some point was sought to be made on behalf of the plaintiff to the effect that Mr Adams was not a party to the agreement because, though he had signed it as a party, the agreement on its first page recorded Miss Assopardi alone as L "the tenant." We did not allow any such issue to be raised. It would have been inconsistent with the plaintiff's own pleading, and it had not been taken in the county court, where it had been accepted or assumed that Mr Adams was a party and no question had been raised below as to any special considerations which might arise with regard to a joint tenancy, and no facts were put in evidence which would give rise to any question with regard thereto. We say no more on that point.

At the hearing in the county court before Judge Curtis-Raleigh the relevant facts were not in dispute. They are clearly and succinctly set out by the learned judge in the first paragraph of his judgment as follows: M

> In this case the landlord plaintiff claims possession of a ground-floor flat, 13 Norland Square, W11, which was let to the defendants under two successive agreements in writing, the first from July 1 1973 for a term of two years expiring on June 30 1975, followed immediately by a tenancy for a fixed term of 11 months from July 1 1975. Both the lettings were lettings of premises which were fully furnished. The landlord was a resident landlord.

When the agreement of July 1 1973 (which we shall call "the first tenancy") was made, the defendants were not protected tenants, because the tenancy was of furnished premises and section 2 (1) (*b*) of the Rent Act 1968, as it then stood, excepted

A such tenancy, so that it was not a protected tenancy. This was so, irrespective of the fact that the plaintiff was what the judge described as "a resident landlord." The Rent Act 1974 came into force on August 14 1974, and was thus in force when the first tenancy ended and when, immediately thereafter, the tenancy brought into being by the agreement of July 1 1975 (which we shall call "the second tenancy") commenced.

The issue, and the only issue, which arises on this appeal is whether, on the facts stated above and on the true construction of certain provisions of the 1974 Act, the defendants were, when the proceedings were brought and at the time of the hearing, protected tenants. Their defence to the claim for possession was
B expressed thus in paragraph 8 of their defence:

> The defendants aver that the said agreement of July 1 1975 was a second and immediately successive fixed-term agreement subsequent to the agreement of July 1 1973 and claim the protection of section 5A (5) (b) of the Rent Act 1968 as inserted by Schedule 2, Part I, paragraph 1 of the Rent Act 1974.

In the county court the learned judge allowed the plaintiff's husband to make submissions on her behalf, since it was said that the plaintiff herself had not a sufficient command of the English language. We allowed the same course to be followed on the hearing of the appeal.

C Judge Curtis-Raleigh held in favour of the plaintiff and granted an order for possession. The second defendant, Mr Adams, appealed to this court. He was represented in this court by Mr Wood, who did not appear in the county court. We would pay tribute to the admirable clarity and fairness of Mr Wood's exposition of extremely complex and highly technical statutory provisions.

Section 1 (1) *(a)* of the Rent Act 1974 extends Rent Act protection to furnished tenancies. That extension applied, not only to tenancies created after the coming into force of the Act, but also to already existing tenancies. However, section 2 (3) provides for the insertion in the Rent Act 1968 of a new section 5A,
D which introduces the concept of a "resident landlord" as a relevant factor in excluding the protected status of furnished tenancies. Its terms are set out in paragraph 1 of Schedule 2 to the 1974 Act. That new section of the 1968 Act, so far as it is relevant in the present appeal, reads:

> 5A—(1) Subject to subsection (5) below, a tenancy of a dwelling-house which is granted on or after the commencement date [August 14 1974)] shall not be a protected tenancy at any time if— . . . (b) the tenancy was granted by a person who, at the time that he granted it, occupied as his residence another dwelling-house which also forms part of that building. . . .

E If it rested there, the defendants in this case would not be protected tenants. The second tenancy, of July 1 1975, was granted after the commencement date of the 1974 Act; and, as appears from the undisputed facts set out above, the landlord was a "resident landlord": that is, he fulfilled the requirements of paragraph *(b)* above. But the provisions of subsection (1) of section 5A are expressly made subject to subsection (5), which, so far as relevant, reads as follows:

> (5) This section does not apply to a tenancy of a dwelling-house which forms part of a building if— . . . . (b) the tenancy is a tenancy for a term of years certain and is granted to a person who, immediately before it was granted, was the tenant under an earlier tenancy of that
> F dwelling-house . . . and, by virtue of this section, that earlier tenancy was not a protected tenancy. . . .

So the exclusion of the tenancy from being a protected tenancy, which section 5A (1) would otherwise achieve in respect of a post-August 14 1974 tenancy, is itself excluded if the requirements of section 5A (5) are fulfilled. There is no doubt but that in the present case all the requirements contained in subsection (5) (b) are met, save only for one possible exception. The second tenancy is a tenancy for a term of years certain. The provision for 11 months' duration, it is not disputed, satisfies that requirement. That second tenancy was granted to a person who, immediately before it was granted, was the tenant under an earlier tenancy (the first tenancy) of that dwelling-house (the G
ground-floor flat). There is no doubt that that first tenancy was not a protected tenancy. The one and only ground on which it could be suggested, and the ground on which the learned judge held, that the requirements of subsection (5) (b) were not fulfilled is that it was not "by virtue of this section" (that is, section 5A) that the first tenancy was not a protected tenancy; but that, instead, it was "by virtue of" something else. The "something else" which the judge held to be relevant and effective was the transitional provisions of the Rent Act 1974 as set out in paragraph 1 of Schedule 3.

The purpose of the transitional provisions was to ensure that H
a resident landlord who before the commencement of the 1974 Act had granted a furnished tenancy, which was still in effect at the commencement of the Act, should be in no worse position than a resident landlord who granted such a tenancy on or after the date of the commencement of the Act. The tenancy granted on or after the commencement of the Act would not be protected because of the new section 5A inserted in the 1968 Act. But in the absence of transitional provisions a pre-commencement date tenancy would be protected because of section 1 (1) of the 1974 Act, coupled with the fact that the new section 5A (1) would not provide an exemption since it was expressed to be applicable to a tenancy granted on or after the commencement J
date.

The relevant transitional provisions in Schedule 3, paragraph 1 (1) and (2), so far as they are material, were as follows:

> 1—(1) In any case where—(a) before the commencement date a dwelling was subject to a tenancy which is a furnished letting, and (b) the dwelling forms part only of a building . . . and (c) on that date the interest of the lessor. . . under the furnished letting—(i) belongs to a person who occupies as his residence another dwelling which also forms part of that building . . . and (d) apart from this paragraph the furnished letting would, on the commencement date, become a protected furnished tenancy, the Rent Act shall apply, subject to sub-paragraph (2) below, as if the tenancy had been granted on the commencement date K
> and as if the condition in paragraph (b) of section 5A (1) of the Rent Act 1968 were fulfilled in relation to the grant of the tenancy.

(Paragraph (b) of section 5A (1) has been set out earlier. It defines what we have called "the resident landlord").

> (2) In the application of the Rent Act 1968 to a tenancy by virtue of this paragraph—(a) sub-section (5) of section 5A shall be omitted; and (b) in section 102A any reference to section 5A of that Act shall be construed as including a reference to this paragraph.

The transitional provisions are plainly relevant and material. If it were not for them, the first tenancy would have had the status of a protected tenancy. It would have been made such by L
section 1 (1) of the 1974 Act. It would not have been deprived of that status by section 5A, because section 5A is concerned with tenancies granted on or after the commencement date. The first tenancy was granted before that date. The way in which the relevant transitional provisions (Schedule 3, paragraph 1 (1) and (2)) achieve their object is by introducing a statutory fiction. The fiction is that the relevant tenancy was granted on the commencement date, though in fact it was granted earlier. It is thus brought within section 5A, and it thus loses its protected status. Schedule 3, paragraph 1 (1), expressly provides that section 5A shall apply. We say "expressly," because there can be no doubt but that the words used, "the Rent Act shall apply," M
in the context mean the Rent Act including the new section 5A. Not only section 5 (3) of the 1974 Act, but also the references, in the immediately following provisions of paragraph 1 (1) and (2) of Schedule 3, to section 5A put it beyond doubt that that is so.

But that does not solve the essential question, which is: is the absence of protected status of the first tenancy "by virtue of this section," within the meaning of that phrase in section 5A (5) (b): or is it "by virtue of" the transitional provisions? If it were not for the statutory fiction created by paragraph 1 (1) of Schedule 3, section 5A would not apply and the first tenancy would thus be protected. Hence it can be said that it is "by

A virtue of" those transitional provisions that the first tenancy does not have a protected status. But the machinery by which the transitional provisions operate to deprive the first tenancy of its protected status is by applying section 5A (less subsection (5) thereof). Hence it could also be said that it is "by virtue of section 5A" that the first tenancy does not have a protected status. Which of the two is the one to choose: for choose one must?

In the absence of any further guidance from the context, we should have inclined to the view that the choice should be the transitional provisions. In some respects it is like a choice of the real or effective cause where there are competing causes. To our mind, it is the statutory fiction, created by the transitional
B provisions, which would be the real or predominant cause of the first tenancy not being protected. If one has to choose between them, as one must, it is by virtue of that statutory fiction, rather than by virtue of the resulting application of section 5A, that the first tenancy does not have protected status. We are confirmed in this view by two elements in the wording of the relevant transitional provisions.

First, the opening words of paragraph 1 (2) of Schedule 3 are: "In the application of the Rent Act 1968 *by virtue of this paragraph*. . . ." The statute thus stresses that the application of section 5A is "by virtue of" these transitional provisions. To
C our mind, that tends to show that the legislature, in using the phrase "by virtue of this section" in section 5A (5) (b), did not contemplate or intend that it should apply to a case where "this section" itself had been made applicable, as the legislature itself has expressly said, "by virtue of" the transitional provisions. The second element is one which is stressed by Judge Curtis-Raleigh in his judgment as being, we believe, the main reason for his conclusion as to the construction of the statute: that is, the wording of paragraph 1 (2) (b) of Schedule 3. It reads: "(b) in section 102A any reference to section 5A of that Act shall be construed as containing a reference to this paragraph." The judge says: "Section 5A should not be read in conjunction with
D Schedule 3, because when it has to be read in conjunction with the Schedule the Act says so—see paragraph 1 (2) (b) of Schedule 3".

In one sense, of course, section 5A has to be read "in conjunction with" Schedule 3, because paragraph 1 (1) of Schedule 3 provides that that section shall apply. But we agree with what we understand to be the reasoning of the learned judge. The new section 102A (1), introduced by paragraph 4 of Schedule 2 to the 1974 Act, contains the words: "If and so long as a tenancy is, by virtue only of section 5A of this Act, precluded from being a protected tenancy. . . ." The legislature, in enacting the trans-
E sitional provisions, which make section 5A applicable in certain cases where it would not otherwise apply, has thought it right, in relation to those words used in section 102A "by virtue only of section 5A of this Act," expressly to provide that those words are to be construed as containing a reference to the transitional provisions in Schedule 3, paragraph 1. There is no such provision—no corresponding words or formula—in relation to the corresponding words "by virtue of this section" where those words occur in section 5A (5) (b). Therefore the legislature did not intend those words in section 5A (5) (b) to be read as "containing a reference to" the transitional provisions. Mr Wood submits that this argument, which appealed to the judge, involves
F placing undue weight on the doctrine *expressio unius, exclusio alterius*. We do not think so. Like the learned judge, we regard it as a powerful indication of the intention of the legislature, which, as is normally the principle of our law, has to be derived from the words used.

Mr Wood criticised the learned judge's careful judgment in two other respects. The first was the passage in which the judge said: "This tenant was the tenant under an earlier tenancy and the only question is whether it became unprotected by virtue of section 5A. The answer is—no, because it was granted before 5A came into force." We do not think that this was an error. At that point the judge was looking at the question without reference to the transitional provisions. The other respect in which Mr G Wood criticised the judgment was the learned judge's "cross-check" of his conclusion with his view of "what the legislature might reasonably be expected to want to achieve." We feel quite unable to express any confident view, one way or the other, as to the construction which would be more likely to fit in with the actual intention of the legislature. In any event, however, it is clear that the judge's "cross-check" was not in any way the basis of his judgment. But in our opinion his decision on the question of construction was right.

We therefore dismiss the appeal.

Appeal dismissed, with costs to respondent (plaintiff) in the sum of £75. Stay of execution of order for possession for 28 H days from date of judgment. Legal Aid taxation of appellant's (second defendant's) costs.

QUEEN'S BENCH DIVISIONAL COURT

October 5 1977

(Before Lord WIDGERY CJ, Mr Justice CANTLEY and Mr Justice PETER PAIN)

GUPPYS PROPERTIES LTD v KNOTT
GUPPYS PROPERTIES LTD v STRUTT J

Estates Gazette March 25 1978

(1977) 245 EG 1023

Fair rents—Evidence of fair rents—Capital values, other registered rents and market values all legitimate evidence—Rent assessment committee not obliged to follow any particular method of valuation—Extent of duty to give reasons—No need to show "arithmetical working" or give reasons in "scientific sense"—Documents which may properly be looked at after hearing considered—No error of law by committee K

These were two statutory appeals under the Tribunals and Inquiries Act 1971 by Guppys Properties Ltd from decisions of a rent assessment committee of the Devon and Cornwall Rent Assessment Panel, the respondents being Mr and Mrs A Knott and Mrs N K Strutt respectively.

R Pryor (instructed by Turner, Peacock, agents for Hollowell & Bollam, of Weymouth) appeared on behalf of the appellants; the respondents did not appear and were not represented.

Giving judgment, LORD WIDGERY CJ said: The court is now required to deal, first of all, with the case of Guppys Properties Ltd against Knott, confining itself to that aspect of the L dispute as relates to the appellants' statutory appeal under the Tribunals and Inquiries Act 1971. In addition to the Knott appeal we have a quite separate appeal to deal with in which the appellants are again Guppys Properties Ltd and where the respondent is a Mrs Strutt. This is a case in which the rent of Honeysuckle Cottage, St Agnes, Cornwall, has been under review by the rent assessment committee, and where the appellants (the landlords) are appealing, as they are entitled to do, assuming that they have a point of law to support their appeal against the decision of the committee. The issue between Guppys and Mr and Mrs Knott relates to a cottage in Cornwall called Chez Nous. The matter had been referred M to the rent officer, that is to say the assessment of a fair rent had been referred to the rent officer and he had produced as his figure of fair rent the sum of £338. Before the rent assessment committee that figure was reduced from £338 to £286. Having failed yesterday to have that reduction set aside by certiorari, Mr Pryor on behalf of the appellants now seeks the same consequence by the slightly different route of an appeal under the Tribunals and Inquiries Act.

I can, therefore, go straight to the decision of the tribunal, producing as it did the figure of £286 as being the fair rent, and see how far it is criticised by Mr Pryor on behalf of the appellants. Before I deal with that in any detail there is one other element of history which has to be brought into this case, and that is that there had been other disputes between the present appellants (that is Guppys) and other tenants of other properties which have often come before the rent assessment committees and been the subject of a judgment. Indeed there was such a case, which was before this court on May 15 1975, which is referred to as *Guppys v Sandoe* [*Guppys (Bridport) Ltd v Sandoe*, reported in full at (1975) 235 EG 689]. I am not going to refer to it in detail at this stage, but it is of importance in the present instance because it will be seen that virtually every point argued by Mr Pryor in favour of the appellants today was argued again by Mr Pryor in the case of *Sandoe* in 1975. In so far as the case of *Sandoe* decided any of these disputed issues, for my part I think we should follow *Sandoe* and apply the same principle to the case before us.

I go back then to the judgment. It is a lengthy judgment. I say that in no sense as a matter of criticism. It deals in considerable detail with the nature of the property Chez Nous and the accommodation it provides; it considers improvements which have been done to the premises by the tenants from time to time, and it goes through the general broad arguments as to the method by which a fair rent should be assessed. It is at this point, in my judgment, that the matters of real consequence in this appeal arise because throughout this litigation (and by that I include litigation affecting other houses beyond those which are the subject of this case today) the landlords have contended as a broad principle, first of all, that as the difference between market rents and registered rents under the Rent Act widens, it becomes more and more necessary, if fairness is to be done to the landlord, to fix fair rents by reference to capital values and not merely by reference to existing registered rents. Fixing rent by capital values of course (sometimes called the "contractors' approach") is a method whereby a modern value is put upon the house as the sale of the freehold. Having ascertained its capital value, an appropriate rate of interest is applied to that capital value, and that gives you a basic figure upon which the fair rent can be ascertained. It would not be disputed, I think, that this is a method which tends to favour landlords, and it is not therefore altogether surprising if the landlords in the present case supported it.

The second matter of principle which has been stressed greatly by the landlords in these cases is that when the rent assessment committee gives its decision it is required by statute to give reasons, and it is contended in one way or another that an assessment committee fails to perform that duty if it does not show what one might call its arithmetical working. What is argued is that it is not enough for the assessment committee to say, "We have considered all the evidence and, doing our best, we have come to the conclusion that £x is the fair rent." It is argued that that will not do, and in some way the figure of £x has to be justified by reasons. In the absence of such reasons, the committee has not performed its function. This was a matter which came very much into evidence in the *Sandoe* case, and in the *Sandoe* case we directed first that there was no obligation on the rent assessment committee to follow any particular method of valuation from those which are accepted as being proper in these cases. In other words, there was no obligation on the committee to use capital values, or registered rents, or current market values, or any particular approach, provided it did its best after consideration of all the relevant matters which had been put before it.

Accordingly, the fact that there seems little attention paid to capital values in the instant case is not an error of law on the part of the assessment committee. The assessment committee is entitled to take any of these courses, and in some instances to take more than one, and mould them together. Assuming that it stays within its jurisdiction and has not acted unreasonably, its conclusions are conclusions which do not raise questions of law and therefore which are not capable of being litigated in this court.

There is only one shadowy dictum in opposition to that principle, but I should refer to it, and I mean no offence by describing it as shadowy. It is found in the case called *Metropolitan Properties Co (FGC) Ltd v Lannon* [1969] 1 QB 577. The actual extract to which I should refer is in the judgment of the Master of the Rolls at p 597 where Lord Denning says this: "The committee are, of course, entitled to use their own knowledge and experience: see *R v Brighton and Area Rent Tribunal, ex parte Marine Parade Estates (1936) Ltd* [1950] 2 KB 410. But that does not mean that they should throw over the evidence altogether. At any rate, they should not throw over the evidence without saying why. That is what troubles me most about their written decision." Lord Denning there is undoubtedly saying that they should give reasons, if they can, for throwing over any evidence, as he describes it. But the learned Master of the Rolls cannot have meant that where the assessment committee is fixing a fair rental on its own experience and instincts, it must give reasons for the figure which it has provided, because in the very nature of things the reasons, in the scientific sense, are not possible in such an instance. It is worth remembering that valuation is not a science: it is an art. There will be many, many cases where all the assessment panel can do is to say, "Doing our best with the information provided, we think the rent should be £x." If they do that, they disclose in my judgment no error of law.

Those being the matters primarily contended before the assessment committee, I ought next to deal with three matters which are more specifically relied upon by Mr Pryor. The first concerns the treatment by the committee of a so-called comparable house called 2 Chelean Cottages at Frogpool. As is common in these cases, a substantial number of existing registered rents were put before the committee for such assistance as those rents might be, and 2 Chelean Cottages is one such rent. The situation in regard to that property was that the actual rent was £494, and in breaking it down to ascertain how much of that figure was attributable to the building and how much to the furniture a complicated calculation had been made, presumably by or on behalf of the assessment panel, which purported to show that £281 was attributable to the building and £213 was attributable to the furniture. Mr Pryor says, and for all I know he may be absolutely right, that that figure of £213 for the furniture was ludicrously large, and by reason of its being large it artificially depressed the figure which was attributable to the house. He says, therefore, it was a very poor comparable if it was relied upon in that way, and indeed I think he says it was misleading.

So be it, but there is nothing in the judgment to discover to what precise extent the committee relied on 2 Chelean Cottages, if at all. Indeed, I suspect they did not because the last reference which is made to that house is in these terms: "We assume he was seeking to prove that 2 Chelean Cottages was in fact worth all or nearly all of the £494 per annum as an unfurnished letting. We cannot accept that proposition, nor the argument on which it was founded." It looks to me as though Chelean Cottages was being disposed of in limbo. Certainly I cannot find anything in that aspect of the case which could conceivably be interpreted as an error of law justifying the allowing of the present appeal.

Then the next specific point made by Mr Pryor was in relation to certain improvements done by the tenants. Section 46 of the Rent Act 1968 contains a special provision to prevent a tenant from being charged twice in respect of improvements. In terms, in section 46(3), it says: "There shall be disregarded— . . . (b) any improvements carried out, otherwise than in pursuance of the terms of the tenancy, by the

A tenant under the regulated tenancy or any predecessor in title of his." In assessing the rent, the committee are required to proceed as though the improvement had never been carried out unless it is an improvement which is made by the tenant in pursuance of the terms of the tenancy. Mr Pryor has battled manfully to try to get us to say that the improvements undertaken by the tenants here were in pursuance of the terms of the tenancy. But there is no tenancy agreement. We have seen nothing in writing to support any such argument, and I just am unable, I fear, despite Mr Pryor's efforts, to understand why he says we should regard these improvements as being in pursuance of the terms of the
B tenancy when we do not even know what the terms of the tenancy were.

The next point to be taken specifically I need not take time over because Mr Pryor accepts that it was not taken below, and that it ought to have been so taken if it was to be used in this court. It is concerned with the question of whether the improvements to which I have recently referred should be disregarded because at the time when the work was done the tenancy premises were not subject to a regulated tenancy. I propose to say no more about the matter because it is clearly not available to Mr Pryor in this court.

C Then there is a reference to another specific matter relied upon as rendering the decision below subject to reversal on appeal, and that is a complaint that the committee were supplied with documents after the hearing, which documents were not placed before the appellants or known to them until the decision of the committee was published.

What seems to have happened is this. The secretary, one L W Weston, of the committee sent on to the chairman of the committee two days after the hearing a number of papers. First of all, there was sent a transcript of the secretary's notes of the hearing. No objection is taken to that. Obviously the chairman in considering the decision would want to have the
D assistance of the notes taken during the hearing. Secondly, there is sent under the same cover some authorities which might be of assistance in regard to the treatment of improvements, which I have already said played a major part in this argument. Thirdly, a copy of the relevant pages of *Hill and Redman* was sent. In other words, the secretary is doing what is clearly a very useful job in giving the chairman of the committee the sort of documents which will help him in his conclusions.

It is only when you come to the fourth and fifth that any real opportunity for criticism arises, because the fourth document or group of documents included under this cover
E was a specific previous decision of a rent assessment panel sent with a view to being used as a comparable if the chairman thought it was of value to him as such. The same goes for the fifth document, which was a copy of the chairman's own decision in a Torquay house, which the secretary thought he might find of assistance.

So far as the practice of sending such documents as these to the chairman is concerned, one must, I think, be a little careful to see that that which is appropriate is sent and that which is not appropriate is withheld. There cannot possibly be anything wrong with sending the chairman a copy
F of the notes of the hearing or of text-books or of authorities. One only touches the point of difficulty, as it seems to me, when one gets down to the specific instances which might be used as comparables because one is there getting into the realms of fact and not of pure law and opinion. It is said that the failure of the committee to see that the appellants were supplied with those documents, or in particular the alleged comparables, is in breach of the statutory instruments which affect this kind of hearing.

The statutory instrument in question is the Rent Assessment Committees (England and Wales) Regulations 1971 (SI 1971 No 1065), and regulation 5 deals with the documents which should be supplied before the hearing. It says: "The G
committee shall, where the reference is to be subject to a hearing, take all reasonable steps to ensure that there is supplied to each of the parties before the date of the hearing," and then certain documents are specified, including a document "which contains relevant information in relation to fair rents previously determined for other dwelling-houses and which has been prepared for the committee for the purposes of that reference."

In so far as documents dealing with comparables are concerned, if they are available before the hearing there is a duty to give copies to the parties. But there is no obligation placed on anybody to send copies of documents which come H
to light thereafter, and there is nothing which I can see in the actions of Mr Weston to infringe the terms of the statutory instrument to which I have referred. In any case I do not think that the terms of regulation 5 are necessarily mandatory, and I am not prepared to accept that the failure to hand over after the hearing copies of material relating to comparable properties would in itself require and entitle the appellants as of right to have the hearing reopened. Whether or not it would be appropriate to have the hearing reopened would depend on the circumstances, and this would be a point properly open to the appellants on an application for J
certiorari. At all events, I can see nothing there which would entitle us to say that there has been an error of law of the kind which would lead to the appeal being allowed.

I propose to conclude my judgment at this point because I am content, if I may say so, with what I said in *Sandoe* as laying down the principles upon which these matters are to be treated, and I do not wish to repeat myself, lest I produce a slightly different conclusion and given counsel material for argument hereafter on the difference. It seems to me that, in summary, as far as the Knott litigation is concerned, there is nothing in the case which is not shown by the *Sandoe* decision to be within the powers and propriety of the committee's K
activities, and therefore it provides no ground for allowing the appeal to this court.

I can deal with the Strutt case even more briefly because this really, as Mr Pryor admits, is a case which reflects the same problem all over again; the same arguments are put forward to the same effect. I can see nothing in the Strutt case which is worthy of being described as a distinguishing feature from either the *Sandoe* case or the Knott case. Accordingly I think that the appeal in regard to Mrs Strutt should also be dismissed.

CANTLEY and PETER PAIN JJ agreed.

The appeals were dismissed. L

COURT OF APPEAL

March 3 1978

(Before Lord Justice STEPHENSON, Lord Justice GEOFFREY LANE and Lord Justice CUMMING-BRUCE)

SOMMA v HAZELHURST

SOMMA v SAVELLI

Estates Gazette April 22 1978

(1978) 246 EG 311 M

Important decision on agreements to grant non-exclusive licences of residential accommodation—Rent Act not applicable to these particular agreements—Separate licences granted to two persons living together to use a double-bed-sitting room in common with the licensor and with such other persons as the licensor should permit, not being more than one at any one time—Agreements conferred non-exclusive licences only—No reason why an owner should not grant a licence—Not a sham merely because it is a device to avoid the consequences of the Rent Act—

Parties intended to be bound by agreements—Impossible to construe as a joint tenancy—Not contrary to public policy—County court judge wrong to decide that the agreements together constituted a joint tenancy—New trial refused—Licensor's appeal allowed—Observations on construction of such agreements and on evidence of surrounding circumstances

These were appeals from decisions of Judge McIntyre at West London County Court concerning the nature and effect of agreements described as "licences" for the occupation of a double room at 4 Cornwall Mansions, London, W14, belonging to the appellant, Miss Immacolata Somma. The room was occupied in consequence of the agreements by Martin Hazelhurst and Miss Rossella Savelli. The application to the county court had been made by Miss Somma under section 105 of the Rent Act 1968 (now section 141 of the Rent Act 1977) for a determination as to the effect of the agreements. Miss Somma's application followed an application on behalf of the occupiers to the rent officer (which he had adjourned in view of the likelihood of county court proceedings) to fix a fair rent. The county court judge decided that the agreements had to be construed together and constituted a joint tenancy.

R Purchas (instructed by W Feldman) appeared on behalf of the appellant; Lord Gifford (instructed by Seifert, Sedley & Co) represented the respondents.

Giving the judgment of the court at the invitation of Stephenson LJ, CUMMING-BRUCE LJ said: In February 1976 two young people Mr Martin Hazelhurst and Miss Savelli (herein called H and S) were looking for accommodation in which to live together in London. He was an educated man employed as a computer programmer, a job involving some mathematical qualifications. She also had a job. They were not married. On Wednesday February 18 they saw an advertisement in the *Evening Standard* in the column headed "Flats and Maisonettes to Let" which read:

ACTON/HAMMERSMITH, West Kensington. Double bedsits and flatlets. All amenities. Near Tubes. £13 to £19 per week 602 5464.

They telephoned the number given and by appointment visited a house at 4 Cornwall Mansions, W14, which belongs to Miss Somma. The house is divided into four flats, subdivided into four rooms and two maisonettes. There they met Mr Ritter, resident managing agent for Miss Somma. Ritter showed them a room 22 ft by 18 ft on the third floor, with two beds in it which he described as a double room. They looked at it and went away. On Friday February 20 they returned, saw Ritter again and said they wanted to take the room and to move in next day. Ritter gave each of them a printed form of agreement into which he wrote the appropriate detail in the blank spaces. They each read the form they were given. H asked a few questions including a query about the clauses which indicated that they would have to share with a third person described as the licensor. They each signed their agreement before they moved in, and, though they had not thought out the legal implications of the contracts, they urgently wanted accommodation and, in the judge's phrase, understood what they were letting themselves in for. The agreements were identical save for the name of the licensee, and we have set forth as an example the agreement signed by S on February 21 1976:

THIS LICENCE is made the 21st day of February 1976. One thousand nine hundred and seventy-six. Between I Somma or Agent of 7 Agate Road, W6 (hereinafter referred to as "the Licensor") of the one part and R Savelli of Milan, Italy (hereinafter referred to as "the Licensee") of the other part.

WHEREAS the Licensor is not willing to grant the Licensee exclusive possession of any part of the rooms hereinafter referred to

AND WHEREAS the Licensee is anxious to secure the use of the rooms notwithstanding that such use be in common with the Licensor and such other licensees or invitees as the Licensor may permit from time to time to use the said rooms

AND WHEREAS this Licence is entered into by the Licensor and the Licensee solely upon the above basis

By this Licence the Licensor licenses the Licensee to use (but not exclusively) all those rooms (hereinafter referred to as "the Rooms") on the 3rd Floor Double B/Sit floor of the building known as and situate at Flat 4, Cornwall Mansions, W14 (hereinafter referred to as "the Building") together with the use of the entrance hall and lift (if any) the staircase outer door and vestibule of the Building and the furniture fixtures and effects now in the Rooms (more particularly set out in the Schedule of Contents annexed hereto) from (Twelve weeks) 21.2.76 until 15.5.76 for the sum of £116.40 on the following terms and conditions:

1. The Licensee agrees to pay the said sum of £116.40 by 4 weekly instalments of £38.80 commencing on the 21st day of February 1976 next and thereafter on Saturday of each 4th week until 15.5.76.

2. The Licensee shall be responsible for the payment of all gas electric light and power which shall be consumed or supplied in or to the Rooms during the Licensee's occupation thereof and the amount of all charges made in respect of the telephone installed therein or in the Building so far as the same relates to his use thereof.

3. The Licensee shall use his best endeavours amicably and peaceably to share the use of the Rooms with the Licensor and with such other licensees or invitees whom the Licensor shall from time to time permit to use the Rooms and shall not interfere with or otherwise obstruct such shared occupation in any way whatsoever.

4. The Licensee shall keep the interior of the Rooms and all fixtures and fittings and fixtures therein in good and clean condition and complete repair (fair wear and tear and damage by accidental fire only excepted) and immediately replace all broken glass.

5. The Licensee shall preserve the furniture and effects in the said Rooms from being destroyed or damaged and make good pay for the repair of or replace with articles of a similar kind and of equal value such of the furniture and effects as may be destroyed lost broken or damaged (fair wear and tear thereof only excepted).

6. The Licensee shall leave the furniture and effects at the expiration or sooner determination of this Licence in the Rooms or places in which they were at the commencement hereof.

7. The Licensee shall pay for the washing (including ironing or pressing) of all counterpanes blankets and curtains which shall have been soiled during the Licensee's occupation (the reasonable use thereof nevertheless to be allowed for).

8. On notice in writing being given to the Licensee by the Licensor or her Agent of all wants of repair cleansings amendments and restorations to the interior of the Rooms and of all such destruction loss breakage or damage of or to the furniture and effects as the Licensee shall be bound to make good found therein the Licensee shall repair cleanse amend and restore or make good the same within two months of the giving of such notice.

9. The Licensee shall not remove any furniture and effects from the Rooms without the previous consent in writing of the Licensor.

10. The Licensee shall not carry on or permit to be carried on in the Rooms any profession trade or business whatsoever.

11. The Licensee shall not do or suffer to be done in the Rooms any act or thing which may be a nuisance cause of damage or annoyance to the Licensor and the other occupiers or users of the Rooms or the Building or of any adjoining premises or which may vitiate any insurance of the Building against fire or otherwise or increase the ordinary premium thereon.

12. The Licensee shall not affix to the windows of the Rooms externally or internally any venetian blinds except of such colour and construction as shall be previously approved in writing by the Licensor or her Agent.

13. The Licensee shall not hang or allow to be hung any clothes or other articles on the outside of the Rooms or the Building.

14. The Licensee shall clean all the windows of the Rooms once at least in every month during his occupation.

15. The Licensee shall not deposit any store of coal elsewhere than in the cellar or other receptacle provided for the purpose and

A shall not keep any combustible or offensive goods provisions or materials in the Rooms.

16. The Licensee shall not cause or permit any waste spoil or destruction to the Rooms or to the Building.

17. The Licensee shall not pull down alter add to or in any way interfere with the construction or arrangements of the Rooms without the previous consent in writing of the Licensor.

18. The Licensee shall not keep any animals or birds in the Rooms Nor shall the Licensee permit any child or children to reside or stay in the Rooms.

19. The Licensor shall not at any one time permit more than one other person to use the Rooms together with the Licensor B and the Licensee.

20. Unless prevented by any cause not under his control the Licensor shall keep the entrance hall staircase vestibule and lift (if any) clean and properly lighted.

21. This Licence is personal to the Licensee and shall not permit the use of the Rooms by any other person whatsoever.

22. Upon the Licensee being in breach of any of the conditions referred to above this Licence shall immediately determine without prejudice to any other remedies of the Licensor and the Licensee shall immediately cease his use of the Rooms and the Building as permitted hereunder.

Signed by the above-named
Licensor or Agent
C S Ritter

Signed by the above-named
Licensee
Savelli Rossella

The Licensee hereby states having received a copy of this contract.

SCHEDULE OF CONTENTS

If an Electric Meter is installed the rent will be reduced by 25 pence per week.

S Ritter
Savelli Rossella

S paid Ritter £77.60 in cash on signing the agreement, being D £38.80 the first four weeks cash in advance, plus £38.80 deposit against breakages or damage. H telephoned his father and asked for a cheque to cover his first four weeks' instalment and his deposit. Ritter received the cheque for £77.60 a few days later. So they moved in and stayed sharing the room with each other until the term of the agreement expired on May 15. Then they each received and signed second agreements in the same terms as the first, and on August 7 they repeated the exercise, signing agreements to expire on October 29. Throughout they continued to make separate payments monthly in advance and received a single receipt.

E In September 1976 H applied to the rent officer to get their monthly instalments reduced, alleging a tenancy or furnished occupation within the protection of the Rent Acts. The rent officer adjourned the applications pending a decision of the county court pursuant to section 105 of the Rent Act 1968 upon the questions whether there was a tenancy and whether Part VI of the Act applied. So Miss Somma instituted these two sets of proceedings by applications for an order determining whether H and S were statutory tenants of the dwelling-house known as Flat 4, Cornwall Mansions, W14.

The grounds of her applications dated December 20 1976 appear there and read:

F The grounds of this application are—By a Licence dated August 7 1976—made between the Applicant and the Respondent the Applicant granted the Respondent a Licence to (but not exclusively) the said Flat consisting of a room on the third floor of a Building known as Cornwall Mansions aforesaid and the furniture therein contained. The names and addresses of the persons on whom it is intended to serve this application are M Hazelhurst. . . .

A word has been left out there I see. It should read, "The Applicant granted the Respondent a Licence to *use* (but not exclusively)" as appears on the application relating to Miss Savelli.

The answers of the respondents H and S read: G

The Respondent denies that the agreement between the parties was anything other than a tenancy agreement. The Applicant well knew that the other person to whom the alleged Licence was granted was in fact living with the Respondent.

On March 9 1977 the applications came on in the county court. They were consolidated and adjourned. On June 17 the hearing took place. The learned judge gave judgment for the respondents H and S. He declared that both respondents were regulated furnished tenants within the meaning of the Rent Acts 1968 and 1974. The order drawn up is in this form: H

And the Court doth declare that the Respondents are tenants of the premises known as Flat 4 Cornwall Mansions, Blythe Road, London W14, within the meaning of the Rent Acts.

If it is upheld, it clearly should be rectified to declare that they are regulated furnished tenants. Against that order Miss Somma appeals.

The submissions of Mr Purchas for the appellant fall under four heads:

(1) On the findings of the judge, the intentions of the parties are to be collected from the two agreements in writing which he conceded should be read together. On their proper construction they were two separate J grants of licences to occupy and not grants of tenancy or tenancies.

(2) The contracts are not within the protection of section 70(1) and (3) of the Rent Act 1968 as the contracts did not grant exclusive occupation of the room that H and S were to share.

(3) The clauses of the contract which provided that H and S were to share with the licensor or another licensee were not illegal or contrary to public policy as the judge held.

(4) As an alternative submission if the court is against him K on any of his first three submissions, there should be a new trial as the hearing was unfair on the ground that the judge wrongly ruled against the application to cross-examine further and call evidence which he made at the end of his closing speech.

Lord Gifford for the respondents submitted that the learned judge's decision was right. There was a single contract, and that contract was a joint tenancy. If not, it was a contract granting to H and S exclusive occupation of their room within the protection of section 70 of the Rent Act 1968. Those clauses in the written agreement which provided that H and S L should be liable to share the room with the licensor or another licensee introduced by Miss Somma were illegal or void as contrary to public policy and should be disregarded in ascertaining the legal consequences of the transaction. He also sought a new trial in circumstances that appear hereafter.

On the issue whether the transaction fell within the protection of the Rent Acts Lord Gifford submitted:

(1) In consideration of questions arising under the Rent Acts the court must look at the substance and reality of the transaction.

(2) In ascertaining the substance and reality of the trans- M action the court is entitled to look both at the documents and at the surrounding circumstances.

(3) Documents which purport to grant licences will be held to grant tenancies if either one of two sets of circumstances apply:

  (a) If upon examining the documents in the light of the surrounding circumstances they are found to be in substance documents granting a tenancy, which he called "the construction route."

  (b) If upon examining the surrounding circumstances the court finds that the documents are a disguise which cloaks the reality of the transaction, which he calls " the disguise route," for such a transaction may disguise the reality of the rights and obligations granted and assumed by the parties without necessarily being fraudulent or attracting the label of a mere sham.

(4) Residential licences should only be upheld as such by the courts in three special classes of case with which we deal hereafter.

(5) The courts must have regard to the policy of the Rent Acts and be astute to prevent their evasion.

In applying those principles to the instant case, Lord Gifford submitted that the learned judge did address himself to the right question, and sought for and found the reality of the transaction in the surrounding circumstances which he admitted in evidence and took into account. Pursuing the construction route, the written contracts are to be construed as parts of a single transaction, granting joint rights and giving rise to joint obligations such as to transfer to the grantees a joint tenancy. Alternatively, pursuing the disguise route, he submitted the true nature of the transaction was, and that the judge's findings can and should be read as finding, that the reality and substance of the transaction was that H and S were granted a joint tenancy disguised by the cloak of the documents drafted for Miss Somma, and that that finding of fact was amply supported by the evidence.

Alternatively, Lord Gifford sought leave to amend his respondents' notice to seek a new trial on the ground that the learned judge failed to appreciate the relevance of the circumstances surrounding the written agreements, stopped Mr Seifert, solicitor for the grantees, from investigating them in cross-examination, and so decided the case without making findings of fact on relevant matters which should have been explored. We gave leave to the respondents to amend their respondents' notice, and entertained their applications for a new trial.

We are confronted with one more attempt by an owner of housing accommodation to provide it at a profit for those in great need of it without the restrictions imposed by Parliament on his or her contractual rights to charge for it and regain possession of it. The attempt which has led to this appeal is made by a document drawn up by one, or a combination, of those who seem to have studied all the efforts, recorded in a welter of cases decided in every court from the county court to the House of Lords, to avoid letting a dwelling-house or part of it by arranging to licence or to share the occupation of it. On the particular facts of this case has this attempt failed as the judge held?

Each document which we have to consider is on its face a licence to each respondent to share occupation—of one double-bed-sitting room for 12 weeks. But the obligation which the document imposes on each respondent is an obligation to share with (1) the owner (described as the licensor), (2) the other respondent (described as a person permitted by the licensor but identified, by reading the respondents' documents with each other, as the other respondent). And the document so repeatedly proclaims itself a licence, and the relationship it creates between the appellant and each respondent as being that of licensor and licensee, that it raises the question why it should be necessary to protest so much and whether so many labels so clearly written all over it give a true or a false description of its real contents. Our first task is to examine closely the course of the proceedings in the county court. The next to observe and analyse findings made by the learned judge.

Sidney Ritter, resident manager of Miss Somma, gave evidence of the circumstances preceding the signature of the agreements. However, the judge observed that he did not think it necessary to go into the background of the agreement because it could not affect the agreements subsequently entered into. But he did not in fact stop Mr Purchas in chief eliciting from the witness the facts that we set forth at the beginning of this judgment.

During cross-examination, the judge again said that he was not concerned with the surrounding circumstances of the agreement and asked to what issue the cross-examination was relevant. Mr Seifert said that he was endeavouring to show that an agreement was entered into before the forms of licence were signed and suggested that H and S had paid a deposit and moved in before they signed the agreements. The judge then observed that unless there was fraud or mutual mistake the parties were bound by what they signed. The witness gave evidence that H and S signed the agreements before they moved in; there was no evidence of a deposit being paid before the agreements were signed, and it appears from the judgment that Mr Seifert did not pursue the allegation of any agreement or bargain antecedent to the written agreements. And indeed he could not do so, because when H gave evidence he did not suggest any such antecedent bargain. Miss Somma gave evidence. She said that she had used the licence agreement for about three years, and that she did not think that she introduced the form of licence because of the Rent Acts. She said that normally if one licensee went, then a friend of the remaining licensee would move in with her approval, and that if another licensee did not move in then she would lose the money and would not be able to recover it from the remaining licensee. We quote the note of the evidence she gave in cross-examination:

> I do not force people to share accommodation with people they do not like. The licence says I can put one other person in the room. I run my business on a humanity basis. I would not let a drunkard into a room in any case. I do not deal with fantasies. My plans are very happy. I do not force anyone to share with those whom they do not like. We both have a right in practice to choose who goes in. If a man was a drunkard I would not let him share. I do not think the licence has anything to do with the Rent Acts. I always let on a sharing basis. I have changed to sharing licences for my flats and rooms. I introduced the forms several years before. It suits people as they are always moving—I use 12-week periods.

H in evidence said that he was not shown the licence agreement on the first visit to Mr Ritter, whereupon the judge observed that there was no point in pursuing that line of questioning as H was an intelligent man and he entered into the agreement and was therefore bound by it, and that he the judge had to construe the documents. H said he read the document for five or 10 minutes at the most, and queried it slightly but was told it was a standard document for letting. He asked about clause 19, as it looked as if three people could share, but Ritter " brushed it off." He said he had moved his possessions into the flat, and that it did not seem a good thing to quarrel with the licence. In cross-examination he said that he and S signed separate agreements and made payments under the licences separately. He said he did not know if he would be responsible for paying both licence fees if S left; that he did not take legal advice because it would have been too expensive; and that he preferred to have the accommodation on the terms it was offered to him than not to have the accommodation at all. That concluded the evidence. The judge did not call upon Mr Seifert until after he had heard Mr Purchas. Mr Purchas invited the judge to construe the documents, and in the course of argument the judge said that everything in the documents was utterly artificial from start to finish. The document was not a licence at all but a joint tenancy, and the landlord could not wrap it up with extraordinary expressions. Mr Purchas said that was at variance with what the judge had said earlier when he said that he had to construe the documents, and asked to pursue further his cross-examination of H. The judge refused to hear further evidence. At the end of Mr Purchas's speech, Mr Seifert was called upon, and reminded the judge of the relevance of surrounding circumstances. Against that background we came to the judgment.

It is clear from the terms of his judgment, from the note to his judgment which he has supplied, and from what he is reported to have said in the course of argument, that he was seeking to ascertain the nature of the agreement solely from the terms of the written documents. Thus, despite the fact that a fair body of evidence was called before him as to the surrounding circumstances, it seems from what the learned judge himself says on the second page of his judgment, that Mr Seifert did not press or at least did not press very strongly that the learned judge should consider anything outside the terms of the documents. This is what the learned judge had to say about it: " at first there was an attempt by the solicitor for the respondents which I am not sure if he abandoned to say that the parties' rights were not circumscribed by these documents "; and a little lower down: " the parties would be entitled to ask the court to consider whether an antecedent bargain was made here, and whether that altered the contract before me. I do not think that side of the argument is now pursued because the male respondent is an educated man. I did not see the girl but I presume she is also educated." It is clear from these passages, and others to which there is no need for us to refer, that the learned judge rejected, inferentially at least, that there was any oral agreement outside the two documents which would be capable of affecting their meaning. Thus we start from the basis that it is to the documents that we must look and to the documents alone. This has an important bearing on the whole of the case.

The next matter is to discover what was the ground of the argument for a joint tenancy vested in H and S. One reads from the judgment again. The learned judge says this, " They understood what they were letting themselves in for. On February 21 1976 there was really a joint licence relating to the flat albeit cloaked around by the documents." There he is saying it is a joint licence. Does he mean joint " interest " of some sort? That is probably the case although it is difficult to say. Then he goes on to say this:

> there is no sign here of the incidents which one finds for a landlord and tenant relationship but heaps of indications which show that the parties if they applied their minds to it conferred on these people a joint tenancy or the rights of joint tenants or the equivalent rights on them. It is not an elaborate matter of this sort. It is trite law that one does not now look for exclusive possession. We are warned off by placing too much emphasis on exclusive possession. The relationship between the landlord and these two young people was one of joint tenants I have no doubt about that whatever.

He then goes on to deal with clause 19 of the documents which permits or purports to permit the landlord to impose another occupant on a remaining grantee whether it be H or S and to say that such a clause must be illegal and is therefore not binding upon the grantees.

It seems to us that the reasoning behind the judge's ultimate conclusion is this:

(1) Although there are on the face of it two documents here which if taken separately would give neither grantee exclusive possession, this is not fatal to the claim of the occupants that this was a tenancy, because the two documents must be construed together and if that is done they are apt to confer on the couple a joint interest of some sort in the room.

(2) The other clauses which appear to make the occupation non-exclusive are illegal possibly on the grounds of public policy and therefore are not binding upon the grantees.

(3) Consequently, one is left with a joint and exclusive occupation.

(4) The nature of the occupation is a tenancy because there are " heaps of indications " which show that the parties if they applied their minds to it conferred on these people a joint tenancy or the equivalent rights. He does not say what these indications may be.

We find ourselves unable to come to the same conclusion as the learned judge. Lord Gifford, basing himself on the judgment of Denning LJ, as he then was, in *Facchini v Bryson* [1952] 1 TLR 1386 at p 1389, and the reasoning in *Marchant v Charters* [1977] 1 WLR 1181 at p 1185 submits that in a " Rent Act " situation, any permission to occupy residential premises exclusively must be a tenancy and not a licence, unless it comes into the category of hotels, hostels, family arrangements or service occupancy or a similar undefined special category. We can see no reason why an ordinary landlord not in any of these special categories should not be able to grant a licence to occupy an ordinary house. If that is what both he and the licensee intend and if they can frame any written agreement in such a way as to demonstrate that it is not really an agreement for a lease masquerading as a licence, we can see no reason in law or justice why they should be prevented from achieving that object. Nor can we see why their common intentions should be categorised as bogus or unreal or as sham merely on the grounds that the court disapproves of the bargain. This matter was expressed most happily by Buckley LJ in *Shell-Mex and BP Ltd v Manchester Garages Ltd* [1971] 1 WLR 612 at p 619 as follows:

> it may be that this is a device which has been adopted by the plaintiff company to avoid possible consequences of the Landlord and Tenant Act 1954, which would have affected a transaction being one of landlord and tenant; but, in my judgment, one cannot take that into account in the process of construing such a document to find out what the true nature of the transaction is. One has first to find out what is the true nature of the transaction and then see how the Act operates upon that state of affairs, if it bites at all. One should not approach the problem with a tendency to attempt to find a tenancy because unless there is a tenancy the case will escape the effects of the statute.

This matter was also touched upon in a New Zealand decision, *Donald v Baldwyn* [1953] NZLR 313 at p 321, where F B Adams J in the course of his judgment said this:

> The law will not be hoodwinked by shams but real and lawful intentions cannot be dismissed as shams merely because they are disliked. It is a sham to say one thing while really intending another, but the court cannot say that a licence is a sham for the reason that the court thinks the parties ought to have intended a tenancy.

There seem to us to be two questions which we have to answer: (1) Did the parties intend to be bound by the written agreement? (2) Can it be said from the words which they used in those agreements that they intended to create a tenancy rather than a licence? As to the first question there is a clear finding of the learned judge that H and S " knew what they were letting themselves in for "; also that they were both educated and that they both signed the document. It is plain from these findings that the learned judge was of the opinion that they knew what they were signing and nevertheless voluntarily signed it. The second question is really the nub of the whole case, namely, do the terms of the two documents show an agreement for a licence rather than an agreement for a tenancy? Immediately one faces the problem of the two separate agreements, one with H and the other with S. If they are truly separate then it must follow that neither H nor S has the necessary exclusive occupation of the room on which to found a tenancy or indeed a Part VI licence. The judge, while saying that he is confined to interpreting the documents themselves, says simply in the passage already quoted, " on February 21 1976 there was really a joint licence relating to the flat albeit cloaked around by the documents." If indeed he was confining himself to the documents, there was no basis for saying that they were joint agreements at all. However,

A if that is too legalistic a way of looking at the situation, as Lord Gifford submits, it is impossible to reconstruct the separate obligations into one joint obligation without doing violence to the obvious intentions of the parties, particularly of the two grantees, H and S. For example, if one takes clause 1, which deals with the consideration which the grantees are bound to pay, it would be necessary, as we see it, either to have a joint interest coupled with a several liability for half the weekly amounts, which is something of a logical inconsistency, or else to redraw the terms of the agreements to make each grantee jointly and severally liable for the full weekly instalment of £77.60, as against the contractual liability B resting on each under the agreement of £38.80. This is not construing the agreement between the parties: it is rewriting it in the absence of any claim for rectification or any allegation of fraud. Lord Gifford was unable to provide any answer to this problem which came near to satisfying us on this point. We find it impossible to say that on the contents of these two documents, even adding the fact that the two grantees made their bid for the room together, a joint interest was created in their favour. That is really the end of the matter, because if the rights of the parties are several as opposed to joint, it is impossible to say that either has the C necessary exclusive possession to found a tenancy.

However, assuming that that conclusion is wrong and that the interest is indeed joint, one must next examine the other clauses to see whether any of them prevents the joint occupation from being exclusive. The recitals and conditions 3, 19 and 21 point very strongly to a licence. Condition 2 imposes an obligation upon the grantee "for the payment of all gas, electric light and power which shall be consumed or supplied in or to the Rooms during the Licensee's occupation thereof . . ." which is a wider obligation than payment for his own proportion of a consumption or supply required or enjoyed also by a second licensee. This points in the direction of a D lease rather than a personal licence. Likewise the obligations to repair and replace in conditions 4 and 5 are wider than an an obligation to remedy damage caused by his own default and are more consistent with a lease than a licence. Condition 8 is the kind of clause which is more appropriate for a lease than a licence for a short term. Conditions 9 and 10 are neutral. Condition 11 by its reference to a duty to other occupiers or users of the rooms is more suitable in a licence than in a lease. Clause 14 is to be read in the knowledge that if another licensee fulfils his obligation the windows will be cleaned twice a month. On balance, though it is not particularly strong, this suggests a joint obligation to clean E once a month. Condition 16 prohibits "waste spoil or destruction to the rooms or to the Building." Waste is a familiar term of art in the law of real property and carries the meaning of damage or prejudice to the reversion. In a document such as this, which is obviously carefully drafted by a hand skilled in property law, the condition points to a lease rather than a licence. Condition 17 prohibits alteration or interference with the construction or arrangements of the rooms. It contemplates a continuing control by the licensor and is more consistent with a licence than a lease.

So some of these obligations are more consistent with a licence than a lease, and some lead to a contrary inference. F The balance comes down in favour of a licence because the document cannot be construed as a lease without substantial rewriting. The recitals, conditions 1, 3, 19, 21 and 22 would all have to be written in substantially different terms or disregarded altogether. But Clause 1 cannot be either rewritten or disregarded, and if it is to the documents that one looks for the terms of the contract the conclusion is that the document as drafted gives rise to personal obligations. The conditions which point to a joint obligation can without violence be explained as a protection required by the licensor who assumes that the licensees who share the use of the rooms from time to time will come to some arrangement of work G or financial contribution between themselves in respect of the obligations which each has severally undertaken.

When the agreements were entered into the unidentified one person in clause 19 was clearly identified as the grantee who signed the other agreement. But if at any time during the currency of the agreement either H or S decided to leave, either by agreement with Miss Somma on surrender of the unexpired portion of the 12-week term, or one of them repudiating the agreement by leaving without paying the balance of instalments as they fell due, Miss Somma has reserved the right to introduce a second licensee. She explained in evidence how she set about it, and was not cross-examined H to suggest that the rights reserved of replacing the second sharing licensee were unworkable or other than of reasonable business efficacy. Such a right to introduce a new licensee as a replacement is wholly inconsistent with a grant of exclusive occupation, much less exclusive possession.

These two clauses also impose on the licensee the obligation to share the room with the licensor as well as with another licensee. The room was that described as being 22 ft by 18 ft, with two beds in it. The question of construction is what are the rights and obligations to be ascertained from the agreement. The nature and frequency of the exercise of J the rights are a different question: see *Goodrich v Paisner* [1957] AC 65. The issue in that case was whether the occupation was of a separate dwelling, but the distinction there made between the rights granted and the exercise of the rights is relevant to the construction of these agreements. Miss Somma has a house. She grants to H and to S the right to use one room, but expressly reserves the right to use that room in common with them. On no construction of such a contract can H and S claim that they have been granted exclusive occupation of the room.

In the second part of his judgment the learned judge considered clause 3 and clause 19. Construing clause 19 he said:

K I pointed out that at first blush it looks as if three people can use the premises at any one time. What was intended but the document does not say so is plain, that the licensee, whether Miss Savelli or Mr Hazelhurst, can share with one other person. It contemplates that the other person shall be whatever other legal person is allowed in. What is laid down plainly is that the landlord is given a complete discretion, putting it mildly, and power in practice if somebody goes so he can say "I have the right to say who is to share and you cannot stop me. You as the remaining licensee can lump it if you do not like it. You can stay or get out." I cannot believe that if anyone applies their mind to the point that is a fair interpretation of the position of the parties that in answer to the question "Are you saying that you L have the sole right to impose any other person or sex, language, class or other taste?" the landlord would give the answer or can give the answer "Yes, I can" as a result of this document. If he imposes someone of three times their age or who snores or plays a musical instrument or has other habits that the surviving person does not like, this means that he can impose such a person on him. Worse still—and this is quite new, I have not heard of it before and the sooner it is dealt with the better—not only can he impose a person of different race or occupation but also of different sex. This is a different, a most revolting different idea. No court known to me as advocate or judge would willingly uphold that. That part of the document is not binding and is illegal in my judgment. That means the parties can carry out the M rest of the agreement without the one party forcing that clause on the other parties' will. I am surprised that counsel cannot find any authority on it. I am sure that in the reports counsel would find a case on this point if not similar to what we have here. If not, I decide it for the first time. It is no part of my decision because it is not necessary for the resolution of the whole case.

In our view the illustrations given by the judge of the kind of uncongenial persons who might be imposed upon H and S by Miss Somma if either of them left only illustrate the risks inherent in the bargain that they made and are no ground for invoking public policy to render unenforceable the right retained by the licensor. The judge did not like the bargain, but the agreements were the agreements of the parties and the fact that the judge envisaged circumstances in which the exercise of Miss Somma's rights would be uncongenial to the remaining licensee does not entitle the court to rewrite the agreement. The judge was deeply offended by the thought that the grantor could introduce a person of a different sex to succeed to the vacancy left after H or S had left. But we can find nothing to indicate that there was such an intention. If S were to depart, and the licensor permitted another lady congenial to H to take her place, a moralist might regard the new sharing arrangement with regret or indignation, but the court would be as reluctant to intervene as it is in the case of the sharing arrangements of H & S. Nor can we construe the clause to find an intention on the part of the licensor to impose on one licensee the company of another licensee of the opposite sex against the wishes of either of them. For those reasons we cannot uphold the learned judge's decision that clauses 3 and 19 are contrary to public policy.

Lord Gifford did not try to support this reason for the judge's judgment but submitted a further objection to these clauses on the ground of illegality as expressing an intention to contravene section 77 of the Housing Act 1957 by overcrowding the room which H and S were given a right to share. If the licensor exercised her right to share the room with H and S by sleeping in the room, the house would be deemed to be overcrowded within the meaning of section 77(1). There are many difficulties in his way. By the subsection a dwelling-house shall be deemed to be overcrowded at any time when the number of persons sleeping in the house either (a) is such that any two of those persons not being persons living together as husband and wife must sleep in the same room or (b) in relation to a house consisting of one room of the size of this particular room, is more than two. By section 87 "dwelling-house" means "any premises used as a separate dwelling . . . or of a type suitable for such use." It is not possible as a matter of construction to hold that clauses 3 and 19 express an intention that the licensor shall sleep in the room when H and S are occupying it, and there was no extrinsic evidence of such intention. This submission fails. It is perhaps only fair to Lord Gifford to say that his own enthusiasm for the point was less than burning.

In the course of the argument we were invited to consider a great number of cases, mostly cases in which the courts have considered the application of different provisions of the Rent Acts. But at the end of the day it was clear to us that the instant case is very much a case peculiar to its own facts. The difficulties, which were real, were largely difficulties in determining the content of the findings of the learned judge. Lord Gifford strongly argued that when properly understood the judge had found that the agreements were a sham, or at least a disguise of the substance and reality of the transaction. We have given our reasons for rejecting that submission, and in the result it has become unnecessary to refer to most of the authorities cited in argument.

So we arrive at the decision that on an analysis of the judgment the result in law is that H and S entered into separate contracts; those contracts were personal licences and not leases or a lease; and by those contracts neither H nor S was given a right of exclusive occupation of the room that they shared. That brings us to the respondents' application for a new trial, leave for which we granted as an amendment to the respondents' notice. Lord Gifford submits that the learned judge approached the case in the wrong way. He confined himself to a construction of the documents when it was his duty, on the authority of the cases which have been concerned with different but in some ways similar problems under the Rent Acts, to admit extrinsic evidence and to consider whether the surrounding circumstances establish that the documents do not record the real intention of the parties at the time when the grantor agreed to let the grantees live in the room. The onus was upon the respondents to establish that the contract was other than that recorded in the documents that they signed. If authority is needed, it is to be found clearly stated in this court in *Buchmann v May* (Bar Library Transcript No 182 of 1976, shortly reported in (1976) 240 EG 49) and the cases therein cited. There Sir John Pennycuick, in a judgment with which Megaw LJ and Stephenson LJ agreed, held that the relevant clause of the agreement in writing was unequivocal and said:

> It seems to me that that provision must stand as evidence of the purpose of the parties unless (the tenant) can establish that the provision does not correspond to the true purpose of the parties. The burden lies upon her to do so.

But Lord Gifford submits that they failed to do so because the learned judge made it clear that he was not interested in hearing evidence of the surrounding circumstances and so either closed his mind to the correct approach or alternatively made it impracticable for the respondents' advocate to present their case adequately. He contrasted the course of the hearing with the proceedings in three other cases in the county courts of London of which he gave us reports of the judgments which he personally could certify. In all those cases the learned judges rightly admitted extrinsic evidence which satisfied them that the written contracts were not the real statement of the intentions of the grantor and two grantees, so that documents, described as licences, almost identical with the present agreements, were held to be the grant of a lease. The judgment of Judge McDonnell in *Walsh and Walsh v Griffiths-Jones and Durant* has been published in the Legal Action Group Bulletin for October 1977.

We say with respect that Judge McDonnell's judgment is a model of the approach to and analysis of an alleged agreement to grant licences to two persons to share residential accommodation, and it deserves wider publicity. Lord Gifford contrasted the hearing before and judgment of Judge McDonnell with the proceedings in the county court with which we are concerned. He pointed to the observations of the learned judge calculated to discourage the advocates for both sides from leading evidence of the surrounding circumstances or from cross-examining in relation thereto. In our view there is great force in Lord Gifford's submission that the hearing had unsatisfactory features, as the judge's interventions point to an inference that he was confused as to the relevance of extrinsic evidence. But with some hesitation we have decided that though there was confusion in the judge's mind while the evidence was being given, in fact his interventions did not have the effect of shutting out the evidence of surrounding circumstances which was relevant to the transaction. In spite of the judge's comments, Ritter, Miss Somma and H appear to have been permitted to give the evidence that they wanted to present of the relevant surrounding circumstances. Mr Seifert did not persist in his contention that H had paid a deposit and moved in before he signed the agreements, and H's evidence was not consistent with the payment of any deposits save the deposits that each paid of £38.80 after the agreements had been signed. Where parties are professionally represented our forensic system requires that parties must persist in examining or cross-examining upon facts, and must seek a ruling from the court if the court indicates an intention to exclude material which the advocate regards as relevant. We do not underestimate Mr Seifert's difficulties, but he did not persist in his cross-examination of Ritter, or his examination in chief of H about the events of the first occasion when he met Ritter, when the judge intervened. We conclude that in this court the case must be scrutinised on the evidence that was given, and that in spite of the criticisms that we have made of the judge's interventions and approach to the case before he came to deliver judgment, it is not a proper case for a new

A trial in order to enable parties to canvas again the evidence of the circumstances in February 1976 which surrounded the signature of the two agreements in writing. Further, having regard to the uncontroverted facts in evidence at the trial, it seems unlikely that a second trial would enable H or S to discharge the onus of establishing that the intentions of the parties are not to be collected from the written documents that they signed, particularly having regard to H's candid evidence that after he had read the document it did not seem a good thing to quarrel with the licence and that he preferred to have the accommodation on the terms that it was offered to him rather than not to have the acccommodation at all.
B This court has today construed those agreements. The respondents' application for a new trial is therefore refused.

The appeal was allowed, with costs in the Court of Appeal and below. The costs awarded against Miss Savelli were limited to £50 in the absence of further order; legal aid taxation of her costs was ordered. The declarations granted were (1) that the respondents were not statutory tenants of the subject accommodation and (2) that Part VI of the Rent Act 1968 did not apply to the contracts dated August 7 1976 (these were the second agreements in the same terms as the first). Leave to appeal to the House of Lords was refused.

C

COURT OF APPEAL

January 19 1978

(Before Lord Justice MEGAW, Lord Justice LAWTON and Lord Justice SHAW)

LLOYD v SADLER AND OTHERS

Estates Gazette May 6 1978

(1978) 246 EG 479

D **Rent Acts—Conditions for acquisition of a statutory tenancy—Important decision likely to affect many tenants, particularly young people joining to share accommodation—Flat let to two joint contractual tenants—One ceases to reside before contractual term expires—Whether remaining tenant qualifies as a statutory tenant on expiry of term—Cases reviewed—Howson v Buxton followed by analogy—Remaining joint tenant entitled to statutory tenancy**

This was an appeal by the plaintiff landlord from a decision by Judge McIntyre at West London County Court in proceedings for possession of a top-floor flat at 22 Cottesmore Gardens, London W8. The landlord claimed that Miss Gloria
E Sadler was not entitled to remain in possession of the flat as a statutory tenant following the expiry of a fixed term because the term had been granted to Miss Sadler and a Miss M Lunt as joint tenants and Miss Lunt had left the flat in order to get married before the expiry of the term. Judge McIntyre rejected the landlord's submission and dismissed the action for possession.

David Neuberger (instructed by Halsey Lightly & Hemsley) appeared on behalf of the appellant; William Birtles (instructed by Tyrer, Roxburgh & Dawson) represented the respondents.

Giving judgment MEGAW LJ said: The relevant facts are
F simple. The question of law which arises on them is of importance, with possibly wide-ranging effects. By an agreement in writing a tenancy of the top-floor flat, 22 Cottesmore Gardens, London W8, was granted by Mr J M Lloyd to Miss G Sadler and Miss M Lunt from December 27 1975 to December 25 1976. The tenancy was a protected tenancy under the Rent Act 1968. About October 28 1976 Miss Lunt left the flat in order to get married. She did not intend to resume, and she did not in fact resume, occupation of the flat. Miss Sadler remained in occupation of the flat during the remainder of the contractual tenancy, and she continued in occupation after December 25 1976 when the contractual tenancy ended. In January 1977 the landlord, Mr Lloyd, G
started proceedings in the West London County Court, claiming possession of the flat. The defendants were Miss Sadler and three other ladies whom Miss Sadler had brought in to live in the flat as her licensees. No separate question arises as to them. Their right to stay in occupation stands or falls with that of Miss Sadler. So much for the relevant facts.

The question of law also can be simply stated: was Miss Sadler a statutory tenant and thus entitled to the security of tenure given to statutory tenants by the Rent Acts, despite the fact that Miss Lunt, before the contractual tenancy had ended, had ceased, permanently, to occupy the flat?

The contractual tenancy was a joint tenancy of Miss Sadler H
and Miss Lunt. The contractual tenancy ended on December 25 1976. The right of Miss Sadler to remain in possession thereafter, if it existed at all, could exist only because she had become the statutory tenant under section 3 (1) of the Rent Act 1968. (That was the statute in force at the time. The corresponding provision of the new Act, section 2 (1) of the Rent Act 1977, is identical.)

Section 3(1)(a) of the Rent Act 1968 provides:

> After the termination of a protected tenancy of a dwelling-house the person who, immediately before that termination, was the protected tenant of the dwelling-house shall, if and so long J
> as he occupies the dwelling-house as his residence, be the statutory tenant of it. . . .

The landlord contended that the contractual, admittedly protected, tenancy was a joint tenancy; and that, as a result of the doctrine of joint tenancy in the law of property, anything that had to be done in connection with the joint tenancy, to have any legal effect, had to be done by the joint tenants: that is, by all the individuals acting collectively. One of the joint tenants, it was submitted (otherwise than when in law one joint tenant could be treated as acting as agent for them all) could not achieve any result recognised by law—he could not act validly—by anything which he himself did K
as an individual. Hence Miss Sadler's continuance in occupation of the flat after the end of the contractual tenancy did not, in law, produce the result that "the protected tenant occupies the dwelling-house as his residence." Hence there was no statutory tenant. Miss Sadler did not achieve that status in law. That submission by the landlord was rejected by Judge McIntyre in his judgment delivered on May 16 1977. The plaintiff landlord appeals to this court.

This case has brought to light two remarkable facts. First, the Rent Acts through all their long history have never made any relevant express provision relating to joint tenants or L
joint tenancies, whether protected or statutory. Secondly, and perhaps even more strangely, there seems to have been no previous case decided by any court, or at least no reported case, in which a question such as falls to be decided in the present case has been raised for consideration on section 3 (1) of the 1968 Act or its predecessors. Yet there must have been, over the years, thousands of instances in which one of two joint tenants has ceased to occupy the dwelling-house before the end of the contractual tenancy, and the other original joint tenant has continued to live there, on the assumption that he has become the statutory tenant. If the landlord's submission is right, that assumption has been wrong. M
Similar instances must have arisen thousands of times where two or more persons have duly become "the statutory tenant," and thereafter one or more of them has, or have, died or ceased to occupy the house as his, or their, residence. What is the effect on the legal status of the person, or persons, remaining in occupation? Does the statutory security of tenure cease?

I do not think that the absence of previous litigation helps one way or the other towards a decision of the issue. But the fact that there are likely to be very many cases with facts similar to those of the present case shows the practical impor-

A tance of this issue. Whatever the answer be, it must affect many people, landlords and tenants, in a matter of great importance to them. If the landlord's submissions in the present case are wrong, the result may be a measure of prejudice to landlords. It would mean that a landlord who looked to two persons as both being liable to him for the rent might find that when the contractual tenancy ended there was only one person who was liable to him for the rent. On the other hand, if the landlord's submissions are right, the potential prejudice to joint tenants would be very much more serious: that is, the value of their legal rights as compared with what those legal rights would be if the landlord's sub-
B missions are wrong.

In view of the importance of the issue, it may be helpful to set out the landlord's submissions as they were conveniently summarised by Mr Neuberger in nine propositions, in the course of his very clear and careful argument:

1. The grant to Miss Sadler and Miss Lunt of a contractual tenancy resulted in Miss Sadler and Miss Lunt being joint tenants.

2. Therefore "the tenant" under the contractual tenancy was Miss Sadler and Miss Lunt.

3. The contractual tenancy was a protected tenancy.

4. Therefore "the protected tenant" was Miss Sadler and Miss
C Lunt.

5. On the determination of a protected tenancy a statutory tenancy can only arise "if . . . he occupies the dwelling-house as his residence."

6. The "he" in the above quotation clearly refers to "the person who . . . was the protected tenant."

7. On the determination of the protected tenancy in this case it cannot be said that Miss Sadler and Miss Lunt were occupying the dwelling-house as *their* residence.

8. Therefore the protected tenant was not occupying the premises in question on the determination of the protected tenancy
D in this case.

9. Therefore no statutory tenancy can arise.

If the ordinary law as to joint tenancy, as it affects rights of property, is to be treated as applicable in all its strictness to section 3(1)(a) of the Rent Act 1968, the logic of Mr Neuberger's propositions appears to me to be unassailable. I agree with the view expressed by Judge McIntyre that section 1(1)(b) of the Interpretation Act 1889 is not of assistance one way or the other.

I have come to the conclusion that, on the true construction of section 3(1)(a) of the Rent Act 1968, the ordinary law
E as to joint tenancy does not have to be, and ought not to be, applied in all its strictness. I base that opinion, primarily at least, on the judgment of Scrutton LJ in *Howson v Buxton* (1928) 97 LJKB 749. It is not a direct authority. There is no direct authority. There are various decided cases to which we were properly referred as providing guidance by analogy, or as illustrating the general rule as to a joint tenancy. Some of them, or dicta in them, certainly lend support to the submissions on behalf of the landlord; but I find the most helpful guidance in *Howson v Buxton*. It appears to me to decide that, where an Act of Parliament refers to "the tenant," and the letting is to two or more persons jointly, it is permissible
F for the court to hold, if so to do makes better sense of the relevant statutory provision in its particular context, that one of those persons, by himself, may for certain purposes be treated as being "the tenant." So here Miss Sadler can be treated, and should on the facts be treated, as having been "the protected tenant" immediately before the termination of the contractual tenancy, and as being "the statutory tenant" thereafter. Hence she has the security of tenure under the Act. I do not propose to discuss at length the cases which I have mentioned. I shall refer to them briefly.

In *Fairclough (TM) & Sons Ltd v Berliner* [1931] 1 Ch 60 Maugham J held that, where there are joint trustees, relief against forfeiture cannot be granted on the application of one G
person only. Otherwise, the joint tenant, who did not join in the application for relief, might find himself saddled with the continuance of obligations under an onerous lease, from which he would prefer to be free.

In *Lloyds Bank Ltd v Jones* [1955] 2 QB 298, a covenant in a lease of a farmhouse and cottages provided that the tenant should "at all times personally inhabit the farmhouse and cottages with his family and servants." "The tenant" was defined to include the named tenant's executors and assigns. The lessee died. His trustees were his widow and another. The widow resided in the farmhouse until her death. The other trustee did not. It was held that this was a breach of H
covenant.

In *McIntyre v Hardcastle* [1948] 2 KB 82 another decision of this court, it was not the tenant, but it was the landlord, who was joint. The joint landlord consisted of two sisters. The case concerned what is now, in the 1968 Rent Act, Case 8 in Part I of Schedule 3: "the dwelling-house is reasonably required by the landlord for occupation as a residence for (a) himself, or (b) any son or daughter of his . . ., or (c) his father or mother. . . ." It was held, following a dictum of Asquith LJ in *Baker v Lewis* [1947] KB 186, at p193, that, where there were joint landlords, "himself" in (a) should be J
read as "themselves"; "any son . . . of his" in (b) should be read as "any son . . . of theirs"; and so on. That was a strict construction of "the landlord", adverse to the interest of landlords. Not unnaturally, Mr Neuberger, for the landlord in this case, says in effect (though he did not put it so inelegantly) "What is sauce for the landlord is sauce for the tenant."

In *Turley v Panton* (1975) 29 P & CR 397 a Divisional Court of the Queen's Bench Division held that a rent tribunal did not have jurisdiction to determine an application made by one person alone, he being one of a number of joint K
lessees under Part VI of the 1968 Rent Act. The possibility of an unfair result to the others of the joint lessees, if such an application could validly be made without their concurrence, was stressed by the court.

Jacobs v Chaudhuri [1968] 2 QB 470 was a decision of this court. It was a decision, not in relation to security of tenure of living accommodation under the Rent Acts, but in relation to security of tenure of business premises under the Landlord and Tenant Act 1954. The letting of the business premises was to two persons who were in partnership. During the currency of the contractual lease the partnership was dissolved, and one of the two partners by agreement between L
them (the landlord not being a party to the agreement) took over the business as his own. He applied, in his own name alone, for a new tenancy under section 26 of the 1954 Act. It was held by a majority, Harman LJ and Winn LJ, Davies LJ dissenting, that the county court had no jurisdiction under the 1954 Act to entertain such an application, made by one individual, when the contractual tenancy had been granted as a joint tenancy to two persons. The majority distinguished *Howson v Buxton*. Davies LJ thought that, though of course *Howson v Buxton* was not direct binding authority in relation to this different statute and these different facts, the principle of Scrutton LJ's judgment was M
applicable by analogy.

The legislature promptly amended the 1954 Act. The Law of Property Act 1969 introduced two new sections into the Landlord and Tenant Act 1954. By section 13 it introduced a new section 43A to overcome a technical defect in the county court jurisdiction, using the draftsman's normal wording to achieve that purpose: the same form of wording (though I do not think this was called to the court's attention in a recent case) as is used in section 1(1) of the Domestic Violence and Matrimonial Proceedings Act 1976. The other new section introduced by the 1969 Property Act was section

41A, which showed that, to put it very broadly, Parliament preferred the view of Davies LJ to the view of the majority. But the fact that Parliament thought it right to include in the new section 41A a detailed code of provisions affecting joint tenants of business premises makes it probable, I think, that it was, indeed, an inadvertent gap which the legislature had left in the earlier legislation, rather than that the court had misinterpreted the meaning actually intended by Parliament to be expressed by the words used in the 1954 Act. If that assessment is right, it tends, in my opinion, strongly to support the view that the absence of express provisions in the Rent Acts as to the effect of joint tenancies is also, equally, an inadvertent gap. If so, and if this court's decision in this case accords with what Parliament would have enacted, had its intention been directed to the particular problem which has arisen in this case, it would, I firmly believe, be far better in the general interest that those parts of the gap which still remain—I suspect that those which remain are neither few nor unimportant, particularly as regards statutory joint tenancies—should be filled as the legislature would wish by an early and comprehensive Act of Parliament. They should not be left to be filled in, piece-meal, by the courts over the years, at random as cases happen to arise and happen to be taken to various stages of appeal. Such a code of judge-made law, however admirable it might prove to be in the end, would be built up slowly, probably erratically (because the various problems would not come up for decision in any logical order), and, most certainly, at no small cost in terms of human stress and anxiety and, in some cases, financial hardship to a large number of individual persons, who, to their misfortune, would happen to be the litigants upon whose troubles the edifice of judge-made law would be built. In my opinion, there can be no doubt which would be the more efficient, sensible and humane way of filling any remaining gaps in the law as to the effect of joint tenancies in the Rent Acts.

The decision in *Jacobs v Chaudhuri* is no longer law. But that does not necessarily mean that the reasoning of the majority in that case is to be disregarded when the court has to interpret other statutory provisions in respect of which it is suggested that there is an analogy. However, in the present case I think it tends rather in the other direction. It is not, I believe, fanciful to suggest that the reaction of the legislature to the decision in *Jacobs v Chaudhuri* tends to show that the legislature's approach to the problem of joint tenants in the 1954 Act, if it had specifically directed its mind to the problem at that time, would have been consonant with Scrutton LJ's approach in *Howson v Buxton*.

Before coming to *Howson v Buxton*, I should mention another case, very recently decided, to which we were referred: *Newman v Keedwell* (1977) 244 ESTATES GAZETTE 469. It was a case arising under section 24(1) of the Agricultural Holdings Act 1948. Fox J delivered, if I may say so, a very clear and careful judgment, in the course of which he discussed *Jacobs v Chaudhuri* and *Howson v Buxton*. But, by reason of the nature of the issues in that case, I do not think that it assists on the issue which we have to decide.

Howson v Buxton was a case which arose under section 12 of the Agricultural Holdings Act 1923, under which "the tenant" of an agricultural holding was given certain rights of compensation in the event of a termination of the tenancy. In *Howson v Buxton* there was a joint tenancy. It was held that, notwithstanding the doctrine of joint tenancy, a claim for compensation founded on a notice given by one only of two joint tenants was valid so as to give jurisdiction to the statutory arbitrator; and that one of the two joint tenants was entitled to recover compensation in respect of a loss which affected him alone, and did not affect also the other of the joint tenants. The relevant provisions of section 12 of the 1923 Act (the emphasis is mine) are:

> (1) Where *the tenancy* of a holding terminates by reason of a notice to quit given by the landlord, and in consequence of such notice *the tenant* quits the holding, then . . . compensation for the disturbance shall be payable by the landlord to *the tenant*. . . .
>
> (6) The compensation payable under this section shall be a sum representing such loss or expense directly attributable to the quitting of the holding as *the tenant* may unavoidably incur upon or in connection with the sale or removal of *his* household goods, implements of husbandry. . . .
>
> (7) Compensation shall not be payable under this section . . . (b) unless *the tenant* has . . . given notice in writing to the landlord of *his* intention to make a claim for compensation. . . .

The subject-matter obviously differs greatly from the subject-matter of section 3(1)(a) of the Rent Act 1968. But the relevance is the statutory reference to "the tenant," "his goods," "his intention," coupled with the same absence of any specific provision for cases where the tenancy is joint. In *Howson v Buxton* it was held by Scrutton LJ, with the concurrence of Sankey LJ, who expressly agreed "for the reasons given by Scrutton LJ," that it was permissible for the court to hold that the statutory requirements which had to be fulfilled by "the tenant" could validly and effectively be fulfilled by the individual act of one of the joint tenants. Thus, one of the two joint tenants could, acting by and for himself alone, give the requisite statutory notice. He could, without the concurrence or co-operation of the other participant in the joint tenancy, become entitled to the compensation which under the statute was payable to "the tenant." Even though it was a joint tenancy, "his goods" could be construed to mean "the goods which belong exclusively to one of the two joint tenants." It did not have to be read as though, whether by the operation of the Interpretation Act or otherwise, it said "their goods." It will be remembered that in proposition 7 of Mr Neuberger's propositions, set out above, it was asserted as a step in his argument that the statutory tenancy could arise only if the dwelling-house could be said to be "*their* dwelling-house"—that is, the dwelling-house occupied by Miss Sadler *and* Miss Lunt. Yet in *Howson v Buxton* "his goods" (the goods "of the tenant") applied to goods in respect of which the other joint tenant had no right of ownership.

In my opinion, the judgment of Scrutton LJ in *Howson v Buxton* shows that, where the strict application of the doctrine of joint tenancy would lead to unreasonable results, or results which the legislature is unlikely to have intended, it is permissible for the court to conclude that the legislature did not so intend: but that, instead, in such a case, the phrase "the tenant," where there is a joint tenancy, is to be read as meaning "the joint tenants or any one or more of them."

There is thus authority that the doctrine of joint participation by joint tenants is not a sacrosanct or immutable doctrine of statutory interpretation, where such phrases as "the tenant" and "the tenancy" are used. Where, then, is the line to be drawn? When does a suggested exception become a heresy? In the present case I believe that, for the purposes of section 3(1)(a) of the Rent Act 1968 the freedom from strict doctrinal restraint should, as in *Howson v Buxton*, be held to apply.

I would therefore hold that Miss Sadler became a statutory tenant on the expiration of the contractual tenancy on December 25 1976, and that she so remains. I would dismiss the appeal.

Agreeing, LAWTON LJ said in the course of his judgment: The object of the Rent Act 1968 was to give security of tenure to persons; and one of two joint tenants might have wanted it and the other might not. The mischief for which Parliament provided a remedy was eviction for reasons other than those which the Act deemed good. If Miss Sadler has to leave the flat because Miss Lunt did not want to stay there any longer that is not a reason which was specified in the Act. I am satisfied that section 3(1)(a) can be construed so that one of two joint tenants can become "the statutory tenant." Any other construction would defeat one of the objects of the Act.

A In deciding as I do, I have not overlooked the fact that under the contractual tenancy the plaintiff could look to both Miss Sadler and Miss Lunt for payment of the rent. On my construction he can only look to Miss Sadler; but if she fails to pay she loses her personal right to security of tenure and the plaintiff will be able to get possession. The Rent Act 1968 took away many of a landlord's rights at common law and was intended to do so for the benefit of tenants. The loss of a right to sue one of two joint tenants after the termination of a contractual tenancy is a minor matter compared to the restriction of the right to possession which the Act imposes on landlords.

B I too would dismiss the appeal.

Also agreeing, SHAW LJ said that if a joint tenant died during the term of a protected tenancy his survivors would unquestionably be entitled to qualify as statutory tenants although the potential sources for payment of rent would have been reduced in number. Moreover the ultimate recourse of the landlord where a statutory tenancy arose and there was default by the tenant or tenants was an action for possession. The prospect of some possible detriment to a landlord was not sufficiently formidable to deflect the interpretation of section 3(1)(a) from its plain objective.

C The appeal was dismissed with costs. Leave to appeal to the House of Lords was refused.

COURT OF APPEAL

November 18 1977

(Before Lord Justice MEGAW, Lord Justice ROSKILL and Lord Justice BROWNE)

HEGLIBISTON ESTABLISHMENT v HEYMAN AND OTHERS

D Estates Gazette May 13 1978
(1977) 246 EG 567

Rent Acts—Landlords' action for possession against statutory tenant—Tenant's wife living in another flat, but visited by him one or two nights per week—Whether tenant residing in subject flat or wife's flat or whether a "two-home" man—Whether suitable alternative accommodation available in wife's flat—Tenant's son, son's girl-friend and a young friend of family living in subject flat—Whether breach of covenants against subletting or carrying on a business by taking in boarders or lodgers—Whether E permitted user for immoral purposes by allowing son and his girl-friend to live as man and wife—Landlords' claims rejected under all headings.

This was an appeal from a decision of Judge Wakley at the Bloomsbury and Marylebone County Court dismissing a claim by the plaintiffs, Heglibiston Establishment, a company registered in Liechtenstein with limited liability, for possession of a flat at 2 Porchester Gate, London W2, against the first defendant, Leonard Alvin Heyman, the tenant, and three other defendants, his son, his son's girl-friend and a young man who was a friend of the family, all of whom were living in the flat. Judge Wakley dismissed the plaintiffs' F claim.

Norman Primost (instructed by Graham Harvey & Co) appeared on behalf of the appellants; David Van Hee (instructed by Wegg-Prosser & Farmer) represented the respondents.

Giving the first judgment at the invitation of Megaw LJ, BROWNE LJ said: This is an appeal from a decision of His Honour Judge Wakley at the Bloomsbury and Marylebone County Court given on October 28 1976. The claim was by the plaintiffs, a company registered in Liechtenstein, for possession of a flat, known as Flat D, 2 Porchester Gate, London W2, against the first defendant, Leonard Alvin Heyman, who was the tenant, and three other defendants G who were living there. The judge dismissed the claim, and also dismissed a counterclaim, which related mainly, or I think in the end entirely, to the disappearance of a cabin trunk. The plaintiffs appeal against the dismissal of their claim for possession. There is no cross-appeal as to the counterclaim.

The first defendant became tenant of the flat by assignment in 1964 of the term under a lease for seven years granted on October 6 1959 by Porchester Estates Ltd to a Miss Pitchford and a Miss Ware. When that lease expired in 1966 he held over and became a statutory tenant entitled to the protection of the Rent Acts. The other defendants are H Preston Ross Heyman, a son of the first defendant, Michael Andrew Newport, a friend of the Heymans, and Miss Mary Dobson; at all material times they were all three living in the flat.

At some time, apparently in 1975, the plaintiffs acquired the freehold reversion. As I have said, the lease is made between Porchester Estates Ltd as the lessor, "which expression shall include its successors in title where the context so admits," and two named ladies "called 'the lessees' which expression shall include their respective executors administrators and assigns where the context so admits." It demises J "all those four rooms with kitchen and bathroom on the top floor of the Mansion Building called or known as 2 Porchester Gate, Hyde Park in the County of London," with certain other rights; and it demises that property to the two ladies as joint tenants from September 29 1959 for a term of seven years at a rent of £400 a year. That rent I think has subsequently been increased by a decision of a rent tribunal.

The relevant covenants by the lessee are these.

Covenant 10:

> The lessee will not at any time during the said term use or permit the said flat or any part thereof to be used for the purpose of any art trade or business whatsoever or for any immoral K purpose or as a private mad-house or receptacle for lunatics or insane persons but shall and will use the said flat as and for a private residence only and in particular will not take in boarders or lodgers.

Mr Primost referred us also to the reference to "business" in clause 11, which he says is an aid to the construction of clause 10, but I do not find it so and I need not, therefore, read clause 11. For the same purpose he relies on a part of clause 13, which binds the tenant:

> Not to do or suffer to be done in the said flat or the said mansion any act or thing that may be or grow to be a scandal L nuisance or annoyance to the lessor or the other tenants of the said mansion or which may tend to injure the character of the said mansion as a private residence but on the contrary will as much as in the lessee's powers lie contribute to the respectability of the building.

In clause 14 there is another reference to user as a private residence.

The other relevant clause is clause 19. It provides:

> The lessee will not without the licence in writing of the lessor assign underlet or part with the possession of the demised premises or any part thereof (such licence not to be unreasonably withheld in the case of a respectable and responsible person) . . . .

I need not read the rest of the clause. M

The two clauses on which the plaintiffs rely in this case are clauses 10 and 19.

The facts are these. The flat, as appears from the lease, has four rooms. There are three bedrooms and a living and dining room, with a kitchen and bathroom. At the material time one bedroom was occupied by the first defendant, Mr Leonard Heyman, when he was there; one was occupied by Mr Preston Heyman, his son, and Miss Dobson, who were living together as man and wife, and one by Mr Newport. Mr Newport was an old school-friend of Mr Preston Heyman and had been living there for about seven

A years at the time of the hearing. Miss Dobson had been living there for about a year or 18 months. The further facts as found by the learned judge, supplemented by the notes of evidence, were these.

> The facts are simply stated. I accept the evidence of L A Heyman which is not seriously challenged. He is an American citizen and has lived at the flat since 1964. During the last few years he has been absent on business from time to time. Four years ago his marriage ran into difficulties. Though on good terms, Mr L A Heyman and his wife agreed that they live better if they do not see too much of each other. Mr Heyman and his wife treat each other in a civilised and decent manner. Mr Heyman bought for his wife a flat in Rotherhithe at 11 Custom House
> B Reach.

Pausing there, the evidence was that he had bought it about four years before the hearing, which would have been some time in 1972, and that he paid £9,000 for a long lease of that flat. The judge went on:

> This was a smaller more modern flat than the one in Porchester Gate and he visits her there one or two nights per week. He spends two or three nights per week on average at Porchester Gate. He spends a further two or three nights per week in Blackburn. He also travels extensively in the United States and Europe and is away for quite long periods.

There was evidence, on which Mr Primost strongly relied,
C that in the telephone directory Mr Leonard Heyman's address was shown as 11 Custom House Reach; and in the annual returns for 1974 of the two companies of which he was a director his "usual residential address" was also shown as 11 Custom House Reach. At Porchester Gate the rent and rates were paid by Mr Leonard Heyman, and he owned the furniture. But a telephone call box had been installed there at about the same time as the Rotherhithe flat was bought and the listed subscriber for that telephone was Mr Preston Heyman. The explanation given by Mr Leonard Heyman was that his son was a professional musician who had to
D be on call all day and night. Mr Preston Heyman, Mr Newport and Miss Dobson pay no rent, but they share the household expenses, including the gas and electricity bills and the food. All of them, including, I gather, Mr Leonard Heyman when he is there, eat together. Mr Leonard Heyman said "Preston works out who has to pay what". Mr Primost suggests that this is evidence that Mr Preston Heyman is the householder; but this seems to me absurd. As my Lord said in the course of the argument, he is simply the treasurer. Mr Leonard Heyman keeps the bulk of his clothes and personal possessions at Porchester Gate, but he has some clothes and a spare razor and toothbrush at Rotherhithe.
E The only person at Porchester Gate who is on the electoral roll is Mr Newport; but, as I have said, Mr Leonard Heyman is a citizen of the United States of America.

The claim is put on a number of grounds by Mr Primost for the appellant plaintiffs. First, he says that the first defendant, Mr Leonard Heyman, is not entitled to the protection of the Rent Acts, because he is not living at the Porchester Gate flat as his home. Secondly, he says that suitable alternative accommodation has been offered to him in the shape of the flat at 11 Custom House Reach in Rotherhithe. He further says that, that alternative accommodation being suitable, it would have been reasonable for the judge to
F make an order for possession. Thirdly, he relies on three breaches of covenant: first, a breach of clause 19 against subletting or parting with possession; secondly, a breach of clause 10 by taking in lodgers and boarders and not using the premises as a private residence and carrying on a business there; and thirdly, again in breach of clause 10, permitting the premises to be used for an immoral purpose.

As to Mr Primost's first ground, that the first defendant is not entitled to the protection of the Rent Acts, he says, rightly, that those Acts only apply when a tenant is residing in the dwelling-house in question in the sense of living there as his home. He says either that Mr Leonard Heyman's home
G is Rotherhithe and not Porchester Gate or that he is a "two-home" man and the judge's decision in this case is an extension of any previous "two-home"-man decision.

The judge dealt with this matter as follows. Having set out the facts which I have already read, he goes on:

> It is therefore said that he is no longer resident in the Porchester Gate flat. I am wholly unable to accept that. In my view, he always intended to remain in Porchester Gate and therefore he has not lost the protection of the Rent Acts. I have heard the authorities cited on the point by Mr Primost.

He then referred to various authorities, several of which have been cited to us, and went on:

> H None of these cover the present case. Mr Heyman is simply a man who uses the Porchester Gate flat as his home base and occasionally goes to visit his wife in Rotherhithe. It is not even right to call him a two-home man. He stays with his wife when he visits her for the night. I do not think that the Rotherhithe flat is his home. It is her home. He keeps a toothbrush, a razor and some casual clothes there, but I do not think it is his home at all. The question has been raised and there is some doubt on whom does the onus of proof lie on this point, as to whether Mr Heyman has given up his residence at Porchester Gate. The question has not been settled. I am content to assume that it is upon the tenant and I am satisfied that the onus has been discharged. There are one or two factors which do not support this view.

J Then the judge referred to the matters of the telephone directory and the annual returns, and to the fact that the gas and electricity bills for Porchester Gate were paid, not by Mr Leonard Heyman, but by his son and Mr Newport and Miss Dobson. He went on:

> In my view this does not displace the evidence that he regards Porchester Gate as his home. He keeps the bulk of his clothes and his personal possessions there. He pays the rent and rates and he owns the furniture. Therefore in my view he is not a two-home man. He only goes to Rotherhithe to visit his wife. He has always resided in Porchester Gate and it is his intention to continue to
> The second issue is the question of the Rotherhithe flat as his residence and is entitled to the protection of the Rent Acts in respect of it. K

Mr Primost says that no reasonable judge knowing the law could have found that Mr Leonard Heyman was living at Porchester Gate as his home. He says that the only reasonable inference is that his home is at 11 Custom House Reach. I do not find the authorities cited by Mr Primost on this part of the case of any help. They were all dealing with problems raised by a "two-home" man, and the question here is whether the judge was wrong in finding that Mr Leonard Heyman was a one-home man, that home being at Porchester Gate. I would only refer to what Lord Evershed MR said twice in *Beck* v *Scholz* [1953] 1 QB 570 at pp 575 and 577— L that the question which the judge had had to decide in that case (and in my view the question which the judge had to decide in this case) was "a jury question." I can see no ground whatever for saying that the learned judge decided this question wrongly. I think he was right.

As to the second ground of appeal, Mr Primost claims that suitable alternative accommodation was available to Mr Leonard Heyman—that is, 11 Custom House Reach—and that it would have been reasonable to make an order for possession. On this point the judge said this:

> The second issue is the question of the Rotherhithe flat as suitable alternative accommodation for Mr Heyman. Apart from M size, it is very far from being suitable. I do not think it is suitable. I do not think it is right to order Mr Heyman to live with his wife there. Though they see each other from time to time, for the sake of their marriage, and their health, it would be absurd for her to live with him as her husband, in that flat. It would be wholly offensive to all ordinary considerations of decency. If I were to make such an order, it would exclude the son, Mr P R Heyman; plainly there would not be room for him.

Mr Primost accepts that he can only challenge this finding if he can show that the judge misdirected himself. I confess that I am not clear what the misdirection is said to be. Mr Primost said that here there was no medical evidence

A that Mr Leonard Heyman and his wife could not live together, and that the judge was wrong in holding that it would not be right to order him to live with his wife in a case where they were regularly living together two nights a week, and there was no evidence that he could not live there four or five nights a week. I think it helpful to refer to one or two passages in Mr Leonard Heyman's evidence. He said that the marriage went wrong and they lived apart. He went on: "There is not room at 11 Custom House Reach. It would not work out." He then referred to certain medical disabilities of his wife and said that stress raises her blood pressure. He says that they had made an arrangement to separate. I
B can see no misdirection. I think this, too, was a question of fact, and there is, in my judgment, no ground whatever for interfering with the judge's decision. There was no misdirection, in my view, and ample evidence to support the judge's finding.

I come now to the alleged breaches of covenant. The judge dealt with the alleged breach of clause 19—that is, subletting or parting with possession. He said:

> The third issue is the question of the breaches of covenant. As to the covenant against subletting, Mr Heyman's son, Preston, lives at the flat as well as his father, together with Mr Newport, who is a family friend. It is quite plain that Mr Newport is
> C regarded as another son by Mr L A Heyman. He has been treated as part of the family ever since he left school seven years ago. It is absurd to suggest that there is any breach of the covenant against subletting. He lives as part of the family and he pays no rent. He shares part of the family expenses.

It does appear from the evidence, as I have said, that he does have a bedroom to himself; but he eats with the rest of the family. Mr Primost submits that there has at least been a subletting or parting with possession of the one bedroom occupied by Mr Newport. He submits that Mr Newport had exclusive possession of that room in exchange for a money consideration in the form of an agreement to share
D the household expenses. I think this submission breaks down at all points. I am not satisfied that there was any intention to create any legal relationship, still less any legal relationship between Mr Newport and Mr Leonard Heyman. I am not satisfied that Mr Newport had either in fact or in law exclusive possession of his room. Mr Primost said that there was some evidence that he had a key of his bedroom, but this does not appear in the notes of evidence. And I cannot possibly regard the agreement to share household expenses as anything in the nature of rent. I entirely agree with the judge's decision on this point.

The next allegation is of a breach of clause 10. Mr Primost
E says that Mr Leonard Heyman had broken his covenant to use the flat as a private residence only and not to take in boarders or lodgers and not to carry on any business there—the "business" suggested being, I understand, the taking in of boarders or lodgers.

The judge said this:

> On the question of the covenant, not to take in boarders or lodgers, of which breach is claimed in respect of Mr Newport and Miss Dobson, I have been referred to *Tendler v Sproule* [1947] 1 All ER 193 on the question of what amounts to a paying guest. That was a decision concerning a few persons living as part of the family. That case was decided upon very
> F different circumstances. I have no difficulty here. Miss Dobson is treated as if she were Mr Preston Heyman's wife and Mr Newport is treated as his brother. Therefore in my view, they are not lodgers or boarders. The test to be adopted is the reasonable man test. I have also been referred to *Segal Securities Ltd v Thoseby* [1963] 1 QB 887. Mr Justice Sachs decided that case on the point of waiver. Therefore, what he said on living with a paying guest was strictly not binding. It is obiter, but nevertheless it is strong persuasive authority. I agree with the approach which he made when he said at p 894 that the taking of a single paying guest who shared the family life would not be a breach of a covenant to use a house as a private residence only. There is no breach here. These are friends sharing accommodation and in consequence there can be no breach. In my view, this covenant has not been broken.

Mr Primost did not suggest that Mr Preston Heyman was G
a boarder or a lodger or a paying guest, but he submits that Mr Newport and Miss Dobson were. He says that *Tendler v Sproule*, the case to which the judge referred, covers this case and binds us to hold that there has been a breach of this covenant. The only statement of the facts of that case appears at page 194 letters B-C, where Morton LJ, delivering the first judgment, said: ". . . all that we know in the present case is that during the tenancy the tenant took in (to quote the judge) 'two paying lodgers'." That is all that we know about the facts of that case, and I cannot regard that as laying down any general principle. Mr Primost relies strongly on what Morton LJ said on the same page, 194, between H
letters G and H, about the previous decision in the case of *Porter v Gibbons and Bissett* (1904) 48 Sol Jo 559. He said:

> I do not think that the decision of Mr Justice Kekewich in *Porter v Gibbons* was well founded. The case is very shortly reported, and it may be that there were matters before the judge which do not appear in the report, but according to the report the defendant, who was the lessee, ". . . had received friends and acquaintances who, with their families, stayed for long periods and contributed to the cost of carrying on the establishment by a fixed weekly or monthly payment. It seems to me that this was a breach of the covenant not to use the said premises except as a private dwelling-house." J

I think that the present case is distinguishable from *Porter v Gibbons*. It seems to me that there is all the difference between a person who makes fixed periodical payments (as was apparently done in that case) and a person who merely agrees to share part of the expenses of a single household. It should be noted that *Porter v Gibbons* went to the Court of Appeal (1904) 48 Sol Jo 814, and the appeal was dismissed; but the court expressly left open the question whether the reception of paying guests was a breach of covenant not to use the house except as a private dwelling-house. I accept the approach of Sachs J (as he then was) in K
the case of *Segal Securities Ltd v Thoseby* [1963] 1 QB 887, particularly at p 894. I need not quote the whole of the passage, which is set out in the headnote; but Sachs J said this: "It is in each case a question of fact and of degree whether the taking in of paying guests is of an order that, having regard to all the circumstances, constitutes a breach of the covenant in question." In my judgment it was a question of fact and degree for the county court judge in the present case to decide whether or not Mr Newport and Miss Dobson were boarders or lodgers. He found that they were not; and again I can see no reason whatever for interfering with his decision. If they were not, no question of L
carrying on a business or not using the premises as a private residence only arises.

Finally, the last allegation is of a breach of another part of clause 10 in that the defendant Mr Leonard Heyman is alleged to have permitted user of the premises for an immoral purpose by Mr Preston Heyman and Miss Dobson, who were living together there as man and wife. Mr Preston relied on the decision of the Divisional Court in *Uphill v Wright* [1911] 1 KB 506. I express no opinion whether or not that case would be decided in the same way today; but it seems to me that the facts of that case are very different from those in the present case and that the cases are clearly M
distinguishable. That was an action for the rent of a flat let to a woman. She said, but the landlord's agent denied, that he knew when he let the flat to her that she was a prostitute, and the county court judge found that he did not know that. But it was found that he did know that she was a "kept woman" and that the rent of the flat was going to be paid by her protector. The Divisional Court held that this was merely a difference in degree. The question in the present case is whether two people who, though not married, live together in a flat as man and wife for a substantial period are "using it for an immoral purpose" within the meaning

of this covenant. In my opinion they are not. I think that the main object of this covenant is to prevent the flat being used for an immoral *purpose*—for example, being systematically run as a brothel or used for prostitution. Mr Preston Heyman and Miss Dobson were (no doubt with the permission of Mr Leonard Heyman) using the flat as and for the purpose of their home, and if in the course of that use they slept together that is not, in my view, enough to constitute using the flat for an immoral *purpose,* nor to amount to a breach of this covenant. As there was, in my judgment, no breach of covenant, the question whether or not it would have been reasonable to make an order if there had been a breach of covenant does not arise.

I would dismiss the appeal.

Agreeing, ROSKILL LJ said in the course of his judgment: I add a few words on the last point—the suggestion that there was a breach of the covenant not to permit the use of these premises for immoral purposes on the part of the respondent Mr Heyman senior because his son, Mr Heyman junior, and Miss Dobson, to Mr Heyman senior's knowledge, were living in that flat together as man and wife. *Upfill v Wright* has been relied on as authority for that proposition. As my Lord has said, the facts there were a very long way away from the facts in the present case. That decision does not bind this court. It is unnecessary to consider whether on its facts that case was wrongly or rightly decided, or whether it would be decided the same way today, or whether the problem there in question—which is not the same as the present problem—would again be resolved by reference to a statute of the 13th year of Edward I and to the Litany in the Book of Common Prayer. We are not concerned with matters of that kind. We are concerned simply with whether or not, as was alleged by amendment to the particulars of claim, the respondent Mr Heyman senior permitted his son and Miss Dobson to use these premises for immoral purposes merely because they were living there as man and wife. All I can say is that if that user is to be interpreted as using premises for an immoral purpose there must be a great many breaches of covenant being committed in London and elsewhere today. In my judgment the point is without substance or merit, and this appeal fails and should be dismissed.

Also agreeing, MEGAW LJ said: I add a few words about *Upfill v Wright.* As a matter of historical interest, the citation, in the judgment of Darling J as a relevant authority in that case, of an alleged statute of the 13th year of Edward 1, statute No 4, is suspect. The learned reporter in that case, in a footnote at the bottom of 511, pointed out that in the 19th year of King Edward III, in a Year Book reporting the cases in that year, it had been said (I should have thought probably correctly) that the purported statute was not a sealed statute at all but was "made by the prelates themselves." That, however, is only a matter of historical interest. So far as *Upfill v Wright* itself is concerned, if, indeed, Mr Primost were right in his submission that the *ratio decidendi* of that case required that, on the facts of this case, it had to be held that there was here a breach of clause 10 of the lease, I would have no hesitation in saying that to that extent, at any rate, the decision in *Upfill v Wright* was wrong and should not be followed hereafter.

ROSKILL LJ expressed agreement with these comments on *Upfill v Wright.*

The appeal was dismissed with costs. An application by the appellants for leave to appeal to the House of Lords was refused.

COURT OF APPEAL

March 16 1978

(Before Lord Justice STAMP, Lord Justice ROSKILL and Lord Justice CUMMING-BRUCE)

LEGAL & GENERAL ASSURANCE SOCIETY LTD v KEANE

Estates Gazette June 3 1978

(1978) 246 EG 827

Important rulings by Court of Appeal of interest to rent officers and rent assessment committees—Appeal from county court judge's interlocutory order in favour of tenant for discovery of information relating to costs of services in action by landlords for rent—Action followed registration of rent—Tenancy agreement provided for a stated rent or such greater rent as should be registered—No reference to a "service charge," but covenant by landlords to provide certain services—Held by Court of Appeal that there was no service charge "payable as part of, or in addition to, the rent" within meaning of section 90 of Housing Finance Act 1972—Held also that discovery should be refused as information required had no relevance to issue in court proceedings—Tenant cannot reopen financial calculations of rent officer or rent assessment committee in county court—Appeal allowed

This was an appeal by the landlords, Legal & General Assurance Society Ltd, plaintiffs in an action for rent against the defendant, Walter S Keane, from an interlocutory order by Judge Llewellyn at Bloomsbury and Marylebone County Court. The order was for the discovery of information relating to the costs during three successive years of providing services at Northways, College Crescent, London NW3, where the defendant was the tenant of flat 39.

P Millett QC and P de la Piquerie (instructed by Lawrence Graham Middleton Lewis) appeared on behalf of the appellants; L Price QC and J C Harper (instructed by Kingsley, Napley & Co) represented the respondent.

Giving the first judgment at the invitation of Stamp LJ, ROSKILL LJ said: Notwithstanding my gratitude for Mr Price's argument, I am clearly of the view that the order made by the learned county court judge cannot be sustained and that this interlocutory appeal must succeed. The matter arises in a curious and unusual way. The appellants, the plaintiffs in the county court action, are the well-known insurance company, Legal & General Assurance Society Ltd; they at the material time were the landlords of a large and well-known block of flats called Northways at Swiss Cottage, NW3. The respondent, the defendant in the action, is a Mr Walter Keane who, at the material time, was tenant of flat 39. As unhappily is not uncommon in London today, problems have arisen between these landlords and their tenant as to rent increases and service charges. Wherever the rights and wrongs of this dispute may lie, it is apparent that Mr Keane feels bitterly aggrieved because of certain rent increases to which he has been subjected and which he objects to paying. He has, it seems, the support of tenants of other flats in that block. The matter has become more acute since this dispute arose because the appellants have sold their reversionary interest in the block to others, who have since sought to impose a further increase upon the tenants.

The dispute arises because the learned county court judge made an interlocutory order that the plaintiffs should give extensive and far-reaching discovery to the defendant in support of an allegation by the defendant that certain rent increases authorised by the rent officer and certain further increases subsequent to the sale of the reversion had been improperly arrived at. I must read some parts of the order made by the learned judge on January 5 1978. He ordered that an application by the defendant dated December 13 1977 —that was a second application—be granted and that "there be specific discovery by the plaintiffs of a summary in writing of the relevant costs of service charges at Northways, College Crescent, NW3, in the accounting year ending (1) December 31 1976, (2) December 31 1975, (3) December 31 1974." The learned judge granted a stay pending appeal and gave leave to appeal.

The learned judge gave his reasons for making this very remarkable order in these terms:

> Although the present action is framed as a claim for rent and does not contain a claim for service charges, it appears that the rent in question was fixed as a fair rent by the rent officer, and that—

and this, with respect, is wrong—

> the defendant (tenant) appealed to the rent assessment committee who then fixed a different fair rent. It is also clear, and indeed conceded by counsel for the plaintiff, that figures relating to service charges were submitted by the plaintiff (landlords) to the rent officer and to the rent assessment committee so that the rent as fixed contained in its composition an element in respect of service charges. The tenancy agreement between the parties refers to the rent, including a reference to "registered rent." It has appeared from the documents produced and disclosed that two widely different figures for service charges were put in by the plaintiff, one to the rent officer and the other to the rent assessment committee. It now appears that a third figure, again significantly lower than either figure given to the rent officer or to the rent assessment committee, has been disclosed by the plaintiff as service charges for the same year, viz 1976. If that third set of figures had been produced to the rent assessment committee it seems certain that the figure for a fair rent would have been lower. What is the position in this court? This court is not a court of appeal from the rent assessment committee. Nevertheless this court is not going to allow the use of the word "rent" to act as a facade to prevent investigation of the service charge element contained in such rent from being investigated in the present action. Sufficient documents have been disclosed to raise a query as to the correctness of the service charge element included in the term "rent" to make it relevant for that matter to be investigated in the present action.

It is against that background that I have to consider the issue the learned judge had before him. The defendant had entered into a tenancy agreement with the plaintiffs which bore the date May 1 1975. The relevant clauses are clauses 1 and 3:

> 1. The Landlord Agrees to let and the tenant agrees to take the premises (hereinafter called "the flat") described in the First Schedule hereto for a term of three years commencing on May 31 1975 at the yearly rent of £830 or such greater rent as shall for the time being be registered as the rent for the flat under the Rent Act 1965 payable in advance in equal instalments on January 1, April 1, July 1, October 1, the first payment to be made on July 1 1975 and to be the proportion thereof computed from May 1 1975 to June 30 1975 and the last payment to be apportioned as necessary.
>
> 3. The Landlord Agrees with the tenant so long as the tenant shall pay the said rent and observe and perform the agreements on the part of the tenant herein contained—

and then follow certain obligations; I can pass over (i) and (ii) and go to (iii):

> To use its best endeavours throughout the tenancy to maintain at all reasonable times an adequate supply of hot water for domestic purposes—

and there then follows an obligation to keep the radiators sufficiently and adequately heated. I need not read the rest except for the proviso:

> that the landlord shall not be liable in damages or otherwise in the event of any interruption of such services

and (iv):

> Unless prevented by circumstances outside its control: (a) To appoint and employ throughout the tenancy porters for attending in and about the entrance hall(s) staircase(s) and passage(s) in the building. (b) Sufficiently to light and clean the said entrance hall(s) staircase(s) and passage(s). (c) Regularly throughout the tenancy to cause the refuse bins to be emptied and cleared.

Those are the express—and the only express—contractual obligations assumed by the landlords towards the tenant.

I go back to clause 1. That clause, which is all-important to my mind, talks about, and only about, rent; there is no reference there to service charges; still less is it the type of clause one often finds, namely that there shall be a separate charge for service charges, which shall be treated as and recoverable as rent; the only liability of the tenant was to pay rent either at the sum stated or as such greater sum as might from time to time be registered as the rent under the relevant sections of the Rent Act.

That agreement is dated May 1 1975. Towards the end of 1976 the plaintiffs wished to get an increased rent registered, and applied to the rent officer. The application form is dated September 22 1976. Attached to it is a calculation of the total expenditure relevant for the purpose of establishing the plaintiffs' claim for an increase. Paragraph 15 of the application form shows that the applicants were claiming that £840 per annum of the new rent sought of £2,000 per annum was attributable to the provision of central heating; one finds in the bundle of papers a calculation designed to justify that figure of £840. The case went before the rent assessment committee, and in due course the new rent was registered. One finds in the papers the entry in the rent register and the plaintiffs' notice dated February 16 1977 that the commencement date of the new rent fixed by the rent officer (which was at the rate of £1,670 per annum) would be January 19 1977.

The first amount the respondent objects to paying is the proportionate part of that increase attributable to the period January 19 to March 31 1977; the second is the proportionate part attributable to the period April 1 to June 30 1977. The figure in respect of the first period is £121 and the figure in respect of the second period is £152. The ground on which it is sought to say that the respondent is not liable to pay that increase is this: since that increase was fixed figures have come to light and have been made available which (it is alleged) show that the rent officer was proceeding on figures which were erroneous (the defendant, in his own defence drafted by himself, uses harsher language, but I ignore that). Different figures have been made available relating to various matters; I do not propose to look at them—it is not necessary to do so. The question is: there being that alleged information showing different figures from those put before the rent officer and assumedly taken into account by him in fixing the rent at £1,670 per annum, is it open to the defendant, by way of defence to or set-off or counterclaim against the plaintiffs' claim for the proportionate part of the increased rent, to say "the rent officer proceeded upon a basis of false or erroneous figures; I am going to ask you, the county court judge, to reopen this matter by investigating my set-off and counterclaim in the hope that I can satisfy you either that I ought not to pay rent at that increased rate at all or, if I am liable to pay it, that I ought to be able to set off, or counterclaim against, that liability a figure representing the shortfall in the value of the services which I claim those figures now reveal."

Reliance is placed on section 90 of the Housing Finance Act 1972, and the learned county court judge appears to have founded a large part of his judgment and the relevant part of the order upon that section. I have already pointed out that clause 1 of the agreement makes no reference to service charges as such; there is but a single liability: to pay rent. The contractual obligation, I repeat, is to pay the specific sum or any increased registered rent. What is said—and Mr Price put this in the forefront of his argument—is that the tenant has an unqualified right to the discovery to which the learned county court judge held him to be entitled under section 90 of the Housing Finance Act 1972.

I ought to read the relevant subsections:

(1) Where the service charges which are payable by the tenant of a flat in any calendar year, or which are demanded from the tenant as being so payable, exceed the amount specified in subsection (2) of this section, the tenant shall, in accordance with this section, be entitled to obtain a summary in writing of the relevant costs in the accounting year ending in or with that year, certified by a qualified accountant as being in his opinion (a) a fair summary of those costs, set out in a way which shows how they are or will be reflected in demands for service charges, and (b) sufficiently supported by accounts, receipts and other documents which have been produced to the accountant, and the certificate shall identify the accounting year to which the summary relates.

I shall not read the next two subsections.

(4) The tenant shall exercise the rights conferred by this section by serving on the landlord a request in writing which states the calendar year to which the request relates, and which is so served not later than 12 months after the end of that year.

Such a notice purports to have been served.

I turn to the all-important provision, and that is the definition of "service charge" in section 90(12):

"Service charge" means any charge for services, repairs, maintenance or insurance, being a charge which is payable as part of, or in addition to, the rent, and which varies or may vary according to any costs (including charges for overheads) incurred from time to time by or on behalf of the landlord or any superior landlord.

I draw attention to the opening words; Mr Price went straight to the third line, but that is no way to construe this definition, one cannot start in the middle; one must follow the definition through from the beginning, and "service charge" is defined as meaning "any charge for services, repairs, maintenance or insurance, being a charge which is payable as part of, or in addition to, the rent." Pausing there, I am unable to construe clause 1 of the present tenancy agreement as creating a service charge which is payable as part of, or in addition to, the rent. As my Lord pointed out, everyone is familiar with the type of agreement which creates a separate obligation on a tenant to pay a service charge; everyone is also familiar with the type of agreement which not only imposes such an obligation upon a tenant but makes it payable as if it were rent, which of course gives the landlord additional rights of recovery; but this definition has no application to the present case, where the tenancy agreement makes no provision for a service charge. It is true that this rent includes provision of particular contractual services by the landlord to the tenant but that does not mean that there is a service charge payable within section 90. As my Lord pointed out, the mischief against which section 90 is aimed is not one which can arise in a case such as this. That mischief was an attempt by landlords to evade Rent Act control by swinging increases on to service charges and then saying that those service charges were free from any form of statutory control by rent assessment committees, rent officers or courts. But an agreement such as clause 1 does not permit such mischief because the relevant rent will, in cases such as the present, be a rent which is controlled by the Rent Acts. As any increase has to be a registered rent, that increase will have to go through the sieve of examination by the rent officer before it can be approved. Therefore the safeguard which section 90 was designed to secure is not something which is necessary for the tenant's protection in the present case. The Rent Acts themselves give him all the protection which he needs or to which he is entitled. Accordingly in so far as Mr Price's submissions and the learned county court judge's order are based on section 90, neither can be sustained. If, therefore, the defendant is to get the discovery he seeks he must show that he is entitled to that discovery under the ordinary rules which govern the disclosure and production of documents in litigation.

Mr Millett, very fairly, did not take the point which would have been unanswerable, that many of the documents sought in the application do not exist; of course it is a condition precedent to an order for discovery of specific documents that the applicant must show that those documents exist and are in the possession or power of the person against whom the order is sought. Mr Millett, rightly as I think, sought to resist this application on wider grounds; he said that even if these documents existed it would be utterly wrong to order their discovery and production because they went to no issue in these proceedings. It must be said at once that when one looks at the defence and counterclaim put in by the respondent it is very difficult to see how these documents go to any issue purporting to be thus raised. Where a litigant in person drafts his own documents the court extends to him the utmost indulgence, and I have read and reread that document in the hope of extracting from it some defence or some possible counterclaim which justify upholding the order of the learned judge; but I am unable to do so, and I think Mr Price at an early stage realised that that must be so. Ultimately counsel evolved a draft amended defence, set-off and counterclaim; one cannot but admire the speed and clarity with which it was produced. The fairest thing for the respondent is to treat this document as properly before the court and then consider whether, if it had been before the court, these documents, discovery of which is sought, would have become relevant under the ordinary discovery rules.

I say at once that in so far as any part of the new defence and counterclaim turns upon section 90 of the 1972 Act, that must fail for the reasons already given. The real question is whether the defendant can get anything from paragraphs 7 and 8 of the draft amended pleading. As I have already said, what he is seeking to say is that he is entitled, by way of set-off or counterclaim, to invite the court to conduct an inquiry into the suggested relationship between the costing figures supplied by the landlords to the rent officer and the contractual obligations of the landlords towards him under the tenancy agreement. To my mind it is at this point that the whole argument breaks down. The draft amended pleading reads thus:

7. Thereafter the plaintiffs could not, consistently with the tenancy agreement and the relevant statutes and the professed basis upon which they had applied for and had obtained an increase in the registered rent, provide services which (a) did not cost what they had professed to the rent officer (and he had accepted) or alternatively (b) did not constitute services equivalent to those which they had so professed. 8. In the periods in question (namely January 19 1977 to March 31 1977 and April 1 1977 to June 20 1977) the plaintiffs have neither spent such sums as would amount to the sums claimed in paragraph 6 of the particulars of claim whether the said premises be considered separately or as a just proportion (by rateable value) of the whole block of flats known as Northways aforesaid nor provided such equivalent services.

Mr Price found himself constrained to say that once there was an increased rent registered, the plaintiffs became under a contractual obligation to provide services equivalent to those costed in the figures supplied to the rent officer. As my Lord put it, this argument involved that once the rent officer had accepted those figures there arose a new implied unilateral covenant by the landlords to provide services equal in value to those to which those figures referred. That involves, as Cumming-Bruce LJ pointed out, that there arises at that moment a new contractual obligation to provide services to the tenant which *ex-hypothesi,* was not part of the original contractual obligation when the tenancy agreement was entered into. That argument is incapable of being sustained. The contractual obligations are those for which the contract provides; they included a contractual obligation to pay rent at the increased rate as registered. But the structure of the relevant statute for the protection of tenants while giving some advantages to landlords, depends upon all questions of cost being resolved through the machinery of the rent officer; once they have been through his sieve, and once the figures have been accepted by him or by the relevant rent assessment committee, it would be driving a coach and horses through the Act if we were to entertain an argument that any tenant sued for the balance of the increased registered rent can seek to reopen in the county court the financial calculations upon which the rent officer or rent assessment committee arrived at their conclusions. They may be right or they may be wrong, the figures they were supplied with may be right or they may be wrong; the place to challenge them is before the rent officer or before the rent assessment committee; it cannot be allowed in the county court. Apart from the error which I think the learned county court judge made with reference to section 90 of the 1972 Act, he appeared to think it was open to him to allow a tenant, by way of set-off or counterclaim or in answer to a claim for the balance of increased rent, to reopen the matter by challenging figures which had been accepted by the rent officer or rent assessment committee. In my judgment that is quite wrong, and this order for further discovery ought not to have been made.

Agreeing, CUMMING-BRUCE LJ said: Under this contract between landlord and tenant there was an obligation upon the tenant to pay such rent as was therein referred to, which was the yearly rent of £830 or such greater rent as shall for the time being be registered as the rent for the flat under the Rent Act 1965. On the facts of this case the landlords applied to the rent officer for an increase of rent; an increase was registered, and the rent officer noted that a specified sum in respect of services provided by the landlords was included in the rent which he had found. Under this contract the rent registered becomes the rent the tenant is liable to pay, and the finding of the rent officer as to the amount of that rent is conclusive. The learned county court judge, while declaring accurately (if I may say so with respect) that his court was not a court of appeal from the rent assessment committee, proceeded to take a course which, if pursued, could only involve first an inquiry into the basis and accuracy of the rent officer's or rent assessment committee's determination and then, if the learned county court judge decided that those calculations were inaccurate and the service charge element included by the rent officer was too much, the only effect he could give to such a finding would be to hold that the rent payable by the tenant was something less than the rent registered by the rent officer. That cannot be an appropriate exercise of the jurisdiction of the county court, because the result has then to be to review the rent which has been conclusively determined by the rent officer.

For the reasons stated by Roskill LJ, I agree the appeal must be allowed. For myself I agree that section 90 of the Housing Finance Act 1972 has no relevance where a rent has been determined by a rent officer. But I decide nothing on the question whether, having regard to the clear wording of section 90, it is possible to invoke that section in aid of proceedings before a rent assessment committee; my view is that the service charge payable contemplated by section 90 is something wholly different from anything contemplated in proceedings before a rent assessment committee, but it is not necessary to determine that for the purposes of this case.

I would agree that the appeal should be allowed.

Also agreeing, STAMP LJ said: I would add only two words of my own because we are differing from the view of the learned county court judge.

It appears that the learned judge came to the conclusion he did because he thought the landlords were somehow seeking to avoid section 90 of the Housing Finance Act 1972; I deduce that from the fact that he remarked "Nevertheless this court is not going to allow the use of the word 'rent' to act as a facade to prevent investigation of the service charge element contained in such rent." Whatever defect, if any—I underline those words—there may have been in relation to the fixing of the new fair rent, there is in my judgment no question of any such avoidance as the learned judge thought there was. He apparently did not appreciate that the purpose of section 90 was to prevent the exploitation of tenants by excessive demands in reliance on a covenant by a tenant to pay a share of the service charges incurred by the landlord. There is no such covenant in the instant case, there being simply a fixed rent. There is no need and no room for that particular form of protection where the rent is of a fixed sum and nothing more can be payable until after a new rent has been fixed by machinery designed to secure an increase to a figure which is a fair rent.

I would accordingly also allow the appeal

The appeal was allowed with costs in the Court of Appeal and below.

COURT OF APPEAL

February 15 1977

(Before Lord Justice CAIRNS, Lord Justice ROSKILL and Lord Justice BRIDGE)

HANSTOWN PROPERTIES LTD v GREEN

Estates Gazette June 10 1978

(1977) 246 EG 917

Rent Acts—Application of doctrine of undisclosed principal—Lease to two named tenants—Whether mother of one of the tenants was entitled also to claim a tenancy on the ground that she was an undisclosed principal—Extent to which parol evidence can be admitted to vary terms of written document—Crucial importance to landlord of knowing identity of tenant, especially when Rent Acts apply and there is a possibility of the emergence of a statutory tenancy—Deputy county court judge wrong to have admitted evidence in support of mother's claim—Appeal allowed

This was an appeal from a decision of Deputy Judge Rose at Brentford County Court refusing an order for possession sought by the plaintiffs, Hanstown Properties Ltd, against the defendant, Mrs Green, who, the plaintiffs claimed, had no rights as a tenant in respect of a flat at 5 Oaklands, Argyle Road, London W13.

Douglas Day (instructed by D J Freeman & Co) appeared on behalf of the appellants; the respondent, Mrs Green, appeared in person.

Giving judgment, CAIRNS LJ said: This is an appeal from a decision of Deputy Judge Rose at the Brentford County Court in proceedings for a possession order. The premises concerned are 5 Oaklands, Argyle Road, W13, a flat which is a dwelling-house within the Rent Acts. It was let by the plaintiffs' predecessors in title under a lease dated August 11 1967 for a period of just over five years to September 29 1972 and the lessees named in the lease were Miss Green and Miss Carrol. After the expiration of the lease the tenants held over as statutory tenants, and it is common ground that Mrs Green, who was the defendant in this action and is the mother of Miss Green, lived with Miss Green and Miss Carrol. There is no doubt that Miss Green and Miss Carrol were statutory tenants. What happened was that early in 1976 Miss Green and Miss Carrol both moved out, and Mrs Green stayed there. In due course an action for possession was brought against her by the plaintiffs, and by her defence she claimed that she was a tenant along with Miss Green and Miss Carrol. That defence was accepted by the deputy judge, who gave judgment for Mrs Green and refused an order for possession. It is from that judgment that the plaintiffs appeal to this court.

There was correspondence before the proceedings were taken, and in the course of that correspondence Miss Green described her mother as "staying in the flat as a guest of Miss Green and Miss Carrol"; and a request was made to the plaintiffs, the landlords, that Mrs Green should be allowed to stay there until she found other accommodation. Mrs Green has conducted her own case in this court, and she has told us that all along she has been most anxious to obtain other accommodation and has no wish to stay in the premises concerned in this action any longer than is necessary. But she says that she has not been able to find any other accommodation, and asks the court to affirm the order of the learned deputy judge allowing her to stay there meanwhile.

It seems to me that the only question which this court has to consider is whether this lease under which two persons are named as lessees can be given effect as a lease to three people on the basis that Mrs Green was an undisclosed principal and, accordingly, that the lease was made by the two who were named on behalf of themselves and Mrs Green. Mrs Green gave evidence at the trial saying: "I obtained the flat for the three tenants. We all moved in together. I had had an accident—I could not travel. Therefore the other two completed the lease. I was, I say, a tenant from the start. The others signed on my behalf." If that had been made clear in the document, then undoubtedly Mrs Green would have been a tenant along with the other two and would have been entitled to continue as a statutory tenant after the lease come to an end. But the question is whether she can contend that, despite the express terms of the lease, she is entitled to say that she was in fact one of the principals on whose behalf the lease was entered into.

There has been considerable litigation with regard to the question of the circumstances in which an undisclosed principal may be a party to a contract in which, being an undisclosed principal, no reference is made to him. A very well-known authority is *Humble v Hunter* (1848) 12 QB 310, a charter-party case in which the headnote correctly expresses the decision of the court:

> In assumpsit on a charter-party executed, not by the plaintiff, but by a third person who, in the contract, described himself as "owner" of the ship. Held, that evidence was not admissible to show that such person contracted merely as the plaintiff's agent.

That is an authority which has been the subject of some criticism.

In a case in the House of Lords, *Drughorn (Fred) Ltd v Rederiaktiebolaget Transatlantic* [1919] AC 203, the headnote reads:

> The description in a charter-party of one of the contracting parties as "charterer" does not, of itself, designate him as the only person to fill that position. Where, therefore, an action for breach of charter-party was brought by persons claiming to be the undisclosed principals of a party described in the contract as "charterer," and objection was taken to the admission of evidence that the plaintiffs were in fact the charterers, on the ground that such evidence would contradict the written contract: Held, that the evidence was admissible.

In the course of giving the leading speech in the House of Lords Viscount Haldane, after referring to some authorities, said at p 206:

> These are authorities for the proposition that evidence of authority of an outside principal is not admissible, if to give such evidence would be to contradict some term in the contract itself.

One question which arises here is whether it can be said that because these two ladies are named as lessees it is inconsistent with that that Mrs Green should be a third lessee.

There is a helpful statement in the case of *Rederiaktiebolaget Argonaut v Hani* [1918] KB 247 by Rowlatt J. He said at p 249 of the report:

> It seems to me that in these cases it is always necessary to look at the document as a whole in order to ascertain whether such words as "as charterers" appearing after a person's name are merely words of description of the parties to the contract, in the same way as the words "of the first part" or "of the second part," or are used for the purpose of describing an essential part to be performed by the party in the transaction which the contract is dealing with.

It so happens that all those cases are on charter-parties. The only case dealing with leases or tenancy agreements to which we have been referred is a decision of Geoffrey Lawrence J in *Danziger v Thompson and Others* [1944] KB 654, where the short headnote reads as follows:

> A person was described in a tenancy agreement as the "tenant" of the premises: Held, that, oral evidence was admissible to show that the person entered into the agreement as the agent or nominee of another.

That, however, was not the case of an undisclosed principal, because it was perfectly well known to all the parties concerned that the person who was not named as the tenant in the written agreement was the person who had authorised the agreement to be entered into on his behalf.

What I think it really comes to is this. Looking at this tenancy agreement as a whole, is the identity of the people named in it as the lessees of such importance as a term of the contract that it cannot properly be taken that some other person could be regarded as an undisclosed principal? I would unhesitatingly answer that question by saying that in an agreement of this kind, containing, as it did, tenant's covenants with regard to repairs, and not to assign or underlet, and so forth, it is a matter of great importance to the landlord to know the identity of the people who are becoming his tenants. It may be said that in this case the tenant's covenants were strengthened rather than weakened by the presence of Mrs Green if she should be allowed to come in. But I see no principle on which it could be said that two lessees can contract on behalf of themselves and another as an undisclosed principal unless they can contract on behalf of an undisclosed principal standing alone. It further appears to me that there is a very real reason in this particular case why the landlords should not find themselves concerned with an additional tenant, because these premises, as I have said, are a dwelling-house within the meaning of the Rent Act, and any additional tenant would add to the number of persons who could become statutory tenants on the termination of the lease, as indeed is the very thing that the defendant seeks to achieve.

I reach the conclusion that Mrs Green is not entitled to claim to be a tenant here. She told us at the beginning of the short argument which she addressed to us that she did not really claim to be a tenant; all that she was asking for was to be allowed to stay there for a little longer until she could get somewhere else to live. I am afraid that that is not a basis in law on which we can decide this case. The only defence that was raised to this action for possession was that she was a tenant, and though one cannot help feeling sorry for Mrs Green, once that point is decided against her it seems to me that the appeal must be allowed.

ROSKILL and BRIDGE LJJ agreed.

The appeal was allowed with costs on Scale 2. The plaintiffs were granted a possession order, the defendant being ordered to give up possession at the end of May 1977.

COURT OF APPEAL

March 22 1978

(Before Lord Justice STEPHENSON, Lord Justice SHAW and Lord Justice EVELEIGH)

TILLING v WHITEMAN

Estates Gazette June 24 1978

(1978) 246 EG 1107

Rent Acts—Recovery of possession by owner-occupier—Case 10 in Schedule 3 to Rent Act 1968 (Case 11 in Schedule

15 to Rent Act 1977)—Joint owners only one of whom in occupation immediately before letting of dwelling-house—House required as a residence by that one only—Whether Case 10 satisfied—Court of Appeal divided—Majority consider that Case does not apply, Eveleigh LJ dissenting—Residence must be required by all joint owners—Leave to appeal to House of Lords

This was an appeal from a decision of Judge Sumner at Canterbury County Court on a preliminary point which concerned the construction of Case 10 in Schedule 3 to the Rent Act 1968 (consolidated as Case 11 in Schedule 15 to the Rent Act 1977). The plaintiffs, who were the joint owners of a dwelling-house, St Leonard's, The Street, Staple, Canterbury, claimed possession from Josephine Whiteman, the defendant, on the expiry of a two-year term on the ground that Case 10 in Schedule 3 to the 1968 Act applied. The judge dismissed the claim, holding that it was not sufficient that one only of the two joint owners required the dwelling-house as a residence. The other joint owner had left the house before the letting to the defendant and had no wish to occupy it. As appears from the judgment of Stephenson LJ, this joint owner, Miss Dossett, although a plaintiff in the county court, withdrew from the appeal, leaving Mrs Tilling, the other joint owner, as the sole appellant.

Christopher Sumner (instructed by Williamson & Barnes, of Deal, Kent) appeared on behalf of the appellant; R Laurie (instructed by Furley, Page, Fielding & Pembrook, of Canterbury) represented the respondent. The second plaintiff, Miss Dossett, did not appear and was not represented.

Giving judgment STEPHENSON LJ said: This is an appeal from an order of His Honour Judge Sumner made in the Canterbury County Court on May 3 1977 deciding on a preliminary point that the plaintiffs have no claim to possession against the defendant. The preliminary issue and the appeal raise a short point of construction. What has to be construed is the language of Case 10 in the Third Schedule to the Rent Act 1968, which is first found—with minor differences—in section 14 of the Rent Act 1965, and has since been repeated word for word in Case 11 of the Fifteenth Schedule to the Rent Act 1977.

The plaintiffs are two elderly ladies who jointly own a dwelling-house known as St Leonard's, The Street, Staple, Canterbury, Kent. On February 19 1975 they made an agreement with the defendant, of Keynes College, the University of Kent at Canterbury, in these terms:

> Whereby it is agreed as follows:
>
> 1. The Landlord lets and the Tenant takes all that furnished dwelling-house known as St Leonard's, The Street, Staple, Canterbury, Kent (hereinafter called " the premises"). 2. There is included in the letting the furniture and effects specified in the Schedule annexed hereto. 3. The premises and the furniture and effects shall be held from the 21st day of February 1975 for a period of two years. 4. The Tenant shall pay for the premises and the furniture the weekly rent of twelve pounds fifty pence four-weekly in advance the first payment to be made on the signing hereof and thereafter every four weeks.

Then there are the usual covenants, and at the foot of the agreement there is this notice typed on it:

> To the tenant the said J Whiteman. Take notice under the Rent Acts 1968 and 1974 that the Landlord may recover possession of the premises under the provisions of Case 10 of Part II of Schedule 3 to the Rent Acts 1968.

Then, under that, comes the sentence: "Dated this 19th day of February 1975," with the signatures of both the landlords, and an acknowledgement: "I acknowledge to have received Notice as above," signed by the tenant. (The reference to the Rent Act 1974 has no relevance to this case.)

On the expiry of the term on February 19 1977 the defendant remained in occupation of the house. On March 4 1977 the plaintiffs began proceedings for possession in the Canterbury County Court. By their particulars of claim they claimed that they were the freeholders and entitled to possession of the dwelling-house known as St Leonard's; the net annual value was £232; the premises, it was alleged, were let on a regulated tenancy by the plaintiffs to the defendant by the agreement in writing which I have just read; the dwelling-house, it is said, was formerly occupied by the plaintiffs as their residence "and is now required as a residence for the first plaintiff." That is paragraph 4 of the particulars of claim. Then they add: " On the 19th February 1975 the defendant was given notice in writing that possession might be required under the provisions of Case 10 of the Third Schedule to the Rent Act 1968." That is paragraph 5; and the plaintiffs claim possession and rent or alternatively mesne profits.

By her defence dated April 21 1977 the defendant raised three defences, one irrelevant and none touching on the issue decided by the learned judge. However, it was agreed between the parties that a question, which is nowhere formulated in the pleadings and nowhere expressly stated in the approved note of the judge's oral judgment, should be tried as a preliminary issue on agreed facts. The agreed statement of facts is this:

> 1. Jointly owned by Miss Dossett and Mrs Tilling. 2. Owners are not related. 3. Property was not occupied by one of the owners immediately before it was let. Property had been occupied by Miss Dossett. 4. Property is only alleged to be required for occupation by Mrs Tilling, the person who is alleged to have occupied immediately before letting.

Made consistent and intelligible, this statement means that Miss Dossett and Mrs Tilling are joint owners of the house; they are not related; Miss Dossett had occupied the house but had left it before it was let to Miss Whiteman; Mrs Tilling had occupied it and was still occupying it immediately before it was let to Miss Whiteman.

The issue which should have been agreed, recorded in writing and submitted to the judge was whether the allegation pleaded in paragraph 4 of the particulars of claim gives the plaintiffs any right to recover possession of the said dwelling-house under section 10(2) of and Case 10 in Schedule 3 to the Rent Act 1968, or whether one of two joint owners of a dwelling-house let on a regulated tenancy who occupied the dwelling-house as his (or her) residence is an owner-occupier entitled to recover possession of it under Case 10 of the Third Schedule to the Rent Act 1968 if the court is satisfied that the dwelling-house is required as a residence for himself (or herself). The judge decided that the plaintiffs were not entitled to recover possession under those provisions of the 1968 Act because it was required as a residence for one of them only.

I think it necessary to read the first three subsections of section 10 before I read Case 10. Section 10 provides:

> (1) Subject to the following provisions of this Part of this Act, a court shall not make an order for possession of a dwelling-house which is for the time being let on a protected tenancy or subject to a statutory tenancy unless the court considers it reasonable to make such an order and either (a) the court is satisfied that suitable alternative accommodation is available for the tenant or will be available for him when the order in question takes effect, or (b) the circumstances are as specified in any of the Cases in Part I of Schedule 3 to this Act. (2) If, apart from the provisions of subsection (1) above, the landlord would be entitled to recover possession of a dwelling-house which is for the time being let on or subject to a regulated tenancy, the court shall make an order for possession if the circumstances of the case are as specified in any of the Cases in Part II of Schedule 3 to this Act. (3) The provisions of Part III of Schedule 3 to this Act shall have effect in relation to Case 8 in that Schedule and for determining the relevant date for the purposes of the Cases in Part II of that Schedule.

A I then turn to the Third Schedule and read Case 10 in Part II, which provides as follows:

> Where a person who occupied the dwelling-house as his residence (in this Case referred to as "the owner-occupier") let it on a regulated tenancy and (a) not later than the relevant date the landlord gave notice in writing to the tenant that possession might be recovered under this Case, and (b) the dwelling-house has not, since 8th December 1965, been let by the owner-occupier on a protected tenancy with respect to which the condition mentioned in paragraph (a) above was not satisfied, and (c) the court is satisfied that the dwelling-house is required as a residence for the owner-occupier or any member of his family who resided
> B with the owner-occupier when he last occupied the dwelling-house as a residence.

Part III of the Third Schedule deals (1) with Case 8, which I shall have to read later, and (2) with the relevant date. Paragraph 2(b) of Part III provides: "Any reference in Part II of this Schedule to the relevant date shall be construed as follows: . . . (b) in any other case," than that which is provided by (a) which I have not read, "the relevant date means the date of the commencement of the regulated tenancy in question."

I next read the note of the judgment in full, because it is short and clear and, in my opinion, with all respect to those
C who are of a different opinion, correct.

> This is a preliminary point. There are no direct authorities. I have been referred to some authority. Two ladies claim possession under Case 10 of Schedule 3 of the Rent Act 1968 as amended, their ground being that basically they propose that one of the ladies, Mrs Tilling, wishes to occupy. Case 10 reads . . .

and then he sets out the provisions of the Case.

> The first matter to be considered is: "Where a person who occupies a dwelling-house as his residence let it on a regulated tenancy." This property was let by two plaintiffs as joint owners. I construe "person" as plural under the Interpretation Act. The plaintiffs only come within Case 10 if they occupied as their
> D residence. It is an agreed fact that they did, Mrs Tilling immediately before the tenancy, Miss Dossett some time earlier. So far so good. The plaintiffs then asked me to adopt a different construction. In order to come within the first limb the owner-occupier had to be construed as both ladies. They then have to argue for sub-paragraph (c) that as only one wants to occupy, "owner-occupier" should be construed as one lady. I do not consider that they satisfy the proviso of sub-paragraph (c). I do not consider that one owner is the owner-occupier as defined. It is inevitably a matter of construction. Should one construe strictly or take a wider view. This was not done in the case of *McIntyre v Hardcastle* [1948] 2 KB 82 in considering Case 8. Here the reference is to person and owner-occupier rather than landlord. Landlord is
> E referred to in Case 8. However, the principle is the same and the overriding consideration is that if one construes person and owner-occupier in the plural to bring the plaintiffs within Case 10 at all, one cannot in the same case construe owner-occupier in the singular. The construction must be consistent. I find the preliminary issue in favour of the defendant. Claim dismissed. Costs Scale 2 taxed or agreed.

Against that judgment only Mrs Tilling, the first plaintiff, appeals. Miss Dossett, the second plaintiff, is living in Oxford and uninterested in any part of these proceedings except their expense. We have in fact granted an application made by Miss Dossett's solicitors for a declaration under Rules of
F the Supreme Court, Order 67, rule 4, that they have ceased to be the solicitors acting for her in this case, and we have also granted an application on behalf of Mrs Tilling for leave to amend her notice of appeal to join Miss Dossett as a respondent.

In *Lloyd v Sadler* ESTATES GAZETTE, May 6 p 479, decided in the Court of Appeal on January 19 1978, Megaw LJ said at page 3G.

> This case has brought to light two remarkable facts. First, the Rent Acts through all their long history have never made any relevant express provision relating to joint tenants or joint tenancies whether protected or statutory. Secondly, and perhaps even more
> strangely, there seems to have been no previous case decided by G
> any court, or at least no reported case, in which a question such as falls to be decided in the present case has been raised for consideration on section 3(1) of the 1968 Act or its predecessors.

It is not so strange that there is no previous reported case in which the question of joint owner-occupiers seeking possession under Case 10 of the 1968 Act has been considered, because its provisions go back only to 1965 and section 14 of the Rent Act of that year. And there are two reported cases in which this court has considered the question of joint beneficial owners who are landlords seeking possession under what was Case 8 in Schedule 3 to the 1968 Act and
is now Case 9 in Schedule 15 to the 1977 Act but goes back, H
beyond paragraph (h) of the First Schedule to the Rent and Mortgage Interest Restrictions (Amendment) Act 1933, to the year 1920. It is therefore even stranger that Parliament has enacted and re-enacted the provisions in respect of the recovery of possession of owner-occupied houses with which we are concerned in this appeal, without making any relevant express provision relating to joint owners or joint owner-occupiers or joint tenants letting dwelling-houses on a regulated tenancy.

The decision which deals with possession of a dwelling-house jointly owned by two landlords is *McIntyre v Hard-*
castle, where this court adopted a dictum of Asquith LJ in J
preference to a dictum of Somervell LJ in *Baker v Lewis,* [1947] KB 186, and decided that two sisters, who sought possession of a house they owned jointly under paragraph (h) of the First Schedule to the 1933 Act on the ground that the house was required as a residence for one of them, could not bring themselves within the paragraph and recover possession. The First Schedule to the 1933 Act read as follows:

> A court shall, for the purposes of section 3 of this Act, have power to make or give an order or judgment for the recovery of possession of any dwelling-house to which the principal Acts
> apply or for the ejectment of a tenant therefrom without proof K
> of suitable alternative accommodation (where the court considers it reasonable so to do) if . . . (h) the dwelling-house is reasonably required by the landlord (not being a landlord who has become landlord by purchasing the dwelling-house or any interest therein after [a certain date] for occupation as a residence for (i) himself; or (ii) any son or daughter of his over eighteen years of age; or (iii) his father or mother.

And then there is a proviso. Case 8 is in very similar terms; Case 8, that is, in Part I of the Third Schedule to the 1968 Act:

> Where the dwelling-house is reasonably required by the landlord for occupation as a residence for (a) himself, or (b) any son or
> daughter of his over eighteen years of age or (c) his father or L
> mother, or (d) if the dwelling-house is let on or subject to a regulated tenancy, the father or mother of his wife or husband, and the landlord did not become landlord by purchasing the dwelling-house or any interest therein after 23rd March 1965 or, if the dwelling-house is let on or subject to a controlled tenancy, after 7th November 1956.

In the leading judgment, Tucker LJ said at p 90 of the report in [1948] 2 KB 82:

> All kinds of difficulties have been suggested to us as likely to follow whichever interpretation is accepted by us. I do not think that the legislature really contemplated this situation when this
> paragraph was framed. Therefore I feel driven to interpret it M
> merely in the light of the actual language used. Looking at it in that way, I feel convinced that the interpretation put upon it by Asquith LJ was the correct one, and I will not attempt to put into better language what he so clearly expressed in the passage just quoted.

In the passage just quoted, Asquith LJ had said that "landlord" in paragraph (h) covered two or more joint beneficial owners and in such a case the dwelling-house must be reasonably required by them for occupation as a residence for themselves (not "himself"), for any son or daughter of theirs (not "his") or for their (not "his") father or mother. That

A decision has been applied by this Court to an application for a new lease of business premises for occupation by two out of three landlords: *Wetherall & Co Ltd v Stone* [1950] 2 All ER 1209.

Mr Sumner has submitted that Parliament had regard to that decision and used words in Case 10 which intentionally escape the impact of that decision and distinguish Case 10 from Case 8. Alternatively, the actual language used enables or compels the court to distinguish the two Cases and to hold that one of two (or more) joint beneficial owner-occupiers can come within Case 10 if he (or she) satisfies the court that the dwelling-house is required as a residence for B himself (or herself). He contends that it is not right to construe "owner-occupier" as referring to both ladies, as the judge did.

For my part, I cannot infer any intention to distinguish Case 10 from Case 8 or its judicial interpretation. I accept that difficulties follow whichever interpretation is accepted and it is the actual language used which must be interpreted and must give Mr Sumner success or justify the judge's adverse decision of the preliminary point. So is Mrs Tilling "a person who" (1) "occupied the dwelling-house as (her) residence", (2) "let it on a regulated tenancy" and (3) requires it as a residence for such a person? If the sentence C is divided into these three parts, each of these questions can be answered affirmatively and Mrs Tilling can be said to be a person who occupied the dwelling-house as her residence, let it on a regulated tenancy and requires it for occupation as her residence. She is one of the two persons who occupied it and let it, and she is the only one of those two who requires it for occupation. But in my judgment it is not correct to say that she let it or is a person who let it within the meaning of those words in their context. The person who let it, and who is referred to as "the owner-occupier," was not Mrs Tilling but herself and her co-owner, Miss Dossett. It is they D two, not Mrs Tilling, who occupied the dwelling-house and were the owner-occupier of it, and it is both of them, not Mrs Tilling alone, who have to satisfy the court that it is required as a residence for them both.

That is a natural construction of the words used even if it is not the only natural construction. It is supported by the difficulty which Mr Sumner's alternative construction introduces for joint owners who disagree on who should occupy the dwelling-house, the tenant or one of the owners. One joint owner can sue for rent or give notice to quit, but all joint owners must sue for possession, and judgment for possession cannot be given in favour of one only. Miss E Dossett's indifference might turn to co-operation or hostility; she might prefer getting rent from Miss Whiteman to letting Mrs Tilling replace her. It is also supported by the general purpose of the Act to protect the security of tenants from eviction by landlords. It is true that Case 10 opens a door to a landlord's exercising his contractual right to possession; but it is one door in a high wall which strictly protects a tenant and it would accord better with the object of the Act if it opens a little than if it opens wide. It was probably both these considerations which led this court in *McIntyre v Hardcastle* to construe "the landlord" in what was in 1968 Case 8 as including the plural and to hold that one of two joint F owners could not recover possession for herself. It was the second consideration which led this court in *Lloyd v Sadler*, to which I have already referred, to refuse to apply *McIntyre v Hardcastle* in reverse so as to deprive one joint tenant of her statutory tenancy when the other had abandoned occupation during the contractual joint tenancy. Following the majority decision of this court in *Howson v Buxton* (1928) 97 LJKB 749, this court held that to apply strictly the doctrine of joint tenancy would have led to results which were unreasonable or unlikely to have been intended by the legislature. But in the instant case the strict application of the doctrine of joint ownership has the opposite effect and leads to a result which is not unreasonable and is much more G likely to have been intended by the legislature than an extension of the right to evict from jointly-owned dwelling-houses. That case and the case which it follows seem to me to give no support to Mr Sumner's interpretation. If they help at all, they help the defendant's case.

I get no help also from the minor changes made in the language of the provision conferring this right to possession in the course of its transfer from the body of the 1965 Act to a Schedule to the 1968 Act (and the 1977 Act). I am satisfied that those changes are due solely to that transfer. Nor do I get help from the definition of "landlord" in section 113(1) of the 1968 Act. But I do get help from section 10 of this H Act. Mr Sumner relies on the fact that in Case 10 the landlord is referred to only in (a), and it is "a person", not "the landlord", who is referred to as occupying and letting the dwelling-house. But it has to be remembered that this Rent Act, like all the other Rent Acts, deals primarily with the landlords and tenants of dwelling-houses. In section 10(1) "the landlord" is not mentioned, but in Case 8, which is linked with the subsection, he is. In section 10(2) "the landlord" is mentioned, but in Case 10, which is linked with that subsection, he is not, except in (a). Case 10, no less than Case 8, states conditions in which a landlord can (in Case 8) or must (in Case 10) be entitled to recover J possession from his tenant. Landlords, or owners, of dwelling-houses may occupy or not occupy the dwelling-house; an owner-occupier is a landlord or owner who is residential, not absentee, at the time when he lets his dwelling-house. If he then occupies, it is strictly correct not to call him a landlord until he actually lets and to call him a landlord when he has let and gives notice that he may recover possession under Case 10. In the instant case these owners gave the notice on the same date and in the same document as the agreement creating the tenancy. But even then it was given at the foot of the agreement and it might in other cases be given on a K later date when the owner who had occupied before he let had indeed become the landlord by letting. These considerations seem to me to account for the use of the words "the landlord" in (a) of Case 10 and of the words "a person" (including the indefinite article) in the opening words of Case 10.

I am therefore unable to find any distinction between the language of Case 8 and Case 10 in any respect material to the decision in *McIntyre v Hardcastle* or to infer from the language of Case 10 a right in one or two (or more) owner-occupiers to recover possession of their dwelling-house for his or her (or their) own occupation. I am of opinion that the L judge's decision to dismiss the plaintiffs' claim was right. I would affirm it and dismiss the appeal.

Agreeing, SHAW LJ said: The circumstances which give rise to this appeal from the decision of His Honour Judge Sumner on a preliminary issue have been recounted in the judgment of STEPHENSON LJ and need no further elaboration. The issue was as to jurisdiction to entertain a claim for possession under the provisions of section 10 and Case 10 of Part II of Schedule 3 to the Rent Act 1968.

The real question was whether the claim as pleaded disclosed a claim within those provisions at all. If it did, the combined effect of section 10 and Part II of Schedule 3 is M mandatory and the plaintiffs must succeed in their claim. The objection taken at the trial on behalf of the tenant was founded on what was averred in paragraph 4 of the particulars of claim. It was there pleaded: "The said dwelling-house was formerly occupied by the plaintiffs as their residence and is now required as a residence for the first plaintiff." The last four words give rise to the controversial issue. The defendant contends that they demonstrate that the plaintiffs' claim is not within Case 10. Condition (c) of that case requires that "the court is satisfied that the dwelling-house is required as a residence for the owner-occupier. . . ." I have omitted the rest

A of paragraph (c) which is irrelevant to the present case. The expression "owner-occupier" is defined, or at any rate described, in the opening words of Case 10 which reads: "Where a person who occupied the dwelling-house as his residence (in this case referred to as 'the owner-occupier') let it on a regulated tenancy." It is apparent that the characteristics which serve to identify an owner-occupier in this context are (i) that he owned the interest out of which the regulated tenancy was granted, and (ii) that he had occupied the dwelling-house as his residence at the time that he let it on a regulated tenancy.

As was the position in the appeal of *Lloyd v Sadler*
B (Estates Gazette, May 6, p 479), it seems clear that the draftsman of the 1968 Act had not in mind the particular situation of joint tenants in relation to statutory or regulated tenancies. In the present case it was contended on behalf of the tenant that "owner-occupier" comprised all those who by virtue of a joint interest comprised the "owner," so that it necessarily included all the joint tenants. It would follow that in regard to paragraph (c) it was an essential element of a Case 10 claim that the dwelling-house be required as a residence for all the joint owners and not merely for one or some of them. This would accord with the view taken in the Court
C of Appeal in *McIntyre v Hardcastle* by Tucker LJ, approving and following the judgment of Asquith LJ in *Baker v Lewis* [1947] KB 186.

Like Tucker LJ—and I quote from his judgment—"I feel driven to interpret (the statute) in the light of the actual language used. Looking at it in that way, I feel convinced that the interpretation put on it by Asquith LJ was the correct one." While it is true that those observations were not made in precisely the same statutory context, I cannot see on what principle of construction any material distinction can be drawn between the grounds for that decision and those which obtain in the present appeal. Counsel for the appellant Mrs
D Tilling (the sole appellant since Miss Dossett has foresworn her erstwhile role of co-appellant) relies on the judgment of this court in *Lloyd v Sadler,* to which reference has been made already. There it was held that one of two joint tenants under a protected tenancy could, on the expiration of that tenancy, claim to be a statutory tenant notwithstanding that the other joint tenant had gone out of occupation and did not intend to return. If, the argument went, one of two joint tenants could claim to be entitled to statutory protection, why should not one of two joint landlords be similarly entitled to statutory relief? The answer to that contention is that the respective legislative provisions fall to be construed
E not only "in the light of the language actually used," but in relation to the object and purpose of those provisions.

The statutory provisions which were the subject of consideration in *Lloyd v Sadler* were designed to give a form of security of tenure to tenants under a protected tenancy. It was a reasonable construction of the relevant provisions that that security of tenure was a personal right of individual tenants notwithstanding that they held under a joint tenancy. Case 10 is to be found as an adjunct of Part II of the 1968 Act which is headed "Security of Tenure," a phrase which is itself followed by the words "Limitations on recovery of
F possession of dwelling-houses let on protected tenancies or subject to statutory tenancies." Thus the theme is the safeguarding of the tenant's right to continue in possession, a theme communicated to Schedule 3, which begins with the heading "Grounds for possession of dwelling-houses let on or subject to protected or statutory tenancies."

The various cases set out under Parts I and II respectively of the Schedule constitute derogations from that theme. Thus they run counter to the main burden of Part II of the Act. This, in my opinion, indicates the need for a strict interpretation of the relaxations of the principle that security of tenure is to be maintained. Accordingly, the problem of construction in the present case has to be resolved from a diametrically G opposed starting point. In this situation I find no difficulty in arriving at the conclusion that "the owner-occupier" in paragraph (c) of Case 10 necessarily involves everyone who was a joint owner at the time the premises were let on a regulated tenancy.

In my view, the learned judge came to the correct conclusion. I respectfully agree with the judgment of Stephenson LJ, and I would dismiss the appeal.

Dissenting, Eveleigh LJ said: In the course of argument great reliance was placed by the respondent upon *McIntyre v Hardcastle* [1948] KB 82. That case, however, was con- H cerned with section 3 of the Rent Act 1933 and the First Schedule to that Act. The words were quite different from those found in Case 10 of the 1968 Act and contained words which are repeated in Case 8 of the 1968 Act. *McIntyre v Hardcastle,* therefore, is a decision upon completely different wording and upon a very different situation, as the separate case, Case 8 in the 1968 Act, can be said to emphasise. Furthermore, there is no general principle to be drawn from *McIntyre v Hardcastle* and certainly not one which governs the present case or the interpretation of Case 10. At p 90 Tucker LJ said: "I do not think that the legislature really contemplated this situation when this paragraph was framed. Therefore, J I feel driven to interpret it merely in the light of the actual language used." In the absence of any reasons to the contrary, that, in my judgment, is the approach which I should adopt to the interpretation of Case 10.

In *McIntyre v Hardcastle* it was necessary, in order for the plaintiff to succeed, to substitute the plural for the singular in the words of the Act. This was of course permissible by virtue of the Interpretation Act, but the plaintiff was not allowed to pick and choose where the substitution should be made but was required to be consistent and substitute the plural wherever the singular occurred. In the present case the appellant claims that she is not asking for the alteration of K any of the words of the Act, not even by substituting the plural for the singular. If this contention is wrong and it is necessary to substitute, then says the appellant "I do no more than substitute consistently."

It has been argued that the appellant's claim runs counter to the policy of the Act. It is said that the intention of the Rent Act is to protect the tenant. That may well be; but it is equally true that the introduction of a new ground for possession as that contained in Case 10 was done for the benefit of the landlord. Furthermore, it is clear that there was an overriding intention to encourage persons in occupa- L tion of their homes to be prepared to let them and not refrain for fear of being kept out for ever. Therefore, if policy considerations are to play a part in the interpretation of the Act in relation to the present claim, it would in my view favour the appellant.

It has also been said that as the Act makes no specific reference to the position of the joint owner as landlord and as *McIntyre v Hardcastle* rejected the claim of one of two joint owners, then this court should proceed upon the basis that the dice is still loaded against one of joint owners. I do not accept this contention. The words of Case 10 are M quite different from those of Case 8; and, while it may be that the legislature did not have joint tenants in mind, I can have no confidence that this was so.

Upon those principles I now turn to consider the relevant words of the 1968 Act actually found in Case 10. Case 10 begins: "Where a person who occupied the dwelling-house as his residence (in this case referred to as 'the owner-occupier') let it on a regulated tenancy." The appellant did occupy the dwelling-house as her residence. (We are not concerned in this present appeal to decide whether the occupation must be immediately before the letting or whether occupation at any previous time will do. However, as the past perfect tense is used for both verbs, there is much to be said for the contention that the occupation has to be the last occupation of the dwelling-house by anyone before the letting.) Further, in my opinion the appellant is a person who let on a regulated tenancy. It is argued that because her joint owner joined in the lease, the appellant cannot be heard to say that she let. Counsel for the respondent says that she is forced to say "they let." I do not accept this argument. Case 10 is concerned to consider the qualifications of a person. The emphasis is upon occupation. A person in occupation should not be deterred from making way for a tenant by the thought that he or she may be kept out of the house permanently. There are two conditions which have to be fulfilled to allow a person back into occupation. That person must have occupied and that person must have let. The benefit of Case 10 is given to one who can stand in the way of freedom of letting not only by virtue of occupation, but also having the power to refuse to let. It seems to me to be wholly truthful for the appellant to say that she let the house on a regulated tenancy. One of two joint owners can in law deal with the property without the concurrence of the other in so far as the rest of the world is concerned. The fact that two people do a thing together does not, in my understanding of the English language, prevent either one claiming that he himself did it. The argument to the contrary entails reading into the Act words something like "on his own." If the appellant had come to an arrangement with the other joint owner that only the appellant's name would appear in the lease, the tenant would have had no answer to the claim for possession. Conscious as I am of the technicalities of the Rent Act, I do not believe that one should encourage a situation where yet another technicality arises unless there is a compelling case.

Notice in writing was given to the tenant, but it is argued that the use of the words "the landlord" is of some significance. It is said that the appellant is not *the* landlord, but only one of two. Reference is then made to section 10(2) of the Act: "If, apart from the provisions of subsection (1) above, the landlord would be entitled to recover possession of a dwelling-house which is for the time being let on or subject to a regulated tenancy, the court shall make an order for possession if the circumstances of the case are as specified in any of the Cases in Part II of Schedule 3 to this Act." Again it is argued that "the landlord" can only mean all those who are the landlords when there are more than one. In my opinion, section 10, subsection (2), is not laying it down as a condition that the plaintiff must be *the* landlord. It is stating what must be shown before the person claiming possession shall be given an order, and what has to be shown is that the case falls within Part II of Schedule 3. It is nowhere stated that the plaintiff must prove that he is *the* landlord. However, if this argument is wrong, I see no difficulty in reading "landlord" to include the plural. If in section 10(2) we read "the landlords," all that consistency requires is that we shall read the plural also when the word "landlord" appears in Case 10. Section 10, subsection (2), would then read: "If, apart from the provisions of subsection (1) above, the landlords would be entitled to recover possession" (in our case they would) "the court shall make an order for possession if the circumstances of the case are as specified in any of the Cases in Part II of Schedule 3 to this Act."

We must now proceed to inquire whether the circumstances are as specified. The relevant case is Case 10. Part II of Schedule 3 is headed: "Cases in which court must order possession where dwelling-house subject to regulated tenancy." As I have said, in my judgment Mrs Tilling is a person who occupied the dwelling-house and let it on a regulated tenancy. The first condition, (a), if we substitute the plural, now reads: "not later than the relevant date the landlords gave notice in writing to the tenant." The landlords did. It is agreed that the next provision, (b), is fulfilled. The final provision, (c), reads: "the court is satisfied that the dwelling-house is required as a residence for the owner-occupier"; in other words, "is required as a residence for the person who occupied the dwelling-house and let it on a regulated tenancy." That person was and is Mrs Tilling. In my view therefore, subject to the facts pleaded being established and to any possible argument as to the date when the owner-occupier is required to have been in occupation, the plaintiff's claim succeeds.

Having done this exercise of substituting the plural consistently, I am far from conceding that joint tenants were not in the minds of the legislature. The change in wording from "owner-occupier" to "landlord" in Case 10 must be deliberate and, in my judgment, achieves the result of giving protection to one of two joint owners. The fact that the expression "joint owners" is not used may well be because there are other forms of shared interests in real property that may also be envisaged and a specific reference to joint owner might be unduly limiting. The Schedule to the Act clearly attempts to deal with situations in a new and simplified manner by setting them out as cases in what no doubt it was hoped would be treated as straightforward language. That language, in my opinion, covers the present claim.

The appeal was dismissed with costs, to be paid by the Legal Aid Fund unless the Law Society gave notice of objection. Leave given to appeal to House of Lords, with liberty to apply as to terms on which leave should be granted.

A **COURT OF APPEAL**

June 20 1977

(Before Lord Justice ROSKILL and Lord Justice CUMMING-BRUCE)

THE PROPERTY & REVERSIONARY INVESTMENT CORPORATION LTD v TEMPLAR AND ANOTHER

Estates Gazette January 28 1978

(1977) 245 EG 307

Application for leave to appeal out of time—Leave sought to appeal from decision against landlords on the operation of a rent review clause in a lease for 21 years—Decision
B **given before the rulings of the House of Lords in United Scientific Holdings Ltd v Burnley Borough Council and Cheapside Land Development Co Ltd v Messels Service Co—Claim that if leave to appeal given the decision appealed against would probably be reversed—Application granted**

This was a motion for leave to appeal out of time from a decision by Judge Fay, sitting as a deputy judge of the Queen's Bench Division, on November 1 1974. The judge had dismissed the plaintiff landlords' action on the ground that the procedure to obtain a rent increase under the provisions of a lease between the parties had not been initiated
C in time.

W Goodhart (instructed by Hicks Arnold Rose Johnson) appeared on behalf of the applicants; D R N Hunt (instructed by Wright, Son & Pepper) represented the respondents.

Giving judgment, ROSKILL LJ said: In 1972 the Property & Reversionary Investment Corporation Ltd issued proceedings against their lessees, a Mr and Mrs Templar, in connection with a rent review clause in a lease entered into for 21 years from March 25 1965 at an initial yearly rental of £1,656. These proceedings came before His Honour Judge Fay sitting as a deputy judge of the Queen's Bench Division, and Judge Fay determined these proceedings in favour of the
D tenants. That decision was given on November 1 1974, when the learned judge dismissed the plaintiffs' action on the ground that the steps to obtain the rent increase had not been previously initiated before March 25 1972. We have not seen a transcript of his judgment, but it would seem that he followed the line of cases then prevailing and took the view that time was of the essence and as the plaintiffs (the landlords) had failed to comply with their obligations under the lease, properly construed, they had lost their entitlement under the rent review clause. That, broadly speaking, was in line with the view taken in two decisions of this court,
E first in *United Scientific Holdings Ltd v Burnley Borough Council* and secondly in *Cheapside Land Development Co Ltd and others v Messels Service Co*. But on March 23 last the House of Lords declared those two decisions to have been erroneous.

Mr Goodhart has therefore claimed, and Mr Hunt has agreed, that we may assume, though he has made no submission, that in the light of the decisions of the House of Lords, Judge Fay's judgment, if appealed to this court, would be held to be wrong. We express no view as to whether that is right or not. It is in the light of that present position, that the plaintiffs have now applied to this court for leave to F
appeal out of time. They are, of course, a long way out of time, for it is now some 2½ years since Judge Fay gave his decision. But it is said by Mr Goodhart that the reason why this court should now give leave to appeal out of time is because this is a case where the parties are in a continuing contractual relationship and that it is wrong that that relationship should still continue and will continue, under the 21 years' lease, until 1986 to be governed by a decision of Judge Fay which, on the assumption I have already mentioned, is to be taken as erroneous.

He relied on a decision of this court, *Re Berkeley* (1944)
171 LT 303. In that case there had been earlier decisions at G
first instance. Then there was a subsequent decision of this court which showed that the earlier decisions were wrong, and it was sought to appeal out of time against the second of those decisions. The Master of the Rolls, Lord Greene, at p 305, said:

> It seems to me that the principle to be extracted is this. It is not sufficient for a party to come to court and say: "A subsequent decision of a superior court has said that the principle of law on which my case was decided was wrong." The court will immediately say to him: "That bald statement is not enough for you. What are the circumstances? What are the facts? What
> is the nature of the judgment? Who are the parties affected? H
> What, if anything, has been done under it?" and so forth. In other words, the whole of the circumstances must be looked at. If the court, in the light of those circumstances, considers it just to extend the time, then it will do so. That seems to me to be the proper principle and it is entirely in accordance with the view taken by this court in the most recent case under this rule of *Gatti v Shoosmith* (1939) 161 LT 208; [1939] Ch 841.

It is therefore plain that it is not enough for Mr Goodhart to say that the recent decisions of the House of Lords clearly show that Judge Fay's decision was wrong. He must show there are special reasons why he should be allowed to argue
that the judgment should not stand. J

At one point Mr Goodhart sought to contend that the plaintiffs might be in a difficult position when next the rent review clause can be invoked in 1979 and said that, notwithstanding the decisions of the House of Lords, they would be bound to comply with Judge Fay's judgment—there being, as he contended, issue estoppel between the parties. Mr Hunt did not accept that, and indeed offered an undertaking that no question of issue estoppel would be raised in 1979 and that the lessees would accept that the plaintiffs would be then entitled to base their claim for rent review in accordance with the decisions of the House of Lords. We cannot
speculate as to the future. The real point here, as Mr Good- K
hart ultimately accepted, is whether it is right that these parties should have their continuing contractual relationship governed by a lease, the terms of which have assumedly been erroneously construed in the court below.

We think that, notwithstanding the way in which Mr Hunt put his case, the plaintiffs should be left to receive the lower rent for the next 18 months or so and thereafter become entitled to claim the higher rent in accordance with the House of Lords' decision and that there are special circumstances which justify appeal out of time. Mr Goodhart, very properly, in the light of *Berkeley's* case, accepted that he could not claim any new rent retrospectively, even if the appeal out of time ultimately succeeds. That is clearly right, and upon his undertaking not to claim any increased rent, if his appeal succeeds, before any date before Midsummer Day next, I take the view that leave to appeal out of time should be given, and accordingly I would allow the motion.

Agreeing, CUMMING-BRUCE LJ said: I would only add this: The court has considered whether there are special circumstances which bring into play the principles described by the Master of the Rolls in *Berkeley's* case, where the Master of the Rolls stated that the court has to look at all the circumstances, and here the context of this application is that there is a lease, whereby the landlord has a right to seek a rent review periodically to keep in step with the market movements of such a lease, which included clauses defining the substantive obligations of the parties and a clause, or clauses, laying down the procedural steps required to enable the lessors to avail themselves of their rights for rent review. What was decided in this case by the learned judge was there had been a procedural default by the landlord in relation to certification in the time limited by the lease for a review of rent, and as a result of that decision, the landlords lost their right to the rent review that they would otherwise have had under the lease.

Now that the House of Lords has decided that the proper construction of the contract is other than that decided by the learned judge, I agree that there are special circumstances here because it does not seem just that future obligations between the parties to the lease should depend upon the construction now shown to be wrong.

Leave given to appeal out of time, both sides' costs to be paid by the plaintiffs in the action.

QUEEN'S BENCH DIVISION

July 13 1977

(Before Mr Justice DONALDSON)

F R EVANS (LEEDS) LTD v ENGLISH ELECTRIC CO LTD

Estates Gazette February 25 1978

(1977) 245 EG 657

Rent review clause in lease—Revision to represent "the full yearly market rental" of the premises (factory buildings and land in Liverpool) let with vacant possession on open market as a whole between a willing lessor and willing lessee—Questions as to assumptions to be made by arbitrator appointed to determine the rent at the end of third year of 21 years' term—Importance of keeping in mind the distinction between the actual landlord and tenant and the hypothetical willing landlord and tenant—The latter abstractions "in the hypothetical life of hypothetical higglers"—Illuminating judgment by Donaldson J (later approved by Court of Appeal: see below)

This case came before the High Court because the arbitrator, Roger Weston FRICS, appointed in default of agreement between the parties to determine the revised annual rent payable in pursuance of a rent review clause in respect of "The Walton Works" in Liverpool, stated his award in the form of a special case raising a question of law for the decision of the court. The landlords were F R Evans (Leeds) Ltd and the tenants English Electric Co Ltd. The facts as to the lease, the contentions of the parties before the arbitrator, the arbitrator's findings, and the question of law for the decision of the court are fully set out in the judgment of Donaldson J. It will be seen from the note at the end of the judgment that it was approved by the Court of Appeal, who dismissed the tenants' appeal against it.

I Glidewell QC and M Barnes (instructed by Lewis Lewis & Co) appeared for the plaintiffs; R H Bernstein QC and E Nugee (instructed by Travers Smith, Braithwaite & Co) represented the defendants. The same counsel and solicitors appeared in the appeal to the Court of Appeal.

Giving judgment, DONALDSON J said: This case is concerned with the meaning and effect of a rent review clause. The parties were unable to agree upon the amount of a revised annual rent and, in accordance with the terms of the lease, the matter was referred to the arbitration of Roger Weston FRICS. He has stated an award in the form of a special case.

The subject matter of the lease is "The Walton Works," which comprises land and factory buildings in Liverpool. The works are exceptionally large premises—60 acres in all, with buildings providing nearly 1m sq ft of floorspace. The lease, dated February 7 1974, is for a term of 21 years from October 1 1973. It gave both parties the right to require a review of the rent at the end of the third and 12th years of the term. The lessee had the further right to determine the lease at the end of the 14th year of the term. This case concerns the review at the end of the third year. English Electric Co Ltd are the original lessees and F R Evans (Leeds) Ltd are the assignees of the interest of the original lessor, the Secretary of State for Defence. I shall refer to them respectively as the tenants and the landlords rather than as lessee and lessor, because the similarity of the latter terminology makes for confusion.

The arbitration lasted for nine working days. The original rent was £90,000 per annum. The landlords contended that this should be raised to about £700,000. The tenants contended that the new rent should be £180,000. On evidence and facts with which I am not concerned, the arbitrator found that on no view should the rent be more than £515,000 or less than £290,000. Which, if either, was appropriate, depended upon the answer to a question of law which appears in the special case. Only a small part of the lease is relevant to a decision on this question of law. That part is as follows:

> 4(2) As from the end of the third year... of the said term the yearly rental hereunder shall if either party so desires... be revised so as to represent the full yearly market rental as hereinafter defined of the demised premises as they shall be at the end of the third... year of the said term... (assuming the demised premises to be fully maintained and repaired in accordance with the lessee's covenants in this lease contained).
>
> 4 (4) The full yearly market rental for the purpose of this lease shall mean the rent at which the demised premises are worth to be let with vacant possession on the open market as a whole between a willing Lessor and a willing Lessee for the remainder of the said term outstanding at the end of such third or twelfth year of the said term as the case may be upon the terms and conditions of this lease (excepting stipulations in this lease as to the amount of the rent but including like provisions as to rent revision from any later date herein specified) without any fine or premium being taken and there being disregarded:
>
> (i) any effect on rent of any tenants or trade fixtures or fittings installed in the demised premises by the Lessee or any occupying tenant or of any improvement or additions carried out to the demised premises not more than twenty-one years before the First day of October one thousand nine hundred and seventy-six otherwise than in pursuance of any obligation under this lease.
>
> (ii) any effect on rent of the fact that the Lessee or its tenant has been in occupation of the demised premises and
>
> (iii) any goodwill attached to the demised premises by reason of the business carried on thereat.

The issue of law between the parties concerns the assumptions which are to be made in assessing this full yearly market rental as defined.

The tenants submit that the arbitrator should assume:

> (i) that in the hypothetical market there is at least one hypothetical tenant bidding for the lease and (if it be found that there would be no other bidder than English Electric Company Limited) that that hypothetical tenant is English Electric Company Limited: and

> (ii) that the rent must be determined at an amount that the hypothetical tenant would agree with the claimants in the light of all the bargaining advantages and disadvantages that would have existed at the 1st October 1976 and in particular on the assumption that, if the parties failed to reach agreement on the rent, the hypothetical tenant was free to occupy two or more smaller properties as an alternative to taking a lease of the subject property.

The landlords submit that the arbitrator should assume:

> (i) that there is a person (who may or may not be English Electric Company Limited) who is willing to take the subject property as a whole on the terms of the lease offered (other than rent) and to enter into meaningful negotiations to arrive at a fair rent, that is to say a rent that is fair in all the circumstances of the case including rental values prevailing in the area: and
>
> (ii) that the rent must be determined at an amount that the claimants would agree with such person in all the circumstances, including in particular the circumstances (if such be found to exist)
>
> (a) that there is no other person willing to take a lease of the subject property on the terms of the lease being offered and
>
> (b) that there is no other property on the market which provides the accommodation and facilities provided by the subject property but disregarding the possibility that the parties would fail to reach agreement on a rent for the subject property.

The question of law is whether either and, if so, which of these submissions is correct.

As I have already said, the arbitrator heard a considerable body of evidence and found facts with which I am not concerned. The only facts material to this question of law have been found by the arbitrator and are set out in the special case. They are that:

> 1. There is no other property on the market which provides the accommodation and facilities provided by the subject property,
>
> 2. If the subject property had been vacant and in hand on the 1st October 1976:
>
> (a) the landlords could have adapted the property for letting in parts and could thereafter have let it in parts and
>
> (b) It is possible but unlikely that the landlords could have let the property on long lease to a single tenant who would thereafter adapt the property for subletting in parts and sublet it in parts.
>
> 3. The Findings set out in 2 above are subject to the following observations:
>
> (a) That in adapting the subject property for letting or sub-letting in parts the landlords or their single tenant would need to demolish parts thereof in order to provide access ways and loading facilities.
>
> (b) That any letting or sub-letting of the subject property in parts would entail considerable delays in finding tenants for all the various parts.
>
> (c) That any single tenant intending to adapt the subject property for sub-letting in parts would require a lease for a term of much longer than 18 years and would expect a rent free period and a reduced rent during the early years of the term.
>
> 4. That apart from the present tenants it is very unlikely on the available evidence and on the state of the market that there would have been a potential tenant in the market for the subject premises on the terms of the lease offered.
>
> 5. That in October 1976 if the present tenants had not been in occupation of the subject premises they would not have made any bid for the lease being offered.
>
> 6. That the present tenants in October 1976 could have split their operation into parts and could have moved each of those parts into smaller alternative premises in the United Kingdom.

Curiously there is no authority giving guidance upon how a rent review clause of this nature is intended to be operated. Despite the diligent research of counsel, all that has been discovered are authorities on rating valuation, compulsory purchase and taxation. These authorities are not irrelevant, but they apply only by analogy. In the circumstances, I propose to start by examining the instructions contained in the lease. The first, and most important, instruction is that the new rent is to be that which the premises are worth for letting on the open market with vacant possession on October 1 1976. It is implicit in this instruction, and is expressed in the later part of clause 4 (4), that the tenants are to be deemed to have moved out or to have never occupied the premises. The second instruction is that the premises are being offered in their actual condition (subject to the assumption as to maintenance and repair) on an 18-year lease, having the same terms *mutatis mutandis* as the present lease. The third instruction is that there is in fact a rent upon which a willing lessor and a willing lessee could and would agree as being the rent which the premises were worth on October 1 1976 for an 18-year term. This is implicit in the clause and is confirmed in fact by the arbitrator's alternative findings.

The first, and perhaps the most important, conflict between the parties is whether, in the application of the clause, the willing lessor is to be identified with the landlords and the willing lessee with the tenants. In a sense the willing lessor must be the landlords because only they can dispose of the premises, but for the purposes of the clause the landlord is an abstraction—a hypothetical person with the right to dispose of the premises on an 18-year lease. As such he is not afflicted by personal ills such as a cash flow crisis or importunate mortgagees. Nor is he in the happy position of someone to whom it is largely a matter of indifference whether he lets in October 1976 or whether he waits for the market to improve. He is, in a word, a willing lessor. He wants to let the premises at a rent which is appropriate to all the factors which affect marketability of those premises as industrial premises—for example, geographical location, the extent of the local labour market, the level of local rates, and the market rent of competitive premises, that is to say, premises which are directly comparable or which, if not directly comparable, would be considered as viable alternatives by a potential tenant.

Similarly, in my judgment, the willing lessee is an abstraction—a hypothetical person actively seeking premises to fulfil needs which these premises could fulfil. He will take account of similar factors, but he too will be unaffected by liquidity problems, governmental or other pressures to boost or maintain employment in the area and so on. In a word, his profile may or may not fit that of the English Electric Co Ltd, but he is not that company.

Against this background I turn to the findings of fact in order to consider their relevance:

(a) The fact that there is no other property on the market which provides the accommodation and facilities provided by the Walton Works is certainly relevant. But its effect upon the negotiations has to be balanced by two other factors. First, while the hypothetical tenant is a willing lessee, he is not an importunate one. He wishes to take a lease of the premises, but he is operating in a commercial field and in deciding what to offer by way of rent, will take account, covertly or overtly, of the alternative of taking a lease of two or more other premises. This is not to say that he would prefer that solution. This will depend upon the level of rent which is under consideration. He is a willing lessee and is quite content to take the Walton Works at the right price. It is just that he is not considering the proposition or negotiating in a vacuum.

(b) The fact that the property could be subdivided and let in parts or sublet and the advantages and disadvantages of that approach are irrelevant. The clause contemplates a letting as a whole.

(c) The fact that it is very likely that English Electric Co Ltd would have been the only potential lessee is relevant, but its relevance is indirect. It does not matter whether the only potential lessee was this company or the XYZ Co Ltd. What matters is that in the state of the market there was not likely to be more than one willing lessee. But the effect of this fact is not decisive because this single potential lessee is to be assumed to be a willing lessee—neither reluctant nor importunate, but willing. Just as the hypothetical lessor cannot rely overmuch upon the fact that no property similar to Walton Works is available on the market, so the hypothetical lessee cannot rely too much upon the fact that he has no competitors—he is, and is known to be, a willing lessee. Furthermore, it is known that he will remain a willing lessee so long as the willing lessor does not press his demand for rent beyond the point at which he is ceasing to act as a willing lessor and at which a willing lessee would cease to be such.

If authority be required for the proposition that monopoly positions on either or both sides do not render hypothetical agreements impossible, it is to be found in *Tomlinson v Plymouth Argyle Football Co Ltd* [1960] EGD 330, a rating case, and in *IRC v Clay* [1914] 3 KB 466, a taxation case.

(d) The fact that if the tenants had not been in occupation in October 1976 they would not have made any bid for the lease being offered is wholly irrelevant. I am quite prepared to accept that the tenants are most unwilling lessees. This does not matter. The arbitrator's concern is with the attitude of the hypothetical *willing* lessee, who is not in occupation of the premises. The clause assumes that there is such a person and it is nothing to the point to prove that there was not.

(e) The fact that the tenants in October 1976 could have split their operation into parts and could have moved those parts into smaller alternative premises in the United Kingdom is also quite irrelevant. It is irrelevant because it confuses the hypothetical lessee with the actual tenants. Certainly the hypothetical lessee would consider this alternative and it would affect the amount of the rent which he would be prepared to offer, but he would remain a willing lessee. The extent of the influence exerted by this alternative would depend upon the extent to which more than one lease and divided premises might be expected to increase his costs. And it is not English Electric's costs which come into the equation, but average and assumed costs of a hypothetical potential lessee.

I now turn to the respective contentions of the parties.

The tenant's contentions

(i) I accept that it is to be assumed that there is at least one hypothetical tenant bidding for the lease. Indeed I would go a little further than the tenants have gone. On the findings of fact, which were of course unknown to the tenants when their contention was formulated, the arbitrator would be justified in proceeding on the basis that, although there was one person who was willing to become lessee of the Walton Works, it was unlikely that there would be more than one.

I reject entirely the proposition that the potential lessee either is, or necessarily has any of the characteristics of, the English Electric Co Ltd. He is a complete abstraction and, like the mule, has neither pride of ancestry nor hope of posterity. He is someone whose needs are such that, in relation to the Walton Works, he is a willing lessee.

(ii) I accept that the rent must be determined at an amount which the hypothetical tenant—the willing lessee—would agree. I do not accept that the person with whom he has to agree that rent is the landlord. Such a concept imports the personal circumstances of F R Evans (Leeds) Ltd. The willing lessee has to agree this rent with a hypothetical landlord—the willing lessor.

The negotiations are assumed to be friendly and fair, but, subject to that qualification, would be conducted in the light of all the bargaining advantages and disadvantages which existed on October 1 1976. But these advantages and disadvantages are those which affected the property and any lessee of that property. The existence or absence of rival potential lessees of the property is merely one indication of the balance of advantage and disadvantage and by no means one which is necessarily decisive.

I do not agree that the hypothetical negotiations—or "higgling of the market place" to use the delightfully archaic phrase which occurs in some of the authorities—are to be conducted on the assumption that if the parties failed to reach agreement on the rent, the hypothetical tenants will be free to occupy two or more smaller properties as an alternative to taking a lease of the Walton Works. The parties *will* reach agreement and the willing lessee will not take up any of the alternative options. However, in the course of the higgling, he will point out to the willing lessor, and the willing lessor will accept, that taking two or more alternative properties is a theoretically available alternative option, the relative advantage or disadvantage of which will be reflected in the rent which is ultimately agreed. In the absence of real higglers, it is for the arbitrator to assess the reality of such an alternative option, its relative advantage or disadvantage and the extent to which the new agreed rent would reflect those facts.

The landlord's contentions.

(i) I accept that there is a person who is willing to take the Walton Works as a whole on the terms of the lease offered (other than rent), but I do not accept that that person may be the English Electric Co Ltd. For the purposes of fixing the new rent, he is not.

I accept that this person is willing to enter into meaningful negotiations, much though I dislike the word "meaningful", which in this context I take to mean neither more nor less than "genuine."

I accept that the negotiations are designed to arrive at the appropriate rent in all the circumstances which in fact affect the property and in theory affect the hypothetical lessor and lessee. I would not myself have used the expression "fair rent" which, like "meaningful," has in recent years been much bandied about in the political arena and may have become distorted in the process. However, I do accept that the appropriateness of the rent is to be judged in the light of all the circumstances of the case—other than those relating to the landlord and tenant as actual juridical persons—and that one of those circumstances is the rental values prevailing in the area. There are, of course, many others. The rent for which each is negotiating is that which is high enough to be acceptable to a willing lessor and low enough to be acceptable to a willing lessee. In the hypothetical life of hypothetical higglers, there is always one rent and never more that one rent which meets these criteria. If the arbitrator is heard to murmer "Oh happy hypothetical higglers" this is only too understandable. He has my sympathy.

(ii) I do not agree that the rent is one which F R Evans (Leeds) Ltd would agree. The person who has notionally to agree the rent is the hypothetical willing lessor.

I accept that the rent has to be agreed in the light of all the circumstances which in fact affect the property and in theory affect the hypothetical lessor and lessee. Any circumstance which affects the actual landlord and the actual tenant, but which would not affect the hypothetical lessor and lessee is irrelevant. I agree that these circumstances include, but stress that they are not limited to, the fact that it is unlikely that there will be more than one willing lessee and that in October 1976, which was the relevant date, there was no other property on the market which provided the accommodation and facilities provided by the Walton Works.

I also agree that the possibility of the parties' failing to reach agreement is to be disregarded—to borrow and adapt an immortal phrase "We are not interested in the possibilities of a failure to reach agreement. They do not exist." But as the negotiations proceed each will be considering whether it would not be better at a given level of rent to break off the negotiations. True it is that they will resist these temptations, but the extent to which they operate upon their respective minds will be reflected in the rent which will, notionally, be agreed in the end.

My comments on the respective contentions of the parties have involved some degree of repetition because these contentions overlap in some respects. For this I apologise. It will be seen that I do not wholly agree with either party's contentions and in some ways my disagreement is as much or more with what is omitted than with what is included.

I have been much pressed to decide in favour of one or other contention as this would avoid a remission to the arbitrator with some degree of further delay and expense. However, what is at stake is £225,000 per annum for a period of nine years—over £2m. I am inclined to think that the views which I have expressed tend to support the landlord rather that the tenant, but I cannot be sure that the arbitrator would adhere to his provisional assessment of £515,000 per annum if he were to read and apply those views. Justice therefore requires that the award be remitted to him for the making of a final award. Let me stress that he is completely free to adhere to the figure which he previously thought was appropriate or to substitute any other figure. He has heard all the evidence and I do not contemplate that he will require further evidence or argument, although I do not wish to fetter his discretion in any respect.

I therefore answer the question of law by holding that it is not correct to adopt wholly the contentions of either party and that, within the scope of disagreement indicated by the special case, the assumptions which should be made are those set out in this judgment.

Note

An appeal from the above decision of Donaldson J came before the Court of Appeal on November 11 1977. The Court of Appeal consisted of two judges only, this being technically an interlocutory appeal. The judgment of the court was given by Browne LJ at the invitation of Stephenson LJ. The court fully approved the decision and reasoning of Donaldson J and rejected criticisms made on behalf of the appellants. The appeal was dismissed with costs in the Court of Appeal and the court below. Leave to appeal to the House of Lords was refused. It is not proposed to report the Court of Appeal judgment. It may be noted that, following the judgment of Donaldson J, the arbitrator, on September 22 1977, made an award determining a rent of £515,000 per annum.

CHANCERY DIVISION

February 3 1978

(Before Sir ROBERT MEGARRY, Vice-Chancellor)

DREBBOND LTD v HORSHAM DISTRICT COUNCIL

Estates Gazette June 17 1978

(1978) 246 EG 1013

Rent review clause—Construction—Notice requiring review given by landlords in time, but notice requiring arbitration on failure to agree on revised rent not served by landlords in time—Whether time was "of the essence" for the arbitration notice—Clause provided for "notice in writing given to the tenant within three months thereafter but not otherwise"—Held that "and not otherwise" applied to both manner and time of service—Intention that time should be of the essence for the arbitration notice—United Scientific Holdings Ltd v Burnley Borough Council considered—Landlords' action for declaration fails

This was an originating summons by lessors, Drebbond Ltd, plaintiffs, to which Horsham District Council, lessees, were defendants, for a declaration in regard to the construction of rent review provisions in a lease for 21 years from November 1 1968 of 1 and 1a East Street, Horsham.

J L Knox (instructed by Bunker & Co, of Hove, Sussex) appeared on behalf of the plaintiffs; R C Pryor (instructed by M J Baty, Solicitor, Horsham District Council) represented the defendant council.

Giving judgment, MEGARRY V-C said: This case largely turns on the meaning and effect of the words "but not otherwise" in a rent revision clause in a lease. At a particular stage in the procedure laid down for rent revision, the clause enables the landlord to require arbitration "by notice in writing given to the tenant within three months thereafter but not otherwise. . . . " Do the last three words apply only to the manner in which the landlord can require arbitration (namely, "by notice in writing given to the tenant"), or do they also apply to the time within which the notice is to be given ("within three months thereafter") and thereby make time of the essence of the contract, so that a notice served after the end of the three months is ineffective? The background to the case is, of course, the recent decision of the House of Lords in *United Scientific Holdings Ltd v Burnley Borough Council* [1977] 2 WLR 806, in which it was held (I put it shortly) that in the time-table of a rent review clause in a lease time was presumed not to be of the essence in the absence of contra-indications in the language of the lease, in the inter-relation of the rent review clauses with other clauses in the lease, or in the surrounding circumstances. At p 822, Lord Diplock expressed the hope that the decision would reduce the number of cases in which it was necessary to resort to the courts to ascertain whether the landlord's delay had deprived him of the right to have the rent reviewed under particular rent review clauses; but this is the second such case to come before me this week, so the hope may still be hope deferred.

The lease was granted on November 11 1968 for a term of 21 years from November 1 1968. The rent was a peppercorn until January 1 1969, and thereafter £1,200 a year for the rest of the first seven years of the term: the £1,200 was called "The initial rent." The lessors were the immediate predecessors in title of the plaintiff company now before me, and the lessees were the defendant council. The assignment of the reversion occurred while the machinery of rent revision was in operation, but nothing turns on that. I shall refer to the plaintiff company and their predecessors in title alike as "the landlord," and the defendant council as "the tenant." Mr Knox appeared for the landlord and Mr Pryor for the tenant.

After the first seven years, the rent during the remaining years of the term was to be "the initial rent or such increased rent as may be payable in accordance with the provisions of clause 2 hereof"; and rent was to be paid on the usual quarter days. That brings me to clause 2. This begins as follows.

> 2. The initial rent payable by the tenant on the grant of this lease shall be reviewed in manner following namely:
>
> (a) The landlord shall be entitled by notice in writing given to the tenant within the last six months before the expiration of the seventh and fourteenth year respectively of the term hereby granted to call for a review of the yearly rent payable under this lease and if upon any such review it shall be found that the commercial yearly rent (as hereinafter defined) of the demised premises at the end of such seventh or fourteenth year as the case may be is greater than the initial rent then as from the end of such year as the case may be or as from the quarter day next following the giving of such notice whichever shall be the later the initial rent shall be increased to the commercial yearly rent provided that in no circumstances shall the rent payable hereunder following any such review be less than the rent payable by the tenant at the date of the landlord's notice calling for review.

Clause 2(b) then defines the "commercial yearly rent" in terms on which nothing turns, and I need not discuss: broadly it is the open market rent for a 21-years' term on the existing terms and conditions.

There is then clause 2(c), on which the dispute centres. It is not subdivided in any way, and of course I shall construe it in its existing form: but for ease of reference I have inserted numerals and letters to indicate its various parts, and I shall read it in this form.

> (c) (1) Such revision as aforesaid shall in the first instance be made by the landlord and the tenant or their respective surveyors in collaboration (2) but if no agreement as to the adjustment (if any) to be made in the yearly rent shall have been reached between the parties hereto or their surveyors within three months after the date of the landlord's notice calling for such revision (3) the question (a) as to the commercial yearly rent of the demised premises at the appropriate date and (b) as to whether there shall be any adjustment in the yearly rent payable and (c) if so what the

A amount of the adjusted yearly rent shall be (4) shall if the landlord shall so require by notice in writing given to the tenant within three months thereafter but not otherwise (5) be referred to the decision of a sole arbitrator to be appointed in default of agreement between the parties by the president for the time being of the Royal Institution of Chartered Surveyors in accordance with the Arbitration Act 1950 or any enactment for the time being amending re-enacting or replacing the same.

It is convenient to refer to the notice which clause 2(a) authorises the landlord to give as the "trigger notice," and the further notice which clause 2(c) (4) authorises him to give as the "arbitration notice." What has happened here is that B the landlord gave the trigger notice in time but failed to give any arbitration notice within the time stated. The last six months before the expiration of the seventh year of the term was the period from May 1 to October 31 1975; and when the landlord gave a trigger notice on May 5 1975, that plainly complied with clause 2(a). Once that notice had been given, the timetable laid down by clause 2(c) began to operate. There were negotiations between the respective surveyors to the parties, but no agreement was ever reached as to the adjustment to be made in the yearly rental. Under clause 2(c) (2), as no such agreement had been reached within three months of the date of the landlord's trigger notice (that is, C within three months of May 5 1975), on August 5 1975 the question of the rent and any adjustment was, under clause 2(c) (3), to be referred to arbitration under clause 2(c) (5), if the landlord gave the tenant an arbitration notice under clause 2(c) (4). Such a notice was, on the wording of the subclause, to be given "within three months thereafter"; and that must be within three months after August 5 1975, that is, by November 5 1975. That date came and went without the landlord giving the tenant any arbitration notice, and on December 22 the tenant wrote to the landlord asserting that the landlord had by then lost the right to a rent review. After taking advice the landlord wrote a letter on D November 19 1976 which appears to be capable of constituting an arbitration notice; but that, of course, was over a year after the date required for such a notice by clause 2(c) (4). The proceedings before me are by way of an originating summons taken out by the landlord, seeking a declaration that the landlord is entitled to require that the revision of the rent should be referred to a sole arbitrator in accordance with clause 2 of the lease.

The starting point is the approach laid down in the *United Scientific* case. Mr Pryor accepted, as he was obliged to do, that under rent review clauses time was presumed not to be E of the essence, and that it was for him to establish that in this case there was enough to rebut the presumption. In the forefront of the contention of each side was, of course, the language of clause 2(c) (4) of the lease. The question of the rent is to be referred to arbitration "if the landlord shall so require by notice in writing given to the tenant within three months thereafter but not otherwise." Mr Knox understandably resorted to the dictionary. "Otherwise," he said, referred to manner or way or respects, and not to time. The phrase "and not otherwise" related solely to the landlord making his requirement by a particular method, namely, by a notice in writing given to the tenant: the landlord must do F it thus, and in no other way. The phrase had no reference to the words "within three months thereafter," for they laid down a period of time rather than a manner of doing something. Mr Knox accepted that if these had been the only words governed by "and not otherwise," that phrase was wide enough in its meaning to be able to operate on them. But whereas here the phrase was preceded by words both of manner and of time, it applied only to the words of manner and not to the words of time.

For Mr Pryor, this approach was a selective disregard of the nearest antecedent. The phrase "but not otherwise" immediately followed the words "within three months thereafter": why should it leap over those words and attach G itself only to the preceding words? In any case, the phrase came at the end of a compound expression, beginning with the word "if," and it was wide enough in its meaning to govern the whole of the compound phrase. The one way in which the landlord could obtain an arbitration was by making a requirement which in all respects (including time) satisfied clause 2(c)(4); he could not do it "otherwise" or "differently." For good measure, Mr Pryor pointed to one of the illustrations of "otherwise" given in the *Shorter Oxford English Dictionary* on which Mr Knox relied. This was "I went at once; otherwise I should have missed him," and this, said Mr Pryor, displayed "otherwise" as doing duty in H relation to time.

Mr Pryor further contended that if the phrase "but not otherwise" governed only manner and not time it had little or no practical effect. If the landlord made a requirement of arbitration orally or by some document not given to the tenant, he could, if the three months' time limit did not apply, cure his default as soon as it came to his notice by giving the tenant a suitable piece of paper. Why insert the emphatic "but not otherwise" merely for this purpose? On the other hand, if those words of emphasis applied to the time limit as well, then there was some point in including them; and their purpose must be to make time of the essence. J

Mr Knox emphasised that once the trigger notice had been given in due time, both landlord and tenant knew that the rent revision machinery would operate so that, if the commercial yearly rent exceeded the initial rent, there would in due course be an increase in the rent. Even if it took a long time to settle the new rent, clause 2(a) made it run from the end of the seventh year or the quarter day next after the giving of the trigger notice, whichever was the later. There was therefore no need for the arbitration notice to be subject to a strict time limit.

Certain other points were discussed, but in the end they did K not seem to me to have any great cogency. I think that I have to decide the case mainly on the basis of the arguments that I have indicated above. So far as grammar and the use of the English language is concerned, I prefer Mr Pryor's approach. The phrase "but not otherwise" seems to me to have the broad sense of excluding all that is not comprehended in what lies between the word "if" and the phrase "but not otherwise," whether it relates to manner or to time. I think that Mr Knox ascribes to the phrase too delicate and selective an import in picking on manner and rejecting time.

There also seems to be considerable force in Mr Pryor's contention that the phrase, with its inherent emphasis, has L little significant function if Mr Knox is right. If it does make time of the essence for the arbitration notice (as I think it does), a sensible scheme emerges from clause 2(c). Once the landlord has given his trigger notice, clause 2(c) (2) allows three months for negotiation. Once that three months has run, the landlord can serve an arbitration notice. He may do it at once or he may leave the negotiations to continue. Much may depend on what the landlord thinks his prospects are from what has emerged from the negotiations. But if he is going to refer matters to arbitration, he must do so within the next three months. The time limit may well concentrate the minds and efforts of the parties and their valuers M so as to avoid the expense and expenditure of time that an arbitration entails. Remove the time limit, and leisurely negotiations may well become more leisurely still, and the tenant may have the prospect of an arbitration being required left hanging over his head indefinitely. I do not for a moment say that without the time limit the clause is unworkable: but I do say that I can see why the parties might wish to impose time limits that mean what they say, and so I can see why the words "but not otherwise" should be inserted in order to produce that result.

I should add a word on the meaning and effect of the

A phrase "and not otherwise" on the footing that, as I have held, it applies to the time limit. Mr Knox said that the "time to be of the essence" was well known to solicitors, and that, of course, is so. Doubtless it is in the same select gallery as "subject to contract," "without prejudice" and many other phrases sanctified by long usage in the law. If the intention was to make time of the essence, said Mr Knox, it would have been easy enough to say so; but that was not done, and the phrase in fact used did not suffice. No authority, I may say, has been put before me on expressions which do suffice to make time of the essence. However, I do not think that the fame of the expression "time to be B of the essence" means that nothing else will do, and I did not understand Mr Knox to contend for so far-reaching a C proposition. As far as the express wording of the document is concerned, all that seems to me to be needed is some expression which shows that the time limit is to mean what it says, and is to be obligatory and not merely indicative. The phrase "and not otherwise" seems to me to do just that. If it does not, then it is difficult to see what it does do; and it is plainly an insertion into the normal flow of the sentence, made for emphasis, and not a phrase which has just slipped in, as it were.

In the result, the landlord's claim fails, and I dismiss the claim for the declaration sought by the originating summons.

The claim was dismissed with costs. D

NEGLIGENCE

A **COURT OF APPEAL**

June 3 1977

(Before Lord DENNING MR, Lord Justice ROSKILL and Lord Justice SHAW)

RUMSEY v OWEN WHITE & CATLIN

Estates Gazette January 21 1978

(1977) 245 EG 225

Negligence of solicitors—Client allowed to sell with vacant possession when tenants in occupation protected by Part II of Landlord and Tenant Act 1954—Assessment of damages —Client to be placed in as good a position as if contract B **correctly performed—Loss likely to flow from breach— Efforts to mitigate loss—Entirety of loss due to original wrong advice—Deputy judge's order varied in quantum**

This was an appeal from a decision of Sir Douglas Frank QC, sitting as a deputy judge of the Queen's Bench Division, in an action for negligence arising from the sale by the plaintiff, Derek Leonard Rumsey, to Portgate Securities Ltd of shops and offices at 60, 60a and 61 High Street, Egham, Surrey, the action being against Owen White & Catlin, solicitors. The proceedings before Sir Douglas Frank were reported at (1977) 241 EG 611.

A L Price QC and H A P Picarda (instructed by Hewitt, C Woollacott & Chown) appeared for the appellants (defendants); R L Johnson (instructed by Rooks & Co) represented the respondent (plaintiff).

Giving judgment, LORD DENNING MR said: This is a tangled story. It arises unfortunately out of negligence on the part of a firm of solicitors. The problem is damages.

In 1973 there were three shops, 60, 60a and 61 High Street, Egham. The owner of them was a Mr Rumsey. Early in 1973 when the property market was exceedingly "buoyant" (as the estate agents describe it) Mr Rumsey was minded to sell the three shops and use the proceeds of sale to set up a factory at Basingstoke. He employed estate agents who intro- D duced a company, Portgate Securities Ltd, as purchasers. They were one of the subsidiaries of the Stern group. Mr Rumsey instructed a firm of solicitors, Owen White & Catlin, to draw up the contract of sale. It was dated April 26 1973. It took this form: the purchase price was £120,000 payable by four instalments of £30,000. The first instalment was payable at once on April 26 1973 (£30,000), the next two instalments on July 26 1973 and October 26 1973. The last instalment was payable on the date of completion. The date of completion was May 10 1974. Note the date. The contract contained the important provision: "vacant possession shall be given on E completion."

The purchasers duly paid the first three instalments of £30,000 each, making £90,000, by October 26 1973. But then a difficulty arose about giving vacant possession. The tenants in the shops were protected by the Landlord and Tenant Act 1954. Each was entitled to a new lease unless there were grounds for getting him out. The solicitors thought that Mr Rumsey had a right to get them out. Section 30(1)(f) of that Act enables a landlord to get a tenant out if he intends to demolish or reconstruct the premises. But the solicitors made a serious mistake. The only landlord who can avail himself of that provision is a landlord who himself intends to F demolish or reconstruct the premises. Here it was plain that Mr Rumsey did not himself intend to demolish or reconstruct the premises at all. So he could not himself get them out. Only the purchasers could get them out—and then only after the purchase had been completed. It was then and then only that they could say that they were landlords who intended to demolish or reconstruct the premises. It was in December 1973 that the mistake was discovered. The opinion of counsel was taken. It was then realised that Mr Rumsey could not implement his obligation to give vacant possession on May 10 1974 unless the tenants could be bought out or got out in some way. One of the tenants refused to be bought out. He G was determined to stay. So Mr Rumsey was unable to implement his contract—owing to his solicitors' mistake.

In an attempt to rectify the mistake the solicitors advised Mr Rumsey to try to make a rearrangement with the purchasers Portgate Securities Ltd so as to convey the shops to them forthwith and enable them to get the tenants out on the ground that they intended to demolish the shops or reconstruct them. Such a rearrangement was made. The solicitors advised Mr Rumsey to enter into a new agreement. It was made on January 28 1974. On that day the title to the property was at once conveyed to Portgate Securities Ltd. H By the agreement it was provided that vacant possession was to be obtained by June 7 1974 and the fourth instalment of £30,000 was thus to be paid. The solicitors no doubt thought to themselves, "If Portgate are the landlords, they can get the tenants out and vacant possession should be able to be given without difficulty by June 7 1974." This rearrangement was beneficial to Mr Rumsey in that he would get the full purchase price from the purchasers. It would also be beneficial to the solicitors, because they would have avoided the consequences of their own mistake. This rearrangement contained, however, some dangers for Mr Rumsey. The purchasers insisted that vacant possession must be given by June 7 1974. To protect the purchasers a clause was included J which provided that if Portgate did not get vacant possession by June 7 1974, then they could call upon Mr Rumsey to repurchase the property for the sum of £90,000. That was the sum they had already paid to him. On the repurchase he would have to pay back the £90,000: and there were stipulations by which he was to reimburse Portgate for the expenses and charges which they had incurred.

That agreement was made on January 28 1974. No doubt the solicitors thought that all would be well. But it was not well. One of the tenants, a Mr Stanley, refused to go. He objected. The matter went to the county court. The judge made an order that he was to go. But at this point it was K discovered that the solicitors had overlooked something else. They had forgotten (in January 1974) that under the Landlord and Tenant Act a tenant is given an additional three months: and, if there is an appeal, he can stay in possession until the appeal is determined. Owing to that oversight, it was impossible for the landlords to obtain vacant possession by June 7 1974. They did not obtain vacant possession by that date. So Portgate Securities, as they were entitled to do, called upon Mr Rumsey to repurchase the property, to pay back the £90,000 and all the costs they had incurred. Now comes the point. By this time in June 1974 when the purchasers required Mr Rumsey to repurchase the shops, the property market had fallen to the very bottom. The Stern Group had collapsed. So had many other companies. The property was now worth only £58,000. So here was Mr Rumsey committed to repurchase the property for £90,000 when it was worth only £58,000. He could not do it. He had not got that £90,000. He had been committed elsewhere to his Basingstoke venture, and the like. So he could not repurchase. Portgate remained the owners and claimed damages from him. The solicitors did not acknowledge their mistakes. They actually sent him their bill for all their costs. In answer he said: "It is you, the solicitors, who have really landed me in all this trouble, and I want damages from you for your negligence." He went to other solicitors and claimed damages from Owen White & Catlin.

The case came before Sir Douglas Frank in July of last year. It is now admitted that the solicitors were negligent in April 1973 in letting Mr Rumsey undertake to sell "with vacant possession," when he could not give it because there were tenants protected by the Landlord and Tenant Act.

Now what are the damages to be? We have had a lot of discussion. The principles are well established. They start with *British Westinghouse Electric & Manufacturing Co Ltd v Underground Electric Railways Co of London* [1912] AC 673. The general principle is that "he who has proved a breach of a bargain to supply what he contracted to get is to be placed as far as money can do it in as good a situation as if the contract had been performed." So I ask myself, if the contract of employment had been performed, that is, if the solicitors had given Mr Rumsey the right advice and told him that he should only sell subject to the tenancies; and that he should let the purchasers get the tenants out as they could, what price would he have got for the property then? There was a good deal of discussion about this before Sir Douglas Frank. Various figures were stated. £102,000 was suggested on behalf of the solicitors. £105,000 was mentioned. Mr Rumsey's advisers said £120,000. Eventually the judge, after hearing all the evidence, came to the conclusion that the figure of £112,000 was the right figure. I will not go into all the arguments and the pros and cons. It seems to me that if £120,000 (payable by instalments over a year) was the price that Portgate were willing to pay for it with vacant possession at the end of a year, it is not unlikely that £112,000 would be the price which might well have been obtainable in April 1973 by a purchaser who was ready to pay that sum down and get the tenants out himself. I would not interfere with the judge's finding that, if the solicitors had given proper advice, £112,000 would have been obtainable in April 1973.

Going on with the story, we know that Mr Rumsey had in fact received £90,000. He had not got the property any longer—the property had been conveyed to the purchasers. So he received £90,000 for the property for which he ought to have got £112,000. So on that simple calculation he has lost £22,000. It seems to me that that is the first head of damage. Then there arises this big problem: it is said by Mr Rumsey that, owing to the fault of the solicitors, much damage has followed because he has been exposed to a liability to Portgate Securities Ltd. They are suing him for damages because he did not repurchase the property for the £90,000 as he ought to have done and paying the costs and so forth. They are suing him for damages because they say that, when he ought to have repurchased, it was worth only £58,000: and they have lost the difference in value and all the costs etc. Portgate have got this big claim against him. We are told that Mr Rumsey has no defence to their claim except as to the amount of damages. Portgate have actually got judgment for part of their claim, amounting to £23,000. They want more and there is to be an inquiry as to damages. So he says he is exposed to this liability to Portgate. The question is whether or not he can recover it against the solicitors.

It seems to me that we have to deal with this in accordance with the ordinary principles as to the measure of damages for breach of contract. They have been stated in *Koufos v C Czarnikow Ltd* [1969] 1 AC 350, and *H Parsons (Livestock) Ltd v Uttley Ingham & Co Ltd* [1977] 3 WLR 990. The damages are those which could reasonably be in contemplation at the time of the breach as being not unlikely to flow as a consequence of the breach.

It seems to me that, if the solicitors had had it in mind that Mr Rumsey could not give vacant possession, they would reasonably have contemplated that the solicitors and Mr Rumsey would try to make some rearrangement with the purchasers. That is exactly what was done in January 1974. They did what seems to me to be entirely reasonable. They made a bona fide attempt to save the situation. If only it had been saved—by getting that other tenant out by June 7 1974—all would have been well. The property would have been purchased for the £120,000, and the deal would have gone through without question. It was a very good arrangement which was reasonably in contemplation. But also it must have been in contemplation that the new agreement might fall down and might not be fulfilled. That is shown by the very terms of it: because it made provision as to what was to happen if vacant possession were not given by June 7 1974. It said that Mr Rumsey was to repurchase for £90,000. So they must have contemplated that it was not unlikely that the agreement would not be fulfilled. Then it is said: "Oh, well; Mr Rumsey ought to have fulfilled the contract himself. His failure to repurchase it for the £90,000 broke the chain of causation. It was his own fault; he ought to have paid back that £90,000." The answer to that is that the solicitors and everyone there knew that that was not a practical proposition. Reading through the correspondence and the matters which have been put before us, it is quite plain that the solicitors knew perfectly well that this money he was getting—this £90,000—was to be used for his other venture in Basingstoke. So it must have been within the reasonable contemplation of the parties that he might not be able to repay that £90,000; and it must also have been within the reasonable contemplation of the parties that it was not unlikely that that would expose him to claims by Portgate.

As to the undertaking in January 1974 (which really has caused a lot of the trouble) it seems to me that it was an entirely reasonable step to take on the part of everyone, but unfortunately it miscarried. In this context I would recall the words in the case of *Banco de Portugal v Waterlow & Sons Ltd* [1932] AC 452 at p 506 of Lord Macmillan. He said:

> Where the sufferer from a breach of contract finds himself in consequence of that breach placed in a position of embarrassment the measures which he may be driven to adopt in order to extricate himself ought not to be weighed in nice scales at the instance of the party whose breach of contract has occasioned the difficulty. . . . The law is satisfied if the party placed in a difficult situation by reason of the breach of duty owed to him has acted reasonably in the adoption of remedial measures and he will not be held disentitled to recover the cost of such measures merely because the party in breach can suggest that other measures less burdensome to him might have been taken.

So there it is. Mr Rumsey was placed in a very difficult position in January 1974 by reason of the unfortunate wrong advice by the solicitors in April 1973. He took measures—very reasonable measures (in which the solicitors were advising him)—to try to overcome the difficult situation which existed. In those circumstances, the consequence of those measures, unfortunate as they are, is that he has been landed with this big liability to Portgate Securities Ltd. It seems to me that it must have been within the reasonable contemplation of the parties that this was not unlikely to happen as a consequence of this initial negligence.

Mr Price has urged upon us very strongly that the January 1974 agreement was a new departure altogether. It was a cut-off point, he said. Any consequences arising from that agreement were the consequences of Mr Rumsey's own failure to repay that £90,000, he said. For the reasons I have given, I do not think it was a cut-off point. I do not think there was a break in the chain of causation. All these consequences flowed one after the other from the initial fault. So it seems to me that Mr Rumsey not only gets his £22,000 but he ought to recover a further sum such as would represent his proper liability to Portgate in the unfortunate situation which has happened. I do not think it is possible to ascertain that sum now. It is the subject of proceedings between Portgate and Mr Rumsey. It may be that the property—I see there is a reference in the papers—has gone up in value much above £58,000. Someone suggested that it has gone up to £100,000. That increase in value may well go to diminish the claim by Portgate against Mr Rumsey. So it may be that Portgate will not be awarded such a large sum against Mr Rumsey as was at first suggested. At all events, I do not think that the amount of the further sum can be resolved until after the proceedings between Portgate and Mr Rumsey are disposed of. When that is done, this case should be mentioned again or dealt with by a master in some other way so that the final figure or judgment against the solicitors can then be ascertained. But at the moment it seems to me that the proper figure should be not £54,000 as the judge thought but £22,000 plus a further sum to be ascertained in a suitable manner after the disposal of the proceedings against Mr Rumsey.

I would vary the order of the judge accordingly.

Agreeing, ROSKILL LJ said: I venture to add to my Lord's judgment because of Mr Price's excellent argument.

There can be no doubt that the defendants failed gravely and sadly in their duty to the plaintiff. It is strange, to my mind, that that was challenged at the trial. The learned deputy judge found that the solicitors had been negligent, as they obviously were. The only question with which we are concerned is what the proper damages are for that undoubted negligence. Mr Price put in the forefront of his argument that there was either no damage or no appreciable damage suffered because he claimed that one looked only at the position on the relevant date and calculated the difference alleged in the value of this property with and without vacant possession. Then one looked at the evidence of the two valuers. One saw that on that evidence the two figures would very nearly meet. So, basing himself on the Sale of Goods Act cases, he submitted that there was no appreciable loss.

It would be strange if that were the position, because the man in the street looking at the facts here could not fail to reach the conclusion that Mr Rumsey has suffered a very serious loss, and it would be curious, to say the least, if, notwithstanding that obvious loss, the law denied him any relief.

Mr Price founded his argument on a number of cases where solicitors have been in trouble for negligent advice given to purchasers of property. For my part, I find no help in those cases. One has in this case, the facts of which are now happily undisputed although somewhat complicated, to resort to first principles in order to ascertain what the right damages are. It has often been pointed out in the House of Lords and elsewhere that there is no certain rule for the assessment of damages which can be universally applied in all cases. Cases of breach of contract vary enormously in their character both in the nature of the contract and in the nature of the breach, and the court must always look at the realities of the position and try as best it can to see what loss it is that the innocent party has suffered subject, of course, to rules as to remoteness of damage and as to mitigation of damage. But it is essential as my Lord has said, to determine this case by reference to first principles of assessing damages for breach of contract.

One finds those first principles stated in classic terms by Lord Haldane, then Lord Chancellor, in *British Westinghouse & Manufacturing Co Ltd v Underground Electric Railways Co of London Ltd* [1912] AC 673 at p 689. He said: "The fundamental basis is thus compensation for pecuniary loss naturally flowing from the breach." Lord Reid said very much the same thing, as my Lord has already mentioned, in *Koufos v C Czarnikow Ltd* [1969] 1 AC 350 at pp 382 and 383, where he said:

> So the question for decision is whether a plaintiff can recover as damages for breach of contract a loss of a kind which the defendant, when he made the contract, ought to have realised was not unlikely to result from a breach of contract causing delay in delivery.

That case, of course, was concerned with delay in delivery. His Lordship went on:

> I use the words "not unlikely" as denoting a degree of probability considerably less than an even chance but nevertheless not very unusual and easily foreseeable.

The question with which we are primarily concerned is one of causation. My Lord has stated the facts, and I only repeat them to the extent necessary to explain what I am about to say. The original agreement was entered into in April 1973. That was when the breach of duty occurred. A subsequent agreement was entered into in January 1974. It seems to me plain that when that second agreement was entered into—I pass over the fact that Mr Rumsey was given no opportunity for independent advice because that is not relevant—it was in the hope of mitigating the damage for which the defendants must by then have known perfectly well, although Mr Rumsey did not, they were going to be liable if this deal fell through, because completion with vacant possession could not be given. It had always been the position under the original April 1973 contract that that £90,000, payable by those three first instalments, would have to be repaid if completion with vacant possession were not given. The test of what is in the contemplation of the parties is an objective test, and it must be taken to have been both in the minds of Mr Rumsey and the solicitors at the time of that April agreement that if vacant possession was not given that £90,000 would have to be repaid. That position seems to me to remain entirely unaffected by what happened the following January. That arrangement in effect was that the date for completion, again with vacant possession, would be postponed, but on the express terms this time that if it were not given the £90,000 would be repaid. Thus it must be taken, I think, to be all part of one continuing transaction which ought never in the first instance to have been entered into having regard to the virtual certainty that Mr Rumsey was not going to be able to give vacant possession. All this trouble arose because of a sensible agreement entered into in January 1974 for the purpose of mitigating the damages for which the solicitors were going to be liable.

Mr Price said that the major part of the loss which arises from Mr Rumsey's undoubted liability to his purchasers, whatever it may be in terms of money, is too remote to recover as damages from the solicitors. The argument runs thus. This liability arose not from the original breach of duty in April 1973 but from Mr Rumsey's failure to comply with the January 1974 agreement in July 1974 because he could not then complete with vacant possession. The argument has been most attractively presented and at first I thought it had force, but the more I have thought about it the more I realise that it contains a fundamental fallacy, in that it ignores that one must regard the whole story as one continuous matter. First, there was a breach of duty committed in April 1974; then an agreement was entered into to mitigate that breach, and if it had enabled the rescue operation to take place it would have been to the benefit of the defendant solicitors as well as to the benefit of the plaintiff; and then there was the final failure to complete with vacant possession in July 1974. I do not think it can be said that there was a break in the chain of causation so that the ultimate liability to Portgate arises from Mr Rumsey's deliberate decision, no doubt brought about by his then financial position, not to pay back the £90,000 and not from the defendants' breach of duty. It seems to me that that loss arose directly from the original breach of duty.

My Lord referred to the well-known speech of Lord Macmillan in the *Waterlow* case, and I will not repeat it. The courts do not look favourably upon claims by contract-breakers that they should not pay for what has so plainly arisen from their own breach of contract merely because a plaintiff has suffered a loss in trying to get himself out of trouble not of his own making. It seems to me that it would be quite unrealistic to try to break this case down into separate compartments and say that one part of the loss flows and the other part of the loss does not. The entirety of the loss, in my view, flows from the original wrong advice and the exposure of Mr Rumsey by that wrong advice to a contract he should never have been allowed to enter into. At the outset of his argument Mr Johnson said that given proper advice Mr Rumsey would have sold the property for £112,000 with the tenants in possession. In fact he has received £90,000 and has been exposed to the Portgate claim. If that argument is right, then the damages are £22,000 plus £X. The " X " is at present not ascertained. That argument has the great merit of simplicity, and the more I have thought about it the more I have realised that there is no answer to it.

There remains, therefore, the problem of how the matter should be dealt with. I do not think, with respect, that the learned deputy judge approached it in quite the right way. He may have intended to have reflected in the figure of £54,000 some part of the liability to Portgate, but he did not express it in a way in which effect could be given to his intention, if that was his intention. It seems to me that the right way of dealing with this, as my Lord has suggested, is to give Mr Rumsey judgment now for £22,000, but the remainder of his entitlement to damages seems to me to be incapable of precise assessment unless and until Portgate's claim against him has been quantified. The position at the moment is that there is interlocutory judgment against Mr Rumsey for the sum of £23,000. For my part, I do not see why that sum is not recoverable from the defendants. I think it plainly is. How much more is recoverable must await the determination of that other case. Therefore, subject to the views of my Lords, I would make an order in some appropriate form that after that claim has been determined in the Chancery Division would be the appropriate time to fix the final sum for which the defendants are liable to the plaintiff. Having regard to the decision of this court in *Trans Trust SPRL v Danubian Trading Co Ltd* [1952] 2 QB 297 the present would not be the right time to make a declaration as to an indemnity.

For those reasons, which I have given at some length in deference to the learned deputy judge, I would vary his order. But subject to that the appeal should be dismissed.

SHAW LJ also agreed.

The appeal was dismissed and the respondent's cross appeal allowed, with costs. Judgment was given for the respondent for £22,000 and interest, together with an additional sum to be ascertained finally after proceedings between Portgate Securities Ltd and the respondent had been determined. Leave for the appellants to appeal to the House of Lords was refused.

RATING

A **COURT OF APPEAL**

February 25 1977

(Before Lord Justice CAIRNS, Lord Justice STEPHENSON and Lord Justice WALLER)

JLG INVESTMENTS LTD v SANDWELL DISTRICT COUNCIL

Estates Gazette January 14 1978

(1977) 245 EG 137

General Rate Act 1967, section 17 and schedule 1—Rating of unoccupied property—Completion of newly-erected
B **building—Service of completion notice by local authority —Work remaining to be done after substantial completion —Partitioning of certain floors not carried out—Period "reasonably required for carrying out the work"—Claim rejected that partitioning could not reasonably be started until tenants were found and consulted—Statute did not mean the period reasonably required for carrying out the work after finding a tenant—Appeal from county court judge dismissed**

This was an appeal from a decision of Judge Davison at West Bromwich County Court who had rejected an appeal by owners of property at 420 High Street, West Bromwich,
C under paragraph 8(4) of Schedule 1 to the General Rate Act 1967 against a completion notice served by Sandwell District Council in respect of an unoccupied recently-erected building. Judge Davison had dismissed this appeal.

C Fay (instructed by Silvermans) appeared on behalf of the appellants; M H Spence (instructed by Geoffrey Albert Hadley, Town Hall, West Bromwich) represented the respondents.

Giving judgment, CAIRNS LJ said: This is an appeal from a decision of His Honour Judge Davison at the West Bromwich County Court relating to a completion notice served by the respondents to the appeal, Sandwell District
D Council, under Schedule 1 to the General Rate Act 1967 specifying a date on which a certain building was to be treated as completed. The building, which was in fact part of a building—it was a building for the purpose of the schedule—consisted of the ground floor, the first floor and the second floor of Kingston House, 420 High Street, West Bromwich, of which the appellants, JLG Investments Ltd, were and are the owners. The notice was served on April 12 1976, stating that the premises could reasonably be expected to be completed within three months of that date—or, rather, actually by July 13 1976, which was no doubt taken as three months from the date of receipt by the appellants of the notice. The learned
E county court judge had to decide whether within the meaning of the schedule the premises could reasonably be expected to to completed by that date. He decided that they could, and accordingly dismissed the appeal which the appellants had brought to him against the notice.

The relevant legislation is contained first in section 17(1) of the General Rate Act 1967, which provides:

> A rating authority may resolve that the provisions of Schedule 1 to this Act with respect to the rating of unoccupied property shall apply . . . to their area, and in that case those provisions shall come into operation . . . in that area on such day as may F
> be specified in the resolution.

Such a resolution was carried in this case.

Under Schedule 1 it is provided in paragraph 1(1):

> Where, in the case of any rating area in which, by virtue of a resolution under section 17 of this Act, this Schedule is in operation, any relevant hereditament in that area is unoccupied for a continuous period exceeding three months, the owner shall, subject to the provisions of this Schedule, be rated in respect of that hereditament for any relevant period of vacancy; and the provisions of this Act shall apply accordingly as if the hereditament were occupied during the relevant period of vacancy by the owner. G

There are certain definitions in paragraph 15 of the schedule: "'building' includes part of a building"; "'owner' . . . means the person entitled to possession of the building"; "'relevant hereditament' means any hereditament consisting of" to put it shortly "a building"; and "'relevant period of vacancy' in relation to any relevant hereditament, means, subject to paragraph 1(4) of this Schedule, any period beginning with the day following the end of a period of three months during which the hereditament has been continuously unoccupied and ending with the day preceding that on which the hereditament becomes or next becomes occupied or ceases to exist." H

Paragraph 7 provides:

> For the purposes of paragraph 1 of this Schedule, a newly-erected building which is not occupied on the date determined under the subsequent provisions of this Schedule as the date on which the erection of the building is completed shall be deemed to become unoccupied on that date.

Paragraph 8:

> (1) Where a rating authority are of opinion—(a) that the erection of a building within their area has been completed; or (b)—

and this is the relevant subparagraph—

> that the work remaining to be done on a building within their J
> area is such that the erection of the building can reasonably be expected to be completed within three months, and that the building is, or when completed will be, comprised in a relevant hereditament, the authority may serve on the owner of the building a notice (hereafter in this paragraph referred to as "a completion notice") stating that the erection of the building is to be treated for the purposes of this Schedule as completed on the date of service of the notice, or on such later date as may be specified by the notice.

Then subparagraph (4):

> A person on whom a completion notice is served may, during the period of twenty-one days beginning with the date of service K
> of the notice, appeal to the county court against the notice on the ground that the erection of the building to which the notice relates has not been or, as the case may be, cannot reasonably be expected to be completed by the date specified by the notice.
> (5) If a completion notice served in respect of a building is not withdrawn and no appeal in pursuance of sub-paragraph (4) of this paragraph is brought against the notice or such an appeal is abandoned or dismissed, the erection of the building shall be treated for the purposes of this Schedule as completed on the date specified by the notice . . .

A Paragraph 9 provides:

> In the case of a building to which work remains to be done of a kind which is customarily done to a building of the type in question after the erection of the building has been substantially completed, it shall be assumed for the purposes of paragraph 8 of this Schedule that the erection of the building has been or can reasonably be expected to be completed at the expiration of such period beginning with the date of its completion apart from the work as is reasonably required for carrying out the work.

It is on that last-mentioned paragraph that the council relied in this case, the notice being to this effect:

> B Whereas you are the owner of the ground floor, first and second floors in the office block known as Kingston House, 420 High Street, West Bromwich, within the rating area of the Metropolitan Borough of Sandwell . . . and whereas the rating authority is of opinion that *the work remaining to be done on the said building is such that the erection of the building can reasonably be expected to be completed within three months* and that the said building is, or when completed will be, comprised in a relevant hereditament as defined in paragraph 15 of Schedule 1 to the General Rate Act 1967. You are hereby notified in accordance with the provisions of the said Act that the erection of the said building is to be treated for the purposes of the said Schedule 1 as completed on July 13 1976.

C It was against that notice that the appeal to the county court was brought in accordance with the provisions in the schedule.

The evidence before the learned judge consisted of the evidence of a surveyor on behalf of the appellants and two witnesses on behalf of the council. Mr Quinney, the surveyor called for the appellants, said that the contractors moved out on December 16 1975; that since then the appellants had been trying to let the property in total or in part; they had successfully let the third, fourth and fifth floors; they had not let the ground, first and second floors; they had expended considerable D sums in advertising the property to let; and it was costing them a good deal of money to have the premises partially unoccupied. There was a carpet on the first floor only, a sort of show office; the light fittings on the first floor only; all the floors had ceilings fitted with acoustic tiles and they would be required to fit the lights; and there was wiring from the plugs in the skirting board. But he said—and this is a matter of great importance in this evidence—

> There is no partitioning on floors 1, 2 and 3. The large open plan office is not often used these days. I can't envisage anyone taking on a floor without any partitioning. The tenants of 3, 4 and 5 are providing their own carpets and partitioning.

E He dealt further with the question of carpets and tiles, and said at the end: "At this moment I expect it to take two months."—by which he must have meant two months to complete the work if it were started then, and, in particular, to complete the partitioning.

. The district rating officer called for the council estimated that the remaining work to be done would take three months. He accepted that partitioning would need to be done. The other witness called for the council was a senior building surveyor. He estimated that it would take two months to complete the building from the state in which it then was, that F is to say, that the physical work would take two months.

The case that was presented on behalf of the appellants was that the work of completion, in particular the partitioning, was work which could not reasonably be expected to be started until tenants were found, because the exact form of the partitioning would be something which the tenants would wish to be consulted about, and it would be inconvenient if partitioning were put in and it was later found that they wanted the partitioning in a different place. The learned judge rejected that argument, taking the view that what was material was the time that would be taken for the actual work of putting in the partitioning and anything else that was still G required to be done, and that it was not relevant to take into account the time that might be taken in finding a tenant. He accepted that there was no deliberate delay on the part of the defendants in seeking a tenant, and that they were making all reasonable efforts to find one. But he said at the end of his judgment:

> . . . the building could be completed from today in three months if there was an intention to do so. I think it could have been completed in three months from the date of the completion notice or when Mr Beasley inspected.

Accordingly, he dismissed the appeal.

H It is contended on behalf of the appellants, in a very careful and interesting argument presented to the court by Mr Fay, that what the schedule contemplates in referring to a reasonable time for completion includes in such a case as this a reasonable time—no more—which might be taken to find a tenant. I may say that Mr Spence, on behalf of the council, concedes that if that is the right test, then on the evidence before the learned judge it could not be said that the completion could reasonably be expected to take place within three months of April 13. So the issue for us is: Was the learned judge right in excluding such reasonable time as might be taken to find a tenant?

J Mr Fay argues, and Mr Spence concedes, that if one looked only at paragraph 8 of the Schedule that might well be the case. But the problem here is whether the same test is to be applied under paragraph 9, having regard to the use of the words "reasonably required for carrying out the work." Mr Fay contends that paragraph 9 is directed only to excluding unreasonable delay, and that there has been no unreasonable delay in any respect on the part of his clients here. He asks us to say that the words "required *for* carrying out the work" are not the same as "required *in* carrying out the work," and points to matters incidental to the carrying out of the K work in addition to the physical activities of performing it. He says that some incidental matters must surely be taken into account, such as, for instance, the time taken for the delivery of goods, and perhaps time taken for the preparing of plans for the work and time taken in obtaining consent under building regulations for it. I express no opinion upon any matters of that kind. They might be regarded as something which was indeed incidental to the carrying out of the work, in the sense that they were an essential part in addition to the actual job of the workmen in putting in partitions and that sort of thing. But it seems to me that it is a very different matter to say that time required for carrying out L the work can include also the finding of a tenant, however convenient it may be to do so, before the work is started.

Mr Fay says that fairness requires the construction for which he asks, because if there has been no dilatoriness on the part of the owners there is no reason why they should be penalised by having to pay rates when the building is unoccupied. As Mr Spence has pointed out, the same might be said in any case where the building has actually been completed, and where undoubtedly if the notice is given the rates would become payable at the end of three months from the notice that the building was completed, although in that case it might well be that every effort had been made to M obtain tenants without success. It does seem to me that both on its literal meaning, and having regard to the purpose which Parliament must have had in mind in enacting paragraph 9 of the Schedule, it ought to be interpreted in the way contended for on behalf of the council and accepted by the learned judge in this case. I think it would be a straining of the meaning of the words "reasonably required for carrying out the work" to say that that expression includes the time taken, however reasonably, in finding tenants willing to occupy the building. As to the purpose of it, I cannot see any answer to the submission made by Mr Spence that unless it has the meaning accepted by the learned judge in this case

A paragraph 9 really provides for nothing more than was already provided for in the earlier paragraphs of the schedule. It is, it seems to me, for just such a case as the present that this special requirement is introduced, that when the building is substantially completed it is to be assumed for the purpose of paragraph 8 that the completion can be expected to take place at the end of the time reasonably required for the carrying out of the work.

I think the learned judge put the right construction upon the provision in this case, and I would dismiss the appeal.

Agreeing, STEPHENSON LJ said: I share the sympathy which the learned judge expressed for the appellants and should have B liked to accede to Mr Fay's submissions, but, in my judgment, for the reasons given by my Lord, the language of paragraph 9 is too plain. In its context "period reasonably required for carrying out the work" does not mean "the period reasonably required for finding a tenant and carrying out the work as directed by the tenant," or "the period reasonably required for carrying out the work after finding a tenant." I would put the construction on the words which my Lord has put, although paragraph 9 was not referred to by the learned judge. I think that that decides this appeal and shows that the learned judge's conclusion was right. I too C would dismiss the appeal.

Also agreeing, WALLER LJ said: Mr Fay's argument depended in part on giving a wide meaning to the word "reasonably" in paragraph 8 of the schedule. He submits that the word is there to safeguard the owner of the property, and that the words "can reasonably be expected to be completed within three months" should, therefore, make allowance for finding a tenant. He says that if completion requires a tenant and you cannot find one, then the building cannot reasonably be expected to be completed until at least you have a tenant; and if paragraph 8 were alone this might be a possible construction. But Mr Spence has submitted, D as my Lords have already said, that where there has been substantial completion, then paragraph 9 applies: and this is such a case. Mr Fay sought to draw a distinction between the use of "reasonably" in paragraph 9. The words, ". . . the erection of the building has been or can reasonably be expected to be completed at the expiration of such period beginning with the date of its completion apart from the work as is reasonably required for carrying out the work," show that there are two different situations to be envisaged, and show, in my opinion, that they cover precisely this case, and if they were not given that meaning there would not be any other reasonable meaning which could be attributed to them.

E The appeal was dismissed with costs. Leave to appeal to the House of Lords was refused.

QUEEN'S BENCH DIVISIONAL COURT

October 6 1977

(Before Lord WIDGERY CJ, Mr Justice CANTLEY and Mr Justice PETER PAIN)

NEWCASTLE-UNDER-LYME BOROUGH COUNCIL v B KETTLE LTD

F Estates Gazette March 25 1978
(1977) 245 EG 1027

Rating—General Rate Act 1967—Important decision on difficult point—Newly-erected hereditament—Proposal by valuation officer for inclusion in list—Action by rating authority to obtain rates more quickly by exercise of powers under section 6 to make amendments in rate—Apparent conflict between section 2(4) and section 6—Provided a proposal by valuation officer is on foot the rating authority may use powers under section 6 to bring new hereditament into charge—Observations on history of rating and "nebulous conception" of action by authority on rate— G Magistrates congratulated but overruled

This was an appeal by case stated brought by Newcastle-under-Lyme Borough Council to challenge a decision by magistrates at Newcastle-under-Lyme refusing an application by the council for a distress warrant in respect of the non-payment of general rates by the respondent company, B Kettle Ltd. The claim for rates was in respect of premises which had been recently erected by the company.

G E Moriarty QC and A Porten (instructed by E Wetherell, Solicitor to Newcastle-under-Lyme Borough Council) appeared on behalf of the appellants; W Glover QC and N J Worsley (instructed by Tinsdill, Hollinshead & Moody) represented H the respondents.

Giving judgment LORD WIDGERY CJ said: This is an appeal by case stated by justices for the Petty Sessional Division of Newcastle-under-Lyme in respect of their adjudication as a magistrates' court sitting at Newcastle-under-Lyme on March 18 1976. They had before them a case of more than ordinary complexity, and I will begin by expressing my appreciation of the efforts which they have made. The case came before them as an application by the present appellants, the Newcastle-under-Lyme Borough Council, for the grant of a distress warrant in respect of non-payment of general rates against the respondent company, who are B Kettle Ltd, in J respect of their premises situate at Newcastle-under-Lyme, where the appellant council claimed that the sum of £9,100-odd was owing to the rating authority in respect of general rates, plus 50p in respect of costs.

An interesting legal argument arose on the facts of this case because those facts were never in any sense in dispute. What had happened was that the premises of Messrs Kettle, which were the subject of the case, had been recently erected and had come into rating recently. That meant of course initially that they did not appear in the valuation list for the rating area concerned. Action was taken by the valuation K officer to bring the new hereditament into the valuation list by making a proposal in the ordinary way. As I understand it, this proposal was made on September 1 1975. If acceptable, or so far as it is acceptable, it would affect the rate in respect of the premises for the year beginning April 1 1975, the rule being that such amendments date back to the beginning of the rating year.

If nothing else had been done, the proposal made by the valuation officer would have been considered in due course by the local valuation court, and in due course an assessment would have been fixed. The respondents would have been liable to pay the rates on that assessment. But perhaps L because the rating authority were anxious to get on and did not think that the machinery was moving fast enough, they took action themselves in order to obtain payment of rates on this substantial new hereditament more quickly than they would have obtained payment had they not taken direct action. As we shall see in a moment, the action which they took is said to be authorised by the early sections of the General Rate Act 1967, and particularly section 6.

It is necessary to have a moment's thought about the history of rating in order to get this case into its proper context. It is not so very long ago that there were no such things as valuation lists and valuation officers and where the M rate book was kept by the local authority and displayed, for all who were interested, the names of the occupiers and particulars of the property and their rateable value. If anybody was dissatisfied with the obligation to pay rates which fell upon him, he could appeal to quarter sessions in cases of that kind. All that has gone. We now have valuation lists, which are a record of the new properties and their value for rating purposes. The valuation list is prepared independently of the rating authority by the valuation officer subject to the two-tier appeal of the local valuation court and now the Lands Tribunal. The function of the rating authority in

A the main (that is to say in the vast majority of new hereditaments) is to fix the poundage which is to be charged having regard, on the one side, to the total rateable value in that area, and, on the other side, to their estimate of costs for the relevant period. It is their function to assess what the rate poundage would be, and when they make that assessment and publish or approve it, they are, in my judgment, making the general rate for the ensuing years. We find in section 2(1) of the General Rate Act 1967 an obligation on every rating authority from time to time to exercise their powers under the Act and make such rates as will be sufficient to provide for the total estimated expenditure.

B That would all be very straightforward and easy were it not for the fact that one finds, floating in a rather nebulous form as a background to this modern legislation, some conception that there exists a rate against which an appeal can be made. Indeed the Act in section 7 specifically provides for the presentation of an appeal to the court. But in section 7 an appeal against the rate goes to the Crown Court. There are other instances as well in which one can see that the draftsman of the Act of Parliament or author of the textbook does recognise this somewhat nebulous conception of an obligation to pay rates which arises from the rate, and that the local authority are the custodians not simply of the
C decision as to what the poundage should be, but also that they can, by altering the rate, make an effect upon the obligation of a particular occupier.

In this case we start by looking at section 2(4) of the General Rate Act 1967 because this is the main provision upon which the respondents' case rested. It provides:

> Subject to the provisions of this Act, the general rate for any rating area . . . (b) shall be made and levied in accordance with the valuation list in force for the time being, except that, where a new valuation list is to come into force for that area, a rate for the year, or any part of the year, beginning with the day on which the new list is to come into force shall be made,
> D and applied in relation to particular hereditaments, by reference to that new list.

So the rate shall be levied in accordance with the valuation list, and it implies it shall not be levied other than in accordance with the valuation list. Here of course the rating authority have sought to include in the rate the substantial sum which has not yet appeared in the valuation list at all.

The justification for departing from the general rule of section 2(4) is alleged to be in section 6, which is the crucial provision. It provides in these terms in subsection (1):

> Subject to the provisions of this section, the rating authority
> E may at any time make such amendments in a rate (being either the current or the last preceding rate) as appear to them necessary in order to make the rate conform with the enactments relating thereto, and in particular may—(a) correct any clerical or arithmetical error in the rate; or (b) correct any erroneous insertions or omissions or any misdescriptions. . . .

That is a somewhat surprising provision to meet in this Act because it clearly contemplates that there is this nebulous conception of the rate which can be affected in its amount by action of the rating authority. So one finds there, without any doubt at all, that Parliament in passing the General Rate Act did contemplate that in the circumstances set out in
F section 6 the local authority could amend the rate—not the valuation list, but the rate—with the result that the rate would then bear the obligations attributable to its new form. In this case it is said that section 6(1) authorises the local authority to amend the rate so as to bring in the new hereditament of Messrs Kettle and the new value for the purposes of the charge.

The section goes on to authorise the rating authority under section 6(1)(c) to:

> make such additions to or corrections in the rate as appear to the authority to be necessary by reason of—(i) the coming into occupation of any hereditament which has been newly erected or which was unoccupied at the time of the making of G
> the rate.

That is not, as I understand it, specifically relied upon by Mr Moriarty, but it is very close to the facts of this case in any event. It authorises the rating authority to amend the rate in consequence of the coming into occupation of a new hereditament. It is very close indeed, as I say, to the facts of our case.

Then in subsection (2) it provides:

> Where the effect of the amendment would be either—(a) to alter, otherwise than by way of correction of a clerical or arithmetical error, the value on which a hereditament is rated;
> or (b) to charge to the rate a hereditament not shown, or not H
> separately shown, in the valuation list, the rating authority shall not make any amendment of the rate unless either the amendment is necessary to bring the rate into conformity with the valuation list or a proposal for a corresponding alteration to the valuation list has been made by the valuation officer; . . .

I pause there for a moment so as not to concentrate too much of the section in one passage of comment. Here we find a substantial restriction on the rating authority's exercise of the powers in section 6(1). They shall not make any amendment to the rate under that power unless certain factors are present, one of them being that there is already on foot
a proposal by the valuation officer. That of course is our J
case. There is a proposal on foot by the valuation officer. Therefore, if the rating authority otherwise have authority to act under this section, they have got over the hurdle presented by that requirement without any question.

Then to finish subsection (2), it provides:

> and if effect, or full effect, is ultimately not given to such a proposal, and the amount of the rate levied in pursuance of the amendment is affected, the difference—(i) if too much has been paid, shall be repaid or allowed; or (ii) if too little has been paid, shall be paid and may be recovered as if it were arrears of the rate.

K

That means, as I see it, that in the end the accountancy will be put right and every liability will lie where it is supposed to lie, because what is contemplated here surely is that, if the assessment goes into the list or is made in the rate pursuant to the application of the rating authority, and the respondents in the present appeal thereupon become liable to pay this substantial sum of rates, then if in the end when the valuation officer's proposal is finally sorted out it is found that the obligation of the respondents is greater than they have paid, they will have to pay out the difference. If it is found that the obligation of the respondents is less,
then they will have the amount returned to them. The effect L
of this somewhat complex provision, as I see it, is to enable the local authority to obtain payment more quickly in cases of this kind, because that would be the consequence if Mr Moriarty's argument for the appellants succeeds.

What is said on the other side? Mr Glover, if I may be allowed to say so, has given us an extremely attractive argument, and one that I have profited by very much. Like all good arguments, it really hangs on two main points. In the first place it is based on section 2(4) and the clear obligation under that subsection not to levy a general rate except in accordance with the values currently shown in the valuation
list. What has happened in the present case, of course, is M
that there has been a flagrant breach of that rule, because the new hereditament going in is not in any sense going in as part of the valuation list. The answer to that which is given by Mr Moriarty is that the section is expressed to take effect subject to the provisions of the Act, and one must regard this as a provision of the Act which enables in special circumstances the section 2(4) rule, if I may so call it, to be ignored.

But more substantial than that is the argument put forward by Mr Glover based on the absence of any resolution of the local authority to do what was done. I have had some difficulty with this, speaking for myself, because this is a case stated and we can only work on the facts which are disclosed, and no word has been said in the case about whether there was a resolution or not. I do not myself consider that any resolution was necessary except such resolution as in the end must accompany all activities of the local authority. In the end, almost everything which a local authority does has to be sanctioned by the resolution of some committee, if not of the authority itself. But leaving that kind of thing aside, I feel quite confident that it was not the intention of Parliament that a specific resolution should be made every time it was desired to bring a new hereditament into the rate under the powers relied upon in this case by the local authority. Section 6 is drafted in clear and specific terms, and I feel confident that it would be said if any further resolution were required.

The justices did not base their decision on the argument which Mr Glover has put before us today. They took another point, which I must say is very ingenious, although it seems to me to be wrong. Their minds were directed at some time to the fact that under section 7(3) of the Act it is thus provided:

> Any rate made by a rating authority shall be leviable and recoverable notwithstanding that notice has been given of an appeal against the rate under this section, except that, if such notice has been given by any person to the rating authority, then until the appeal has been determined or abandoned no proceedings shall be commenced or carried on to recover from that person any sum greater than that at which the last effective rate was charged in respect of the hereditament to which the appeal relates.

The magistrates, no doubt thinking that this would be a very fair way of disposing of the case, relied on that provision and said that on that account the local authority's action was unjustified. I think, again with respect, that they were wrong, because this was not an appeal against a rate, as I have pointed out, and I do not think that the provisions of section 7(3) have any relevance today whatever. That is not to depart from the words of appreciation which I gave to the magistrates earlier on on having to deal with a case which has not been easy for us, or any of us.

In the result I would say that the case for the local authority is made out, and I would answer the question which is attached to the case in the affirmative by saying that the justices were wrong in law in their interpretation of section 6(2). I do not think as at present advised that this court would make any other order which would assist.

CANTLEY J agreed.

Also agreeing PETER PAIN J said: I must say that the earlier sections of the General Rate Act with which this case is concerned seem to me to be extremely difficult to interpret, especially when they relate to the making of a rate and the amendment of a rate. But, doing the best I can, I agree ultimately with what has fallen from the Lord Chief Justice.

Appeal allowed, with costs. Leave to appeal to the Court of Appeal granted.

COURT OF APPEAL

March 21 1978

(Before Lord Justice MEGAW, Lord Justice LAWTON and Lord Justice BROWNE)

CAMDEN LONDON BOROUGH v HERWALD

Estates Gazette June 3 1978

(1978) 246 EG 823

Distress for rates—Appeal from Divisional Court—Preliminary question as to Court of Appeal's jurisdiction—Jurisdiction accepted—Not a "criminal cause or matter"—Doubt set at rest—Substance of appeal whether it is a defence to application for distress warrant that ratepayer was in occupation of part only of hereditament—Correct principle is that distress warrant should not be issued if description in valuation list cannot be satisfied without inclusion of property which person rated does not occupy—Cases reviewed—Appeal from Divisional Court allowed

This was an appeal from a decision of the Queen's Bench Divisional Court allowing an appeal from Hampstead magistrates, who had refused to authorise the issue of a distress warrant on a complaint laid by Camden Borough Council against the present appellant, Mr Theodore Herwald. The case concerned the non-payment of the general rate in respect of premises at 63 Loveridge Road, London NW6. The Divisional Court had remitted the case to the justices with a direction to authorise the issue of the warrant.

Sir Peter Rawlinson QC MP and Desmond Keane (instructed by Walford & Co) appeared on behalf of the appellant; Guy Roots (instructed by Francis Nickson, Camden Borough Council) represented the respondent authority.

Giving the judgment of the court at the invitation of Megaw LJ, BROWNE LJ said: This is an appeal by Mr Herwald from a decision of the Divisional Court of the Queen's Bench Division given on April 8 1976 and reported in [1977] 1 WLR 100. The Divisional Court allowed an appeal by way of case stated from a decision of the Hampstead Magistrates' Court and refused leave to appeal, but leave was given by this court on March 28 1977.

The case relates to distress for rates. On March 13 1975 the Camden Borough Council laid a complaint in the magistrates' court against Mr Herwald for a summons for non-payment of the general rate in respect of premises at 63 Loveridge Road, London NW6, for the period from October 1 1972 to March 31 1975, amounting to a total of £1,019.80. On May 1 1975, the magistrates heard the complaint and refused to authorise issue of a distress warrant. The Divisional Court, having allowed the appeal, remitted the case to the justices with a direction to authorise the issue of a distress warrant.

When the appeal came on in this court, the court raised the question whether we had jurisdiction to hear it, in view of the decision of this court (consisting of Lord Esher MR, Kay LJ and A L Smith LJ) in *Seaman v Burley* [1896] 2 QB 344. It was held in that case that there was no appeal to this court from a decision of a Divisional Court on a case stated by justices relating to the grant of a distress warrant to enforce a poor rate under a local Act, because it was an appeal in a "criminal cause or matter" within section 47 of the Judicature Act 1873, inasmuch as the proceedings before the magistrates might end in imprisonment (see now section 31 (1)(a) of the Supreme Court of Judicature (Consolidation) Act 1925 and section 1 (1) (a) of the Administration of Justice Act 1960). Sir Peter Rawlinson, for Mr Herwald, wished us to deal with the case; and Mr Roots for the Camden Borough Council did not take any objection to our jurisdiction, and indeed advanced a helpful argument in support of it.

The statutory provisions as to distress for the general rate are now contained in Part VI of the General Rate Act 1967, beginning with section 96 (1).

Section 96 (1) provides that

> . . . if any person fails to pay any sum legally assessed on and due from him in respect of a rate for seven days after it has been legally demanded of him, the payment of that sum may, subject to and in accordance with the provisions of this Part of this Act, be enforced by distress and sale of his goods and chattels under warrant issued by a magistrates' court; and, if there is insufficient distress, he may be liable to imprisonment under the provisions of this Part of this Act in that behalf.

Section 97 (1):

> The proceedings for the issue of a warrant of distress under this Part of this Act may be instituted by making complaint before

a justice of the peace and applying for a summons requiring the person named in the complaint to appear before a magistrates' court to show why he has not paid the rate specified in the complaint.

Section 98:

The justices may state a case under the Magistrates' Courts Act 1952 when called upon to issue a warrant of distress under this Part of this Act.

Section 99 (1):

A warrant of distress under this Part of this Act may be directed to the rating authority, to the constables of the police area in which the warrant is issued and to such other persons, if any, as the magistrates' court issuing the warrant may think fit, and the warrant shall authorise the persons to whom it is directed to levy the amount which the person against whom the warrant is issued is liable to pay by distress and sale of his goods and chattels.

Section 102 (1):

If the person charged with the execution of a warrant of distress for levying a sum to which some other person has been rated makes a return to the magistrates' court that he could find no goods or chattels (or no sufficient goods or chattels) on which to levy the sums directed to be levied under the warrant on that other person's goods and chattels, a magistrates' court may, if it thinks fit, and subject to the provisions of section 103 of this Act, issue a warrant of commitment against that other person.

Section 102 (5):

The order in the warrant of commitment shall be that the said person be imprisoned for a time therein specified but not exceeding three months, unless the sums mentioned in the warrant shall be sooner paid. . . .

and the sub-section goes on to provide for a remission of the term of imprisonment proportionate to the amount of any later payment.

Section 103 (1):

Section 102 of this Act shall have effect subject to and in accordance with the following provisions: (a) on the application for the issue of a warrant for the commitment of any person, the magistrates' court shall make inquiry in his presence as to whether his failure to pay the sum to which he was rated and in respect of which the warrant of distress was issued was due either to his wilful refusal or to his culpable neglect; (b) if the magistrates' court is of opinion that the failure of the said person to pay the said sum was not due either to his wilful refusal or to his culpable neglect, it shall not issue the warrant.

Section 103 (2):

Where on the application no warrant of commitment is issued, the magistrates' court may remit the payment of any sum to which the application relates, or of any part of that sum.

Seaman v Burley was decided on the provisions of a local Act relating to the parish of Paddington (5 George IV, cap cxx (vi). Section cxx of that Act contained provisions generally corresponding with those of Part VI of the 1967 Act, with one important exception: it contained no provision corresponding to section 103.

In *Ryde on Rating*, 13 ed p 878, the authority of *Seaman v Burley* in the present context is doubted. In two later cases this court has heard and decided appeals from the Divisional Court relating to distress for rates, though it does not appear that in either of them *Seaman v Burley* was cited or any point taken as to jurisdiction (see *Associated Cinema Properties Ltd v Hampstead Borough Council* [1944] 1 KB 412 and *Bexley Congregational Church Treasurer v Bexley London Borough Council* [1972] 2 QB 222). In *Re McGreavy* [1950] Ch 269 this court disagreed with what it regarded as an *obiter dictum* of Lord Esher in *Seaman's* case (see p 279-280), and in *China v Harrow UDC* [1954] 1 QB 178 the Divisional Court (we think) impliedly distinguished *Seaman*, which was cited in argument. But we do not think we need refer further to those cases, because two years after *Seaman v Burley* was decided it was distinguished by this court on a ground which in our judgment equally distinguishes it from the present case. In *Southwark and Vauxhall Water Co v The Hampton UDC* [1899] 1 QB 273 it was held that an application to a court of summary jurisdiction for an order to enforce a general district rate under the Public Health Act 1875 was not a "criminal cause or matter" and that an appeal lay to this court from a decision of the Divisional Court on a case stated. The leading judgment was given by A L Smith LJ, who had been a party to the decision in *Seaman v Burley*, and Rigby LJ and Collins LJ concurred. The ground on which *Seaman's* case was distinguished was that in that case the Summary Jurisdiction Act 1879 did not apply but in the *Southwark and Vauxhall Water Co* case it did. The 1879 Act contained provisions in section 35 the effect of which was, in our view, substantially the same as the effect of the provisions of section 103 (1) of the 1967 Act, and it was held that the proceedings were therefore not a "criminal cause or matter." The proceedings in the present case are, therefore, in our view, equally not a "criminal cause or matter." This distinction also avoids the startling conclusion that every judgment summons in a county court under Order 25 rule 33 of the County Court Rules and section 5 of the Debtors Act 1869 is a "criminal cause or matter." In our judgment, this court has jurisdiction to hear this appeal. We hope that our decision will end the uncertainty referred to in *Ryde*.

We come, therefore, to the substance of the appeal. It is well established that an application for a distress warrant to enforce payment of rates can only be resisted on certain limited grounds. It is also well established that one of such grounds is that the defendant is not in occupation of the hereditament in respect of which it is sought to rate him. The question in this case is whether, and if so in what circumstances, it is a defence for him to show that he is only in occupation of part of that hereditament.

The case stated, having found that the relevant rates were made, published and demanded, finds the following facts:

Paragraph 4(ii):

That the valuation list shows 63 Loveridge Road as a single hereditament as a workshop and store.

Subparagraph (iii):

That Mr Stephenson, rates inspector, London Borough of Camden, visited 63 Loveridge Road on several occasions, the first time being on March 11 1974.

Subparagraph (iv):

That Mr Herwald had been a subtenant of a Mr Jellineck from 1960 until six or seven years ago and then of a Miss Marcotics to whom he had paid a portion of the total rent and rates.

Subparagraph (v):

That during Miss Marcotics' lifetime he had occupied certain parts only of 63 Loveridge Road, namely an office on the first floor of the house, with use of the kitchen, and area marked "Factory A" (see attached rough sketch plan), and that subsequent to her death he continued to occupy the same parts of the premises as he had always occupied.

The sketch plan attached to the case and the legend on it show that 63 Loveridge Road comprises a house of three floors consisting of a shop and office on the ground floor, four rooms on the first floor (including a kitchen) and four rooms on the second floor; a one-storey factory (Factory A); a two-storey factory (Factory B); and two yards separating the buildings. The case contains no findings about the use made by Mr Herwald of the parts which he occupies, nor as to the use or occupation (if any) of the other parts. Sir Peter Rawlinson emphasised that the effect of the findings was that during the lifetime of Miss Marcotics Mr Herwald only paid (to her) the rates in respect of the parts he occupied; that after her death he continued to occupy the "same parts of the premises as he had always occupied"; and that he was now being called upon by the rating authority to pay the rates on the whole.

A The conclusion of the justices and their question to the court were as follows:

Paragraph 5:

> We were of opinion that as Mr Herwald, the respondent, had at all times occupied only certain parts of 63 Loveridge Road, he would not be liable for the rates for the whole premises and accordingly refuse to authorise issue of a distress warrant.
>
> *Question:* The question for the opinion of the High Court is whether it is correct that Mr Herwald is liable only for rates in respect of that part of the property in which he has been found to be in sole occupation when that part of the property is con- B tained in the valuation list only as part of a single hereditament.

In view of some criticisms of the Camden Borough Council which were made or implied during the argument and of the finding in paragraph 4 (iii) of the case stated, we think we should emphasise the division of responsibilities in rating matters which is now embodied in the General Rate Act 1967. We think this division of responsibilities and procedures is also important in relation to the substance of this case. Mr Roots gave us an interesting and helpful summary of the history of the development of rating procedure from the Poor Relief Act 1601 to the Lands Tribunal Act 1949. The scheme of the 1967 Act (re-enacting earlier legislation) is a division C between responsibility for assessment or valuation and responsibility for collection. Valuation is the responsibility of valuation officers appointed by the Inland Revenue and collection is the responsibility of the rating authorities. The provisions as to valuation lists and the duties of valuation officers are contained in Part V of the Act. It is the duty of valuation officers to prepare a valuation list for each rating area (section 67 (1)) and to insert in the list the prescribed particulars " with respect to every hereditament in the rating area and the value thereof " (section 67 (2)). By section 67 (6) the valuation list is made " conclusive evidence for the purposes of the levying D of [the] rate of the value of the several hereditaments included in the list." It is not made conclusive evidence of anything else, but it is obvious that before making a valuation of any hereditament the valuation officer must decide what the hereditament is which he is valuing. The identification of hereditaments is an essential part of the process of valuation and so within the province of the valuation officer (see the definition of " hereditament " in section 115 (1) of the Act). Section 69 and the following sections contain provisions for the alteration of the valuation list by means of proposals and for appeals to local valuation courts and to the Lands Tribunal if there are objections to a proposal. If the occupier of part E of premises included in the list as a single hereditament thinks that his part should be shown as a separate hereditament, he can and should make a proposal for the alteration of the list under section 69 (1) (a) or (c) or both. We were told that since the hearing in the Divisional Court this has been done and that 63 Loveridge Road is now shown in the list as two hereditaments. When a proposal takes effect, its effect relates back to the beginning of the rating period in which it was made (section 79), but the rates claimed in this case are in respect of earlier periods. Collection of the rate is the responsibility of the rating authorities. They have nothing to do with assessment or valuation nor with the identification or F definition of hereditaments. We think that this is emphasised by their duty under section 85 (1), which is only to give information to the valuation officer. By section 2 rating authorities are under a mandatory duty to make and levy the general rate, and by section 2 (4) (b) the general rate "shall be made and levied in accordance with the valuation list in force for the time being. . . ." So long as the entry in the valuation list remained unaltered, it was therefore the duty of the Camden Borough Council to levy the rates in respect of 63 Loveridge Road in accordance with the list. The rating authority had no power to split up the hereditament or apportion the rates. We agree with the Divisional Court that G Mr Herwald's proper remedy was to make a proposal for the alteration of the list (see [1977] 1 WLR at p 104 (B-C)).

But this is not the end of the matter. The foundation of Sir Peter Rawlinson's argument was the incontrovertible proposition that liability to pay rates depends on the occupation of land (or certain other types of hereditament); a person is liable to be rated in respect of his occupation of land. He referred us to section 16 of the 1967 Act, to *Allchurch v Hendon Union Assessment Committee* [1891] 2 QB 436, and to *Westminster City Council v Southern Railway Co Ltd* [1936] AC 511, especially Lord Russell of Killowen at p 529. He submitted that: (a) Mr Herwald is not liable to be rated H in respect of land of which he is not in occupation; (b) on the findings of fact he was not in occupation of part of this hereditament; (c) no rate can lawfully be levied in respect of the part of which he is not in occupation; (d) therefore, no distress warrant can lawfully be issued for rates which include rates on the part he does not occupy.

As we have said, there is no doubt that on an application for a distress warrant it is a defence for the defendant to show that he is not in occupation of the hereditament at all (see, for example, *Associated Cinema Properties Ltd v Hampstead Borough Council* [1944] KB 412). But where the defence is that the defendant is only in occupation of part of J the hereditament the position is different. The rating authority relied on the decision of the Divisional Court in *Manchester Overseers v Headlam* (1888) 21 QBD 96, and the Divisional Court in the present case accepted their submission and directed the distress warrant to issue. If we may respectfully say so, we think that the effect of that decision was stated with complete accuracy in the judgment of Robert Goff J, with which the Lord Chief Justice and O'Connor J agreed. I quote from [1977] 1 WLR begining at p 102 (H):

> It is right that a person is only liable to be rated in respect of property of which he is the occupier: see section 16 of the Gen- K eral Rate Act 1967. But it does not follow that, merely because he can show that he does not in fact occupy part of premises in respect of which a rate has been made, a distress warrant should not be issued. To resist the issue of a warrant, he must show that the description of the rated property in the valuation list includes on its face property which he does not occupy. The principle was stated by this court in *Overseers of the Poor of Manchester v Headlam and London and North Western Railway Co* (1888) 21 QBD 96, at p 98 in a passage which has since been frequently cited and applied: ". . . if one entire assessment be made in terms upon property which he does occupy, and upon other property which he does not occupy, so that upon the true state of facts being ascertained it is impossible to satisfy the L description in the rate book without including property which he does not occupy, the rate will be bad and ought not to be enforced."

That is the end of the quotation from *Headlam.* Robert Goff J goes on:

> In that case, property occupied by the railway company had been assessed as " offices and land with rails," but in assessing the amount of the rate the overseers had included certain buildings which were not occupied by the company. It was held that, since the property in fact occupied by the company satisfied the description in the rate book, the rate was good on the face of it and a distress warrant must be issued. The proper remedy of the M company in such circumstances was to appeal against the assessment; not having appealed, they could not resist the issue of a warrant. By way of contrast, in *Langford v Cole* (1910) 102 LT 808, where a single assessment of poor rate was made on property described in the rate book as " mansion house and grounds " and it was established that the mansion house itself was unoccupied at the date when the rate was made, it was held that the rate made in respect of the whole property could not be enforced and that a distress warrant should not therefore be issued.
>
> The position is therefore as follows. If the person rated is in occupation of premises which fulfil the description in the valuation list, that is sufficient for the issue of a warrant: but if the description in the valuation list cannot be satisfied without includ-

A ing property which the person rated does not occupy, the rate cannot be enforced against him and a distress warrant should not be issued.

Applying that test, the Divisional Court held that the part occupied by Mr Herwald did fulfil the description in the valuation list and that non-occupation of even a substantial part did not prevent him from being liable for the rate in respect of the whole: accordingly, he could not resist the issue of a distress warrant (see p 104 (A-C)).

Sir Peter Rawlinson submitted (i) that we should overrule *Headlam* and the later cases which have followed and app-
B lied it; (ii) alternatively, that *Headlam* should not be rigidly applied to all cases but only to "trivial matters," (iii) in the further alternative, that even if the principle of *Headlam* should be applied, the Divisional Court was wrong in holding that the part occupied by Mr Herwald fulfilled the description in the valuation list.

As to (i), Sir Peter submitted that the decision in *Headlam* was inconsistent with the fundamental principle that liability to be rated depends on the occupation of land and with the decision of the Court of Appeal three years later in *Allchurch v Hendon Union Assessment Committee*. *Headlam* is not binding on us, and Mr Roots did not feel able to submit that
C the decision of the Exchequer Chamber in *Crease v Sawle* (1842) 2 QB 862 was binding upon us because of later changes in procedure. But *Headlam* has stood for 90 years, and was followed and applied in at least the three later cases referred to in the judgment of the Divisional Court, in 1910, 1922 and 1956: we should therefore be very slow to overrule it. But, in our judgment, it is not inconsistent with the general principle on which Sir Peter relies. We agree with Mr Roots that the question in this case relates to procedure and jurisdiction rather than to liability; so it did in *Headlam*, where the proper remedy would have been to appeal against the rate. *Allchurch v Hendon Union Assessment Committee* is
D not, in our view, inconsistent with *Headlam*, because *Allchurch* related to the assessment process (under the procedure then applicable) and not to the collection process; it was concerned with the process now governed by Part V of the 1967 E
Act. On an application for a distress warrant the magistrates have no jurisdiction to alter the valuation list or to apportion the rates on any such basis. It does not appear to have been suggested in any of the later cases in which *Headlam* has been applied that there was anything in *Allchurch* which threw any doubt on it. We therefore reject Sir Peter's first submission as to *Headlam*. Nor can we see any reason to limit that decision to "trivial matters." No such suggestion seems to have been made in any of the later cases.

But we accept Sir Peter's third submission that, applying the principle of *Headlam*, these rates cannot be enforced against Mr Herwald. F
The description in the valuation list at the relevant times was "a workshop and store." "Factory A," which he occupies, can fairly be described as a "workshop," but on the findings of fact he does not occupy anything which can be described as "store." Further, he does occupy the office in the house, which is not covered by the description in the list. We doubt whether we are entitled to take into account the later alteration in the list, but it is interesting to note that we were told that 63 Loveridge Road is now entered as two hereditaments, the part occupied by Mr Herwald described as "Factory and Office" and the rest as "Stores and premises—void." We do not find any help in the case G
on which the Divisional Court relied—*Vernon v Castle* (1922) 127 LT 748; it seems to us that each case must depend on its own facts and on the construction of the particular entry in the list.

The appeal will therefore be allowed. The question stated in the case is not the right question; on our decision, the position is not that Mr Herwald is liable only for rates in respect of the part of the property which he occupied, but that he is not liable for any part of the rates claimed. The actual decision of the magistrates in refusing to authorise issue of a distress warrant was, however, right.

The appeal was allowed with costs in the Court of Appeal H
and below. Legal aid taxation ordered, the appellant being legally aided.

REAL PROPERTY AND CONVEYANCING

QUEEN'S BENCH DIVISION

July 29 1977

(Before Sir DOUGLAS FRANK QC sitting as a deputy judge of the division)

NICHOLSON v SECRETARY OF STATE FOR ENERGY AND ANOTHER

Estates Gazette January 14 1978

(1977) 245 EG 139

Opencast Coal Act 1958—Acquisition of Land (Authorisation Procedure) Act 1946—Authorisation to work coal by opencast methods—Public inquiry—Refusal by inspector to allow plaintiff to cross-examine local authority witnesses—Whether breach of rules of natural justice—Inspector wished to avoid repetition and irrelevancies and considered that points could be made in plaintiff's examination-in-chief—Plaintiff substantially prejudiced—Limits of inspector's discretion—Authorisation quashed and inquiry to be reconvened

In these proceedings Michael Burrows Nicholson, of The Manor, South Hiendley, near Barnsley, Yorks, applied under paragraph 13 of Schedule 1 to the Acquisition of Land (Authorisation Procedure) Act 1946, as applied by paragraph 10 of Schedule 1 to the Opencast Coal Act 1958, to quash an authorisation by the first respondent, the Secretary of State for Energy, to the second respondents, the National Coal Board, for the working of coal by opencast methods at Cawker, near Royston, South Yorkshire, on the ground that the public inquiry preceding the authorisation was invalid in that the plaintiff was refused the right of cross-examination.

Michael Barnes (instructed by Conway Kremer & Co, of Leeds) appeared for the plaintiff; Harry Woolf (instructed by the Treasury Solicitor) represented the first respondent; Lord Colville (instructed by the solicitor to the National Coal Board) represented the second respondents.

Giving judgment, SIR DOUGLAS FRANK QC said that section 1 of the Opencast Coal Act 1958 prohibited the working of coal by opencast methods except in pursuance of an authorisation granted by the Secretary of State for Energy, who, in turn, could not grant an application for authorisation unless it were made in accordance with the provisions of Schedule 1 to the Act. Those provisions included the hearing of any objections at a public inquiry.

In the present case the plaintiff was one of 16 objectors. He prepared his case with considerable care and cross-examined at the inquiry at least one of the witnesses called on behalf of the National Coal Board. But the inspector refused to allow him to ask questions of two of the witnesses who gave evidence on behalf of the West Yorkshire County Council and one each on behalf of the Wakefield Metropolitan District Council and the South Yorkshire County Council. In his affidavit the inspector said he was anxious to avoid unnecessary repetition and irrelevancies and, in addition, considered that any points which the plaintiff wished to develop were more properly developed when he called evidence and presented his case.

For the plaintiff it was now contended that there was a breach of the rules of natural ju tice. He referred to Professor S A de Smith in *Judicial Review of Administrative Action* at p 188 where it was said, " Refusal to permit cross-examination of witnesses at an administrative hearing will usually be a denial of natural justice. Seldom can such a refusal be justified if a witness has testified orally and a party requests leave to confront and cross-examine him—the fact that the proceedings may be inquisitorial and informal is inconclusive—but there may exceptionally be valid grounds for disallowing questions to a witness on a particular matter." In *Halsbury's Laws of England* (4th ed vol 1 para 76) the same author said, " Rejection of a request to be permitted to cross-examine witnesses who appear at a hearing for the other side will normally be construed as a breach of natural justice. . . ."

The plaintiff particularly relied on *Wednesbury Corporation v Ministry of Housing and Local Government (No 2)* [1966] 2 QB 275 and, in particular, on the judgment of Diplock LJ, as he then was, in which he said, " He [the inspector] must permit persons making such representations a reasonable opportunity of adducing relevant material in support of their representations, and if such material includes oral statements of witnesses of fact or of opinion I am inclined to think he must permit a reasonable opportunity of cross-examination of such witnesses by persons appearing at the inquiry who wish to dispute what witnesses have said." That case, like the present, concerned an inquiry into objections. The plaintiff also relied on the speech of Martin B in *Osgood v Nelson* (1872) LR 5 HL 636 at p 645 in which stress was laid upon the right to cross-examine: and on *Marriot v Minister of Health* (1935) 154 LT 47 at p 50 in which it was held that at a public local inquiry into a compulsory purchase order the strict rules of evidence did not need to be observed provided that substantial justice was done between the parties. Swift J, referring to the speeches in *Board of Education v Rice* [1911] AC 179, said " Nothing should be done which could offend against that instinct which every citizen possesses, to offend against which is said to deny something which is contrary to natural justice . . . if one party is heard without the other having the opportunity of knowing what he has said or of testing what he has said by cross-examination . . . then clearly there has been an infringement of the rules of natural justice "

The plaintiff had also relied on *R v Deputy Industrial Injuries Commissioner, ex p Moore* [1965] 1 QB 456 and contended that that case was clear authority for the proposition that a person at an inquiry had a right to cross-examine anybody who gave evidence contrary to his case because only in that way could the inspector fulfil his function of inquiring into objections.

The policy behind the authorisation procedure was to merge economic and town planning considerations and make the first respondent the final arbiter. The pre-inquiry procedure ensured that those directly affected because of their ownership or occupation of the land concerned and those representing broader interests could require a public inquiry to be held. Paragraph 7(2) of Schedule 1 to the 1958 Act originally provided, " If any objection duly made by any such local planning authority or other local authority or by any such owner, lessee or occupier, is not withdrawn, the minister shall cause a public local inquiry to be held, and shall consider the objection and the report of the person who held the inquiry before determining whether to grant the authorisation; and if he determined to grant it, he may do so either in accordance with the board's application or subject to such modification as he may determine." The Coal Industry Act 1975 amended that paragraph so that the words "the objection" were amended to "all objections to the grant of the authorisation which are duly made and not withdrawn."

Mr Woolf now contended that the effect of the amendment was that "all objections" included objections by whomever made. Lord Colville, however, contended that there were two classifications of persons—specified objectors and others —and that the first respondent's primary duty was to the first class. He also contended that the obligation to the latter class was no more than that their representation should be heard. He said that under rule 7(3) of the Compulsory Purchase by Public Authorities (Inquiries Procedure) Rules 1976 (SI 1976 No 746), the right of cross-examination was denied to any non-statutory objector. He went on to argue that it would offend commonsense for objectors at inquiries where such rules did not apply—such as in the present case— to have greater rights. The rules did not apply, but it did not follow that because a certain objector did not have an express right to cross-examine he might not have such a right as a rule of natural justice, depending upon whether his objection was bona fide. It was clear that all objectors had equal status and were entitled to the benefit of the rules of natural justice. The question was to what extent those rules required an objector to be permitted to cross-examine. Mr Woolf submitted that the inspector had a discretion to disallow cross-examination by a person intending to give evidence and that discretion had been properly exercised.

It had been said that the phrase "contrary to natural justice" was sadly lacking in precision. A helpful starting point was the judgment of Kerr J in *Lake District Special Planning Board v Secretary of State for the Environment* [1975] JPL 220 where he said, "In administrative procedures the question [of natural justice] had to be approached on broad lines. The test was whether a reasonable person, viewing the matter objectively, and knowing all the facts, would consider that there was a risk that injustice or unfairness had resulted. It was a question of fact and degree in administrative procedures whether the rules of natural justice had been broken."

The views of the "reasonable person" varied according to the climate of public opinion. Undoubtedly public opinion in relation to administrative acts had altered considerably over the years. Its development had taken the form of increasing vigilance. The recent case of *Performance Cars v Secretary of State for the Environment* [1977] JPL 585, at p 586, showed how ready courts were to impute a rule of natural justice. In that case Lord Denning MR said: "People should not go away from an inquiry feeling 'I've not had a fair deal.'"

He (his Lordship) had been led to the conclusion that a reasonable person, viewing the matter objectively, would consider that there was a risk that injustice or unfairness would result if a person considering himself to be directly affected by a proposal was denied the opportunity to cross-examine a witness who had given evidence contrary to his case. Cross examination was not to be used for the purpose of a "fishing expedition" but the plaintiff had been denied the right to cross-examine the very evidence upon which his own case turned. An inspector was entitled to refuse to hear, or at least to stop cross-examination by objectors who intentionally engaged in irrelevancies and repetition, and also by those whose objection was not to the proposed development as such but to any development notwithstanding that it was authorised by statute.

It was common ground that the plaintiff was a person aggrieved and, being deprived of cross-examination, he lost "a chance of being better off in relation to the proposed authorisation." Therefore, having regard to the test propounded by Browne J in *Hibernian Property Company Ltd v Secretary of State for the Environment* [1974] P & CR 197 at p 218, he was substantially prejudiced. The authorisation should be quashed and the matter remitted to the Secretary of State for Energy so that he might reconvene the inquiry and allow the plaintiff to cross-examine at least the witnesses called by the West Yorkshire and Wakefield local authorities.

The defendants were given leave to appeal.

QUEEN'S BENCH DIVISION

November 10 1977

(Before Sir Douglas FRANK QC sitting as a deputy judge of the division)

HARLOW v HARTOG

Estates Gazette January 14 1978
(1977) 245 EG 140

Restrictive covenant—No building other than private dwelling-house of a certain annual value—Proposal to enclose existing swimming pool in ground of private dwelling-house—Injunction sought by neighbour—Whether "building scheme" in force—Whether proposed enclosure a breach of covenant—Object of covenant to prevent erection of factory or possibly a block of flats—Injunction refused

This was a claim by Richard Harlow, owner of 21 Sibsey Road, Boston, Lincolnshire, for an injunction to prevent the erection of a building to cover in an existing swimming pool at adjoining premises, 23 Sibsey Road, owned by William Hartog. The plaintiff relied on a covenant which provided that "No building other than private dwelling-houses shall be erected on the said land and each of such dwelling-houses shall be of an annual value not less than thirty-five pounds."

Patrick Ground (instructed by Collyer-Bristow & Co, agents for Roythorne & Co, of Boston) appeared for the plaintiff; G H Newsom QC and Robert Wakefield (instructed by Thompson Quarrell, agents for Frost, Gunning & Co, of Boston) represented the defendant.

Giving judgment, Sir Douglas Frank QC said that the plaintiff sought an injunction restraining the defendant from constructing a building to enclose an existing swimming pool. The issues were (1) whether there was in force a building scheme so as to enable the plaintiff to enforce certain restrictive covenants and (2) if so, whether the proposed building was a building other than a private dwelling-house.

The land on which the plaintiff's and defendant's houses were built was part of about seven acres of undeveloped land that was sold off in plots to six different purchaers in 1927. Planning permission was granted to the defendant, who bought his home in 1976, for the pool to be covered, but the plaintiff objected because of its proximity to his own house and the effect it would have on his outlook. On March 22 1977 he was granted an *ex parte* injunction and on March 31 the plaintiff gave undertakings as to damages and the defendant undertook not to continue to construct the building until trial of the action.

On behalf of the plaintiff it was submitted that the essentials of a building scheme as expressed in *Elliston v Reacher* [1908] 2 Ch 374 and set out in *Halsbury's Laws*, 4th ed, vol 16 para 1355 were satisfied because: (1) there was a common vendor; (2) there was a defined area with lots as shown on the defendant's conveyance; (3) that as a matter of inference the purchasers of each of the plots were intended to benefit from the covenants; and (4) that, again as a matter of inference, the purchasers bought from the common vendor on the footing that the restrictions imposed on the land purchased by them were to enure for the benefit of the other lots. He referred to *Bell v Norman C Ashton Ltd* (1956) 7 P & CR 359, *Baxter v Four Oaks Properties Ltd* [1965] Ch 816, *Brunner v Greenslade* [1971] Ch 993, *Texaco Antilles Ltd v Kernochan* [1973] AC 609, *Lund v Taylor* (1973) 31 P & CR 167 and *Nottingham Patent Brick & Tile Co v Butler* (1885) 15 QBD 261.

The defendant contended that if there was a building scheme here it could only be an implied scheme and would have to satisfy the four tests referred to by Mr Ground for the plaintiff. He further submitted that only the first ground was met, namely that there was a common vendor. It was common ground that if the plaintiff was to succeed he had to show that at least there was an estate plan defining the area within which the scheme was to be operative. He (His Lordship) found no evidence that there was an estate plan, and still less that it was brought to the attention of any of the purchasers. The plan in the 1927 conveyance defined only the land being conveyed and showed an adjoining rectangle of similar size which, it stated, was contracted to be sold. The plan fell a long way short of being a plan of a defined area of a building estate.

On the question of breach of covenant he (his Lordship) thought that it was a matter of fact and degree whether a building was other than a private dwelling-house. It was beyond question that the erection of a building for recreational purposes or otherwise, used ancillary to a house such as a greenhouse, would not be in breach of the covenant unless it could be said that looking at the hereditament as a whole the offending building was not ancillary or incidental to the dwelling-house, but a separate building used independently. The swimming pool in question was a somewhat large one for a private dwelling, but there could be no question but that its use was part of the enjoyment of the dwelling-house, and it followed that the fact that it was to be enclosed would not alter that status. His Lordship thought that the purpose of the covenant on the land was to prevent the erection of, say, a factory on the land or possibly a block of flats.

Judgment was entered for the defendant.

COURT OF APPEAL

May 17 1977

(Before Lord Justice STEPHENSON, Lord Justice BROWNE and Sir JOHN PENNYCUICK)

CALABAR (WOOLWICH) LTD v TESCO STORES LTD

Estates Gazette February 11 1978
(1977) 245 EG 479

Claim for a declaration that the conduct of a freezer centre at certain premises was a breach of a covenant in a lease which restricted the use of the ground floor to the business of a supermarket for the sale of a named range of articles and other things ancillary thereto—Whether premises not used as a supermarket because of sales of freezer cabinets—Definitions of "supermarket"—Not all goods sold need be portable so as to be the subject of self-service—Sale of freezer cabinets not a breach of covenant—Appeal dismissed

This was an appeal from an order made by Judge Thomas, sitting as a deputy judge of the Chancery Division, on an originating summons in which Calabar (Woolwich) Ltd sought a declaration that Tesco Stores Ltd were in breach of a covenant in a lease which restricted the use of part of premises at 190-194 Eltham High Street to the business of a supermarket. Judge Thomas refused the declaration sought, but, as will be seen from the judgment of Sir John Pennycuick, the question raised in the Court of Appeal differed from that which was before the judge in the court below.

L Caplan QC and J A Speed (instructed by Tobin & Co) appeared on behalf of the appellants (plaintiffs); Edwin Prince (instructed by Berwin Leighton) represented the respondents (defendants).

Giving the first judgment at the invitation of Stephenson LJ, Sir John Pennycuick said: We have before us an appeal from an order made by Judge Thomas, sitting as a deputy judge of the High Court, on July 22 1976. The case arises upon an originating summons in which Calabar (Woolwich) Ltd is the plaintiff and Tesco Stores Ltd is the defendant. The question at issue is whether, in conducting a freezer centre at certain premises—190-194 Eltham High Street—Tesco is in breach of a covenant contained in a lease between Calabar and Tesco's predecessor in title. The learned judge decided that there was no breach and Calabar appeals from that decision.

The facts can be shortly stated. By a lease dated May 1 1961 Calabar demised these premises, 190-194 Eltham High Street, to a company known as Fellows Foodtown Ltd for the term of 21 years at an annual rent of £5,600. The lease contains a covenant upon which the whole of the issue in the present case turns. The covenant is contained in subclause (q) of the lessees' covenant and runs as follows: "Not at any time during the term to use or permit the demised premises" etc; then:

> . . . nor to use or permit or suffer the ground floor of the demised premises to be used except for the purpose of a retail shop for the purposes of the lessees' business of a supermarket for the sale of groceries provisions garden produce fresh meat confectionery domestic and hardware and toilet requisites and also as ancillary thereto for the sale of such other articles or things as are usually sold by and in supermarkets.

The whole case turns upon the construction of these words and their application to the facts of this case. There is no evidence as to the precise nature of Foodtown's business, but that is not of practical importance, since we know from the terms of the covenant itself the nature of that business, namely, the business of a supermarket for the sale of the specified articles. On July 13 1971 Fellows Foodtown Ltd assigned the lease to Tesco. Some few years later Tesco decided to fit out and use the ground floor of the premises as a freezer centre.

It will be convenient at this stage to quote a few paragraphs from the affidavit sworn in these proceedings by D N Edgerton on behalf of Tesco. Mr Edgerton is a chartered surveyor and is employed as property manager by Tesco; his affidavit contains the following explanation of the nature of a freezer centre and how Tesco came to use these premises for that purpose. Mr Edgerton's affidavit is not contradicted on these points.

Paragraph 17 of Mr Edgerton's affidavit says:

> It became apparent during the early part of 1974 that the plaintiffs were unwilling to agree to a change of use and the defendants resolved in March 1974 to use the store as a centre for the sale of frozen foods, groceries, provisions and freezers ("the proposed use").

Then, in paragraph 20:

> After the exchange of correspondence in early June 1974 . . . the defendants resolved to reopen the store as a "Cadena Freezer Foods Centre."

"Cadena" being a trading name of the Tesco Group.

Paragraph 24 confirms that a plan produced by Mr Tibbatts in an affidavit on behalf of Calabar is a fair reflection of the layout of the store, and he (Mr Edgerton) goes on:

> It is difficult to attribute percentages of floor space to the different types of goods sold at the said store but the turnover of the store is represented by the following . . .

and then under the headings "Types of Goods" and "Approximate Percentage of Turnover" we see: Frozen Meat, 10.65 per cent; Other Frozen Food Products (immensely the greatest) 59.58 per cent; Non-Frozen Food Products, 14.72 per cent; Freezer Cabinets, 15.05 per cent. As the case has developed, it is with freezer cabinets that we are primarily concerned.

In paragraph 26, Mr Edgerton says:

> It is important to bear in mind that the frozen food industry is a comparatively new industry in this country and it is only in the past five years that there has been a marked development in the supply of deep-frozen foods for storage in home freezers and in that period represented one of the fastest-growing categories of total food expenditure.

In paragraph 33, after having given certain particulars with regard to the layout and the goods sold in the freezer centre, Mr Edgerton says:

> In all other respects the said store is like a more traditional Tesco supermarket. The range of non-frozen food items is similar to that sold at other Tesco supermarkets although more limited in range.

Then, in paragraph 34, he says:

> The said store uses the carrying baskets and wheeled trolleys used in other Tesco supermarkets and the opening hours are the same. The average purchase amounts to £1.95 which is not very far different from that at other Tesco supermarkets.

What is meant by "freezers"? They are cabinets for sale to customers who take the freezers back to their own homes and use them for the storage of the deep-frozen goods. As I understand it, they serve the same purpose as refrigerators, except that they are designed to keep food frozen for a much longer time. There is no precise evidence as to their size and weight; indeed, the size and weight vary, some obviously being larger and heavier than others, but I think it is fair to say that by and large they are not portable in the sense that a customer could pick one off the shelf and wheel it through the exit, paying at the check-point as he goes.

Calabar had refused its licence to use the ground floor as a freezer centre, and on March 20 1975 Calabar issued the originating summons which commenced these proceedings. The summons asks for a declaration that on a true construction of subclause (q), the defendants are not entitled to use the ground floor of the premises comprised therein as a freezer food centre, and an injunction restraining the defendants from using the ground floor for any purpose other than that authorised by the subclause. On the same day Calabar gave notice of motion for an injunction restraining Tesco from using the ground floor as a freezer food centre. Upon the motion affidavits were sworn by M S Tibbatts, the chartered surveyor who is employed as the managing agent of Calabar—that was in support of the summons—and then by Mr Edgerton on behalf of Tesco, the affidavit from which I have already quoted; then there is a short affidavit in reply by Mr Tibbatts, and finally a very lengthy affidavit on behalf of Tesco by a Mr Corina. I think it will be useful at this stage to quote a number of passages from the affidavit of Mr Corina.

Mr Corina is the industrial editor of *The Times*, and for 12 years he has written, lectured and broadcast extensively on the distributive trades; he was formerly the commercial correspondent of the *Financial Times*. He deposes, *inter alia*, at paragraph 3 that "There has been no more profound development in British retailing than the introduction of self-service trading methods since the end of the second world war." Then, in paragraph 7, he says: "What can be stated is that self-service has contributed to a dramatic change in the structure of retailing"—and he elaborates that. Then he says, at paragraph 20:

> The first supermarkets (ie self-selection stores over 2,000 sq ft) contained only a few cabinets. Today, a large area is devoted to long runs of open cabinets, offering dairy produce, meats, fish, vegetables, pies, speciality foods and cakes.

That is referring to the cabinets in the shops, not the cabinets for sale to customers.

Then he says:

> A constraint in the 1950s and early 1960s was the limited size of the domestic refrigerator and even the supply of new thin wall home refrigerators with larger freezing sections did not meet many family requirements. Consequently, the manufacture and sale of special deep-freezer cabinets for household use had enlarged. These vary in size, from modest box-like units for the small kitchen to large cabinets kept in a garage or some other convenient place.

Those are the freezer cabinets with which this appeal is primarily concerned.

Then, at paragraph 21:

> The trend has posed the food retailer with the problem of supplying frozen items on a larger scale.

Then, at paragraph 27:

> Because of innovation, there can be no standard definition of a supermarket, which was a pre-war American phrase coined to describe retail stores attempting to create the street market atmosphere within premises not relying on counter service.

Paragraph 37:

> Food retailers will state there is no fixed commercial formula for a supermarket. Outside the grocery trade, there are other traders who also use the term for their particular systems of selling motor accessories, hardwear and do-it-yourself materials (to cite some examples).

Paragraph 38:

> The supermarket is what a retailer wants it to be, with an important proviso. That is that their operation should depend on a substantial element of self-selection by consumers in making their purchases. The word supermarket describes a system of selling within premises. In short, it is a principle. A shop is somewhere shopping takes place. A supermarket shop is one based on the principle of self-service.

Paragraph 41:

> The modern freezer centre is engaged in the sale of quick-frozen foods by the supermarket principle. That the goods may be purchased in bulk does not conflict with the essential aim of supermarket operations to maximise sales of stock to compensate for lowered profit margins.

Then finally, paragraph 47:

> The only way to cut through all this is to find the common basis for classifying premises engaged in retailing as supermarkets. This has to be by acknowledging the one underlying principle—the sale of goods through self-service means.

No further evidence was filed; there was no oral evidence or cross-examination on the affidavits, which of course is in accordance with the usual procedure on a motion in the Chancery Division.

The hearing took place on July 21 1976, the motion being treated as the trial of the action. It appears that the case made by Calabar upon that hearing was basically that this freezer centre could not be described as a retail shop at all. That contention was abandoned before us; I need not elaborate upon it, except to say that the learned judge said that the contention was plainly ill-founded. The judge gave a very short judgment, of which we have his rather summary note, but it is important to bear in mind that he was addressing himself to a very different question from that with which we are concerned. He said: "A supermarket is a large self-service shop selling a variety of goods. At the date of the lease traditional supermarkets carried on selling groceries and provisions to customers of all sections of the public for immediate consumption." Then he says: "Permitted use was retail shop, which this is; for supermarket, which it is; for the sale of groceries etc; these are things sold and it seems that the current use falls within the use permited by subclause (q)." The notice of appeal is relatively uncommunicative.

Upon this appeal Mr Caplan, for Calabar, having jettisoned the contention that this was not a retail shop, relied, and relied entirely, on a different point, namely, that for the reasons which he advanced the business carried on by Tesco in this shop is not the business of a supermarket. The basic ground for that contention, to which I shall return in more detail, is that among other things there were offered for sale in this shop freezer cabinets, which are not a proper subject matter of self-servicing.

It will be convenient at this point to construe the restriction in subclause (q). In order to comply with the exception, the following requirements have to be observed: first, the premises must be used for the purpose of a retail shop; then that retail shop must be a retail shop for the purposes of the lessees' business of a supermarket. One may at this stage elide the word "lessees" in so far as it relates to Foodtown. The purposes are the business of a supermarket. Then the business of a supermarket must be a particular business of a supermarket, namely, the business of a supermarket for the sale of groceries, provisions, garden produce, fresh meat, confectionery, domestic and hardware and toilet requisites. So far, it has to be a supermarket for the sale only of that range of articles. Then follow the important words: "and also as ancillary thereto for the sale of such other articles or things as are usually sold by and in supermarkets."

To avoid misunderstanding, I think it is really clear that in putting a meaning upon the word "supermarket," one must see what was the natural meaning, in its context, of that word at the date of the lease, namely 1961. But obviously, a person carrying on the business of a supermarket, as that word stood in 1961, might, while still carrying on that business, vary the methods by which he carried on that business. That becomes apparent if one thinks of any other trade, for example a covenant not to carry on any business other than that of an hotel or a chemist. Again it is clear that a person carrying on the business of a supermarket might vary, and add to, the articles within the specified range of articles, but not outside that range. So two issues arise in this case: first, is the business carried on by Tesco on these premises today the business of a supermarket? Second, are the cabinets within the specified range of goods, including the extending words, "such other articles or things as are usually sold by and in supermarkets"?

I shall consider these points separately, although, in view of certain concessions made by Mr Caplan, only the first issue is now a live issue. (1) Is the business of Tesco the business of a supermarket? I would apply the uncontradicted evidence of Mr Corina. He deposed that "The word supermarket describes a system of selling within premises. In short, it is a principle ... A supermarket shop is one based on the principle of self-service," and he says that the one underlying principle is the sale of goods through self-service means. Mr Corina nowhere states—and his evidence is to the contrary—that in order to constitute the business of a supermarket, every single item sold on the premises, or sold in the course of that business, must independently be the subject matter of self-service. And it seems to me, the question must be one of fact and degree in each particular case. To state the extreme example, if 95 per cent of the articles sold are the subject matter of self-service, then I should have thought, *prima facie*, that the business would be nonetheless a supermarket by reason that 5 per cent were not the subject of self-service; but as the percentage of articles which are not the subject of self-service increases, then it becomes more difficult to say that the business is that of a supermarket, and if the percentage rises to 95 per cent then one may say fairly confidently that the business would not be that of a supermarket. The percentage which these cabinets form of the whole of the sales in the supermarket in the present case is 15 per cent. It seems to me that the plain and sensible conclusion corresponds to the evidence of Mr Corina; that is, that this business is the business of a supermarket.

Mr Caplan contended that the evidence of Mr Corina (and indeed, any evidence on the meaning of the word "supermarket") is wholly inadmissible. He quoted a statement from the case of *Lovell & Christmas Ltd v Wall* (1911) 27 TLR 236 which includes these words after certain other qualifications: "If it contains technical terms an expert may explain them," and then it goes on to deal with trade custom and secondary meaning.

> But unless the case can be brought within some of these exceptions it is the duty of the court, which is presumed to understand the English language, to construe the document according to the ordinary grammatical meaning of the words used, and without reference to anything which has previously passed between the parties. It is not in dispute that the court, in making use of its knowledge of the use of the English language, may refresh itself by reference to dictionaries.

Mr Caplan contended that the word "supermarket" is indeed an ordinary word in use in the English language, so that evidence is inadmissible, though one may look at a dictionary. I say "is" an ordinary English word, but strictly speaking the question is whether it was an ordinary English word in 1961. Mr Caplan then referred to three modern dictionary definitions. *The Concise Oxford Dictionary of Current English*, 1976, defines "supermarket" as a "Large self-service store selling foods and some household goods"; *Osborne & Grundage, Concise Commercial Dictionary*, 1966 (not a dictionary with which I am personally familiar but I am told that it is of authority), defines "supermarket" as "A retail shop selling a large range of goods to customers who collect their own purchases and pay at a central point"; and *Webster's Third New International Dictionary* (which I understand is an American work) defines "supermarket" as "A departmentized self-service chain or independent retail market that sells food, convenience goods and household merchandise arranged in open mass display."

From his contention that "supermarket" is an ordinary English word and from those dictionary definitions, Mr Caplan derives these propositions: A supermarket is a retail shop with three characteristics: (1) the goods are on display within the shop for self-service, with a certain exception which is not now material, as to goods which have to be weighed or measured; (2) such goods, following selection, are taken away from the shop by the customer himself, after first taking the goods to a paying point; and (3) the goods sold are of a wide variety falling within a general description of portable goods for domestic and personal use or consumption. The effect of (2) and (3) is that the goods displayed for self-service in the shop must be confined to goods which are portable and are capable of being taken to the checkpoint by the customer.

I do not accept the contention that the word "supermarket" is, even now, an ordinary English word such that the court must be taken to be acquainted with its meaning, subject only to refreshing itself by reference to dictionaries. It is clear that the word "supermarket" started as a fancy trade name in the distributive food trade, and as such it was a technical term. It may well be that the word "supermarket" is in the course of development into an ordinary English word; but even today it seems to me that that word is no more than passing through an intermediate area between a technical term and an ordinary English word. For myself, apart from the evidence in the present case, I should not be certain of its precise meaning, and I am not prepared to pretend that I am certain of its exact meaning only to the assistance of dictionaries. It seems to me that evidence of the meaning of the word "supermarket" is admissible today. I would add that in any event the very short definitions in the various dictionaries do not seem to me sufficient to

A establish that the word "supermarket" denotes a place in which only self-service can take place and in which no single article, not being the subject of self-service, could be sold. I do not accept the three characteristics advanced by Mr Caplan; it seems to me that the true view is that a supermarket is a shop in which substantially the greater part of the articles on display are for self-service, but not a shop in which exclusively the articles are on display for self-service. I conclude, then, on this first issue, that the business of Tesco in these premises is, upon the evidence, a supermarket.

The second issue is whether these cabinets fall within the specified range of goods. They can be said to do so under B either of two heads, namely (1) that they constitute domestic requisites, and (2) that they fall under the head "such other articles or things as are usually sold by and in supermarkets." I do not propose to elaborate on this issue, because Mr Caplan very fairly accepts that if he is wrong on his construction of the word "supermarket" there is no evidence that these cabinets are not usually sold in supermarkets.

I conclude, then, that in carrying on the business which it does carry on upon the demised premises, including the sale of these freezer cabinets, Tesco is not in breach of the covenant contained in subclause (q) and I would dismiss this C appeal.

Agreeing, BROWNE LJ said: I agree that this appeal should be dismissed for the reasons given by Sir John Pennycuick. As my Lord has said, in this court the only point argued by Mr Caplan was that these premises are not being used for the purposes of the business of a supermarket because of the sale of the freezer cabinets. As my Lord has pointed out, this was not a point raised by the originating summons or the notice of appeal and it appears from the note which we have of the judgment of Judge Thomas that it was not really raised before him. My Lord has already read Mr Caplan's submission that a place cannot be a supermarket unless it has D the three characteristics to which my Lord has referred. I agree with my Lord that the word "supermarket" is not yet established as an ordinary English word with a clearly defined meaning. For this reason, and also perhaps because it is a technical term, I agree that evidence is admissible to help us to give a meaning to that word.

My Lord has already referred to the judge's definition in one sentence: "A supermarket is a large self-service shop selling a variety of goods." He has also referred to a number of passages in the affidavit of Mr Corina, which seems to me to be of very great help and importance. I am not going to read them again; I will only read from one paragraph in E which he says that in effect the essential feature of a supermarket is "that their operation should depend on a substantial element of self-selection by consumers in making their purchases." It is not saying that it must depend entirely on self-selection. There are two other definitions in the evidence; first of all, Mr Edgerton in his affidavit filed on behalf of the defendants said in paragraph 6 on page 52: "The defendants define a supermarket as a self-service store of over 2,500 sq ft net space, selling groceries, provisions, fruit, meat and toiletries, wines and spirits, and since the early 1960s increasingly including a limited range of convenience and durable goods not normally sold in association with F foodstuffs, such as menswear, ladieswear, childrens wear, hardware, domesticware, electrical goods and furniture." Mr Tibbatts, in his second affidavit, filed on behalf of the plaintiffs, said in paragraph 6 at page 80: "I do not dispute the defendants' definition in paragraph 6 of the said affidavit," but he said that he would add two qualifications, on which Mr Caplan did not rely in this court. My Lord has also referred to three dictionary definitions which have been put before us, and I need not repeat them.

Mr Caplan's submission to us was that premises cannot be a supermarket if anything is sold there (to any extent beyond what may be regarded as *de minimis*) which cannot be carried by the customer to the pay desk and then carried G away from the premises by him or her. I cannot accept this. Mr Caplan concedes that if no freezer cabinets were sold here, the premises would be a supermarket, as on the evidence they clearly would be. When Mr Tibbatts visited these premises on March 11 1975 he observed that there were at least six deep-freeze units on offer for sale to the public—see paragraph 7 of his affidavit of March 24 1975—and the sale of freezer cabinets represents 15 per cent of the turnover; see the passage from Mr Edgerton's affidavit which my Lord has already read. That does not in my view prevent the premises from being a supermarket. The plan exhibited to Mr Tibbatts' first affidavit, at page 50, though not to scale, H indicates that only a small part of the premises was then used for the sale of freezer units.

Whether one takes the definitions of the judge, or of Mr Corina, or of the dictionaries, these premises in my view are within them—they are substantially used as a supermarket; I shall come back a little later to this point. But whatever may be the general meaning of "supermarket," the decision in this case must of course ultimately depend on the construction of this particular covenant. Plainly the parties contemplated and accepted, when the lease was granted in 1961, that the business was at that time that of a super- J market. Mr Caplan accepts that the covenant would not be broken merely because the lessees had changed their method of carrying on that business; for example, by beginning to supply different ranges of goods. I understand that it is for this reason that Mr Caplan does not rely on the reference to *fresh* meat in the covenant.

For the reasons given by Sir John Pennycuick and for those that I have tried to give, the premises are in my view still used for the purposes of the business of a supermarket. They are still used substantially for that purpose, even on Mr Caplan's definition. As Sir John Pennycuick has said, the question whether or not the premises are used for the K purposes of a supermarket may in some cases be a question of degree; as a question of degree, it seems to me that in the present case they clearly are still used for the purposes of a supermarket. The words of the clause after "the business of a supermarket" exclude any suggestion, for example, that a supermarket can only be used for the sale of food or groceries. In my view, as my Lord has said, these freezer cabinets are domestic requisites. I would also hold that the sale of the freezer cabinets is "ancillary" to the sale of deep frozen meat. It is true that "ancillary" is qualified by the words "such other articles and things as are usually sold by or in a supermarket." But Mr Caplan accepted this morn- L ing, in answer to my Lord, that those words do not refer only to the articles which are usually sold by or in supermarkets in 1961, but also cover articles which are usually sold by or in supermarkets in 1977. As it is the landlords who are alleging a breach of covenant, the burden of proving that freezer cabinets are not now usually sold by or in supermarkets is in my view on them, and Mr Caplan very fairly accepted that there is no evidence before the court to that effect.

Accordingly, for I think substantially the same reasons as those which Sir John has already given, I agree that this appeal must be dismissed. M

Also agreeing, STEPHENSON LJ said:

These premises are admittedly still used for a retail shop and for the sale of many of the articles or things specified in the covenant. It is also conceded that the sale of frozen meat and other frozen food products, which are now the greater part of what is sold on the premises, would not prevent the premises from continuing to be used for the purposes of the respondent's business of a supermarket. That concession would seem to destroy the case made by the appellants in correspondence, in their originating summons and notice of

A motion, in their affidavits, in argument before the learned judge and even in their uncommunicative notice of appeal.

But it is submitted by Mr Caplan in this court that the sale of freezer units or cabinets prevents these premises from being premises used for the purposes of the respondent's business as a supermarket. Why? Because they destroy the character of self-service which we should take judicial notice is essential to a supermarket and is limited to providing purchasers with things they can collect and carry past the paying point without help from the respondents' staff or from too many helpers of their own. That seems to me to be equivalent to submitting that we ought to know without evidence that
B such articles or things as freezer cabinets are not usually sold in supermarkets (though Mr Caplan submits that they are usually sold in freezer food centres) and that there is indeed an essential incompatibility between a supermarket and the sale of freezer cabinets, even if they constitute a relatively small part of its trade.

I regret to have to admit that I do not know anything of the kind, and Mr Caplan has not persuaded me that we must be presumed to know precisely what a supermarket is. I consider that a supermarket is a recent and still growing concept, and about its essential nature the courts need evidence. I see no reason, and I find no evidence, to contra-
C dict the evidence of Mr Corina, to which my Lord Sir John Pennycuick, has referred. I accept the judge's definition, which has also been read. On the assumption that a freezer unit cannot be purchased, to quote Mr Corina's words, "through self-service means" truly so called, in my judgment to use these premises for the sale and purchase of freezer cabinets to the extent specified in the evidence does not contravene this covenant.

Accordingly I agree that the appeal should be dismissed.

The appeal was dismissed with costs.

D

CHANCERY DIVISION

November 24 1977

(Before Mr Justice SLADE)

P B FROST LTD v GREEN

Estates Gazette February 11 1978

(1977) 245 EG 485

Jurisdiction of High Court—Mortgagees' claim for possession —Land in question two separate hereditaments—First
E **hereditament with net annual value for rating of £742—Second hereditament agricultural land not liable to be rated—Question whether the High Court had jurisdiction to hear case or whether it fell within exclusive jurisdiction of the county court because not above the £1,000 limit—Agricultural land treated as having a net annual value for rating for this purpose "equal to its value by the year"—Aggregate net annual value thus sufficient to give High Court jurisdiction**

This was an originating summons under which P B Frost Ltd claimed possession as mortgagees of certain land against the defendant, William James Green. A question
F arose as to whether the action was within the jurisdiction of the High Court.

C P F Rimer (instructed by Burchell & Ruston, agents for Gepp & Sons, of Chelmsford) appeared for the plaintiffs; Stephen Hunt (instructed by Blacket Gill & Langhams, agents for Peter M Cox & Co, of Melton Mowbray) represented the defendant.

Giving judgment, SLADE J said that the plaintiffs' title arose under a legal charge dated January 19 1977 whereby in consideration for a loan of £40,000, the defendant charged two separate hereditaments at Pickwell, Leicestershire, namely, a manor house with 12 acres of adjoining land and 218 acres of freehold agricultural land comprising no G
dwelling-house. It had been stated that the net annual value for rating of the manor house was £742. The 218 acres were not liable to be rated or included in any valuation list for rating purposes because they comprised solely agricultural land. It was, he thought, common ground that the plaintiffs' title to possession had been made out subject only to the question of jurisdiction and to any question of suspending the operation of any order or of postponing the date for delivery of possession.

Section 37(1) of the Administration of Justice Act 1970 provided:

> Where a mortgage of land consists of or includes a dwelling-house and no part of the land is situated in Greater London, then, subject to subsection (2) below, if a county court had jurisdiction by virtue of section 48 of the County Courts Act 1959 or section 38 of this Act to hear and determine an action in which the mortgagee under that mortgage claims possession of the mortgaged property, no court other than a county court shall have jurisdiction to hear and determine that action. H

There was no dispute on the facts that the first two of the three conditions were fulfilled here. The question arose in relation to the third. Mr Hunt, for the defendant, claimed that the county court had jurisdiction to hear and determine the present proceedings by virtue of section 48 of the County J
Courts Act 1959 and accordingly, that the third condition was fulfilled, just as much as the first and second, so as to deprive the High Court of jurisdiction. Section 48(1) of the County Courts Act 1959 provided that a county court should have jurisdiction to hear and determine any action for recovery of land where the net annual value for rating of the land in question was not above the county court limit. The county court limit referred to was fixed at £1,000 by the Administration of Justice Act 1973, section 6 and Schedule 2. But the answer, in his (his Lordship's) judgment, was to be found, as Mr Rimer for the plaintiffs had submitted, in K
section 200 of the County Courts Act 1959.

Section 200(2) provided the answer to the problems that would otherwise arise in attempting to apply section 48 in the untidy case where the property sought to be recovered did not consist of one or more separate hereditaments, each having a separate net annual value for rating. Paragraph (a) of the subsection covered the case where the relevant property or part of it did not have a separate net annual value for rating, but formed part of a larger hereditament which did have such a value. Paragraph (b) then went on to cover any other case where the property sought to be recovered did not consist of one or more separate hereditaments each L
having a separate net annual value for rating. In particular, in his judgment, it covered the case where the property sought to be recovered was or included property which did not have a net annual value for rating at all, because it was not liable to be rated. He concluded that section 200(2)(b) was apt to cover a case such as the present.

Applying this to the facts of the case, the net annual value for rating of the 218 acres fell to be treated as being "equal to its value by the year." This phrase was not defined in the 1959 Act, but the judge thought that the meaning to be attributed would be one very similar to the formula provided by section 19(3) of the General Rate Act 1967 in relation M
to hereditaments other than houses or other non-industrial buildings. But he did not think it was necessary to pursue this point. The plaintiffs had deposed to the fact that the value of the 218 acres "by the year" must be "well in excess of £258." This allegation had not been denied in any evidence on behalf of the defendant.

Accordingly he found that the net annual value for rating for the purpose of the two sections of the lands which were the subject of this application totalled in the aggregate more than £1,000. He therefore concluded that the county court did not have jurisdiction to hear the application under section 48 of the County Courts Act 1959. He therefore rejected the defendant's submission that the jurisdiction of the High Court was excluded by section 37 of the Administration of Justice Act 1970. In his judgment this court had jurisdiction to hear the case.

QUEEN'S BENCH DIVISIONAL COURT

December 12 1977

(Before Lord WIDGERY CJ, Lord Justice CUMMING-BRUCE and Mr Justice PARK)

HOOPER v EAGLESTONE

Estates Gazette February 18 1978

(1977) 245 EG 572

Caravan Sites Act 1968—Appeal from justices' dismissal of information alleging deprivation of caravan dweller's right to electricity—Right dependent on site being a "protected site"—Site protected although landowner not eligible for a site licence—Appeal allowed and case remitted to justices

This was an appeal, by way of case stated by justices sitting at Newquay, by William David Hooper, clerk to the borough of Restormel, against the justices' dismissal on February 22 1977 of an information preferred by him against Leslie George Eaglestone alleging that he, being the owner of land at Trebarber Farm, St Columb Minor, Newquay, did, with intent, cause Orton Swandley, the occupier of a caravan on a protected site, to suffer the loss of electricity supply.

Michael Burrell (instructed by Sharpe, Pritchard & Co, agents for the solicitor to the borough of Restormel) appeared for the appellant; F H S Gilbert (instructed by Tozers, of Dawlish) represented the respondent.

Giving judgment, LORD WIDGERY CJ said that the information, preferred under section 3(1) of the Caravan Sites Act 1968, alleged that the respondent had cut off the electricity to the occupant of a caravan on his land. The sole question was whether the justices were right in accepting a defence submission that the site was not protected within the meaning of the Act.

The Act defined a protected site as "any land in respect of which a site licence is required under Part 1 of the Caravan Sites and Control of Development Act of 1960. . . ." Turning to the 1960 Act one found that the occupier of any land used for caravans must be the holder of a site licence. The effect of that was that land could not be used for a caravan site without a site licence. A local authority could issue a site licence if, and only if, an applicant had the benefit of permission to use the site for caravans.

In the present case the owner could not get a site licence because of the unauthorised presence of caravans on the land. But he could not argue that because he could not get a site licence the land was not a "protected site." A site licence was *required* in respect of the land and therefore it was protected within the Caravan Sites Act. The case should go back to the justices for them to continue the hearing.

CUMMING-BRUCE LJ and PARK J agreed.

Appeal allowed with costs. Case remitted to the justices for them to continue the hearing.

COURT OF APPEAL

November 17 1977

(Before Lord Justice MEGAW, Lord Justice BRIDGE and Lord Justice WALLER)

SWORDHEATH PROPERTIES LTD v FLOYDD AND OTHERS

Estates Gazette February 25 1978

(1977) 245 EG 657

Possession order against squatters—Part 1 of Order 26 of the County Court Rules—No power to suspend execution without the owner's consent—County Court and High Court rules the same

This was an appeal by Swordheath Properties Ltd from an order of Judge Llewellyn at Bloomsbury and Marylebone County Court granting the company possession of the ground, first and second floors of a house in St Anns Villas, Notting Hill, London, pursuant to Part 1 of Order 26 of the County Court Rules, but suspending its execution for 14 days.

David Neuberger (instructed by Harold Stern & Co) appeared for the company. The defendants did not appear and were not represented.

MEGAW LJ reading the judgment of the court, said that Order 26 corresponded with Order 113 of the Rules of the Surpreme Court. Some of the defendants were present at the originating application and the requirements of Order 26 were therefore fulfilled. The judge asked counsel for the company whether he would agree to execution of the order being suspended for 14 days. Counsel said that on his instructions he could not. The judge nevertheless made an order for possession not to be executed for 14 days.

The company applied for the appeal to be expedited. The motion came before the Court of Appeal on November 10. None of the defendants was present, but the court was satisfied that they had been served with the notice. The court allowed the appeal and deleted the suspension. It had been held in *McPhail v Persons, Names Unknown* [1973] Ch 447 that the High Court had no power under Order 113, in the absence of the applicant's consent, to suspend an order for possession. That decision was binding on their lordships regarding Order 113. It was plain that there was no distinction between that order and Order 26 relevant to the question before the court. In the absence of consent, the judge had no power or discretion to suspend execution. Hence the appeal had to be allowed.

The court was satisfied that the judge had not deliberately made an order which he knew to be outside his powers when he made it. Nor did he deliberately misconduct himself in respect of an appeal.

The court had been told that in a number of county courts suspensions had been granted in Order 26 orders without the applicants' consent. As the order at present stood, that was not within the courts' powers. It would seem that in not a few cases consent was given by applicants to a short suspension. It might be that those responsible for considering such matters might wish to consider whether some power, possibly a strictly limited power, of suspension should be made available, despite the absence of consent. It might be, however, that there were practical reasons to the contrary. The court expressed no view.

CHANCERY DIVISION

April 19 1977

(Before Mr Justice SLADE)

ENGLISH v DEDHAM VALE PROPERTIES LTD

Estates Gazette March 4 1978 and March 11 1978

(1977) 245 EG 747 and 839

Action for rescission of conveyance and damages, alternatively for damages only, for misrepresentation, alternatively for an account of profits by reason of non-disclosure—Sale of a property consisting of under 4 acres containing a bungalow and garage—Application for planning permission for a limited development of the property made in the name of the vendors but without their knowledge or permission by purchasers' agent before exchange of contracts—Purchasers liable to account for profits—Categories of fiduciary relationship not closed—Order for payment of sum due on taking of account

A This was an action begun by Thomas English and his wife, Elsie English, and continued by the latter after the death of her husband, for relief on the grounds of misrepresentation and/or non-disclosure in connection with the sale of a property known as Rusty Tiles, Coggeshall Road, Dedham, Essex. The cause of action was based on an application for planning permission made in the names of the plaintiffs, but without their knowledge, before the exchange of contracts. The detailed allegations in the statement of claim and the history of the events are dealt with fully in the judgment of Slade J.

A Scrivener QC and T Thomas (instructed by H T Bowles
B & Co, of Manningtree) appeared for the plaintiff; L Joseph (instructed by David Cohen & Goldsobel) represented the defendant company.

Giving judgment SLADE J said: At the beginning of 1971 the two persons named as plaintiffs in this action, Thomas English and his wife, Elsie English, were the joint legal and beneficial owners of a property known as Rusty Tiles, Coggeshall Road, Dedham, Essex, which I will call "the property." They were an elderly couple, who had owned it since about 1955 and had lived there since about 1957. On July 21 1971 they exchanged contracts for the sale of the
C property at the price of £7,750 to the defendant company, Dedham Vale Properties Ltd. This is a property-developing company which has at all material times been controlled and managed by a Mr Harrington, who is a qualified architect with long experience in planning matters. By a conveyance on sale dated December 9 1971 Mr and Mrs English conveyed the property to the defendant.

By the writ in this action, which was issued on March 30 1972 Mr and Mrs English claimed, as the primary head of relief sought, rescission of the conveyance and damages, and, alternatively, damages for misrepresentation. Mr English subsequently died on January 6 1974, but, the cause of action
D being a joint one which survived to Mrs English, she has continued the action on her own. The statement of claim had been amended before the trial began. I propose to read it as originally amended and then to refer to the reamendments made at the trial, because its precise form may be of some significance for certain purposes. As first amended, it read as follows:

1. Until the conveyance referred to hereafter the plaintiffs were the beneficial owners of the property known as Rusty Tiles, Coggeshall Road, Dedham, in the County of Essex.

2. Between February 1971 and July 1971 the defendants by their servants or agents were negotiating with the plaintiffs for
E the purchase of the said property.

3. During the said negotiations the defendants' servant or agent one Harrington made the following oral representations to the plaintiffs:

(a) That he the said Harrington had found out that there was "No planning permission going" in respect of the said property.

(b) That there was no immediate possibility or prospect of planning permission being granted for housing in respect of the said property.

4. In reliance upon these said representations the plaintiffs agreed to accept the sum of £7,750 from the defendants for the said property instead of a higher sum had there been
F permitted development potential for the same.

5. The said sale was concluded by an exchange of written contracts on July 21 1971 and completed by a conveyance dated December 9 1971.

6. On July 14 1971 the defendants' servant or agent one Mead without the authority, consent or knowledge of the plaintiffs made an application for planning permission in respect of the said property. The said application was false in the following particulars:

(a) The said Mead falsely held himself out as being the authorised agent of the plaintiffs for the purpose of making the said application.

(b) The said Mead falsely gave a certificate under section 16 of the Town and Country Planning Act 1962 to the planning authority purporting to show that he was acting as the authorised agent of the plaintiffs and therefore by implication that they knew of and consented to the said application being made. G

7. On October 20 1971 the planning authority, the Lexden and Winstree Rural District Council, gave permission for the development of the said property by the building of a dwelling house and garage thereon. The document giving the said permission was sent to the said Mead who received the same as the defendants' servant or agent. The defendants did not inform the plaintiffs of the said application or the said permission being granted. H

8. In the aforesaid premises the representations set out in paragraph 3 thereof were made falsely to the knowledge of the said Harrington and thereby to the knowledge of the defendants.

9. By reason of the said false representations and/or the said failure to disclose the matters set out in paragraph 7 hereof, the plaintiffs have suffered damage and loss in that they were induced in selling the said property for £7,750 whereas the true value of the same was in excess of £18,000.

10. Further or in the alternative by making the said application for planning permission of the plaintiffs' said property as aforesaid and/or by receiving notification of the grant of planning permission as aforesaid the defendants became trustees for the plaintiffs and were under a duty to inform the plaintiffs of the said application and of the said grant of planning permission. J

11. In breach of the said duties the defendants failed to inform the plaintiffs of the said application and of the said grant of planning permission thereby causing the plaintiffs damage and loss in that they did not know of the true value of their said property and sold the same for £7,750 instead of at least £18,000. And the plaintiffs claim:

(1) Rescission of the said conveyance and damages.

(2) Alternatively, damages under paragraph 9 or alternatively under paragraph 11 hereof.

(3) A declaration that the planning permission granted on October 20 1971 by the planning authority in respect of the said property is null and void. K

(4) Costs.

(5) Further or other relief.

At the trial I gave leave for the reamendment of the statement of claim, so as to introduce a new cause of action, though one substantially based on the facts alleged in the statement of claim in its original form. A new paragraph 12 was added, reading as follows:

12. Further or in the alternative by reason of the matters set out in paragraphs 6 and 7 hereof, the defendants were able to make profits out of the purchase and subsequent development of the said plot and are accountable to the plaintiffs for such profits as are found to have been made. L

The reamendments also involved the incorporation of the following two new paragraphs in the prayer:

(4) An account of the profits accruing to the defendants by reason of the non-disclosure to the plaintiffs of their application for and subsequent grant of planning permission in their name.

(5) An order for the payment by the defendants to the plaintiffs of all moneys found to be due to them on the taking of such account. M

I now proceed to a more detailed recital of the facts of the case. The property consists of a site comprising a little under 4 acres of land. The width of the site towards the road frontage is substantially narrower than the width at the rear. On the narrow front portion there has at all material times been situate a modest bungalow and garage. The wider rear portion has at all material times been rough grassland. Adjoining the site, on its northern side, is a site of similar size and shape known as "Shorelands."

In 1964 the Essex County Council approved a non-statutory "Dedham Village Plan," embracing Dedham Village itself and the settlement of Dedham Heath, which is situated about a mile to the south and includes the property. On that plan the property was within one of three areas allocated for residential development. Such development was, however, dependent upon the provision of adequate drainage and sewage disposal facilities. Mrs English at all material times knew of the substance of the plan and that the property had thereby been zoned as a site for some 30 houses. She had in her possession a copy of an article in a newspaper dated February 21 1964 in which the relevant map and proposals were reported and reproduced. In or about November 1968 the Essex County Council withdrew the Dedham Village plan because it was thought that the development as originally proposed would be detrimental to the amenity and rural character of the area, part of which was designated as being of great landscape value.

The owners of " Shorelands " have at all material times been a Mr and Mrs Coe. In 1968 Mr English and a Mr McCarthy, who was the son of Mrs Coe, jointly applied to the Essex County Council for planning permission for residential development of the property and " Shorelands," which together comprised about 7½ acres. On February 17 1969 the council refused this permission. The refusal, however, was expressed to be based on the following single ground: " The site is at present without means of sewage disposal and the development of the site at the present time is therefore considered to be premature." The very terms of this refusal thus gave some grounds for inferring that, if means of sewage disposal were made available thereafter, development might perhaps be permitted.

Early in 1970 Mrs English, realising the importance of sewerage if there were to be any hope of developing the property, paid a visit to the advice bureau of her Member of Parliament, Mr Anthony Buck, with regard to the sewerage scheme at Dedham Heath. Mr Buck made certain enquiries with the Lexden and Winstree Rural District Council and on March 4 1970 sent her a copy of a letter which he had received from the clerk of that council, in which he said that tenders for the sewerage scheme would be opened within the next two months and it remained to be seen whether they would be accepted. A tender was in fact accepted in June 1970 and work on the scheme began on September 1 1970 on the basis of an 80-week contract. On September 7 1970 following some intermediate correspondence with her, Mr Buck wrote to Mrs English to say that he had received a letter from the council telling him that work on the sewerage scheme was to start at the beginning of that month and that it was scheduled to take 80 weeks to complete. Accordingly Mrs English learned from this letter that if the work were done according to schedule it would be completed in March 1972 and that she would have a chance of obtaining planning approval for her desired development at any time after that. She accepted in her evidence that she appreciated that planning permission might become obtainable in respect of her property in 1972.

Meantime, however, she and her husband had decided that they must try and dispose of the property. He had unfortunately become ill in 1969 or 1970. In view of this and in view of her age, Mrs English felt that she could not cope with looking after him and so large a property at the same time. The correspondence before me reveals that a number of persons were interested as potential purchasers of the property. Between August and October Mr and Mrs English were negotiating with a company known as Hey & Croft Ltd and negotiations reached an advanced stage for the possible sale of the property at a price of £14,000. By October 1970, however, this proposed sale, which had been negotiated in conjunction with the proposed sale of " Shorelands " to the same purchaser by Mr and Mrs Coe, had fallen through. It may be mentioned that, during the course of negotiations for this sale, namely on August 13 1970, Fenn Wright & Co, estate agents acting for Mr and Mrs English, wrote to Mrs English saying that, because the contract had been placed for the new sewerage scheme, they had visited the Lexden and Winstree Rural District Council offices to discover whether any planning application for residential development of the property, if submitted at that time, would be likely to be granted. They reported that the official whom they had seen had given his opinion that a further application at the present time would be considered premature, but that the council would have no objection to the land being developed when the new sewage works were in operation, though he could not say whether the county planning authority would take the same view. It is in fact common ground between the parties in the present case that it was at all times the Essex County Council which presented the main opposition to the development proposals, rather than the local council, which was sympathetic.

By August 1970 Mrs English was also negotiating with a Mr Noonan acting on behalf of a company known as Gough Cooper (South East) Ltd. By November 1970 she was also negotiating with a Mr N H Norrington, who has given evidence in these proceedings. He is an estate agent but was acting at that time on his own behalf, as he was interested in purchasing the property for his own occupation. On November 2 1970 he wrote saying that he had made enquiries with the local council and that, since they would give no indication as to the possibilities of gaining planning permission for the property, he could not offer her asking price of £16,000. However, he made two alternative propositions to her; either he would pay her the current market valuation of the property to be assessed by an independent valuer, would further undertake to apply for planning permission and, if that application were successful, would make up the purchase price to a total of £18,500; alternatively, he would pay her the current market value of the property plus £2,000 for the possible potential and complete the deal on that basis.

On December 11 1970 Mrs English, who had not accepted either of Mr Norrington's alternative offers, wrote to the Department of the Environment, which, on December 21 1970, wrote to her saying that it understood that permission to develop the land two years ago had been refused because of the absence of main drainage in the area and suggested that she should reapply for planning permission to the rural district council.

On January 19 1971 Mr Norrington wrote to Mrs English asking if she was in a position to reply to the offer previously made by him, this referring to his letter of November 2 1970. By that date, she had received an offer of £13,500 from Gough Cooper (South East) Ltd through Mr Noonan, but had not accepted it. Mrs English did not reply to this letter. By the end of January 1971 Mr Norrington, as he told me in evidence, had for the time being lost interest in the property. He therefore told Mr Harrington about it. Mr Norrington had previously acted as an estate agent for the defendant company in various transactions, and had by then become a personal friend of Mr Harrington. As Mr Harrington put it, Mr Norrington passed on the information to him as a " tip."

On January 29 1971 Mr Harrington went to see the property and spoke to Mrs English, with whom he dealt throughout the subsequent negotiations, in which her husband took no part. He was naturally very interested in the planning position. She herself showed him the press cutting to which I have referred and the correspondence with Mr Buck. At a very early stage he visited the offices of the rural district council to investigate the planning position. He discovered that the Dedham Vale proposals referred to in the newspaper cutting were non-statutory. He also discovered from the planning list that an application for the development of the whole site had already been made and refused, on the grounds that it was premature, and that there had been a reference to the sewage situation. He also discovered that the sewerage scheme had been started and was due to be completed in 1972. He gained the impression that, while the rural district council were sympathetic, the county council were unlikely to give permission. He considered, however, that there was some chance of obtaining permission on appeal to the minister, because of the use of the word " premature " in the previous refusal.

Mr Harrington had subsequent meetings with Mrs English on February 26 and April 24 1971. It is common ground that on one or more of the three meetings to which I have referred he made her an unconditional offer of £9,000 for the property, not linked to obtaining planning permission, and that she refused this offer. She says that at one of the meetings he also made her an unconditional offer of £10,000, but he denies this. He says that at one of these meetings he made her an offer of £18,000 conditional on the obtaining of planning permission, but she denies this. I think nothing really turns on these two disputed points.

On March 9 1971 Mr H W Whybrow, of Fenn Wright & Co, wrote to Mrs English informing her that Gough Cooper (South East) Ltd were not prepared to go ahead with the purchase and that they had also decided not to proceed with the acquisition of " Shorelands," because they appeared to be " less confident of obtaining planning consent for building than they were some months ago."

On March 29 1971 Mr Norrington wrote to Mrs English in effect repeating the offer made in his letter of November 2 1970, but nothing came of this and he faded out of the picture.

On April 16 1971 Mr Whybrow wrote to her informing her that Hey & Croft Ltd were no longer interested in making an offer for the property or for Shorelands. He therefore advised her to submit to the rural district council a fresh planning application, in conjunction with Shorelands, similar to that made about two years before. Mrs English did not take up this suggestion. It appears that by the middle of May the best offer she was left with was one of £7,500, because all other proposals had fallen through.

On or shortly before May 21 1971 Mrs English telephoned Mr Harrington asking him to go and see her at her house, which he did on May 21. Mrs English's version of this crucial meeting, as given in her evidence in chief and under cross-examination, was substantially as follows. He told her that there was " no building obtainable " for the land at the moment, or words to that effect, having obviously recently visited the planning authority. They did not have a long conversation. He offered her £7,750 for the property, representing a little bit more than the only other offer that was available to her at the time, of which she had informed him. She accepted this offer. He told her that he contemplated altering the bungalow by converting the loft into a room and by putting in dormer windows. He also said he would put the 3 acres at the back of it under rye to get thatching material for it. There was some discussion as to the date of completion and he agreed that this should be deferred for six months in order that she and her husband might find somewhere else to live. This was, I think, substantially Mrs English's account of this conversation.

Mr Harrington's version of it was rather different, being substantially as follows. She told him that the best offer left to her was £7,500, that she was most anxious to sell and leave because her husband was ill and she wanted to move to the coast to live and that the land was more than they could manage. He accordingly offered her £7,750, which she accepted. Though they had on previous occasions discussed the planning position, the situation was probably discussed that day as well. She knew as well as he did (according to his evidence) that the Essex County Council was not in favour of development of the land at the time. He told her that he was of the opinion that in the long run permission for development would be granted. He told her in particular that he thought the minister might in the end reverse the county council on appeal, but he thought all this would take some time. She said she was not prepared to wait. He did not recall specifically saying anything about altering the bungalow, but he saw no reason to dispute that he had said it, even though that was not his motive for buying the property. He did not remember saying anything to Mrs English about setting the 3 acres with rye and denied the precise words attributed to him by further and better particulars of the statement of claim, which alleged that he had said that he would do this to get the " hay " for thatching, since he would have referred to it as " straw." He could not, however, remember even referring to straw. He accepted that they discussed the date of completion and said that they agreed that the time should be six months, with an option for her to call for completion at any time after three months. He also undertook to help her find alternative accommodation. She asked him to suggest a firm of solicitors whom she might instruct, and he suggested Elwes Witty & Co. Then, according to his evidence, after the agreement had been reached, he told her that he would like, on behalf of the defendant company, to make an application for planning permission for a single plot, as opposed to residential development of the whole site, on the small narrow strip of land adjacent to the road frontage and to the existing bungalow on the property, which I will call " the small plot." According to his evidence, he told her that he would like to make the application in the plaintiffs' names, since they would continue to be the theoretical owners of the property for some period of time, and this she readily agreed to. Mr Harrington explained in evidence that he particularly wished to make the application in their names rather than his own or that of the defendant company, because he had had a brush of personalities with a person who at that time held a responsible position with the Essex County Council, and he believed that this brush had previously led to 12 consecutive refusals of applications for outline planning permission made by him or his company, most of which were in the event subsequently allowed on appeal. He explained in evidence, though he did not suggest that he told Mrs English this, that it was this pattern of refusals, which he regarded as unreasonable, and the belief that he was the victim of discrimination which made him particularly wish to use Mr and Mrs English's names. At the meeting of May 21 1971, according to his evidence, being delighted to have concluded the deal and not anxious to delay further, she readily agreed to his making the application in their names, using words to the effect, " Do what you like; I am not interested."

Mrs English absolutely denies that, at this meeting of May 21 1971, Mr Harrington made any mention to her whatsoever of any intention on his part to make a planning application in respect of the small plot. She said that she never agreed to lend her name to such an application and that, if he had mentioned it, she would have wanted more money. As she put it, " That would not have gone past unnoticed."

There is thus a stark conflict between her evidence and that of Mr Harrington on this very important point. They were the only two persons who could give evidence as to what occurred at this meeting, and this will be a convenient moment to state my assessment of them as witnesses. Mr Harrington, no doubt assisted by his experience in giving evidence in planning applications and such like, gave his evidence very clearly and well. I do not think there is any point of straightforward fact on which it can be shown to have been demonstrably incorrect; nor did I gain the impression that he was attempting to mislead the court in any respect. Mrs English, on the other hand, was a much less good witness. She is a lady of 82 years and, though I think her mental faculties seem essentially unimpaired, she appeared in the witness box somewhat excitable, angry and abrasive. Her bitterness over the whole transaction, and in regard to

Mr Harrington in particular, led her into a number of fairly clear, if generally trivial and pointless, misstatements of fact on matters of detail. She could hardly bear to admit anything that might be thought to show Mr Harrington even in a slightly favourable light. I therefore accept the submission, made by Mr Joseph on behalf of the defendant company, that her evidence, coloured as it was by obvious rancour in some instances, should be treated with some reserve. I would also accept that, where her evidence conflicted with that of Mr Harrington on detailed questions of fact, his evidence should, in general, be preferred. Nevertheless, I think due allowances must be made for her age; and I certainly did not get the impression that she was a person who was deliberately fabricating the principal points of her evidence for the purpose of bolstering up her case. On a number of important points, as will appear, I thought her evidence carried conviction, even allowing for the fact that some elderly persons may come sincerely to believe that certain things have been done and said in the past, when such belief has no foundation in truth.

One point was freely admitted by Mr Harrington in relation to the meeting of May 21 1971. Though he and Mrs English had had at least three meetings before that day and had discussed the question of planning permission at one or more of these earlier meetings, they had never previously discussed the possibility of a limited application being made for permission in respect of the building of one house on the small plot. I am prepared to accept from his evidence that at the meeting of May 21 he did very briefly mention to her that he was proposing to put in an application in respect of the small plot in the names of her and her husband. As will appear, however, he did not mention that he was even considering doing this before exchange of contracts. Furthermore, since the proposition mooted involved not only the use of their names but also a completely different form of application from that ever discussed between him and Mrs English before, I think it was incumbent on him to make it quite clear to her at that meeting, first, precisely what form of application he had in mind and, secondly, that he intended to use the names of Mr and Mrs English for the purpose of making the application. I do not think he did make these two matters sufficiently clear. All the evidence in this case indicated that Mrs English has at all material times been a lady with a keen business sense, who would assuredly have asked for some additional *quid pro quo* beyond a price of £7,750 if she had been made fully aware that Mr Harrington was contemplating a different, more limited, form of planning application, which inherently had a greater chance of success than the wider form previously discussed between them, and that he particularly wished to use the names of her husband and herself for this purpose. She herself said in evidence that she would have done so, and on this point I believe her. Mrs English, I am satisfied on the evidence, gave no true consent to the use of the plaintiffs' names in regard to this application, because Mr Harrington had not sufficiently explained to her what he wanted to do. The fact that he did not do so, coupled with his admitted failure subsequently to inform her or her solicitors either of the making or of the result of the application, has, I think, become the principal reason for the intense bitterness which she obviously now feels in relation to the transaction.

However, I entirely reject the allegation that at this meeting of May 21 1971 Mr Harrington made any representations to her in relation to the planning position which he knew to be untrue. It was common ground between the two of them in their evidence that, in the course of one or more of the four meetings between them to which I have referred, Mr Harrington had told her, in effect, that he had formed the view, as a result of his visit to the council offices, that while it might take a considerable time before any planning permission would be granted for any housing in respect of the property, he himself thought that permission would be granted in the end. In the light of the planning situation as I have already described it, I do not think statements to this effect misrepresented the factual position in any way. Nor, in my judgment, did Mrs English's oral evidence reveal any other respect in which Mr Harrington had misrepresented either the current planning position or the prospects of obtaining planning permission. Nor do I think she was actually misled in any way by anything that was said to her. As a result of her long previous experience and interest in the subject, she knew the position and the prospects in relation to the obtaining of permission for a full-scale development of the property more or less as well as Mr Harrington. If he said anything as to the prospects of obtaining permission on the limited application—and the evidence does not suggest that he did—she would not have been misled, because, as I have found, she did not realise that a limited application was under contemplation. For these reasons, in my judgment, the claim that representations were made to her at the meeting of May 21 which at that date were untrue and were known by Mr Harrington to be untrue, fails. I wholly acquit Mr Harrington of any charges of fraud in relation to this incident, and indeed in relation to any other incidents in this case. This is, however, far from the end of the matter and I must revert to its history.

After the meeting of May 21 1971 the plaintiffs' solicitors proceeded to draw up a formal contract. A draft was sent to the defendant's solicitors on May 25 1971. Preliminary inquiries were made and answered. The draft contract was returned by them, in amended form, on June 29 1971. On July 12 1971 the defendant's solicitors wrote to the plaintiffs' solicitors saying that they had engrossed the contract for signature by the defendant. On July 13 the plaintiffs' solicitors wrote to them saying that they agreed their amendments to the draft contract and that they had their part engrossed and were ready to exchange. Then on July 14 1971 before contracts had been exchanged Mr Harrington caused an employee of the defendant company, a Mr J C Mead, to sign and submit to the Essex County Council an application to carry out development on the property by the erection of one detached house with a garage on the small plot. On this application form, the applicants were shown as Mr and Mrs English, their address being given as Rusty Tiles; Mr Mead purported to sign the form as the "agent" of Mr and Mrs English and the form, as completed, requested that the decision notice should be sent to his address. In the usual way, the form had to be accompanied by a certificate of the nature referred to in section 16(1) of the Town and Country Planning Act 1962. The form as completed by Mr Mead, on Mr Harrington's instructions, was in the form specified in section 16(1)(a), namely that "The applicant is the estate owner in respect of the fee simple of every part of the land to which the accompanying application dated July 14 1971 relates." If this form of certificate had not been given, then, in order to comply with the requirements of section 16, it would have been necessary for the application to be accompanied by a certificate in the form specified in section 16(1)(b), certifying that the requisite notice of the application had been given to all persons who, at the beginning of the period of 21 days ending with the date of the application, were owners of the land and setting out the names of those persons, their addresses and the date of service of such notice.

Mr Harrington was quite frank in telling me that he asked Mr Mead to sign the application form, rather than signing it himself, because, for reasons which I have already mentioned, he did not wish the council planning authority to know that he or the defendant company were concerned with it. Though the defendant's own expert witness, L J Trott, a chartered surveyor, said under cross-examination that he would not have been a party to a planning application which

A concealed the identity of the true developer, I express no opinion as to whether it was a proper course, either in relation to planning law or to the planning authority, for Mr Harrington to procure Mr Mead to submit an application in this particular form in these particular circumstances. In relation to Mr and Mrs English, however, Mr Harrington was in my judgment at fault in causing this application to be submitted in this form at this time. As he admitted in his oral evidence, at the meeting of May 21 1971 it had at all times been contemplated by both sides that there would in due course be an exchange of formal contracts and the arrangement reached was to be subject to such exchange. B Mr Harrington accepted in evidence that he never informed Mrs English whether he would be making the planning application before or after the exchange of formal contracts. In these circumstances, I think, as he himself more or less accepted, that even if, contrary to my view, he had made it sufficiently clear that he did intend to make a planning application in her name, Mrs English could reasonably have assumed that he would not do so until after contracts had been exchanged. It is one thing for A to make a planning application in B's name in relation to a house after he has actually contracted to buy the house from B and thus become the beneficial owner of it in equity; it is quite another thing C for A to do this at a time when A and B are merely in an advanced stage of negotiation and A as yet has no legal or equitable interest whatsoever in the house. Mr Harrington explained in evidence that he thought he was justified in making the application before exchange of contracts because his solicitors had heard from Mr and Mrs English's solicitors saying that they were ready to exchange contracts. This explains what he did and absolves him from any charges of dishonesty in this particular context; but I do not think it justifies the action he took before the exchange of contracts without her consent. If, contrary to my conclusion, D he did make it sufficiently clear to her that he wished to make a planning application in the names of Mr and Mrs English in respect of the small plot, any consent which she gave to the making of such application must, in my judgment, be regarded as having been given subject to contract, as much as all other consents given by her at the meeting of May 21 1971.

In these circumstances, I think that, on any footing, Mrs English cannot be regarded as having consented to the making of the application at the time and in the circumstances in which it was made, before exchange of formal contracts. Mr Harrington should, I think, on any footing, have either E waited until contracts were actually exchanged or obtained her express consent to the making of the application before exchange. He could very easily have sought confirmation through the respective solicitors that she did consent. Whether the consent would have been given is another matter; I think it would not, at least if the contract were to proceed on the basis of a £7,750 purchase price. It is common ground that no one told the plaintiffs or their solicitors that the application had been made before contracts were exchanged and that such exchange took place on July 21 1971, although the part of the contract which I have seen, signed on behalf of the defendant, inexplicably bears the date July 23. The F contract, of course, contained no mention of the planning application, which, unknown to the plaintiffs, had already been made in the name of Mr and Mrs English. The date for completion specified in the contract was on or before October 23 1971.

On September 7 1971 Mrs English asked Mr Harrington to go to see her. She told him that she had changed her mind, that her husband was too ill to move and that she wanted to cancel the contract. Mr Harrington refused to cancel it. By that time he was in active negotiation with Mr and Mrs Coe for the purchase of "Shorelands." It seems not unlikely that Mrs English would have been aware of these negotiations. On September 21 1971 the defendant's solicitors G wrote to the plaintiffs' solicitors saying that if Mr and Mrs English needed a deferment of the completion date the defendant would consider it, but there could be no question of their being released from the contract. They also referred to certain assistance which Mr Harrington had given her in attempts to find alternative accommodation. On October 12 1971 the defendant's solicitors offered on behalf of the defendant what they described as an "ex gratia" payment, upon completion, of £150 towards the plaintiffs' legal and removal expenses, and said that, without prejudice to the provisions of the contract, the defendant would agree to defer completion for the time being. H

On October 20 1971 outline planning permission was granted in respect of the proposed building on the small plot. A copy of the permission was sent addressed to Mr Mead at the address given by him on the application form; no copy was sent to Mr and Mrs English in view of the form which the application had taken.

On November 8 1971 the plaintiffs' solicitors wrote acknowledging the offer of £150 and offering to complete at the end of that month. A completion date was subsequently arranged for December 9.

Between October 20 and December 1 1971 Mrs English and Mr Harrington did not meet. Then on December 1 he visited J her. By this time, according to his evidence, which I accept on this point, she was very bitter, because he had in September insisted on holding her to the contract. At that meeting, also according to his evidence, she used words in conversation which made it clear that she knew of the grant of the planning permission which had been given in September. Mrs English, on the other hand, was adamant in her evidence that this was not so, because she did not hear of the permission until January or February 1972. If she had known, she said, she would not have completed the transaction as she did a few days later. Once again, there is a direct conflict of evidence K between Mrs English and Mr Harrington. In the end, however, I have not been satisfied that Mrs English in fact knew of the planning permission at that time. It is common ground that neither Mr Harrington nor anyone else on his behalf actually told her of the permission either at this meeting or at any time before. Though he could suggest no other way in which she might have learned it, he suggested in his evidence that she could have discovered by means of a direct inquiry of the planning authority. This is no doubt theoretically so, but I can see no good reason why she should have made such inquiry, and I think it inherently improbable that she did so. Mr Harrington told me that at the meeting of L December 1 she was sarcastic and used words to the effect "You are doing very well out of this, aren't you?" He could not, however, recall in any further detail what she said. The inference which I draw is that he wrongly attributed her bitterness and sarcasm to a knowledge of the permission, whereas in fact it was at that time due to her general dissatisfaction with a transaction which she had already come to regret for other reasons. Having seen her in the witness box, I accept her evidence that, if she had known of the permission, she would not have completed the transaction a few days later without any further demur or complaint. I think that at very least she would have instructed her solicitors to raise the point M in correspondence, but this she obviously did not do at this stage. Significantly, though there are in evidence a quite large number of letters written in 1971 by her solicitors or by her personally, or by the defendant's solicitors, not one of them refers to the planning application in any way. On this point as to knowledge, or lack of it, her evidence, though inaccurate in other details, carried conviction.

If, however, as is possible, I have drawn an incorrect inference as to this point, or, indeed, on the question whether Mr Harrington made it sufficiently clear to Mrs English at the meeting of May 21 1971 what he proposed to do in relation to applying for planning permission, then I think to some extent he has himself to blame. This was an important part of the transaction as far as he was concerned. It would have been so easy for him, for the avoidance of any possible doubts or misunderstandings, to instruct his solicitors to inform the solicitors instructed on behalf of Mr and Mrs English precisely what he proposed to do in relation to the application, what he had done, and, indeed, what the result of it was. If misunderstandings have occurred, his failure to put anything in black and white for the benefit of an old lady, such as Mrs English, must be partly responsible.

I therefore find that at the date when completion took place, namely on December 9 1971 Mrs English and her husband still did not know that an application for planning permission had been made or that it had been granted. Then in January or February 1972, according to her evidence which I accept on this point, Mrs English learned for the first time that a house was to be built on the property. She consulted her solicitors, who obtained a copy of the planning application, and on March 7 1972 wrote to Mr Mead saying that Mr and Mrs English had never heard of him and had given him no authority to make an application or to certify on their behalf. This letter was handed to the defendant's solicitors who, on March 14 1972 replied saying that, at the time of the application, Mr Mead was an employee of the defendant. They continued:

> Shortly before exchange of contracts, our client, Mr Harrington, asked your clients whether they would object to an application for planning consent being made by him on their behalf. Your clients confirmed that they had no objection to this and Mr Mead was accordingly instructed by our clients.

On March 21 1972 the plaintiffs' solicitors wrote to the defendant's solicitors saying, *inter alia*, that Mrs English absolutely denied having given permission to Mr Harrington to apply for planning permission. This letter was answered on March 22 1972 by a letter in which the defendant's solicitors said, *inter alia*:

> Our client reiterates that he sought and obtained your client's permission to make a planning application. The matter was discussed between our respective clients in or about July 1971, as your clients had requested an extended completion date. The application for planning consent was in respect of one dwelling-house by the side of Rusty Tiles.

It is thus to be noted that the two letters of March 14 and March 22 1972 from the defendant's solicitors referred to the conversation in which Mrs English is alleged to have given her consent to the planning application, as having taken place either "shortly before exchange of contracts" or "in or about July 1971"—a rather different version from the defendant's present case, which is that everything was arranged on May 21 of that year. It is obvious that the defendant's solicitors had taken Mr Harrington's instructions before writing those two letters. Since it is possible that they misunderstood his instructions, I do not attach very great significance to this point. It is, however, one further indication that the evidence that Mrs English gave permission for the plaintiffs' name to be used is neither clear nor conclusive, and one further reason why, in my judgment, she should be given the benefit of the doubt on this point.

The letter of March 22 1972 was followed by the issue of the writ in the present action on behalf of the plaintiffs on March 30 1972. The only other facts to which I think I need refer at the present stage are these. In January 1972 the defendant had entered into a conditional contract for the purchase of "Shorelands." In March 1972 Mr Harrington, on the defendant's behalf, applied for development of the whole site, that is to say the property and "Shorelands." As anticipated, there was a refusal by the county council. There was then an appeal to the minister which, after some delay, was allowed. On April 23 1974 planning permission was granted. Long before that, the sewerage scheme had, of course, been completed. In the event, there can be no doubt that the transactions entered into by Mr Harrington or his company, the defendant, in relation to the property and "Shorelands" have proved very profitable for them. Various figures were mentioned to me in evidence, but I do not think I am concerned with the details. It was accepted by both sides that if I were to hold that Mrs English was entitled to damages or an account of profits, the question of quantum should, in default of agreement, be remitted to an inquiry in chambers.

Mrs English is not now seeking rescission of the conveyance, nor have her counsel pursued before me the claim in the prayer to her statement of claim for a declaration that the planning permission granted on October 20 1971 is null and void. Her claim now is limited to a claim for damages and/or an account of profits. It should also be observed in passing that she has at no time sought rescission of the conveyance on the grounds of alleged undue influence or alleged unconscionable bargain. Her claim for rescission was at all times made simply on the footing that Mr Harrington had positively misled her or alternatively had failed to disclose to her what he should have disclosed.

The grounds upon which Mr Scrivener, on behalf of Mrs English, claims that she is entitled to damages and/or an account of profits may, I think, fairly be summarised under five headings, though I do not reproduce the submissions in precisely the same order as that in which they were made.

First, it was submitted that the defendant is liable to Mrs English for damages for fraudulent misrepresentation on the grounds that the representations referred to in paragraph 3 of the statement of claim were false to Mr Harrington's knowledge at the date when they were made and that Mrs English acted on them to her detriment. I have already dealt with this submission and unhesitatingly rejected it. I do not think that, at or before the meeting of May 21 1971, Mr Harrington misrepresented the planning prospects in respect of the property. Still less do I think that he deliberately misrepresented them or that Mrs English was misled by anything he said in relation to such prospects. As she said in her evidence, she "knew it all already."

Secondly, and alternatively, it was submitted that the defendant is liable to Mrs English in damages for fraudulent misrepresentation on the grounds that, even if the representations made to her by Mr Harrington in relation to the prospects of planning permission at or before the meeting of May 21 1971 were not false to his knowledge at the date when they were made, he acquired knowledge that they were false either when he made the planning application in July 1971 or, alternatively, when planning permission was obtained in October 1971; he thus, it was submitted, became under a duty to disclose the change of circumstances. In this context, Mr Scrivener referred me to the decision of the Court of Appeal in *With v O'Flanagan* [1936] 1 All ER 727. This case illustrates that if, during the course of negotiations for a contract, one of the parties has made a statement which is at the time false in fact, but which he believes to be true and which is material to the contract, and during the course of the negotiations he discovers the falsity of the statement, he becomes under an obligation to correct his erroneous representation. It further shows that if during such negotiations a party makes a material statement which was true at the time it was made, but which before the conclusion of any contract becomes untrue, then if the person who made it discovers before conclusion of any contract that it has become untrue, he becomes under an obligation to disclose to the other the change in circumstances. In either situation, his wrongful failure to disclose may be treated as a deemed misrepresentation.

I do not, however, think that the principle of *With v O'Flanagan* has any relevance to the facts of this particular case. Mrs English does not claim or suggest that Mr Harrington ever made any representation to her of any kind in relation to the prospects of obtaining planning permission for building *on the small plot.* The representations which she says he made related to a quite different matter, namely, planning permission for the development of the three acres at the back of the property. These were the only representations which she claims to have induced her to enter into the contract. The making of the planning application in relation to the small plot on July 14 1971 did not affect the truth of the latter representations in any way. They did not cause the statements which Mr Harrington had made in relation to the prospects of obtaining planning permission for the whole plot to be any less true than they had been at the time when they were made. In these circumstances, the principle of *With v O'Flanagan* did not, in my judgment, itself impose any obligation on him to disclose to her the making of the planning application, though, as I have already indicated, I think he was under an obligation to obtain her consent to the making of the application before contract for other reasons. *A fortiori,* in my judgment, the principle of *With v O'Flanagan* did not itself impose on Mr Harrington or the defendant any obligation to disclose in October 1971 the fact that planning permission had been obtained; for by that time Mrs English was already committed to the contract.

Thirdly, it was submitted on Mrs English's behalf that the defendant is liable in damages on the grounds that, since it placed itself in the position of a self-appointed agent in making the planning application, without the plaintiffs' authority, an implied contract must be deemed to have arisen between the defendant and the plaintiffs, and the defendant was in breach of the duties arising from such implied contract in failing to disclose the making and result of the planning application. If this duty had been fulfilled, it was submitted, Mr and Mrs English would not have concluded the contract or, alternatively, would with proper justification have refused to complete it and would thereby have been saved from a transaction which caused them substantial damage. However, no reported case was cited to support the proposition that where A, without B's authority or knowledge, purports to act as agent for B, a contractual relationship may thereby arise between A and B. It seems to me that, just as an express contract must be founded on the express mutual consent of the contracting parties, so an implied contract must be founded on the implied mutual consent of the parties interested. Mutuality is *ex hypothesi* lacking in a case where the relevant acts of A are done without B's knowledge or consent. In these circumstances, I reject the argument based on any implied contract in relation to the application for planning permission.

The fourth proposition put forward on behalf of Mrs English by Mr Scrivener is in my view of much the greatest substance. This is that in all the circumstances of the case the defendant became and is accountable in equity to Mrs English in respect of the profit derived by it from the planning application made in July 1971. Mr Scrivener advanced this submission in a number of alternative ways. His basic points, however, were I think these: The defendant had, without the plaintiffs' authority, put itself into the position of a self-appointed agent for them in making the planning application. It thereby placed itself in a fiduciary relationship with the plaintiffs; in the course of this relationship it used their names and their property without their authority in making the planning application; also in the course of this relationship it acquired information which it should have disclosed to them, namely, the making and the grant of the planning application; as a result of this unauthorised use of the plaintiffs' property and names, and of the defendant's breach of duty in failing to disclose to them the information to which I have referred, the defendant made a substantial profit through the obtaining of the planning permission and is liable to account to Mrs English in equity for that profit.

These submissions, which are reflected in the reamendments to the statement of claim, raise some difficult and to some extent novel points of law. Before considering them further, it will be convenient to rehearse my interpretation of the basic relevant facts of this case. On May 14 1971 Mr Mead, by the direction or authority of the defendant, but without the plaintiffs' consent or knowledge, made an application for planning permission in respect of the plaintiffs' property. The form of application showed the plaintiffs as the applicants, Mr Mead as their agent, and made no mention whatsoever of the defendant. The application was made at a time when contracts between the plaintiffs and the defendant had not yet been exchanged and the defendant thus had as yet no legal or beneficial interest in the property. Pausing there, it seems to me that, although the form of application had been presented by the authority or direction of the defendant or Mr Harrington, it was nevertheless the application of the persons on whose behalf it purported to have been made, namely the plaintiffs, in the sense that, at least at any time before exchange of contracts, the plaintiffs (on ordinary principles relating to agency) would have been free to ratify what Mr Mead had done purportedly on their behalf and to take over the full benefit of the application for themselves, to the complete exclusion of Mr Mead, or the defendant, or Mr Harrington. I infer that, if they had known of the making of the application, this is precisely what they would have done. I also infer that they would have declined to proceed with the exchange of contracts, at least at a price of £7,750, because they would have wished to see what became of the application. This course they were, in the event, unable to adopt, because neither Mr Harrington nor Mr Mead had informed them of the application. In the result, this non-disclosure, coupled with the making of the application, has resulted in the defendant obtaining a substantial profit which it would not otherwise have obtained.

In the circumstances, it is hardly surprising that Mr Scrivener has placed non-disclosure in the forefront of his reasons for submitting that the defendant should be made to account for the profit obtained by it from the permission. Apart from other relevant considerations, he pointed to section 16 of the Town and Country Planning Act 1962 which, in ordinary circumstances, precludes a local planning authority from entertaining an application for planning permission unless it is acompanied by either (i) a certificate stating that, in respect of every part of the land concerned, the applicant is either the freeholder or is entitled to a tenancy thereof, or (ii) a certificate stating that the applicant has given the requisite notice of the application to all the persons (other than the applicant) who, at the beginning of the period of 21 days ending with the date of the application, were owners of any of the relevant land. Section 16(5) imposes penalties in respect of false or misleading certificates, and section 17 obliges a local planning authority, in determining applications for planning permission for development, to take into account representations received by them within a stated period. One of the clear purposes of section 16 is to confer protection on an owner of land, in respect of whose property an application for planning permission is being made by another person, by providing for notice of the application to be given to him. Mr Scrivener did not submit that the section itself operated to confer on Mrs English the statutory right to receive a notice, enforceable by damages in the present case. He did however submit, in my judgment correctly, that if the ordinary and proper procedure had been adopted which should be adopted under the statute, in a case where an application is being made in respect of the land of a person other than the signatory of the application and the owner has not consented to the signatory making the application as his agent, Mrs English would have received notice of the application before it was made on July 14 1971 in the present case. If she had received such notice, as I have indicated, she would not, on my analysis of the facts, have proceeded to exchange contracts in the form which they actually took as to price and other matters.

I now proceed to consider whether all or any of these various points operate to confer on Mrs English the right to an account of the profits received by the defendant from the relevant planning permission. In this context it was emphasised on her behalf that she is not seeking recovery of the profit obtained by the defendant from the wider permission which it subsequently obtained in respect of the whole property and Shorelands. Her claim is in respect of the permission obtained in her name without her knowledge by means of an application made before exchange of contracts.

On behalf of Mrs English, strong reliance was placed on a passage from the judgment of Lord Denning MR in the Court of Appeal decision in *Phipps v Boardman* [1965] Ch 992. In that case a beneficiary under a will trust claimed an account of profits made by the appellants as the result of purchasing shares in a company in which the trust had a substantial holding. One of the appellants was solicitor to the trustees and the other was a beneficiary. They were dissatisfied with the company's accounts and, on behalf of the trust, attended a meeting of the company in December 1956 with proxy forms signed by the other two trustees. Early in 1957 they decided to make a take-over bid personally for further shares in the company so as to obtain control and by a liquidation of assets make a repayment of capital to the shareholders. During the negotiations for the purchase, they referred to their representative capacity and made use of the information which they had received at the meeting as representatives of the trustees. Further detailed knowledge of the assets of the company and their value was obtained during negotiations, the information being acquired upon the basis of their representation of the shareholding of the trust. Eventually, the proposed purchase was effected and it proved profitable to the appellants personally. The Court of Appeal held that they were accountable for this profit, substantially on the grounds that they had assumed the character of self-appointed agents for the trustees and had used the knowledge of the company which they had acquired and which was the property of the trust to make a profit for themselves. Lord Denning said this at p 1017 of the report:

> I think the judge put the position right when he said that "in 1956 they assumed the character of *self-appointed agents* for the trustees, for the purpose of extracting information as to the company's business from its directors." The word "self-appointed" was criticised by Mr Bagnall, but its meaning is clear enough. These two gentlemen took upon themselves an authority which they did not truly possess: and, by virtue of this assumed authority, they obtained information and knowledge which they would not otherwise have got. This sort of thing has happened often enough before. There are many cases in the books where a person has assumed to have authority when in truth he has none. It has always been held that he is accountable just as if he had in fact the authority which he assumed. The classic instance is an executor *de son tort.* If a person intermeddles with the assets of an estate in such a way as to denote an assumption of the authority of an executor, he is accountable just as if he were an executor, see *Stamford's* case, *Read's* case. Likewise with a man who assumes to act as bailiff without authority. Thus in *Gawton v Lord Dacres,* Anderson CJ said: "If one becomes my bailiff of his own wrong, without my appointment, he is accomptable to me." So with a person who assumes to act as trustee. In *Rackham v Siddall,* Lord Cottenham LC held a lady who took on herself to act as trustee was just as liable as if she were in truth a trustee; similarly in *Lyell v Kennedy,* the House of Lords held that a person who took upon himself to receive the rents of property which to his knowledge belonged to others was chargeable in a fiduciary character with the rents he had received. In most of these cases the person had already some position or connection which gave him the opportunity of assuming authority to himself. So here Boardman was the solicitor to the estate. Tom Phipps was one of the chief beneficiaries. This gave them the opportunity of assuming authority to act as agents to the estate. They did assume it in their negotiations with Smith and are accountable just as if they were agents. That is how the judge regarded them. . . .
>
> Treating them as agents, to what extent are they accountable? We have been through once again the cases where an agent makes a profit for himself out of his agency. It is quite clear that if an agent uses *property* with which he has been entrusted by his principal, so as to make a profit for himself out of it, without his principal's consent, then he is accountable for it to his principal, see *Shallcross v Oldham.* So also if he uses a *position of authority,* to which he has been appointed by his principal, so as to gain money by means of it for himself, then also he is accountable to the principal for it, see *Reading v Attorney-General.* Likewise, with *information or knowledge* which he has been employed by his principal to collect or discover, *or which he has otherwise acquired,* for the use of his principal, then again if he turns it to his own use, so as to make a profit by means of it for himself, he is accountable, see *Lamb v Evans, Regal (Hastings) Ltd v Gulliver,* for such information or knowledge is the property of his principal just as much as an invention is. . . .

Pearson LJ (at p 1030 *ibid*) said:

> It was contended that the defendants could not appoint themselves agents. . . . It seems to me, however, that as the defendants assumed the character of agents and acted in that character over a long period with resulting profits to themselves, they can be held accountable on the basis of agency: *Lyell v Kennedy.*

Mr Scrivener submitted that the passage from the judgment of Lord Denning which I have read is an accurate summary of the relevant rule and precisely covers the present case. The defendant assumed an authority which it did not truly possess, namely, the authority as the plaintiffs' agent (through Mr Mead) to make the application for planning permission. By virtue of this assumed authority, it obtained information, namely, information as to the making and result of the planning application, which it would not otherwise have got. It also obtained a benefit which it would not otherwise have got, namely the permission itself; therefore, it is submitted, it is accountable to Mrs English just as if it had in fact made the application solely and simply as the plaintiffs' agent.

Mr Joseph, on behalf of the defendant, submitted in effect that there is no general principle of English law which obliges a self-appointed agent, acting without authority from his purported principal, to account to such principal for any profit acquired by him by virtue or during the course of the assumed authority. He pointed out that the Court of Appeal's decision in *Phipps v Boardman,* though upheld by the majority in the House of Lords (under the name *Boardman v Phipps* [1967] 2 AC 46) was upheld by the majority on somewhat different, broader grounds, namely, the rule of equity that a person in a fiduciary capacity must not make a profit out of his position which is part of the wider rule that a person in a fiduciary capacity must not place himself in a position where his duty and his interest may conflict. If a person, without any preceding authority, purports to enter into a contract with a third party for and on behalf of an alleged named principal and the named principal later ratifies and adopts the contract, the relationship of principal and agent will usually be treated as arising between the two parties concerned with all the normal consequences attaching to such relationship: see *Bowstead on Agency* 14th ed at pp 37 *et seq.* In the present case, however, the defendant cannot be said to be a self-appointed agent of the plaintiffs in respect of any *contract* entered on their behalf. In these circumstances, Mr Joseph, in the course of a helpful analysis of the authorities, in effect submitted that the defendant could be made liable to account to the plaintiffs if, but only if, the case fell within one or more of three well-recognised categories where a liability to account falls upon a self-appointed agent:

A (1) Mr Joseph accepted that if, at the date when the planning application was made, there already subsisted between the plaintiffs and the defendant a relationship of a fiduciary nature and the defendant had used this relationship to obtain the planning permission, then it would be accountable to the plaintiffs for any profit so derived. By way of illustration he referred me to *Reading v Attorney-General* [1951] AC 507, where a soldier was held to be accountable to the Crown for money obtained by him by the use, or rather the abuse, of his military status. Mr Joseph, however, pointed out that in the present case there was no fiduciary relationship B which existed before the date when the planning application was made and enabled it to be made.

(2) Secondly, he recognised that the defendant could be made liable to account as self-appointed agent, if it could be said that it had meddled with a pre-existing trust or with trust assets with knowledge of the existence of the trust. Examples of this principle are *Rackham v Siddall* (1849) 1 Mac & G 607 and *Soar v Ashwell* [1893] 2 QB 390. This principle clearly has no application on the facts of the present case.

(3) Thirdly, Mr Joseph recognised that the defendant could C be liable to account if, at the date when the planning application was made, there already existed some kind of fiduciary relationship between the plaintiffs and the defendant, and the defendant, in making the application, could be said to have been placing himself in a position where its interest and its duty conflicted. One example of such a situation would be that in *Phipps v Boardman*. Another would be that in *Parker v McKenna* (1874) 10 Ch App 96, where trustees were held accountable for profits made by them personally out of dealings with trust property on the grounds that "no agent in the course of his agency in the matter of his agency can D be allowed to make any profit without the knowledge and consent of his principal": (*ibid* at p 124 per Sir W M James LJ).

Mr Joseph, however, submitted that these were the only three situations in which the defendant could have been made accountable on the present facts and that the present case fell within none of the three categories. He referred to the sentence of Lord Denning's judgment in *Phipps v Boardman* in which it had been said, in relation to self-appointed agents, "in most of these cases the person had already some position or connection which gave him the opportunity of assuming E authority to himself." Mr Joseph submitted that the sentence could more aptly have read "In *all* of these cases. . . ." He submitted that, because the defendant at the time that it made the application did not already have any position or connection which gave it the opportunity of assuming authority to itself, this was really fatal to any claim of the plaintiff to an account of profits in equity.

In my judgment, in the end the question of the liability, if any, of the defendant to account must depend on the view which the court takes as to the nature of the relationship subsisting between it and the plaintiffs at the date when the planning application was made. The liability to account F would, in my judgment, arise if, though only if, the relationship was in the eyes of equity a fiduciary one in the sense that it imposed relevant fiduciary duties on the defendant towards the plaintiffs. In this context, I regard one point as being of great significance. As I have indicated, if the plaintiffs had learned of the application which had been made in their names and in respect of their property before contracts had been exchanged, they would have been entitled to say to the defendant and Mr Mead "We ratify and adopt this application, which has been made in our names, as our own application and neither of you are to have anything more to do with it." Furthermore, this is, I think, effectively the course which the plaintiffs would have actually adopted, G unless a more favourable form of contract had then been offered to them. This being so, I do not think that a court of equity should or will allow the defendant to be in a better position than that in which it would have found itself if, before exchange of contracts, the defendant or Mr Mead had told the plaintiffs of the application which had been made in their names, as I think it or he should have told them. I do not think that the categories of fiduciary relationships which give rise to a constructive trusteeship should be regarded as falling into a limited number of strait-jackets or as being necessarily closed. They are, after all, no more than formulae for equitable relief. As Ungoed-Thomas J said in *Selangor* H *United Rubber Estates Ltd v Cradock (No 3)* [1968] 1 WLR 1555 at p 1582:

> The court of equity says that the defendant shall be liable in equity, as though he were a trustee. He is made liable in equity as trustee by the imposition or construction of the court of equity. This is done because in accordance with equitable principles applied by the court of equity it is equitable that he should be held liable as though he were a trustee. Trusteeship and constructive trusteeship are equitable conceptions.

Mr Joseph drew my attention to the judgment of Buckley LJ in *re Coomber* [1911] 1 Ch 723 at p 730 in which it was observed that the doctrine of equity which induces a court J of equity to set aside a gift does not rest upon the existence of a fiduciary relationship whatever be its nature. He said: "It rests upon the existence of such a fiduciary relationship as will lead the court to infer undue influence or knowledge in the one party concealed from the other or other circumstances into which I need not go."

It has not been submitted that, but for the making of the planning permission, any fiduciary relationship could be said to have arisen between Mrs English and the defendant or Mr Harrington before exchange of contracts. Mr Joseph in effect submitted that the mere making of a planning application could not by itself have given rise to any such relationship, K when none would have otherwise existed. I see the force of this submission but am not in the end convinced by it. My reasons may be in the form of two general propositions:

(1) Where during the course of negotiations for a contract for the sale and purchase of property, the proposed purchaser, in the name of and purportedly as agent on behalf of the vendor, but without the consent or authority of the vendor, takes some action in regard to the property (whether it be the making of a planning application, a contract for the sale of the property, or anything else) L which, if disclosed to the vendor, might reasonably be supposed to be likely to influence him in deciding whether or not to conclude the contract, a fiduciary relationship in my judgment arises between the two parties.

(2) Such fiduciary relationship arises in the sense that there is a duty on the proposed purchaser to disclose to the vendor before the conclusion of the contract what he has done as the vendor's purported agent, and correspondingly, in the event of non-disclosure, there is a duty on him to account to him for any profit made in the course of the purported agency, unless the vendor con- M sents to his retaining it. In such circumstances, the person who, for his own private purposes, uses the vendor's name and purports to act as his agent cannot reasonably complain if the law subjects him to the same consequences *vis-à-vis* his alleged principal as if he had actually had the authority which he purported to have.

On my analysis of the facts of the present case, the plaintiffs never consented to the defendant, or Mr Harrington, or Mr Mead purporting to make the planning application as their agent before contract; the fact that this had been done was never disclosed to them before the exchange of contracts;

A and they never consented to the defendant's retaining the profit ultimately received by it as a result of the making of the planning application. In these circumstances, it is, in my judgment, accountable for such profit.

Mr Scrivener submitted that Mrs English would at her option be entitled to damages in lieu of an account of profits. In this context he referred me to the decision of the Court of Appeal in *Seager v Copydex Ltd* [1967] 2 All ER 415. This was a case where the defendant company had made use, albeit honestly, of information which had been received in confidence and was not available to the public. The remedy ultimately granted by the Court of Appeal was B not an account of profits nor an injunction, but damages to be assessed by the master. It is I think clear that, since the Judicature Act, each division of the High Court has full power to grant either an injunction or damages. I read this case as being an instance where the court granted damages in lieu of an injunction. I am not convinced that it illustrates or establishes any general principle that the court has power to grant damages in lieu of an account of profits; and no other case has been cited to me which suggests that the court has any such jurisdiction.

Whether or not such jurisdiction exists, however, I am satisfied that the present case would not be an appropriate C occasion on which to exercise it. The whole essence of my reasoning, if it be correct, is that, while the defendant and Mr Harrington are absolved from the charges of fraud made against them, the defendant has received a profit which, in all the circumstances of the case, equity requires that it should repay. In legal terms, it seems to me it is the profit to the defendant which is the relevant factor, rather than the plaintiff's loss.

In these circumstances, I propose to grant relief in the terms of paragraphs (4) and (5) of the prayer to the reamended statement of claim, that is to say an account of the profits accruing to the defendant by reason of the D non-disclosure to Mrs English of its application for, and subsequent grant of, planning permission in the plaintiff's name, and an order for the payment by the defendant to her of all moneys found to be due to her on the taking of such an account. I will remit such account to be taken in chambers in default of agreement, but I think it will be desirable that, so far as possible, the relevant principles upon which it should be taken should be determined at the present stage. I will, therefore, hear any further submissions which counsel may wish to make in this context and as to the form of my order generally.

E [The sum of £3,000 was later agreed and an order made accordingly.]

QUEEN'S BENCH DIVISION

November 22 1977

(Before Mr Justice TALBOT)

MASTERS v BRENT LONDON BOROUGH

Estates Gazette May 6 1978

(1977) 246 EG 483

F **Damage caused by tree roots—Lime tree planted by local authority on ground outside house—Foundations fractured —Subsidence of bay at front—Father's interest in house transferred to plaintiff—Whether plaintiff could claim for damage occurring during father's occupation—In cases of continuing nuisance the person in possession may recover for damage suffered whether damage began before or after he acquired his interest—As to quantum of damage, builder's evidence accepted on necessity of demolishing and rebuilding bay and underpinning wall**

This was an action by John William Masters against the London Borough of Brent for damage caused by nuisance consisting of the undermining of the foundations of his house G at 23 Wendover Road, London NW10, by the spreading roots of a lime tree planted on adjacent ground by the defendant authority or their predecessors in title.

M F Harris (instructed by Parsons, Evans & Francis) appeared on behalf of the plaintiff; Ian Goldsworthy (instructed by Barlow, Lyde & Gilbert) represented the defendants.

Giving judgment, TALBOT J said: The plaintiff brings this action claiming damages for nuisance. He is the owner of leasehold premises, 23 Wendover Road, London NW10. He claims that by virtue of a lime tree planted by the defendants, or their predecessors, in the vicinity of his house, the roots of that tree undermined the foundations of the premises and H thereby caused him loss and damage. Two principal points arise for decision in this action. Logically the first is: has the plaintiff a right to claim for the loss incurred by the undermining of the foundations and the necessary building works in consequence thereof? Secondly, if he has a right so to claim, what are the damages to which he is entitled?

At the outset of this case certain matters were accepted by the defendants. 1. It is accepted by them that the roots of the lime tree, which is in the pavement outside the plaintiff's house, did encroach upon the land of 23 Wendover Road. 2. It is accepted that that encroachment of the lime tree roots constituted an actionable nuisance. 3. The defendants accept J that the lime tree roots substantially caused the subsidence of the land, though there was left in issue the amount of subsidence and the works which would be required to make good the damage.

The necessary facts in order to determine the first question, namely, whether or not the plaintiff has any right to claim for the loss incurred, are these: The plaintiff, John William Masters, lives at 23 Wendover Road with his father, John George Masters. In addition to those two the plaintiff's wife also lives there. This property is leasehold and the leasehold interest belonged to the plaintiff's father. In the autumn of K 1971 the plaintiff noticed a crack in the passage wall in the hallway of the house. There had been some crack visible before, but he noticed then it was getting worse. He asked the advice of a friend of his, who was a young builder. His advice was that the work which would be required to put the house in order was too large for him and he recommended a firm, a company of builders, T R Atkins Ltd. The plaintiff therefore consulted Mr Atkins, a director of that company, who made an inspection shortly before November 17 1971. As a result of that inspection the plaintiff was told that the bay at the front of the house was definitely subsiding. Mr Atkins then prepared an estimate of the cost of the necessary L works. That was given on November 17 1971 and amounted to £876. Neither the plaintiff nor his father, the owner of the leasehold interest, had at that time the money available to pay for the necessary repairs. They therefore consulted their solicitors.

As a result of the advice of the solicitors, on November 24 1971 the plaintiff's father transferred to the plaintiff his leasehold interest in the premises. That was done in order that the plaintiff, who then became the holder of the leasehold interest, might raise money by way of mortgage. He went to the Chelsea Building Society and obtained a loan on mortgage of £4,000. One of the conditions of the grant of M the loan was that the works necessary to put this house in order would be carried out within a period of six months. In the months immediately succeeding the plaintiff also negotiated the purchase of the freehold interest in the premises and paid for that out of the mortgage loan, the freehold transfer being completed on June 14 1972.

In April 1972 Mr Atkins' company began work. The works were eventually completed—and I shall have more to say about them later on—and on October 4 1972 the plaintiff received an account for £1,240.93, which he promptly paid out of the moneys advanced to him on mortgage. I say

A in passing that, in addition to the first estimate of £896, there had been a second estimate for additional necessary work dated December 3 1971 in the sum of £135.64.

Coming, therefore, to the first point taken by Mr Goldsworthy, it is shortly this, that the plaintiff, who acquired the leasehold interest on November 24 1971, is not in a position to claim, in this action, for damages based upon the cost of repair of this house, the damage occurring before he acquired the leasehold interest and when his father was the leaseholder. Mr Goldsworthy submitted that, as a matter of law, the plaintiff was in no better position, or worse position, whether he was a volunteer, or a purchaser for B consideration, or for consideration of natural love and affection. He would be in no better position than would a complete stranger if he had taken over the house and the premises. Such person, he submitted, would have no right of action against the defendants. He cited and adopted for the purpose of his argument that which is set out in *Clerk and Lindsell on Torts,* 14th ed para 1413. The paragraph is headed: "Who can sue for nuisance?" The opening sentence reads: "Apart from rights arising from public nuisance the person who can sue for a nuisance is the person in possession or occupation of the land affected."

C One of the authorities cited in the text is that of *Malone v Laskey* [1907] 2 KB 141. In that case the wife of a licensee occupier of a house who was injured by the fall of a tank caused by vibrations set up by the defendant was held to have no right of action for her injury either in nuisance or in negligence. So far as that decision relating to negligence is concerned, the decision of the House of Lords in *Billings (AC) & Sons Ltd v Riden* [1958] AC 240 held that that part of the decision was wrong. Their Lordships expressed no views, other than to limit their reversal to that part of the decision on negligence, on the question of nuisance. It is, therefore, Mr Goldsworthy's submission that this plaintiff here D is in no better position than was the wife of the occupier in *Malone v Laskey* in that at the time of the damage he had no interest in the property. It does not fall to me to decide whether or not that is a proposition that I ought to follow, for reasons which will be seen later.

The next authority that I was referred to was *Thompson v Gibson* (1841) 7 M & W 456. That was an action for nuisance arising out of the placing of a building in a market place whereby the public use of that part of the market place was prevented. At p 460 of that report Mr Goldsworthy relied upon this part of the judgment of Parke B. I will read a little further back than the passage which Mr Goldsworthy E drew my attention to. It begins as follows, half-way down p 460:

> Is he, who originally erects a wall by which ancient lights are obstructed, to pay damage for the loss of the light for the first day only, or does he not continue liable so long as the consequences of his own wrongful act continue and bound to pay damages for the whole time, and if the then owner of the market might have maintained an action against the defendants for the injury to his franchise, for the whole period during which the defendants' act continued to be injurious to him, his lessee must be in the same condition as to subsequent injuries; for it is clearly established that he has a right of action for every continuing F nuisance.

There are further passages from that judgment to which I will return later. But Mr Goldsworthy stressed the words "subsequent injuries" and pleads here that all the plaintiff can do is to claim for any loss, or injury, subsequent to his acquisition of the premises and, therefore, not permitting him to add to his claim, a claim in respect of the damage which occurred during his father's ownership of the legal interest. He can only, said Mr Goldsworthy, sue for damage which accrued during his occupation and a purchaser cannot inherit any right to damages which accrued to his predecessor.

He next dealt with the measure of damages, namely, the G difference in value of the interest in the land between the time before and the time after the damage was done and submitted the plaintiff never enjoyed the land in its uninjured state. It was then in his father's ownership. It was conceded by Mr Goldsworthy that in assessing that type of damages it is normally, in these circumstances, permissible to look to the cost of reinstatement.

I was then referred to another paragraph in *Clerk and Lindsell,* 1393. I read from that paragraph:

> A private nuisance may be, and usually is, caused by a person doing on his own land something which he is lawfully entitled to do. His conduct only becomes a nuisance when the conse- H quence of his acts are not confined to his own land but extend to the land of his neighbour by (1) causing an encroachment on his neighbour's land, when it closely resembles trespass, (2) causing physical damage to his neighbour's land or buildings or works or vegetation upon it.

I need not refer to the first category.

A little later on in that paragraph the learned editor writes as follows:

> Nuisances of the first kind in the nature of encroachments occur when a man builds on to his house a cornice which projects over his neighbour's garden so as to cause rainwater to flow thereon. J when his trees overhang his neighbour's land and when the roots of his trees grow into his neighbour's land.

Then he goes on to deal with nuisances of the second kind causing physical damage to land.

A further submission involved a reference to the authority of *Darley Main Colliery v Mitchell* (1886) 11 App Cas 127. I need not deal in detail with this quite complicated decision, but merely extract from it what Mr Goldsworthy desired me to do, namely, that in that case there had been, as a result of the working of coal, a subsidence of land and injury to houses in 1868. The working of coal ceased, but in 1882 K a further subsidence took place, causing further injury. The decision of their Lordships was that that second subsidence gave rise to a cause of action in the hands of those who suffered the second injury. The principle that Mr Goldsworthy desires me to extract from that case is that there there were two separate claims in respect of the same nuisance and he likens that to this case, namely, there are two separate claims here, namely, a claim by the father and a claim by the plaintiff, if the plaintiff has suffered damage since his acquisition of the premises. Mr Goldsworthy conceded that the plaintiff may have a cause of action for encroachment if he is in a position to show damage. The burden of his L submission on this point was that the plaintiff cannot use the damage which occurred during his father's ownership to support his cause of action for the nuisance which continued during his own occupation.

Another authority was *Clegg v Dearden* (1848) 12 QB 576. There the defendants in working a mine had excavated wrongfully into their neighbour's mine. That gave rise to a claim for damages in trespass. Mr Goldsworthy adopted the principle to be derived from that case, namely, that there was a once-for-all loss and that in this case occurred during the ownership of the plaintiff's father.

A more modern authority relied upon was *Sparham Souter* M *v Town and Country Developments (Essex) Ltd* [1976] 1 QB 858. Principally that decision of the Court of Appeal dealt with the time at which a cause of action accrues. For instance, the court held that in the case of negligence the cause of action accrues not at the date of the negligent act or omission, but at the date when the damage is first sustained by the plaintiff. Again the question of the Statutes of Limitations was involved in this decision. But there is a passage in the judgment of the Master of the Rolls which Mr Goldsworthy submitted was very much in point. It occurs at p 868 between letters "G" and "H." The Master

A
of the Rolls said this: "One word more: the only owner who has a cause of action is the owner in whose time the damage appears. He alone can sue for it unless, of course, he sells the house with its defects and assigns the cause of action to his purchaser." Thus, it is submitted, that, in effect, covers this case. Those were the principal submissions made to me on behalf of the defendants.

Mr Harris points out, somewhat wryly, that if the defendants are right, had the plaintiff's father sued, they would have said: where is his loss in that the cost of reinstatement has fallen upon his son? In the same breath, or in the next breath, they say: the plaintiff, the son, can't sue, because
B
the damage did not occur during his ownership of the interest. That, submits Mr Harris, is palpably wrong. I must confess, looked at that way, there is great force in that submission. He analysed the matter in this way: the defendants have committed the tort of nuisance, that tort has damaged the premises, as a result the value of the premises has been diminished by the damage and the measure of that diminution is the cost of remedying the damage and, finally, that cost has been borne by the plaintiff who, at the time of the commission of the tort, had sufficient interest to found an action in nuisance.

He then referred me back to *Thompson v Gibson*. I go
C
back to that authority at p 460 and continue reading the judgment of Parke B, who referred to Roll Abr, Nuisance, K2:

> If one is seized of land near a river, and another stops it with loads of earth, and the tenant of the land adjoining leases to another for years, and then the stoppage continues, by which the land of the lessee is surrounded, the lessee shall have an action on the case against him; for though the stoppage was in the time of his lessor the continuance was a wrongful damage to the lessee, for his land was surrounded.

Then a little further on the learned judge said this:

D
> In the case of *Rosewell v Pryor*, which was an action against the defendant, who erected an obstruction to the ancient lights of the plaintiff, and then aliened, Lord Holt lays it down, that it "is a fundamental principle in law and reason, that he that does the first wrong shall answer for all consequential damage; and here," he says, "the original erection does influence the continuance, and it remains a continuance from the very erection, and by the erection until it be abated." And he adds, "that it shall not be in his power to discharge himself by granting it over." It is true that Lord Holt afterwards says, "that if the alienee of the land brought an action against the erector and the erection had been before any estate in the alienee the question would be greater, because the erector never did any wrong to the alienee."

E
The essence of the case of *Thompson v Gibson*, as it was dealt with by the court, Mr Harris argues, indicates that it was a continuing nuisance, and where there is a continuing nuisance it indicates that a successor in title can sue in respect of it. The case of *Clegg v Dearden*, he points out, is really quite distinct from the present case. *Clegg v Dearden* was an action for trespass, possibly for nuisance, but it was not expressed to be a continuing wrong. He further argues that the defence, by accepting that the plaintiff can sue for nuisance, accepts that he can only do that because he is in possession of land subject to a continuing nuisance.

F
The final submission that he makes upon this is that here is a case where there is a continuing actionable nuisance affecting the land both during the time of the father's ownership and during the plaintiff's ownership, that there was in fact, not in theory, continuing damage and that the plaintiff, who has remedied the damage at his own expense, is entitled to recover that cost as his loss for the continuing nuisance. In my judgment, that is a correct submission. Where there is a continuing nuisance inflicting damage upon premises those who are in possession of the interest may recover losses which they have borne whether the loss began before the acquisition of the interest, or whether it began after
G
the acquisition of the interest. The test is: what is the loss which the owner of the land has to meet in respect of the continuing nuisance affecting his land?

Having decided that point, therefore, in the plaintiffs favour, I now turn to the question of damages. There were two witnesses whose evidence I have to consider. On behalf of the plaintiff I had the evidence of the building contractor concerned, Mr Atkins. On behalf of the defendants I heard the evidence of a Mr Stevens, a chartered architect of considerable experience, but who had the misfortune, so far as assisting this court is concerned, in not having seen the property until the remedial works were being carried out. A
H
number of points were taken by the defendants based upon Mr Stevens' evidence, one of which was that the works undertaken by Mr Atkins were really wholly unnecessary.

I will turn, therefore, to Mr Atkins' evidence. He was a builder of considerable experience, having been in business for over 40 years. Upon his inspection about two days before November 17 1971 he found the bay at the front of the house seriously out of line, with the brickwork fractured, and he came to the conclusion that it was beyond the point of restoration and would have to be demolished. There was a fracture in the ceiling on the ground-floor room, a crack which ran diagonally right across the partition wall—that is
J
the wall in the hall—the sills and lintels were all fractured and were not reusable and it was only sensible, in his view, to renew the foundations as the subsiding of those foundations would have been the cause of the damage that he saw. Indeed, the foundations were tilted and fractured. As the result of his survey he felt that if left very much longer the main structure might collapse and the main wall—that is looking at the house from the front on the right—had some serious cracks showing. It was difficult, he said, to say that it would have got worse, but, from the structural point of view, he would expect something serious might happen to
K
the bay owing to the fractured sills and lintels. During the course of the works the plaintiff expressed a preference for a rectangular bay instead of what is called a cant or splayed bay which had previously existed. Mr Atkins was prepared to rebuild the bay in this rectangular form as it would not in any way add to the cost which he had already quoted for the restoration and rebuilding of this bay. Work commenced, as I have said, and in the course of the work, as a result of his instructions, he was called to the site and there he saw a number of lime tree roots under and around the foundations, which confirmed his original suspicions.

L
One of the points taken by the defendants is that at the front of this house there was, what is considered by Mr Stevens to be, a substantial privet hedge, and the roots from that privet must have contributed to the damage which occurred to the house. Mr Stevens conceded that substantially the damage was caused by the lime tree roots. But it was argued by the defendants, on his evidence, that some part of the loss ought to be attributed to these privet roots. Mr Atkins' inspection did not reveal to him anything in the nature of a large or substantial hedge. It was rather sparse, so far as he recalls. His actual inspection of the roots, which he was called to see by his workmen, did not reveal
M
to him anything particular that was related to privet hedge roots. He did not notice any, he said, at the foundation level. He could then see at that point of the excavation that the foundations were in fact fractured and tilted. His view as to whether the privet roots could have added anything to the effect of the lime tree roots was that it was purely a matter of speculation. The privet roots were nearer to the surface and he had never, in his experience, come across a case where privet roots had contributed to this kind of damage. His conclusion was that the main substantial cause here was the effect of the lime tree roots. So far as the ceiling in the room was concerned, that had lost its key to

the laths and the only proper way of dealing with it was to replace it.

Mr Stevens made an inspection on May 12 1972 and, as has been pointed out in his report dated November 18 1977, which was a document admitted for the purposes of being used by either side, Mr Stevens indicated that they were not able to spend very much time at the site and they were not able to see anything of the old foundations of the bay. There was partly to be seen the excavation trench, because it had not been completely backfilled. It was a matter of conjecture, though it was his view, that the privet roots would have affected the foundations, and, after much consideration, he put it at the quantity 15 to 20 per cent. He queried the need to demolish the bay. Only in extreme cases does one need to demolish and rebuild. He gave various reasons as to why he thought it was probably unnecessary in this case to have undertaken complete demolition and rebuilding. Again, by reason of the fact he had never had an inspection of the damaged bay at the time when a decision had to be made, it seems that what he can contribute to this is largely a matter of conjecture.

With regard to the underpinning which Mr Atkins carried out of the wall on the right, Mr Stevens considered that that was not necessary unless there were very serious conditions, and one of the reasons that it was not usually done unless conditions were very serious was because it was not always very beneficial. He did not consider again that the underpinning was necessary in this case. In respect of this matter, of course, he was in as good a position as Mr Atkins to reach a judgment. He could see the position, as indeed Mr Atkins could see it. He made one or two remarks about the windows and whether it was necessary to replace the windows. So far as the ceiling is concerned, he was not able to say anything about the keying, because he was not able to see that. But it was his impression that it might have been possible to repair it rather than take it down and replace it. With regard to the hall and the staircase, redecoration was indeed necessary, but the actual cause of the damage might have been due, he said, to movement either from the front or from the back of the house, the movement at the front of the house, of course, being caused by the roots of the lime tree. He then was asked what was the unnecessary amount spent on underpinning. He gave two amounts £250 to £270. Referring to the final bill submitted on September 4 1972 by Atkins Ltd he gave amounts which he said ought to have been expended upon the various items therein set out.

I went through those with Mr Goldsworthy. Therefore, two questions ultimately have to be decided. One is: is it right to take the view that the plaintiff's contractors carried out unnecessary work and, therefore, the cost should be whittled down by the amounts spoken to by Mr Stevens? Secondly, ought I to reduce whatever amount is due to the plaintiff by the possible effect of the privet roots? As I indicated at the outset of this part of my judgment Mr Stevens was in the difficulty of not having seen the works that had to be carried out. I find no difficulty, in those circumstances, in accepting the word and the evidence of Mr Atkins, the experienced builder, as to what in fact was necessary. Subject to any possible reduction, in my judgment the plaintiff ought to have that which a competent builder said was the necessary cost of putting right the damage caused by the encroachment.

Finally, therefore, ought I to reduce it by a possible effect of the roots of the privet hedge? Again it was bound to be a matter of conjecture, 15 to 20 per cent. It is not possible to measure it. I can find no link in the facts here to point to any such figure. Furthermore, in that case, too, Mr Stevens did not have as fair an opportunity of judging the matter as Mr Atkins, and when they both say the substantial cause here was the lime tree roots I do not see any reason but to depart from that and to attribute this damage to the encroachment of the lime tree which was the responsibility of the defendants.

The plaintiff, therefore, is entitled to judgment in respect of the sum of £1,133.33.

TAXATION

A **COURT OF APPEAL**

March 22 1978

(Before Lord Justice STAMP, Lord Justice ROSKILL and Lord Justice CUMMING-BRUCE)

WATKINS v KIDSON (INSPECTOR OF TAXES)

Estates Gazette June 24 1978

(1978) 246 EG 1113

**Capital gains—Land owned on April 6 1965—Computation of chargeable gain—Time-apportioned basis or formula based on April 6 market value—Latter basis applicable if
B case fell within paragraph 23(1)(b) of Schedule 6 to the Finance Act 1965—Land without planning permission for development, but with substantial hope value, at date of sale—As development would in any case have been unlawful without permission before disposal, could it be said to have "become" unlawful within the statutory hypothesis in paragraph 23(1)(b)—Held that the hypothetical absolute prohibition of development applied—Market value formula applicable—Decision of Fox J upheld**

This was an appeal from a decision of Fox J, who had upheld a ruling by the special commissioners for income tax dismissing an appeal by the taxpayer, William Henry
C Watkins. The appeal related to 22 acres of farmland, part of a larger area which the taxpayer had inherited known as Westhill Farm, Brackley, Northamptonshire, on the death of his father in 1962. The value of the 22 acres on the death of the taxpayer's father in 1962 was £1,581. The taxpayer sold the 22 acres in 1972 for £260,000. At the date of the sale no planning permission for development had been obtained, but there was a hope or expectation of permission and permission was in fact subsequently obtained. The issue in the case was as to whether the computation of the capital gain should be in accordance with paragraph 23 of Schedule 6 to the Finance Act 1965 or whether, as the taxpayer contended, the time-
D apportionment basis should be applied.

D C Potter QC, A E Park and D G Robins (instructed by Speechly Bircham) appeared on behalf of the appellant; J E Vinelott QC, A Fletcher and B J Davenport (instructed by the Solicitor of Inland Revenue) represented the respondent.

Giving judgment, STAMP LJ said: The judgment I am about to read is the judgment of the court.

This is an appeal from an order of Fox J dated February 25 1977, whereby the judge, being of the opinion that the determination of the Commissioners for the Special Purposes of the Income Tax Acts was correct, dismissed an appeal by way of the case stated by the taxpayer, Mr Watkins.

E The case is reported in [1977] 3 All ER 545. The learned judge there sets out the facts in full and it is unnecessary therefore to recite them. The proceedings relate to an assessment to capital gains tax for 1971-72 in respect of a gain of £258,961. It relates mainly to the gain accruing on the disposal of 22 acres of land under a contract of January 26 1972. As appears from the learned judge's judgment the 22 acres are to be treated as having been acquired by Mr Watkins in January 1962 for £1,581. It was sold for £260,000. The price paid reflected the hope or expectation that the planning authorities would be willing to grant planning permission F
authorising residential development of the land. No planning permission authorising such development had, however, been obtained.

Section 19(1) of the Finance Act 1965 charged the capital gains tax on "chargeable gains computed in accordance with this Act and accruing to a person on the disposal of assets." When this new capital gains tax was introduced by the Finance Act 1965 it was not the policy to tax gains which had accrued prior to April 6 1965 (see, eg section 22(10) of the Act and paragraph 24(2) of Schedule 6). Accordingly on a disposition made after April 6 1965 of property which had been acquired prior to that date provision had to be made G
for ascertaining how much of any profit had accrued prior to that date and how much of it subsequently. Only the latter part was made taxable.

Two methods of achieving this result were laid down by the Act. Except where otherwise provided the profit was to be treated as accruing at a uniform rate over the whole period of the taxpayer's ownership: see paragraph 24 of Schedule 6. By the effect of a rising rate of inflation, or because subsequent to April 6 1965 there has been a sharp rise in the development value of the land disposed of, this method of ascertaining the taxable gain favours the tax- H
payer, for it throws back part of the gain which in truth accrued after April 6 1965 to the earlier non-taxable period. It is this method of ascertaining the post-April 6 1965 gain which in the instant case is favoured by the taxpayer. There can be no doubt, on the facts of the case, that the value of the land rose far more sharply after the relevant date than it did before.

The other method of ascertaining the post-April 6 1965 gain—and it is that for which the Revenue contends in the instant case—is to ascertain what was the market value of the asset on April 6 1965, which is a very tiresome process except in relation to such assets as quoted investments, and J
compute the post-April 6 1965 gain accordingly: see paragraph 23(2) and the following subparagraphs of Schedule 6. Where the case fell within paragraph 23 it was mandatory to adopt the last-mentioned method of computing the post-April 6 1965 gain. Whether paragraph 23 did or did not apply depended on whether the case fell within subparagraph 1 of paragraph 23, and that in turn depended on whether the consideration for the disposal exceeded what the market value of the land would be upon the hypothesis specified in paragraph 23(1)(b).

Paragraph 23(1) was in force in relation to disposals at the relevant time. It appeared under the cross-heading "Sales K
of land in United Kingdom reflecting development value," and provided as follows:

> 23(1) This paragraph shall apply in relation to a disposal of an asset which is land in the United Kingdom, or an estate or interest in land in the United Kingdom—
>
> (a) if, but for this paragraph, the expenditure allowable as a deduction in computing under this Schedule the gain accruing on the disposal would include any expenditure incurred before April 6 1965, and

A (b) if the consideration for the asset acquired on the disposal exceeds what its market value would be if, immediately before the disposal, it had become unlawful to carry out any development in, on or over the land other than development of the kinds specified in Schedule 3 to the Town and Country Planning Act 1962 (for land in England and Wales or Northern Ireland) or Schedule 3 to the Town and Country Planning (Scotland) Act 1947 (for land in Scotland).

In this paragraph "development" has, in relation to land in England or Wales or Northern Ireland, the meaning given by the Town and Country Planning Act 1962 and, in relation to land in Scotland, the meaning given by the Town and Country Planning B (Scotland) Act 1947.

(2) For the purposes of this Part of this Act, including Part 1 of this Schedule, it shall be assumed in relation to the disposal and, if it is a part disposal, in relation to any subsequent disposal of the asset which is land in the United Kingdom or an estate or interest in land in the United Kingdom that that asset was sold by the person making the disposal, and immediately reacquired by him, at its market value on April 6 1965.

We have not hitherto mentioned paragraph (a) of paragraph 23(1) because it is common ground that in the instant case it was satisfied.

C Mr Potter, on behalf of the taxpayer, fastening on the words "it had become unlawful to carry out any development . . .," submits that paragraph (b) is only intended to apply to a case where, at the date of the disposition, it was lawful to develop the land. What the legislature had in mind was the case where the value of land has been dramatically increased by the effect of any actual planning permission which made it lawful to develop it and not to a case where, as here, no planning permission had been obtained at the time of the disposal and it was, as well before as after the disposal, unlawful to develop the land. The expression "had become unlawful," so the argument runs, imports a change of D condition—that is, the entry into a new state or condition by a change from some former state or condition. Here, it is submitted, because it was unlawful as well before as after the disposition to develop the land there was no such change. E

There would be much to be said for Mr Potter's submission were it not for the fact that paragraph (b) is drawn upon the assumption that the arithmetical exercise of deducting the hypothetical value from the actual consideration is one that can be performed: for only by performing it can you ascertain whether subparagraph (2) does or does not apply to the case which you are considering. Paragraph 23 in terms directs that the paragraph *shall apply* if the two conditions set out in paragraphs (a) and (b) are satisfied, and you can only find out whether the condition in paragraph (b) is satisfied by doing the arithmetic there contemplated. And so the language of the paragraph must, if possible, be construed so F as to enable the exercise to be done.

We find no difficulty in so construing it. When one comes to analyse the wording of the hypothesis it is not a change in the law whereby some particular development had become unlawful which is supposed but a change in the law by the effect of which it had become unlawful to carry out *any* development in or over the land. The judge in the court below thought that the words "any" and "unlawful" required, for the purpose of the hypothesis, the absolute prohibition of all development of the land with or without planning permission except what has been referred to as Schedule 3 development, and we agree with him. Here it G was lawful, at the time of the disposition, to carry out a development subject to obtaining planning permission, and there is no difficulty in supposing a situation when it had become unlawful to do so. And so the expression "had become" is satisfied. So read, paragraph (b) will always operate to determine whether a particular case did or did not fall within paragraph 23(2).

The appeal is dismissed.

The appeal was dismissed with costs. The court directed that three counsel on each side should be allowed (a planning junior counsel was added to each side as planning law was H involved in addition to tax law). Leave to appeal to the House of Lords was refused.

TOWN AND COUNTRY PLANNING

A **QUEEN'S BENCH DIVISION**

May 25 1977

(Before Sir DOUGLAS FRANK QC, sitting as a deputy judge of the division)

HAYNS v SECRETARY OF STATE FOR THE ENVIRONMENT AND ANOTHER

Estates Gazette January 7 1978

(1977) 245 EG 53

Application under section 245 of Town and Country Planning Act 1971 to quash decision of inspector dismissing appeal against refusal of planning permission—Permission sought B for provision of means of access to applicant's land across grass verge forming part of a common—Planning authority's requirement of proper visibility splays—Question whether applicant had such rights over verge as to enable him to provide splays—Applicant unable to establish such entitlement—Refusal of application upheld

In these proceedings, the applicant, Andrew Bickford Hayns, the owner of land at Chipperfield, Herts, sought to have quashed the decision of an inspector appointed by the Secretary of State for the Environment to hear and determine an appeal against the refusal of the planning authority to grant permission for the construction of a means of access C to the applicant's land from a highway now called The Common, formerly known as Chipperfield Callipers Hall Road.

Norman Wise (instructed by L O Glenister & Sons, of Eastcote, Middlesex) appeared on behalf of the applicant; Harry Woolf (instructed by the Treasury Solicitor) represented the first respondent, the Secretary of State for the Environment. The second respondents, the planning authority, were not represented and took no part in the proceedings.

Giving judgment, Sir Douglas Frank said: This is an application under section 245 of the Town and Country Plan- D ning Act 1971 to quash a decision by an inspector appointed by the first respondent to hear and determine an appeal against a refusal of a planning permission. The applicant is owner of the land at Chipperfield in Hertfordshire. Between the frontage of his land and the highway, a road now known as The Common [*sic*], there is a grass verge which is not part of the highway but forms part of Chipperfield Common. The applicant was minded to have a means of access from The Common across the grass verge to his land in preference or in addition to other presumably less commodious existing means of access. He applied for planning permission which was refused on grounds which are not before me.

E The first respondent appointed A J Bingham, one of his inspectors, to hear and determine the appeal, and he held an inquiry on June 9 1976. From his decision letter, which is dated June 29 1976, it is clear that he came to the conclusion that there was no objection to the means of access on amenity grounds, but he found that the planning authority was right in requiring proper sight visibility splays in accordance with the recommended standards. These are 15 ft by 300 ft (4.5m by 91m) triangles in both directions. These are standard requirements for means of access in this type of situation. So the question arose whether the applicant was F in a position to provide those splays, that is to say whether he was in a position to ensure that the visibility from 15 ft back and 300 ft in both directions would be unimpeded. The planning officer had asserted that though the sight line required would be unimpeded by any obstruction such as trees or shrubs or any other means that would interfere with the sight line, the land concerned was not under control of the applicant.

After the end of the hearing the inspector had asked to see copies of the title deed upon which the planning officer had based his assertion that the second respondents were the owners of the verge, with the effect that it was not within G the control of the applicant. During the course of the inspection which followed the hearing the inspector was handed a copy of a conveyance of the whole of Chipperfield Common made on October 1 1936 and he took care to ensure that the chartered surveyor representing the appellant was supplied with a copy and that was done within two days. It is right that I should point out that nowhere, so far as I have been able to ascertain, was it contended on behalf of the applicant that he had the control of the land required for the visibility splays. And although the copy of that title deed was sent to the chartered surveyor there was no rejoinder by him or by any other person representing the applicant to H the effect that that assertion by the planning officer was wrong. The inspector concluded that the conveyance confirmed the contention that the land concerned was under the control of the second respondents.

I now consider Mr Wise's argument based on the 1936 conveyance. It was in fact a conveyance of the whole of Chipperfield Common to the predecessors of the second respondents, but the conveyance was subject to and there was reserved out of the grant

> a right for the vendors jointly and severally and for any person or persons authorised by them or either of them and J for his or their successors in title and assigns to construct repair and use from time to time with or without horses carts carriages and motor cars any road or pathways and to lay thereunder sewers drains electric light and gas and water mains required for the use or development of any property now belonging to the vendors or either of them adjoining the said Common across such portions of the said Common as lie between the Chipperfield Callipers Hall Road and such property then belonging to the vendors as aforesaid and between the Chipperfield Belsize Road and such property as aforesaid.

I am told, and I accept, that Chipperfield Callipers Hall Road is now known as The Common, which is the highway to which the appellant requires direct access. Mr Wise argues K that the reservation out of the grant of the right-of-way, the regrant as he prefers to call it, was such as to carry with it such ancillary rights as would be necessary in order to exercise that grant in today's conditions. He says one such ancillary grant would be the control over the land required for visibility splays, because he says without that control the grant cannot today be exercised. He cites a number of cases which he said support that conclusion. For my part, however, I do not think that there was any such ancillary grant implied with the reservation or regrant which I have mentioned. I accept that as a general proposition where there is a grant of an easement there are deemed to be also granted such rights as are normally incidental to the easement, but such rights are normally incidental and are such as to be physically necessary to exercise the easement. I do not think that this is any such case. In my judgment where there is a grant of a right-of-way there is not to be assumed as incidental to that right any control to be given to the owner of the dominant tenement for the purpose of complying with modern town planning standards. I would say that that is true today. It was certainly true at the time of this grant in 1936 at which time such requirements of visibility splays were unknown.

Mr Wise referred to *VT Engineering Ltd v Richard Barland & Co Ltd* (1968) 19 P & CR 890. There the landlord granted to his tenant a right-of-way at all times and for all purposes over a roadway leading to the tenant's premises. The question was whether the landlord was entitled to erect such a new building as would limit vertically the size of vehicle which could be used. On the facts it was held that while to require lateral swing space of indefinite dimensions would be unjustifiable the tenant's entitlement to vertical swing space in a right-of-way previously open to the skies would be severely curtailed in a tunnel 9 or 10 ft high. Accordingly an injunction was granted. That seems to be a long way from the present case. I draw particular attention to the judgment at p 895 where the learned judge says:

> I can well see that on the grant of a right-of-way the grantor must accept that over the way granted he cannot thereafter exercise rights which materially interfere with the enjoyment of the easement. It may perhaps be that he must allow some degree of tolerance for wide loads, so that he cannot fence or build up to the very edge of the way granted, but must leave a free-board of a foot or two, particularly if there are bends in the way. . . . But that is very different from saying that the grant of a way may in effect sterilise a strip of land of indefinite depth on each side of the way, depending on the loads and methods of loading from time to time adopted. This seems to me to go far beyond any necessary or reasonable expectation or implication; and quite apart from the absence of authority on the point, it seems to me that such a right would or might subject the grantor to a quite unjustifiable burden. I accordingly reject such a claim.

It seems to me that in the present case what the applicant is saying is that that grant of easement carried with it a sterilisation of a very considerable area of other land outside any right-of-way, and would certainly place an unjustifiable burden on the grantor. I appreciate that this is common land to be used for open-space purposes, but the control would effectively prevent, as is the intention of a visibility splay, the planting of any trees or any shrubs of more than a very limited height, and I think that would go far beyond any incidental right to be annexed to an easement of that kind. So I have little doubt that a grant of a right-of-way does not carry with it an implied grant of such rights as are necessary for the grantee to exercise the right by the provision of visibility splays.

That would be sufficient to dispose of the matter, but the question arises, if I had decided otherwise, whether there was evidence before the inspector that the present applicant (to quote from the easement) "was the owner of any property now belonging to the vendors," that is to say whether he, the applicant, is the owner of property which belonged to the vendors in 1936. Mr Wise properly concedes that at the inquiry the appellant did not prove any title as successor to the vendor of 1936, but he said that there was presumptive evidence to put the inspector on the alert, and he should have seen that the applicant owned adjoining land and looked at the proposed access in the context of the 1936 conveyance and at all the circumstances; and from that he should have concluded that the applicant was successor to the 1936 vendor. I do not agree. There was nothing at the inquiry which could have enabled the inspector to take the view that the applicant was entitled to the benefit of the 1936 conveyance, and indeed it does not seem to have been represented to him that the applicant was so entitled. I would have thought that, if there were a burden here, the burden was on the applicant to make out his entitlement and indeed he would have no doubt had a further opportunity after the inquiry having been sent a copy of the conveyance. There is no evidence before me, and there was no evidence before the inspector, that the land now owned by the applicant was then owned by the vendor. On that ground in any event I would hold that the inspector's decision was right. In fact, it seems to me that he went out of his way to try to get it right. If there were any failure here it was on the part of those representing the applicant.

Finally I should say this. Mr Wise made a plea *ad misericordiam*, saying that in effect there was no objection on the grounds of amenity. All that was required was this visibility splay, and the second respondents, they being in a position to provide it, are being difficult. Of course I do not know sufficient of the facts, whether they are justified in adopting an obstructive attitude, if indeed they are. They would be quite wrong, I would have thought, to refuse to co-operate unless they have good reason for not preserving the sight lines, but really it is nothing to do with me. It is a matter for the second respondents as owners of the land concerned.

In the circumstances the appeal must be dismissed.

The appeal was dismissed with costs.

QUEEN'S BENCH DIVISION

July 5 1977

(Before Mr Justice WILLIS)

H SABEY & CO LTD v SECRETARY OF STATE FOR THE ENVIRONMENT AND OTHERS

Estates Gazette February 4 1978

(1977) 245 EG 397

Town and Country Planning Act 1971—Appeal against refusal to permit gravel extraction—Secretary of State's decision challenged—Rules of natural justice—Audi alteram partem—Decision based on conclusion that removal of gravel would adversely affect supply of moisture to the detriment of crop yields—Applicants not given opportunity to deal with this issue—Breach of rules of natural justice—Decision quashed

In these proceedings H Sabey & Co Ltd applied under section 245 of the Town and Country Planning Act 1971 to the court to quash a decision by the Secretary of State for the Environment (APP/4401/A/73/18173, dated July 8 1976) dismissing, after a public inquiry before an inspector, the applicants' appeal against a refusal of permission for gravel extraction from land at Sipson, Middlesex.

M Fitzgerald (instructed by Braby & Waller) appeared on behalf of the applicants; Harry Woolf (instructed by the Treasury Solicitor) represented the first respondent. The other respondents, the Greater London Council and the London Borough of Hillingdon, were not represented and took no part in these proceedings.

Giving judgment, WILLIS J said: This appeal under section 245 of the Town and Country Planning Act 1971 raises a short point, whether the decision of the Secretary of State dated July 8 1976 dismissing the applicants' appeal against a refusal to permit gravel extraction was contrary to the rules of natural justice and so of course outside his powers.

The land in question consists of 66 acres classified by the Ministry of Agriculture as grade I agricultural land, albeit at present used only for pony grazing. The applicants applied in February 1973 to extract sand and gravel by dry working and refilling. Objection was raised by the Ministry of Agriculture among a number of other objectors, and after a long inquiry which ranged over many topics the inspector finally based his recommendation to dismiss the appeal on what can broadly be described as agricultural grounds. The remaining issues have been resolved in favour of the applicants. The Ministry of Agriculture's objection was reflected in the second ground of refusal of the Greater London Council which stated:

> Extraction and subsequent filling would be likely to reduce seriously the agricultural quality of the land. This would in turn reduce the agricultural productivity of the land and there is consequently an objection on agricultural grounds to the proposal.

There was no dispute at the inquiry that the land was rightly classified grade I and its capacity for supporting intensive horticultural operations was recognised by the applicants (see report paragraph 60). However, it was said by them that agricultural or horticultural operations appropriate to such quality land were quite uneconomic. The applicants devoted a great deal of their evidence towards meeting the objection that the agricultural quality of the land would be seriously reduced as alleged in ground 2 of the refusal. Evidence was given to support the expected achievement of grade II status after restoration, and, as the report shows in paragraphs 84 to 114 inclusive, there was detailed evidence of the proposed method of working, equipment and restoration.

Reference is made in the report to the evidence on behalf of the Ministry of Agriculture. There appear the following references germane to the question at issue in this case.

Paragraph 182:

> The underlying gravel by virtue of its structure is an important asset in the production of ground moisture.

Paragraph 184:

> Movement, storage and subsequent replacement of the soil in accordance with the appellant's proposals . . . would present major practical problems of compaction to soil structure by the passage of heavy site machinery. The moisture content of soil is a critical factor and no matter how well the site be operated . . . there would inevitably be occasions when the structure of wet soil would be damaged by heavy equipment during the all-year-round general site operations; damage which could not be rectified. . . .

Paragraph 185:

> The ministry is monitoring restoration works in the area and while some results are encouraging it has not yet been proved that land of this type can be restored to its former quality.

After setting out his findings of fact the inspector came to his conclusions of which I set out the material parts.

Paragraph 213:

> . . . the major issues to be balanced . . . are those of the undisputed local gravel demand against the potential loss of high-quality grade I agricultural land. I have no reason to doubt that the company could restore the land to a condition at least equal to that demonstrated on nearby workings and return it to an agricultural use.

Paragraph 214:

> It appears to me, however, that an essential feature governing the quality of the unique overlying brick earth is the ability of the underlying gravel to provide through its structure, water essential to intensive market gardening operations. The removal of the gravel and its replacement with various materials must put at risk the supply of moisture compatible with the demands of high yields which could be expected from this grade I land. While I accept that an agricultural crop could be obtained, experiments now in their early stages have not yet demonstrated that high yields or several annual market garden crops could consistently be obtained from restored land.

The Secretary of State in his letter of decision set out the inspector's conclusions verbatim, and in paragraph 3, after agreeing with the conclusions and accepting the recommendation to disallow the appeal, said (*inter alia*):

> . . . the agricultural objection is of major importance, having regard to the Government's policy of protection of agricultural land of the highest quality. The inspector's view that the removal of the underlying gravel from the land and the consequential alteration in its fundamental characteristics would be likely to result in a decline in its quality, even if carefully restored, is accepted.

Mr Fitzgerald submits that there was no evidence before the inquiry upon which the inspector could properly reach the conclusion that the removal of gravel and its replacement with various materials must put at risk the supply of moisture compatible with the demands of high yields which could be expected from this land. He said it formed no part of the case of the second respondent or the Minister of Agriculture, that no questions were put to his witnesses on what turned out to be this all-important topic, and that the applicants have not been given the opportunity which they should have had to deal with it if they could and so have been prejudiced. He sought to put in affidavits by the appellant's managing director, Mr Penfold, and Dr Lloyd, his agricultural expert, to the effect that no evidence was given by the Ministry of Agriculture witness or anyone else which showed or tended to show that removal of gravel and its replacement by other materials would have adverse effects upon or put at risk the necessary supply of moisture, nor as to the sufficiency or otherwise of the moisture supply after the land's restoration.

Mr Woolf objected to the admission of this evidence on lines for which there is authority, but conceded that the words of Lord Denning MR in *Ashbridge Investments Ltd v Minister of Housing and Local Government* [1965] 1 WLR 1320 at p 1327: "The court can receive evidence to show what material was before the minister" were to some extent against his submission in the present case, and that there was no authority to support the exclusion of evidence designed to show that there was no evidence before the minister.

Mr Fitzgerald's submission is of the second or negative kind, and while I fully understand Mr Woolf's insistence that the court must beware of turning itself into a Court of Appeal on fact, I concluded that, in the exceptional circumstances of this case, Mr Fitzgerald would be put into the position of not being able properly to present his point of law unless I allowed the introduction of affidavit evidence limited to the negative proposition of there having been no evidence to support the inspector's conclusion. I therefore did so, and also granted leave to Mr Woolf to introduce an affidavit by Mr Loveday, the inspector, who, while agreeing that K H Gordon, the ministry witness, gave no evidence on the "moisture question," as I call it, during examination, cross-examination and re-examination, recalled asking Mr Gordon a question at the end of re-examination couched in general terms, in answer to which he confirmed that "the underlying sand and gravel strata provide water through capillary attraction to the overlying top soil."

Mr Fitzgerald submits that, even if it is accepted that in this way there was some evidence as to the "moisture question" which might be said to provide some basis for the inspector's conclusions in paragraph 214, nevertheless it came in very much by a side wind and an opportunity should have been given to the applicants to deal with the issue before the inspector reached a conclusion which was the basis for his decision. On this matter he says the applicants have not had "a fair crack of the whip": see *Fairmount Investments Ltd v The Secretary of State for the Environment* [1976] 1 WLR 1255 per Lord Russell of Killowen at p 1266.

Mr Woolf submits that the applicants are really trying to have a second bite of the cherry and to be put in a position to call evidence now which they should have called at the inquiry as part of their case. He submits that, looking at the report as a whole, there was sufficient material from which the inspector could draw the inferences which led him to the crucial words in the conclusion at paragraph 214. As to the affidavit evidence, Mr Woolf says it was plain that the "moisture question" was of critical importance in the consideration of the future agricultural quality of the land, and even if Dr Lloyd had tried to limit the issues to be debated in discussions with the Ministry of Agriculture before the inquiry, the onus was on the applicants to lead evidence to show, if they could, that the moisture content of the land would not be adversely affected by their operations, and if they had not done so in chief, they should have realised the importance of the question during the inquiry and applied to lead such evidence at a later stage, possibly after adjournment.

I have not found this question altogether easy, and my mind has wavered during the argument. I think there is force in Mr Woolf's submission that the applicants might have been expected to have included as part of their case a claim that the supply of moisture through the replacement materials would not adversely affect the quality of the land. That they did not do so I think was probably due to the earlier discussions between Dr Lloyd and the ministry designed to limit the issues. In the result neither the ministry nor the applicants said a word, or at least no more than appears in paragraphs 182 and 184 of the report, on the question which critically affected the inspector's views and the decision of the Secretary of State. I do not think anything which emerged during the hearing, including the answer referred to in Mr Loveday's affidavit, can fairly be said to have put the applicants on the alert so that they should have applied to call further evidence. I would not wish anything I have said to be thought to be critical of the inspector. He had a long and complicated inquiry to conduct and has produced a careful and lucid report. In it, although the importance of the underlying gravel is referred to, any risk posed to the moisture content of the land by its removal and replacement by other materials really finds no place, in my view. No doubt the inspector considered that he had the material, perhaps principally by way of inferences which he could legitimately draw, to justify the view he expressed in paragraph 214.

On the whole, however, I have come to the conclusion that if there was any evidence upon which to base paragraph 214 and the Secretary of State's decision, as to which I feel very considerable doubt, to have acted upon it as he did without giving the applicants the opportunity to deal with it if they could, the Secretary of State acted in breach of his duty *audi alteram partem*.

In the circumstances of this case I do not think anyone is to blame, but I am left with the strong feeling that justice will not be done unless the applicants are afforded an opportunity to present their evidence on the "moisture question" before the Secretary of State reaches his final conclusion. Accordingly the decision must be quashed. I only add that while, of course, it is a matter entirely for the Secretary of State, there being no longer any criticism of the decision save in this one particular, one can naturally hope that any rehearing can be limited to this single question.

The Secretary of State's decision was quashed, with costs.

QUEEN'S BENCH DIVISIONAL COURT

December 13 1977

(Before Lord WIDGERY CJ, Lord Justice CUMMING-BRUCE and Mr Justice PARK)

HOOPER v SLATER

Estates Gazette February 18 1978

(1977) 245 EG 573

Enforcement notice—Failure to remove caravans—Plea that storage of caravans was within "warehouse" use accepted by justices—Town and Country Planning (Use Classes) Order—"Warehouse" implied covered storage—Appeal against justices' decision allowed

This was an appeal by William David Hooper, by way of case stated by justices sitting at Newquay, from their dismissal on February 15 1977 of an information preferred by the appellant on behalf of the Restormel Borough Council against Peter Richard Slater, alleging his failure to comply with an enforcement notice requiring him to remove two caravans on land at St Columb Road, Newquay, on the ground that such use constituted a material change of use for which no planning consent had been given.

J Sullivan (instructed by Sharpe, Pritchard & Co, agents for the solicitor to the borough of Restormel) appeared for the appellant. The respondent did not appear and was not represented.

Giving judgment, Lord Widgery CJ said that in 1971 the respondent was granted planning permission for use of the land in question for warehousing and as a storage depot. Proceedings had subsequently been taken against him for failing to remove from the site two caravans in accordance with the terms of an enforcement notice.

The argument before the justices, and in the present appeal, was that the presence of the caravans represented a breach of planning control. The justices, however, appeared to have been misled by argument to the effect that the planning permission granted in 1971 would cover the use now complained of. Just how they got hold of that idea was not clear, but it was quite obvious that the use permitted by Class X of the Town and Country Planning (Use Classes) Order—use for warehousing and as a storage depot—was not to be equated with the use of *any* kind of storage.

It was quite clear that the use to which the land was put was not a use sanctioned by Class X. "Warehouse" use implied covered storage. The justices, however, took the view that the presence of the caravans meant that the land was being used for "storage" and that was good enough. But it was not good enough, because it was not a use of the kind allowed by the permission granted. The case should go back to the justices for them to continue the hearing.

Cumming-Bruce LJ and Park J agreed.

Appeal allowed, case remitted to the justices with a direction to continue the hearing. Appellant awarded costs against the respondent.

QUEEN'S BENCH DIVISION

October 14 1977

(Before Mr Justice GRIFFITHS)

BRADFORD METROPOLITAN DISTRICT COUNCIL v SECRETARY OF STATE FOR THE ENVIRONMENT AND OTHERS

Estates Gazette February 18 1978

(1977) 245 EG 573

Town and Country Planning General Development Order 1977—Whether the facing of dwelling-houses with stone cladding "permitted development" within Class I in Schedule 1 to the Order—Submission that as a "cosmetic alteration" it was not covered by permission given by Order—Secretary of State's ruling that it was covered upheld by court—Stone cladding did not project beyond the wall, the moulded surrounds and window sills being regarded as part of the wall—Application for planning permission not required

This was an appeal by Bradford Metropolitan District Council against a decision of the Secretary of State for the Environment that an application for planning permission was not required by householders who wished to face their houses with stone tiles, this being an action covered by Class I in Schedule 1 to the Town and Country Planning General Development Order 1977 as permitted development.

M G V Harrison (instructed by chief solicitor, Bradford Metropolitan District Council) appeared on behalf of the appellant council; M D Kennedy (instructed by the Treasury Solicitor) represented the first respondent, the Secretary of State. The second and third respondents, Robert Paterson and Frederick French, whose houses were the subject of the dispute, did not appear and were not represented.

Giving judgment, GRIFFITHS J said: Mr Paterson and Mr French, who are the second and third respondents in this application, decided that they would like to face their houses at nos 2 and 6 Redcar Road with stone tiles. They applied for planning permission to the Bradford Metropolitan Council, but those responsible for planning decisions in the council refused them permission because in their judgment it would be bad planning, as it would be aesthetically unacceptable to have a road of semi-detached red-brick houses turned into a hotch-potch by sporadic stone cladding. The householders, being aggrieved by that decision, appealed. The inspector decided that the Bradford Council had jurisdiction to entertain the planning application and that they came to the right decision. That inspector's report then went for review by the Secretary of State. By this time, as I understand it, Mr Paterson and Mr French had got so fed up with all the delays that they decided to paint their houses, and have no further interest in this appeal. However, I am told that there is a divergence of view between various councils and the Secretary of State as to whether or not planning permission is required if a householder of a dwelling-house wishes to put stone cladding on the face of his house. Some councils apparently take the view that planning permission is not required, others, including Bradford, that planning permission is required.

The Secretary of State in this matter has given it as his opinion that planning permission is not required, and the question now that I have to determine is whether or not planning permission is required, and this depends upon the construction of Schedule 1 to the Town and Country Planning General Development Order of 1977. First I should read Article 3 of the order, which provides:

> Subject to the subsequent provisions of this order, development of any class specified in Schedule 1 to this order is permitted by this order and may be undertaken upon land to which this order applies, without the permission of the local planning authority or of the Secretary of State.

I now turn to Schedule 1. Schedule 1 is divided into various classes, and the two classes with which I shall be concerned are Class I and Class II. Class I provides: "Development within the curtilage of a dwelling-house. (1) The enlargement improvement or other alteration of a dwelling-house so long as," and there then follow a number of provisions. Giving those words their ordinary meaning, to add stone cladding to the face of a building is clearly at least an alteration of the dwelling-house. Presumably, in the mind at least of the occupier it is an improvement of the dwelling-house and perhaps it is arguable that it is an enlargement, only that I doubt. For the purpose of the appellants' argument, however, it is submitted that these words should not be given their ordinary general meaning but that in the context in which they appear in this order, and against the general background of planning legislation, they must be given a more restricted construction. The restricted construction for which the appellants contend is that they are limited to the physical structure of the building as opposed to any alteration to its cosmetic appearance. I say at once that I find this a difficult concept because they are not mutually exclusive concepts. You may well have a structural alteration which significantly affects the cosmetic appearance of the building. Indeed, the whole object of a structural alteration may be to affect the cosmetic appearance. I give just as one example—altering the shape of a window from a rather ugly ill-proportioned window perhaps to a more attractive round window, or something else of that nature. So if one has to try to draw a line between structural alteration and cosmetic alteration I can see areas of great difficulty that would immediately arise.

In support of the argument the appellants rely on the three provisos set out under sub-paragraphs (a), (b) and (c): (a) limits the size of the alteration, so that the dwelling-house is not made larger than an additional 50 cubic metres or one-tenth of the capacity of the building, subject to a limit of 115 cubic metres; (b) places restrictions on the height to which the building can be enlarged, and, finally, (c) is in the following terms: "No part of the building as so enlarged, altered or improved projects beyond the forwardmost part of any wall of the original dwelling-house which fronts on a highway." These provisos, the appellants say, show that the permitted development is aimed solely at what they call "structural features." I find it very difficult to infer that. True, they are placing a limitation upon the size to which you can enlarge your building, but I do not see why, because there is a limitation imposed upon the size to which the building can be enlarged, I should conclude from that that it was intended to rule out cosmetic alteration.

The other matter upon which the appellants rely in support of a restricted reading of the wording of Class I is to be found in Class II. Class II is of general application. It is headed "Sundry minor operations." It is not intended to apply only to dwelling-houses but it is of general application to all buildings and structures, and it specifically provides under paragraph 3 for "The painting of the exterior of any building or work otherwise than for the purpose of advertisement, announcement or direction." So you do not have to have planning permission to paint the exterior of any building or work, and the argument is that that was incorporated in Class II to enable a householder to paint his house because he would not have been able to paint it under the general words of Class I because they were not intended to apply to cosmetic alteration. Now I should see some force in that argument if Class II were limited to dwelling-houses, but Class II is not limited to dwelling-houses. Class II is of general application, and because a provision is made that allows any building or work to be painted without planning permission, I cannot see why I should infer from that that the very wide words in Class I dealing with dwelling-houses would not have included painting. I think they would. So I get no help from Class II, paragraph 3.

So the first rung of the appellants' objection in my view fails because I see no warrant in either argument to cut down the ordinary general scope of the words of Class I, which are clearly wide enough to cover such an operation as stone cladding. But the appellants have a second string to their bow. They point to proviso (c) of the first paragraph of Class I, and contends that the stone cladding will project beyond the wall of the dwelling-house. The relevant measurements are as follows: The dwelling-house is constructed of red brick. It has windows with window-sills, and it also has a moulded surround attached to the wall around the lower windows. These project 3 in beyond the brickwork. The stone cladding when it is attached to the brickwork will be of a total thickness of 1¼ in, so it will be within the building line, if one takes as one's building line the edge of the sills and the moulded surrounds to the lower windows, but it will project 1¼ in beyond the original brickwork. The argument for the appellants is that as that will be in breach of sub-paragraph (c) it is therefore not covered by the development permission granted under Class I. The minister rejected that contention because he gave it as his opinion that on the facts of this case the sills and the surrounds should be regarded as part of the wall, and that as the proposed stone cladding would not project beyond them it fell within Class I development.

I have had my attention drawn by Mr Harrison to the way in which this proviso was first introduced into the legislation in 1963. The proviso in Schedule 1 to statutory instrument 1963 no 709 provided that: "No enlargement

A improvement or alteration shall project beyond the forwardmost part of the front of the original dwelling-house," and he says that this was an intentional alteration, so that any additions should not be entitled to project out as far as any projections that might have been incorporated on to the wall of the building and is to hold any alteration to the wall line and not to the line of any projection.

Mr Kennedy says that is reading far too much into the alteration, and the real purpose of the alteration of the wording is to make quite clear that the limitation was aimed at the building line that fronted on to the highway. I was impressed by that argument, but it is weakened by the fact B that that effect could have been achieved by redrafting subparagraph (c) in the present Class I of Schedule 1 omitting the words "of any wall." I do not think I get very much help either way from the alteration of the wording, and the question is, in my view, whether it was permissible for the Secretary of State, looking at this matter in a commonsense way, to regard the moulded surrounds and the window-sills as part of the wall, and I come unhesitatingly to the conclusion that it was. I have no doubt that the real object of 1(c) is to hold any alterations or enlargements to the original building line, and it seems to me that it would be a most nit-picking approach to say that although it is attached to C the brickwork and within the building line as established by the sills and mouldings it nevertheless fails to be protected because it is projecting beyond the wall viewed solely as the brickwork.

I was referred to the decision of the House of Lords in *Brutus* v *Cozens* [1973] AC 855, and to a planning appeal in which that decision was applied, *LTSS Print & Supply Services Ltd* v *Hackney Borough Council* [1976] QB 663, and following the guidance to be found in those decisions, it is, in my view, permissible on the facts of this case for the Secretary of State to construe that word "wall" in the way in which he did, and to regard the mouldings and the D sills as part of the wall.

Accordingly, in my view, the Secretary of State came to the right conclusion and this appeal fails.

The appeal was dismissed with costs.

QUEEN'S BENCH DIVISIONAL COURT

June 2 1977

(Before Lord WIDGERY CJ, Mr Justice O'CONNOR and Mr Justice WATKINS)

KENTUCKY FRIED CHICKEN (GB) LTD v SECRETARY E OF STATE FOR THE ENVIRONMENT AND ANOTHER

Estates Gazette March 11 1978

(1977) 245 EG 839

Town and Country Planning Act 1971—Appeal against enforcement notice—Appellants required to close down hot food take-away shop after complaints of smell, noise etc—Expert evidence called at inquiry before inspector to show that nuisance could be alleviated—Inspector decided against appellants—Submission that inspector was bound to accept opinion of appellants' experts in the absence of contrary evidence—Inspector held entitled to use his commonsense and not bound to accept experts' evidence

F This was an appeal by Kentucky Fried Chicken (GB) Ltd under section 246 of the Town and Country Planning Act 1971 against the decision of an inspector dismissing their appeal against an enforcement notice served by Lambeth Borough Council as planning authority. The Council was the second respondent to the present appeal to the Divisional Court. The first respondent, the Secretary of State, did not appear and took no part in the proceedings.

N Wise (instructed by Reimann & Cummins) appeared on behalf of the appellants; C Whybrow (instructed by S J G Smith, Director of Administration and Legal Services, Lambeth) represented the second respondent.

Giving judgment, Lord Widgery C J said: This case comes G before the court as a purported appeal under section 246 of the Town and Country Planning Act 1971 by the appellants, who are known as Kentucky Fried Chicken (GB) Ltd, against the decision of an inspector appointed under the Town and Country Planning Acts, which decision was given on May 6 1976 and referred to premises in part occupied by the appellants at 36 Brixton Road in South London.

In this instance, although the matter is in every sense a planning appeal, it has not taken the more common course of being heard by an inspector at an inquiry, the inspector making a report and the Secretary of State basing his decision on that report. This is a case in which the inspector is himself charged H by delegated authority with the duty of reaching a conclusion upon all the facts which the Secretary of State would normally have to deal with on the more conventional method of appeal.

The cause of all the trouble is that on the ground floor at 36 Brixton Road there is a shop operated by the appellants, which is described as a hot food take-away shop. In other words, as that description implies, food is cooked in this shop and is handed over the counter in a hot state to customers who then take it away and eat it. Not altogether surprisingly, the value of the shop is apparently good and the customers numerous.

The collection of people and their activities resulting from the J opening of this shop has given rise to complaints both from the local authority and from the occupiers of adjacent buildings. The sort of complaints which have been made are that there is smell of fumes and from the cooking, noise from the pan, noise from people who come as customers, and noise from people who slam doors in the early hours of the morning and foul the area with waste paper and other things.

On discovering that this change of use had been made the planning authority served a notice in brief complaining of the development occasioned by the opening of this shop and requiring it to be shut down. In answer to that enforcement notice, the appellants (as they were perfectly entitled to do) have K brought this appeal, and it has come to us, as I have indicated, from the inspector who conducted the inquiry and who reached a conclusion.

The gist of what happened before the inspector is this. The appellants called a quantity of expert evidence, no doubt of very high quality, containing recommendations as to what should be done to alleviate the nuisance created by the activities of this shop to which I have referred. References were made to British Standards and so on in demonstration of the adequacy of the proposed remedial measures, and I am prepared to approach this case on the basis that the appellants made out a strong case for saying that they could alleviate some, if not all, of the L nuisance which had been proved to exist.

The council thought differently, and so did one or two of the adjoining occupiers, and the inspector's conclusions come to this. He said:

> On the evidence and from my own observations I am satisfied that fumes and smells from cooking in the appeal premises penetrate to the residential accommodation above and that this is seriously damaging to the amenities that the occupants of that residential accommodation can reasonably expect to enjoy. Although it has been stated on behalf of the appellants that improvements can be made which would substantially reduce the smell and fumes, in my view this would not entirely eliminate or reduce to a satisfactory degree the damage to the amenities M now being caused.
>
> I am also satisfied that the amenities of the occupants of the residential accommodation above the premises are being seriously damaged by noise from the air duct and from the general activities within the premises. By the nature of the building and the close proximity of the residential accommodation to the seat of the commercial activities, it does not appear to me that it would be feasible to reduce the noise levels sufficiently, whatever structural alterations are made, so as to eliminate this justifiable source of complaint.

Finally, he says:

> While I accept that the appellants are unable to control the activities of their customers outside the premises, it seems to me that this shop,

A which is open seven days a week until a very late hour at night, inevitably attracts customers who are likely to cause disturbance outside the premises by slamming of car doors, playing of radios and other activities. Such noise, in my opinion, is likely to be more disturbing to the amenities of residents than the fairly constant traffic noise to which they are accustomed. Moreover, I do not accept the argument that the emanation of additional noise is permissible because of the already high level of noise in the area; and I am not convinced that it would be practicable to control the activities outside the premises by the imposition of conditions on a planning consent.

At that point the court was prompted to ask counsel appearing for the appellants where the point of law was, because section 246
B authorises an appeal to this court on a point of law, and so far we have only considered the merits. The answer given by counsel was, in substance, if he will forgive me for paraphrasing it, that the appellants having called high-level experts who had expressed opinions about this, the inspector, without any evidence to contradict those opinions, was bound to accept them.

That just is a complete and total fallacy. The inspector (who is a man of experience, and, above all, specialised qualifications, who is sent to assess a problem of this kind) is supposed to use his own knowledge and, if I may say so, commonsense as well. He is intended to use his commonsense and he is not bound to accept the evidence of experts. It is exactly the same
C situation that justices and juries find themselves in when experts of great distinction go into the witness box before them. The inspector is no more bound to accept the evidence of the experts than are they.

Accordingly, the plea that there was no evidence to support the conclusions of the inspector is a plea which is totally inapt for the circumstances. He was entitled to reach his own conclusion and unless it could be shown that he had taken into account something which he should have ignored, or ignored something which he should have taken into account, his decision must stand.

I would dismiss the appeal.

D O'CONNOR and WATKINS JJ agreed and the appeal was dismissed with costs.

QUEEN'S BENCH DIVISION

December 13 1977

(Before Sir DOUGLAS FRANK QC sitting as a deputy judge of the division)

NIARCHOS (LONDON) LTD v SECRETARY OF STATE FOR THE ENVIRONMENT AND WESTMINSTER CITY COUNCIL

Estates Gazette March 11 1978
E (1977) 245 EG 847

Appeal against decision of Secretary of State for the Environment upholding refusal of planning consent for use of 37 and 41-43 Park Street, Mayfair, as offices—Minister's decision quashed—Failure to take into account cost of converting premises for residential use

These were two applications by Niarchos (London) Ltd seeking orders quashing a decision of the Secretary of State for the Environment dismissing appeals against the refusal of planning permission to continue office use of the company's premises at Park Street, Mayfair.

F David Widdicombe QC and Matthew Horton (instructed by Coward Chance) appeared on behalf of the applicants; Harry Woolf (instructed by the Treasury Solicitor) represented the Secretary of State.

Giving judgment, SIR DOUGLAS FRANK said that the premises were made up of a basement, ground floor and five other floors with a mezzanine floor. The applicants did not occupy the three upper floors which were sublet. Temporary planning consents were granted for office use after the last war. They had been renewed in the past, but when they came up for renewal in December 1973 consent was refused.

There was an appeal to the minister and after an inquiry in 1975 the inspector reported that schemes produced by Westminster Council and Niarchos (London) Ltd for conversion back to residential use were not economically viable. The inspector, noting that the premises were in an area allocated primarily for residential use, concluded that a sufficiently strong case had not been made out for renewal of office use, but that the alternative, in the prevailing economic circumstances, if the premises were not used as offices, was for them to stand empty. In his opinion, therefore, it could not be said that the buildings could be reasonably adapted for residential use at that time (1975). He recommended that a temporary permission for five years should be given to allow time for the economic situation to improve and for Niarchos (London) Ltd to find alternative accommodation. G H

However, the minister rejected the inspector's recommendation and took the view that financial consideration which made conversion unprofitable were not a factor of such importance as to persuade him to permit an exception to the council's policy of not renewing temporary permissions for office use. He dismissed the Niarchos appeal.

Niarchos now claimed that the minister had failed to apply the tests in the development plan. They complained that he had asked himself the wrong question, namely whether the financial considerations which made conversion unprofitable were of such importance as to warrant making an exception to the policy for the area. His Lordship agreed. The policy was contained in the development plan and the question that the minister should have asked himself was whether or not the premises could reasonably be adapted for use as residences. J

It was argued by Mr Woolf that it was entirely a matter for the minister how he interpreted the development plan, and further, that the minister was entitled to change his policy from day to day. It was trite law, in his Lordship's view, that the minister could change his policy and, accordingly, he was not bound to apply the policy in the development plan. However, when he expressed himself as deciding a case in accordance with a stated policy it must follow that if he decided otherwise than in accordance with that policy he misdirected himself and acted in excess of his powers. K

His Lordship also rejected the submission that the minister was entitled to put his own interpretation on the provisions of the development plan.

The provision that premises should not be capable of being reasonably adapted was not confined to physical practicability. The minister was bound to take into account the financial implications of adapting the premises and the inevitable conclusion was that the premises could not reasonably be adapted for residential occupation. L

The present case was the first town planning decision which had been quashed for being unreasonable, but his Lordship believed he was doing no more than applying an established principle to particular facts.

The applications were allowed with costs.

QUEEN'S BENCH DIVISIONAL COURT M

November 28 1977

(Before Lord WIDGERY CJ, Mr Justice PARK and Mr Justice MAY)

ST ALBANS DISTRICT COUNCIL v NORMAN HARPER AUTOSALES LTD

Estates Gazette March 25 1978
(1977) 245 EG 1029

Town and Country Planning Act 1971—Appeal by planning authority against dismissal by magistrates of information alleging failure to comply with enforcement notice—Building 2½ ft too high—Question whether information laid out of time under

Magistrates' Courts Act 1952—Offence a once-and-for-all, not continuing, offence—Authority's power to extend time for compliance not to be exercised after expiry of time-limit to detriment of owners—Justices acted correctly in dismissing information

This was an appeal by St Albans District Council, by case stated, against the dismissal by magistrates at St Albans of an information laid by the council complaining of failure by Norman Harper Autosales Ltd to comply with an enforcement notice. The breach of planning control alleged in the notice was the erection of a building above the maximum height allowed without special planning permission.

Alistair Dawson QC and John Furber (instructed by J D R Jeffery, Solicitor, St Albans District Council) appeared on behalf of the appellants; Dennis Naish (instructed by Murgatroyds, of St Albans) represented the respondents.

Giving judgment, LORD WIDGERY CJ said: This is an appeal by case stated by justices for the County of Hertford sitting as a magistrates' court at St Albans, who, on July 13 1976, had before them an information laid by the present appellants, the St Albans District Council, against the present respondents, Norman Harper Autosales Ltd, alleging that the respondents, having been served with an enforcement notice under the Town and Country Planning Act 1971, had failed to take the step required by the notice within the period allowed for compliance therewith, namely, by January 26 1976. It was said that that was contrary to section 89 of the Town and Country Planning Act 1971.

The facts and history of this matter are straightforward enough. The respondents as owners of the property in question erected a building at the property, 318 Watford Road, St Albans, and the aspect of the size of the building which is relevant and material is that it was 12 ft high. Under the regulations then current with regard to the height of such buildings in that area the maximum height permissible without special planning permission would have been 9 ft 6 in, and so it was $2\frac{1}{2}$ ft too high. This fact attracted the attention of the planning authority, and they, as they were perfectly entitled to do, served an enforcement notice under section 87 of the Act with a view to having the breach of planning control remedied.

The relevant passage is subsection (1) of section 89 of the Town and Country Planning Act 1971, which provides:

> Subject to the provisions of this section, where an enforcement notice has been served on the person who, at the time when the notice was served on him, was the owner of the land to which it relates, then, if any steps required by the notice to be taken (other than the discontinuance of a use of land) have not been taken within the period allowed for compliance with the notice, that person shall be liable on summary conviction to a fine not exceeding £400 or on conviction on indictment to a fine.

There one has the development alleged to infringe control, the serving of the enforcement notice and non-compliance by the respondents with the notice. But, as they were perfectly entitled to, they took the matter by way of appeal to the Secretary of State, and the Secretary of State found against them. He confirmed the enforcement notice, but he prescribed a new date for compliance with the notice because much, if not all, of the compliance time previously allotted had been taken up in the machinery of appeal. Fixing that new date, the Secretary of State chose September 29 1973, and so the enforcement notice required the removal of these buildings by that ultimate date. It did not happen like that, and the local authority in due course brought proceedings in the magistrates' court alleging a breach of section 89 (1), to which I have referred.

The case came on for hearing on May 13 1975 and the information was dismissed. The reason seems to have been that the information was laid out of time because reliance was placed on section 104 of the Magistrates' Courts Act 1952 in these terms:

> Except as otherwise expressly provided by any enactment, a magistrates' court shall not try an information or hear a complaint unless the information was laid, or the complaint made, within six months from the time when the offence was committed, or the matter of complaint arose.

What seems to have happened in May of 1975 was that, their attention having been drawn to that provision, the justices thought that the matter of complaint had arisen more than six months before the information, and they dismissed the proceedings. Four days before this hearing the respondents had put in an application for planning permission, the purpose of which was to modify these buildings, among other things, and no doubt to comply with the law. But I do not find it necessary to go into that in detail, except to mention it merely as a part of the background. We have reached the stage where the justices have dismissed the information in May of 1975, and the planning authority, in order to cure the difficulty produced by their being said to be out of time, laid down a new period for compliance with the enforcement notice. They said that such notice should be required to be complied with within three months of June 29 1975, or, to look at it the other way, by September 30 1975. That was a purely unilateral action (we have been shown the letter, but I do not find it necessary to look at it again) and its purpose was obviously to extend the time during which the respondents could carry out the stage required by the aforesaid notice.

The authority for this action on the part of the local authority is said to be section 89 (6), which provides:

> Any reference in this section to the period allowed for compliance with an enforcement notice is a reference to the period specified in the notice for compliance therewith or such extended period as the local planning authority may allow for compliance with the notice.

No doubt most people reading that subsection would regard it as a power for the local authority to extend a concession to the landowner if they thought fit. In this instance the attempted purported exercise of that power, on the face of it, was not for the benefit of the landowner at all but for the benefit of the authority, as we shall shortly see. The extension so made was renewed again, so that finally the expiry of the period for compliance with the enforcement notice had been advanced to January 26 1976. Within six months of that date, though only just within six months of that date, the planning authority brought their second proceedings alleging breach of the enforcement notice, and it was of course from that information and its related hearing that the matter comes before this court.

The way the argument has gone is thus. Mr Dawson for the local authority (the appellants) takes us to the justices' reasons in the case, and they are found to be these. First of all, the justices took the view, contrary to the submissions of the local authority, that the offence created by section 89 (1) was not a continuing offence. Furthermore, the justices held that the extension of the period for compliance could not be done in a unilateral way without regard to the interests or concern of the landowner. Putting it in their own language, they said that "any power to extend the period for compliance with an enforcement notice must be exercised within a reasonable time and in particular before the expiration of the six months' time limit provided by section 104 of the Magistrates' Courts Act 1952."

It is common ground that the appellants did not exercise the power within that time limit. Against that background, the argument goes thus. Mr Dawson says that this is a continuing offence. If it is a continuing offence, then he has no trouble because such an offence is renewing itself, as it were, every day, and there are plenty of instances within the last six months which may be relied upon as being the basis for the prosecution. He says that it is a continuing offence and that that is enough to get him home. The argument on the other side is that the justices are right and this is not a continuing offence. Furthermore, it is argued on behalf of the respondents that the justices were also right in substance, if not in detailed reasoning, when they held that the period for compliance could not be extended at the whim of the local authority and without regard to the interests and needs of the landowner.

We must deal with those two points separately. Counsel, with great frankness, has said that it is not easy to direct our attention to any principle or authority which shows beyond any doubt whether we are dealing here with a continuing offence or with a single offence. In the end, everybody seems to agree that it is a matter of construction of the statute, and very largely a matter of impression at that. So I go back to look at section 89 with that in view. Subsection (1) seems to me to have all the signs of a once-and-for-all offence. It is not without importance that under section 91 of this Act the local authority, on the default of the owner, can go in and do the work, and I think that the draftsman of section 89 (1) had in mind that this was a once-and-for-all offence. The authority's sanction to get rid of the offending building was section 91, and not some remedy whereby the offence was kept alive for that purpose.

Suffice it to say that, in my judgment, when one reads section 89 (1) it has the smack of a single offence about it. Compare subsection (5) of the same section, which is dealing with unauthorised development which has taken the form of a change of use, and which provides:

> Where, by virtue of an enforcement notice, a use of land is required to be discontinued, or any conditions or limitations are required to be complied with in respect of a use of land or in respect of the carrying out of operations thereon, then if any person uses the land or causes or permits it to be used, or carries out those operations or causes or permits them to be carried out, in contravention of the notice, he shall be guilty of an offence, and shall be liable on summary conviction to a fine not exceeding £400, or on conviction on indictment to a fine; and if the use is continued after the conviction he shall be guilty of a further offence and liable on summary conviction to a fine not exceeding £50 for each day on which the use is so continued, or on conviction on indictment to a fine.

I say nothing about the first half of the section because it is not necessary for us today to decide whether the initial offence under subsection (5) is a continuing offence or not. But what is perfectly clear, in my judgment, is that the offence constituted by the second half of subsection (5), which can only arise when there has already been a successful prosecution on the first half, is a continuing offence. I say that not for the first time, because in the case of *R* v *Chertsey Justices, ex parte Franks* [1961] 2 QB 152 this court had to decide on the then current legislation the substantial question of whether what is now section 89 (5) created a continuing offence in its second half. Giving the judgment of the court, I gave the opinion of the court as being that a continuing offence was created. Now one asks oneself what is the justification for that decision, and the answer is that the language is clearly language, in my judgment, which points to a continuing offence. There is the daily penalty. All these matters are relevant and powerful in directing one's attention in that manner.

In default of any other authority, and in a situation in which I think we have to use our commonsense and our best ability and produce an answer consistent with the language of the statute, I have come to the conclusion that the offence charged under section 89 (1) is *not* a continuing offence, as the justices themselves had found.

One still has to deal with the adjustment of the date for compliance, because there is the argument that, even if this is a once-and-for-all offence, that offence is committed on the first day after the expiry of the time prescribed by the notice for compliance with the requirements of the notice. That means, of course, that in the instant case one begins with the last day as being September 29 1975—the date fixed by the Secretary of State as already described. That period has long since expired, but it is said that the effect of it has been continued or resuscitated by the unilateral directions to which I have already referred whereby the planning authority purported to extend the time for compliance. Ultimately that period, it will be remembered, was extended to January 26 1976, and the argument is that the planning authority can bring these proceedings, even on the basis of a one-offence section, as long as the proceedings are taken within six months of January 26, as they were.

I do not agree with that either. It seems to me that it cannot possibly have been the intention of Parliament that this simple little power, so obviously designed in the first instance to enable the local authority to help a landowner, should be capable of being used to the landowner's detriment without any special provision, and without any sort of application by the landowner to bring the matter before the authority or the court. I think that the power of the local authority to extend time under section 89 (6) is a power which must be exercised before the time had expired. In other words, if the initial time, or any lawfully extended time, has expired, it then becomes too late for the authority to extend time again.

For those reasons I think the justices were correct in their conclusion and right in their reasoning, and I would dismiss the appeal.

Park J agreed.

May J also agreeing, said: I would merely add that, in my judgment, the single-offence nature of subsection (1) of section 89 is made even clearer by the presence of subsection (4), which, as I think, does for subsection (1) the same as the second half of subsection (5) does for the first part of subsection (5).

The appeal was dismissed with costs.

QUEEN'S BENCH DIVISION

December 9 1977

(Before Sir Douglas FRANK QC sitting as a deputy judge of the division)

THANET DISTRICT COUNCIL v SECRETARY OF STATE FOR THE ENVIRONMENT AND ANOTHER

Estates Gazette April 15 1978

(1977) 246 EG 229

Town and Country Planning Act 1971—Listed building consent—Refusal by planning authority upheld by court and decision of Secretary of State quashed—Old farmhouse with 14th- and 15th-century works owned by Oxford college—Secretary of State differed from inspector on a finding of fact without giving parties the opportunity to make representations under the rules governing procedure at local inquiries—Finding as to value of building if restored—unintelligibility of Secretary of State's reasons not raised in appeal—"Temporary comfort" only for appellants

This was an appeal against the decision of the Secretary of State by Thanet District Council. The Secretary of State had allowed an appeal by Corpus Christi College, Oxford, from the council's refusal to consent to the demolition of a listed buiding owned by the college, namely, the Old Farmhouse, Netherhale Farm, near Birchington, Kent, which contained works dating from the 14th and 15th centuries. The college were the second respondents to the present appeal.

C Cochrane (instructed by Sharpe, Pritchard & Co), appeared on behalf of the appellant council; Harry Woolf (instructed by the Treasury Solicitor) represented the first respondent, the Secretary of State for the Environment; J M J Burford (instructed by Howell, Peel & Gamlen, of Oxford) represented the second respondents, Corpus Christi College.

Giving judgment, Sir Douglas Frank QC said: This is an appeal under section 245 of the Town and Country Planning Act 1971 whereby the appellant asks for an order quashing a decision by the first respondent allowing an appeal against their refusal to give listed building consent to the demolition of a listed building on land at Netherhale Farm, near Birchington, Kent.

The building concerned is largely composed of works of the 14th and 15th centuries but had been unoccupied for 18 years, although the second respondents have tried to find a purchaser for it. It has been owned by the second respondents for about two centuries, who, after considering the various alternatives, reluctantly decided that the building should be demolished in such a way that the parts of special interest could be dismantled and removed to a suitable site.

Consent having been refused, the second respondents appealed to the Secretary of State, whose inspector held an inquiry on October 21 1975. The main issue at that inquiry was whether it would be practical to restore the building for use as a dwelling. The second respondents gave evidence that the cost would be of the order of £100,000 and that the value as restored would be not more than £60,000, and that the sale of the house would result in interference with the farm on which it was situated and would therefore reduce the value of the farm. On the other hand the appellant's witness estimated that the cost of restoration would be about £25,000 plus VAT, although he could not name a sum which he could guarantee would not be exceeded and at the end of the work " the result would not be a modern house but one that would attract special clients."

The inspector, in his findings of fact, having described the defects in the building, said: " There is a conflict of evidence between the appellants and the council about the necessary repairs and the likely costs of works necessary to bring the building into a habitable condition. I prefer the evidence of Mr A W Swaine, the honorary investigator, that the defects are mainly superficial." He also said that " the separate ownership of the old house would have a negligible impact on the working of the farm." In his conclusions he said that he was of the opinion that every effort should be made to preserve the building, and then said: " The building is now in a sad state of disrepair but as the honorary investigator states that the defects are mainly superficial then they can be easily remedied. The defects have been exaggerated by the appellants and the cost of necessary repairs to render the building habitable is likely to be much less than estimated by them. The value of the restored building would largely exceed the cost of the repairs."

The letter conveying the decision of the Secretary of State set out the inspector's conclusions and recommendation and then stated:

> The Secretary of State has given very careful consideration to the inspector's conclusions, but in his view the present state of disrepair of the building and its location in the surroundings of a farmyard make it unlikely that, if restored, it would be disposed of and beneficially occupied for any worthwhile period. In these circumstances he thinks the best course is to grant permission for the demolition of the building so that any parts of special interest can be dismantled and suitably preserved. Therefore the Secretary of State hereby allows the appeal and grants listed building consent for the demolition of the Old Farmhouse, Netherhale Farm, nr Birchington.

In their notice of motion the appellants set out one ground for challenging that decision, namely that the first respondent relied upon findings of fact which were different from those made by the inspector, without notifying the appellants. From the affidavit sworn in support of the application, and in the arguments advanced by Mr Cochrane for the appellants, the appellants' case is that the Secretary of State differed from the inspector's finding of fact that the defects were mainly superficial.

I am bound to say that I find the first respondent's reasons difficult to construe and I think the way in which he has expressed them is open to criticism both on the ground of brevity and lack of clarity. However, the decision is not challenged on these grounds, so that I do not have to consider the application of the principle enunciated in *Givaudan & Co Ltd v The Minister of Housing and Local Government* [1967] 1 WLR 250.

Both Mr Woolf and Mr Burford, appearing respectively for the first and second respondents, were prepared to accept that the inspector's so-called conclusions, which I have quoted, were in truth findings of fact, but, argued Mr Woolf, there was no reason to assume that the first respondent was not accepting the inspector's findings of fact and it was not competent for me to infer otherwise. I agree that I must not infer that the first respondent was differing from the inspector, unless the inference is unavoidable, and I have come to the conclusion, doing the best I can with reasons which I do not really understand, that the inference is unavoidable. The inspector's finding that the value of the restored building would largely exceed the cost of repairs inherently advanced two findings of fact, namely (a) the value of the building as restored, and (b) the cost of repairs. If, therefore, the first respondent found that it was unlikely to be disposed of, then by necessary inference he is saying the restored building would have no value, and that, in my judgment, would be differing from his inspector upon a finding of fact. I appreciate, as Mr Woolf contended, that the words " disposed of " are apparently used conjunctively with " and beneficially occupied for any worthwhile period," but I find that addition meaningless in the context. Moreover, the reference to the present state of disrepair seems to me to be irrelevant having regard to the fact that the first respondent was apparently resting his decision on what would happen if the buildings were restored. Indeed Mr Burford in his argument for the second respondents apparently was forced to rely on the evidence given for his clients. All in all I find it irresistible to conclude that there was a difference in finding of a fact or facts and, accordingly, I quash the decision.

However, I may say that this appeal should have been brought on the ground that the reasons are unintelligible and therefore are not reasons at all. How can the present state of disrepair have any bearing on the question of whether the premises can be disposed of? Why would they not be occupied for any worthwhile period? And where is the evidence in support of that assumption?

I should add, perhaps, that the council may take only temporary comfort from this judgment, for it seems to me that the inspector's conclusions fly in the face of the evidence.

No order was made as to costs.

COURT OF APPEAL

January 26 1977

(Before Lord DENNING MR, Lord Justice BROWNE and Sir John PENNYCUICK)

PERFORMANCE CARS LTD v SECRETARY OF STATE FOR THE ENVIRONMENT

Estates Gazette May 13 1978

(1977) 246 EG 563

Town and country planning—The applicant company, having been refused planning permission by planning authority to use land as a car park, requested copies of documents, such as letters and plans, for the purpose of the public inquiry following an appeal to Secretary of State—Authority refused to provide documents, but, possibly on counsel's advice, handed a bundle of them to the company's managing director at opening of inquiry—Request for adjournment to study documents rejected except for an extension of luncheon break—Inspector's refusal of substantial adjournment a failure to comply with rule 10 (5) of Town and Country Planning (Inquiries Procedure) Rules 1974 and also a breach of rules of natural justice—Secretary of State's decision dismissing appeal quashed.

In this case the applicant company, Performance Cars Ltd, applied under section 245 of the Town and Country Planning Act 1971 to quash the decision of the Secretary of State upholding the refusal of planning permission for the use of land near Layton Road, Brentford, as a car park. The application was

A dismissed on January 23 1976 by Douglas Frank QC sitting as an additional judge of the Queen's Bench Division ((1976) 240 EG 51), whereupon the applicant company appealed to the Court of Appeal. The relevant facts are fully stated in the judgment of Lord Denning MR.

N Primost (instructed by Loxdales) appeared on behalf of the appellants; Harry Woolf (instructed by the Treasury Solicitor) represented the respondent.

Giving judgment, Lord Denning MR said: Not far from the Great West Road in Brentford there was in the old days a set of railway sidings. They ran parallel to a road called Layton Road.
B They covered a site of two acres. In 1970 they ceased to be used: afterwards the two acres of land were bought by the council, the London Borough of Hounslow. They let it to a firm called Performance Cars Ltd. It was on a lease which was determinable by six months' notice. The terms of the lease were that Performance Cars were to be able to use it as a car park, storage of contractors' plant, for machinery and parking vehicles. For a time Performance Cars used the land contrary to the lease for dismantling cars, but, on complaints being made, they stopped so using it. In February 1972 they made an application for planning permission to park cars and lorries there, but the local authority refused on the ground that "the general activities,
C including noise, associated with the present use of the site, are considered detrimental to the amenities of nearby residential properties." The local council sent round a circular to the people who lived near this site. These lodged a petition against the use of it as a lorry park. It was signed by 77 of the local residents, mostly people from Layton Road. It asked for "the removal of the lorry park which is causing loss of sleep and is generally disturbing the peace at all times of the day and night."

The planning request was refused. There was an appeal to the minister. He sent an inspector down to hold an inquiry. Much of the neighbouring property is used as an industrial area with factories and builders' yards. But Layton Road is an exception.
D It is a residential road with terraced houses built about 18 years ago.

In order to prepare his case for the inquiry, the managing director of Performance Cars, acting in person, wrote to the local council and he sent a copy of his letter to the Department of Environment on July 26 1974. It said: "With reference to the appeal hearing on September 10 1974 . . . we would be obliged if you expedited the delivery of all documents and statements" wanted to prepare the case. He asked for: "(1) Photostat copies of all letters and details of an alleged petition from 77 people of Layton Road as mentioned in the 'Evening
E Mail' on Monday October 9 1972, and (2) copies of applications, plans and approvals, letters and circulars, including objections, in respect of a petrol station now undergoing completion in Layton Road," and other documents. He did not get a reply at that time. On August 21 1974 he wrote again.

Eventually on August 22 1974 the borough council by their solicitor's department replied: "The planning officer has sent me copies of your letters. In my opinion the council is not required to supply you with the information and documents you requested and accordingly I regret that I am unable to accede to your wishes." So there it was. On August 22 1974 the solicitor's department of the borough refused him the documents he wanted
F in order to prepare his case. The judge said: "I feel that the council's failure to respond to that request is to be deplored."

The inquiry came on on September 10 1974. This time the borough council were represented by counsel. The company was represented by Mr Goldsmith in person. I expect that counsel told the council that they were wrong in refusing to let Mr Goldsmith see the documents. The position was governed by Rule 6, sub-rule (4) of the Town and Country Planning (Inquiries Procedure) Rules 1974 which govern these inquiries and which says:

> Where the local planning authority intend to refer to, or put in evidence, at the inquiry documents (including maps and plans), the authority's statement shall be accompanied by a list of such documents, G
> together with a notice stating the times and place at which the documents may be inspected by the applicant and the section 29 parties; and the local planning authority shall afford them a reasonable opportunity to inspect and, where practicable, to take copies of the documents.

It is quite plain that if local authorities intend to put documents in evidence, they should let the other party see them beforehand, and the council did not fulfil the terms of that rule. When the case was coming on for hearing, counsel probably said: "You had better let Mr Goldsmith see them."

So at the beginning of the inquiry on September 10 1974 Mr Goldsmith was handed this bundle of documents. Whereupon he said: "I would like an adjournment to deal with this." He H
said that these documents had not been provided; he had not had time to study them in detail; it was not possible for him to prepare his case; and he wanted 30 days' adjournment. The inspector was in a very difficult position. He and everybody else had given a day up for the case. He said: "You may have an extended lunch hour and you can study them during the lunch hour." There was an adjournment of an hour and 40 minutes over the lunch break. During that time Mr Goldsmith no doubt read the documents. In the afternoon Mr Goldsmith made some quite sensible comments on those letters. He did the best he could. He did not make any further application for an adjournment because he thought that that had been ruled upon. So it J
had.

After the inquiry was concluded the inspector had a view. He made his report on September 30. In his findings of fact the inspector said "there is an urgent need for a lorry park in the Brentford area and none is at present available". He also said: "Present usage involves about 8 to 10 movements of lorries to or from the appeal site in 24 hours." In his conclusions the inspector said: "I consider that the continued use of the appeal site as a lorry park on the present scale would be detrimental to the well-being of the occupants of the dwellings in Layton Road, particularly in the eastern half. An intensification of this use would be likely to follow from the granting of planning permission and K
this would be intolerable for local residents. I consider that these objections outweigh the need, urgent and important as it is, for a lorry park in this area." So there it was. The inspector thought, if permission were granted, the noise would become much worse and would become intolerable.

The minister upheld that decision. Now the company apply to this court to set the decision aside. Mr Goldsmith appeals under section 245 of the Town and Country Planning Act 1971 on the ground that his interests have been substantially prejudiced by a failure to comply with any of the relevant requirements in relation to that Act. Alternatively, he appeals on the L
ground that there has been failure of natural justice.

Under the Town and Country Planning (Inquiries Procedure) Rules, rule 10, sub-rule (5), says that the inspector may allow documents in evidence if they have not been disclosed to the applicant previously, but he shall if necessary during the inquiry give the applicant or the local planning authority, as the case may be, "an adequate opportunity of considering any such fresh submission or document."

The whole question really comes down to this: did the inspector give Mr Goldsmith an adequate opportunity of considering these new documents? I think the inspector was put in a very difficult position. The local authority had, however, made a bad mistake M
in not disclosing the documents previously as they should have done. In view of the mistake I think that the inspector should have given Mr Goldsmith more opportunity of considering them and dealing with them. I can understand Mr Goldsmith saying: "If I had been able to have these documents before, I could have dealt with them. I asked for them by letter both to the council and to the ministry, but I did not have them." In those circumstances he would go away with a sense of injustice. He had asked for those documents expressly a long time before the hearing. He was not given them. He then had them given to him at the last moment. It seems to me in those circumstances that he would

A have a grievance. He would not feel that justice had been done. It is that feeling, I think, which in all our proceedings we should try to avoid. People should not go away from any inquiry feeling "I've not had a fair deal." It is for this reason that I feel we must, although one regrets it, let the order go and set the matter aside.

This point is similar to the point Browne LJ had in the case of *Hibernian Property Co Ltd v Secretary of State for the Environment* (1973) 27 P & CR 197. He said: "When the inspector gave way to this temptation I have no doubt at all that she was acting from the highest motives . . . I am not satisfied that in fact the result would have been different if the questions had not been asked. In view, however, of the authorities to which I have B referred and of the vital importance of ensuring the strictest observance of the rules"—he set aside the decision.

In this case I feel the same way as Browne L J did in that case. I think that we ought to allow this appeal and set aside the order.

Agreeing, Browne L J said: It is now accepted by Mr Woolf, and it is clearly right, that the local planning authority was wrong in its refusal to produce these documents at an earlier stage. It was, in my judgment, clearly a breach of rule 6, sub-rule (4) of the Town and Country Planning (Inquiries Procedure) Rules 1974. It may well also have been a denial of natural justice. But I agree with Mr Woolf in the construction which he puts on C section 245 (4) (b) of the Town and Country Planning Act 1971. which says: " . . . if satisfied that the order or action in question is not within the powers of this Act, or that the interests of the applicant have been substantially prejudiced by a failure to comply with any of the relevant requirements in relation thereto, may quash that order or action." The present case relates to an "action" as defined by section 242 (3) (b), that is ". . . any decision of the Secretary of State on an appeal under section 36 of this Act." I agree with Mr Woolf that the "action in question" referred to in section 245 (4) (b) is the decision on the appeal. Equally, I agree that the word " thereto" refers back to the decision on the appeal. Accordingly, there is apparently no D direct sanction, at any rate under that subsection, for a previous failure by a local planning authority to comply with the rules.

But there is a sanction in fact because the position seems to me to be this. The council, having wrongly refused to produce these documents, the matter then comes before the inspector and is covered by rule 10 of the rules. The question here is how the inspector dealt with the situation with which he was presented on September 10 when the inquiry opened, by reason of the previous failure of the local planning authority to do what they should have done. If the inspector unfortunately got it wrong, the Secretary of State of course is responsible, because the inspector is simply representing him.

E The case therefore, I think, comes down to the question whether, in view of the local planning authority's failure to produce the documents until the morning of the hearing, the inspector's refusal to adjourn the case was either a denial of natural justice or a breach of rule 10, sub-rule (5), or both. I am myself very doubtful indeed whether in fact the letters and the petition made any difference to the recommendation of the inspector or the decision of the Secretary of the State. I am also very doubtful whether, even if the applicant had had an adjournment, he could have got any useful rebutting evidence. But, when the letters and the petition were produced at the beginning F of the hearing, it should, I think, have been plain to the inspector that they *might* be relevant. It seems to me also that it should have been clear to the inspector that the applicant might be able to get some rebutting evidence.

It is not clear on the information before us whether the inspector read the letters and the petition before making his decision only to grant an adjournment by giving a longer luncheon hour. If he did read the letters, he should have realised that they might be relevant and that the applicant might be able to get rebutting evidence and that it would only be fair to give him the chance of doing so. If he did not read them before giving the decision, his decision was clearly wrong. I agree that G the inspector was put in an extremely difficult position, and I think my Lord has already made it clear that we make absolutely no criticism from a moral point of view of the conduct of the inspector. I am quite satisfied that he did what he thought was right in all the circumstances. But, in my judgment, that being the position, there was a breach of natural justice by reason of the decision to refuse any substantial adjournment and to go on with the hearing before the applicant had had any substantial chance of considering the documents. It also follows, I think, that the inspector did not give the adequate opportunity for consideration which he is required to do under rule 10, sub-rule (5) of the Inquiries Procedure Rules.

H As I have said, I have strong doubts whether the refusal of the adjournment really made any difference; but, having decided that there was in my judgment a breach of natural justice, I do not think my doubts matter. I hesitate to refer to a previous judgment of my own—I am not going to read it—but in the *Hibernian Property Co* case (1973) 27 P & CR 197 I dealt with this matter at pp 212 to 213, and said that I thought the question was not whether the evidence which had been, as I held, obtained contrary to the requirements of natural justice had in fact prejudiced the applicant's case, but whether there was a *risk* that it had prejudiced the applicant's case. I quoted the decision of the Judicial Committee of the Privy Council in which the opinion J was delivered by my Lord in the case of *Kanda v Government of Malaya* (1962) AC 322 and I went on to say that it was obviously impossible for the court to investigate what weight the various relevant factors had had in inducing the Secretary of State to make the decision he did, but that there was clearly a risk that the evidence which I held to be wrongly obtained might have influenced his decision.

In view of my doubts to which I have already referred, I am reluctant to quash this decision, but my Lord has already quoted what I said at the end of the *Hibernian* case, and I will not repeat it. It seems to me, for the reasons I there gave, that the K importance of maintaining the observance of the rules which the courts have laid down in the interests of natural justice outweighs any other consideration, and I agree that we should quash this decision.

Also agreeing, Sir John Pennycuick said: The local authority was admittedly at fault in not giving Mr Goldsmith an inspection of the letters before the opening of the inquiry. Once this is accepted, it seems to me to follow that the inspector ought to have allowed an immediate adjournment at least sufficient to enable Mr Goldsmith to read the letters and consider his position. I do not think that it was right for him merely to extend the luncheon adjournment by 40 minutes in order that Mr Goldsmith L might consider the letters then and, meanwhile, direct him to open his case and call his evidence before considering the letters. I think this was a breach of natural justice which invalidates the recommendation of the inspector and the decision of the Secretary of State based upon that recommendation.

A parallel position would arise in a civil action if the defendant failed to produce at the opening of the trial particulars which he had been bound to produce at an earlier stage and if the plaintiff then applied for an adjournment.

It seems to me that Mr Goldsmith might have renewed his application for an adjournment—that is to say, an adjournment M of some days at least—after considering the letters. No doubt it was difficult for him, especially as he appeared in person, to do that in the existing circumstances. The fact remains that he did not make any further application for an adjournment and, for myself, I rather doubt whether there could be said to have been a failure of natural justice by reason of the omission of the inspector to grant a further adjournment on his own initiative.

The initial and serious fault in these proceedings was of course that of the local authority, and I have every sympathy with the inspector who had to make this difficult procedural decision and was in no other respect at fault. I think, nonetheless, that the inspector was at fault in the course which he took over the adjournment. The rules of natural justice, as my Lords have pointed out, must be strictly complied with, and I think it would be right to quash this decision.

The application was granted and the decision of the Secretary of State quashed. The appellant company was awarded costs before the Court of Appeal and below.

QUEEN'S BENCH DIVISION

March 10 1978

(Before Sir Douglas FRANK QC sitting as a deputy judge of the division)

RKT INVESTMENTS LTD v HACKNEY LONDON BOROUGH

Estates Gazette June 10 1978

(1978) 246 EG 919

Planning permission—Alleged imposition of ultra vires condition—Permission granted for office use incorporating a condition in an office development permit restricting the office use to that of a specified public body—Statement in the permission document indicating that if the office use of another building owned by plaintiffs was abandoned in favour of residential use the above condition might be removed—Whether condition or whole permission vitiated by ulterior object—Admissible evidence—Ulterior object not established—Plaintiffs' action for declaration fails

This was an action by RKT Investments Ltd for a declaration that the incorporation of a condition in a planning permission restricting office use to that of the Inner London Executive Council of the National Health Service was *ultra vires*. The permission in question related to a five-storey building, of which the plaintiffs were the leasehold owners, at 32-43 Chart Street, London N1, built in 1966 for warehousing, but never used as such.

D M W Barnes (instructed by Biddle & Co) appeared on behalf of the plaintiff; A R Vandermeer (instructed by H F W Wilson, Greater London Council Solicitors' Department) represented the defendants.

Giving judgment, Sir Douglas Frank QC said that on March 25 1971 the Secretary of State for the Environment issued under Part I of the Control of Office and Industrial Development Act 1965 to the Inner London Executive Council of the National Health Service an office development permit (ODP) permitting the office use of the premises. The permit was expressed to be subject to a condition that "the office premises created by this development shall be used by the Inner London Executive Council of the National Health Service only." On April 23 1971 the defendants issued a planning permission permitting the use of the premises as offices subject to a number of conditions, including that contained in the ODP the reason for which was stated to be "as required as a condition of the ODP pursuant to Part I of the Control of Office and Industrial Development Act 1965." There then followed these words:

> The local planning authority certify that the condition no 4 [that is the condition quoted] is a condition subject to which this planning permission is granted in accordance with section 6 of the Control of Office and Industrial Development Act 1965, which the local planning authority could have imposed, apart from the provisions of Part I of the said Act of 1965 and sections 85 and 86 of the Town and Country Planning Act 1968, and would have imposed if that part of the Act of 1965 and those sections of the Act of 1968 had not been enacted.
>
> I would inform you that:
>
> (1) When considering the application, the Greater London Council understood that before Insurance House, Insurance Street, Islington, is offered for reletting for offices, the leaseholders will explore with the freeholders the possibility of redeveloping the site for residential purposes in conformity with the zoning; and
>
> (2) Should a scheme for residential development of Insurance House be carried out then the Greater London Council would consider an application to modify any permission, granted by the Borough Council, in respect of the condition now certified under the Control of Office and Industrial Development Act 1965.

It was common ground that the defendants were bound to impose the condition concerned, but it was said by the plaintiffs that the defendants could not otherwise lawfully have imposed the condition. The practical consequence was this: conditions imposed solely as required as a condition of the ODP have a limited life (see section 18 of the Act of 1965), whereas the effect of the certification is to give the condition the status it would have had if it had been imposed independently of the ODP.

Before moving to Chart Street the Executive Council had two sets of offices, one being Insurance House which they held under a lease having about 40 years to run. Thus the above-quoted information was to the effect that if the office use of Insurance House were abandoned in favour of residential use, then the user condition in the Chart Street permission might be removed.

Mr Barnes, for the plaintiffs, submitted that the defendants, in deciding to impose the certification, took into account as a consideration and as a relevant factor the desirability of persuading and inducing the appellants to give up their existing office-use rights at Insurance House and to develop that property residentially and that was a matter which could not properly be taken into account and, alternatively, the certification did not fairly relate to the development permitted. Mr Barnes said that this raised three issues, namely:

(1) was the desirability of persuading the giving up of the residential use of Insurance House a considera- in the certification?;

(2) if so, was this certification *ultra vires*?; and

(3) if so, is the certification severed so as to leave the remainder of the planning permission intact?

His Lordship said that before he dealt with these issues he had to determine whether documents other than the planning permission itself were admissible in evidence and whether oral evidence could be given. In *Slough Estates Ltd v Slough Borough Council (No 2)* [1971] AC 958 that question was considered and the consensus seems to have been as expressed by Lord Pearson at p 967, namely, that where a planning permission is on its face a complete and self-contained document extrinsic evidence should not be taken into account in considering it save where other documents are incorporated by reference.

Here, however, both counsel seem to be agreed that the *Slough* case was distinguishable in that it was concerned with the construction of the permission, whereas in the instant case the question was what was the motive for imposing what was in effect a condition. Mr Barnes submitted that the court was entitled to receive any evidence to decide whether any matter was or was not taken into account. Mr Vandermeer agreed but submitted that if documents were looked at then the maker of them ought to be allowed to speak to them, and he referred to *Poppetts (Caterers) Ltd v Maidenhead Corporation* [1971] 1 WLR 69.

In his (his Lordship's) judgment, where reasons were given for a condition, as they were required by statute to be, albeit the provision was directory and not mandatory, then the court should not go behind those reasons unless their bona fides was challenged or it was said that there could be no evidence to support them. In the instant case, however, no reasons were given and Mr Vandermeer for the defendants conceded that that was an omission, and so his Lordship thought it proper to ascertain the reasons by extrinsic evidence. Once that principle was established, there was no reason for restricting the evidence to documents, particularly having regard to *Poppetts* case, but of course evidence of intention by an officer or member was not admissible as evidence of the collective mind of the defendant council. The court was, therefore, entitled to look at all the surrounding circumstances in which this permission was granted in

A so far as they assisted in discerning the defendants' reason for the certification.

Although the London Borough of Hackney ostensibly were the defendants in this case, in truth they were no more than the representatives of the Greater London Council. The reason was that by section 24 of the London Government Act 1963 the GLC had power to direct that the certification should be imposed and did so direct. In fact the history of this case showed a schism between the two authorities in that since 1969 it had been the borough council's view that planning permission for office use should be granted, whereas that of the GLC had been to the contrary, and on
B April 28 1970 the latter directed that planning permission for that use on the ground and first floors by the Trustee Savings Bank should be refused. When the present application was first made the defendants decided that permission should be granted without the certification. However, when the matter came into the hands of the GLC they explored the possibility of the conversion of Insurance House to residential use and it was beyond question that the GLC saw a gain in planning terms if that conversion took place. Mr Barnes argued that that was a wrongful consideration in deciding the planning application before the council. It was that which caused the council to write the condition into the cer-
C tification and for that reason it was *ultra vires*. He referred to *Hanks v Minister of Housing and Local Government* [1963] 1 QB 999, and in particular to p 1020 where Megaw J (as he then was) said:

> I am prepared to assume, for the purposes of this case, that, if it be shown that an authority exercising a power has taken into account as a relevant factor something which it could not properly take into account in deciding whether or not to exercise the power, then the exercise of the power, normally at least, is bad. Similarly, if the authority fails to take into account as a relevant factor something which is relevant, and which is or ought to be known to it, and which it ought to have taken into account, the exercise of
> D the power is normally bad.

Mr Barnes also referred to the oft-quoted dictum of Lord Denning MR in *Pyx Granite Co Ltd v Ministry of Housing and Local Government* [1958] 1 QB 554 at p 572:

> The principles to apply are not, I think, in doubt. Although the planning authorities are given very wide powers to impose "such conditions as they think fit," nevertheless the law says that those considerations, to be valid, must fairly and reasonably relate to the permitted development. The planning authority are not at liberty to use their powers for an ulterior object, however desirable that may seem to them to be in the public interest.

E He (his Lordship) accepted that the defendants would have violated the principles propounded in the two cases cited if the facts established that the purpose or motive for the certification was to coerce the applicants into changing the use of Insurance House into residential use. However, he did not think that that was established by the facts. There could be no doubt that the office use of the Chart Street premises was contrary to the provisions of the approved development plan in that they are shown in an area zoned for general industry and, further, the office policy of the development plan was to restrain further office development
F in the centre of London. The evidence established that this permission was granted only because of the special needs of the Executive Council as was made plain in a report by the GLC's Joint Director of Planning and Transportation in which he said:

> Because such a large proportion of the Council's staff live in the Hackney/Islington area, there could be a serious dislocation of the Council's work if they were forced to find accommodation outside that area.

It was clear from oral evidence given that permission would not have been granted but for the Executive Council's special case. It seemed to follow that the certification would have been made irrespective of Insurance House, because without it the premises might become occupied by somebody G not having a special case and the development plan provisions would thereby be prejudiced. It was possible, as was said in evidence, that but for the potential residential use of Insurance House permission might not have been granted at all, but that did not vitiate the certification. The conclusion was that the only motive for the certification was to safeguard the development plan and that the statement added for information was no more than an encouragement to somebody to bring about the change to residential use of Insurance House so that there would be a planning gain in that use to set off against the planning loss at Chart Street.

Accordingly it followed that the action failed and it was H not necessary to deal with the third question in issue.

QUEEN'S BENCH DIVISIONAL COURT

March 20 1978

(Before Mr Justice MELFORD STEVENSON, Mr Justice CANTLEY and Mr Justice CROOM-JOHNSON)

KENSINGTON AND CHELSEA ROYAL BOROUGH v ELMTON LTD

Estates Gazette June 17 1978 J
(1978) 246 EG 1011

Town and Country Planning (Control of Advertisements) Regulations 1969—Section 109 (2) of the Town and Country Planning Act 1971—Conviction for unauthorised advertisement—Offending advertisement replaced after a short interval—Complaint of continuing offence—Appeal by borough against justices' dismissal—Held new offence, not continuing offence—Appeal dismissed

This was an appeal by the Royal Borough of Kensington and Chelsea against the decision of London (South Western) K magistrates on September 6 1976, dismissing an information alleging that Elmton Ltd, having been convicted of displaying an advertisement board outside premises in Beauchamp Place, London SW3, without planning consent, did continue to display the board between May and August 1976, contrary to the Town and Country Planning (Control of Advertisements) Regulations 1969 and section 109(2) of the Town and Country Planning Act 1971.

Douglas Day (instructed by the borough solicitor) appeared for the appellants. The respondents did not appear and were not represented.

Giving judgment, MELFORD STEVENSON J said that the L respondents, having been convicted of displaying an advertisement outside premises in Beauchamp Place, London SW3, without planning consent were summoned by the appellants alleging that they had continued to display the advertisement and were therefore liable for a daily penalty.

That information was heard by justices for the South Western Metropolitan Petty Sessional Division on September 7 1976, when they dismissed it on the ground that there had been no continuing offence following the original conviction. The justices, by their case stated, found that after the respondents' appeal against conviction had been dismissed the advertisement was taken down but, a week afterwards, had M been replaced. The court was now asked to say that there had been a continuing offence of displaying the advertisement.

As he (his Lordship) saw it, it was not a case of a continuing offence but of a new offence for which a fresh information could have been issued. The appeal should therefore be dismissed.

Agreeing, CANTLEY J said that the removal of the advertisement had been a bona fide attempt to abide by the appellants' order to remove it and it would be an abuse of language to regard the replacement as being a continuing offence.

CROOM-JOHNSON J also agreed.

The borough's appeal was dismissed.

VENDOR AND PURCHASER

A

COURT OF APPEAL

November 8 1977

(Before Lord Justice MEGAW, Lord Justice BRIDGE and Lord Justice WALLER)

BATTY AND ANOTHER v METROPOLITAN PROPERTY REALISATIONS LTD AND OTHERS

Estates Gazette April 1 1978

(1977) 246 EG 43

House built on unstable hillside land "doomed" from outset—Actions by purchasers against vendors (a development company), (B) builders and local authority—No appeal from judge's dismissal of action against local authority —Liability of vendors and builders affirmed on appeal—Vendors liable both in contract for breach of warranty and in tort for negligence—Builders liable in tort for negligence on Donoghue v Stevenson principle— Important analysis of duties and review of case law

These were appeals by Metropolitan Property Realisations Ltd (developers) and Trippier Construction Ltd (builders) from a judgment of Crichton J, who found the former liable in contract and the latter liable in tort in respect of a house at Rawtenstall, in Rossendale, North-East Lancashire, which (C) the judge held to be "doomed" as a result of movement of the strata on the hillside adjacent to the house. Mr and Mrs R Batty had acquired from the developers a 999-year lease of the house, which had been built by the builders. The judge dismissed the claim, based on alleged negligence or breach of statutory duty in inspecting the foundations, against Rossendale District Council. There was no appeal against the dismissal of this claim, but Mr and Mrs Batty cross-appealed against the refusal of the judge to find the developers liable in tort as well as in contract.

Douglas Brown QC and Andrew Gilbert (instructed by Tobin & Co) appeared on behalf of the appellants (Metro- (D) politan Property Realisations Ltd); J M Collins QC and John Hoggett (instructed by Bannister, Preston and Ormerod) appeared for the appellants (Trippier Construction Ltd); Benet Hytner QC and A C Jolly (instructed by Gregory Rowcliffe & Co, agents for Whiteside & Lord, of Rawtenstall) represented the respondents (Mr and Mrs Batty).

Giving judgment, MEGAW LJ said: In August 1971 the two plaintiffs, Raymond Batty and his wife, Helga Batty, took from the first defendants, Metropolitan Property Realisations Ltd, on a 999-year lease, a house and garden. The house had just been built, being one of a number of houses on the same estate, at Rawtenstall, in the district of Rossendale, in North- (E) East Lancashire. They paid £5,250 as consideration for what was in all respects other than a legal technicality a purchase by them of the house. The house became known as 33 Redwood Drive.

The house had been built by the second defendants, Trippier Construction Ltd, under arrangements made by them with the first defendants, who are a development company. The land on which the house was built, along with neighbouring land on which other houses were built at about the same time, was sold by the second defendants, who had earlier bought it from the local authority, the third defend- (F) ants, the Rossendale District Council. It was sold by the second defendants to the first defendants. Under their contract with the first defendants, the second defendants proceeded to build this house and other houses. When this house had been completed and leased to the plaintiffs on a 999-year lease, the first defendants sold the reversion to the second defendants. Thus, in effect the building of the house and its disposition was a co-operative effort by the first defendants and the second defendants, the second defendants doing the building, the first defendants providing the finance and being the party with whom the purchasers of the house, the plaintiffs, entered into contractual relations. Both the first (G) defendants and the second defendants, through their representatives, had inspected the site, and also had walked over the surrounding territory in 1969 and 1970, before the decision to build was taken and, therefore, before the building had started. The object of such inspection was to decide, by reference to various considerations, including the safety and suitability of the site for housebuilding, whether or not they or either of them would undertake, or become involved in, housebuilding in that area. They decided to build, on the basis, as between them, which I have just very summarily outlined.

(H) The house was built on a sort of plateau, as it has been described, at the top of a steep slope which fell down at a gradient of about 1 : 3 to a stream, called the Balladen Brook, which ran in the bottom of the valley below the house. The front of 33 Redwood Drive faced on Redwood Drive, with a small front garden between the house and the road. At the back of 33 there was a piece of land, bought by the plaintiffs along with the house, which was intended to provide a garden at the back. It was somewhere down that garden that the relatively level area (it was only relatively level) which I have called "the plateau" changed its inclination to the 1 : 3 slope which was the general inclination of the slope falling (J) from there towards the stream. The total area covered by the house 33 Redwood Drive and the land sold with it was about 100 ft from front to back—from Redwood Drive to the bottom of the back garden—and about 40 ft from side to side in width. Coloured photographs which we have seen give a reasonably clear impression of the general nature of the ground in and about the house and its neighbourhood. They show the steepness and roughness of the hillside on both sides of the valley. I shall not attempt to describe the topography further, as it is unnecessary to do so for the purpose of this appeal.

(K) That was in 1971. The plaintiffs bought their house and moved into it and made it their home. In 1974 there was a severe slip of the natural strata of the hillside. It did not directly or immediately damage the plaintiffs' house or its foundations, though it did cause direct damage to a part of the back garden. As a result, urgent investigations were undertaken and legal proceedings were begun by the plaintiffs against three defendants—the two already mentioned and, as the third defendant, the Rossendale District Council, the local authority.

The gravity of that litigation—its seriousness for the parties—will become apparent when I recount that it was held by Crichton J, from whose judgment this appeal is brought, that the plaintiffs' house is doomed; and the finding on that issue, though it was the subject of much conflicting evidence at the trial, is not now disputed or challenged. At some time not later than 10 years from the date of the trial, possibly much earlier, the movement of the strata on the hillside on the slopes adjacent to the plaintiffs' house will cause the foundations of that house to slide down the hill and the house will be in ruins. Already, of course, in those circumstances, the house is unsaleable. The reason, or at least the principal reason, as found by the learned judge, for this very grave and disastrous prognosis is the presence in the boulder clay, which is the principal constituent of the hillside, of a layer of what is described technically as varved clay. I do not need to go into the technical or geological details. They are not relevant for the purposes of the decision of the appeal.

The plaintiffs' claim against the first defendants was in tort, for negligence, and for breach of contract. Against the second defendants the claim was in tort for negligence, on what I may call *Donoghue v Stevenson* principles. Against the third defendants, the local authority, it was for negligence or breach of statutory duty, in respect of the local authority's duties with regard to inspection of the foundations. It was, however, held by the judge—and it does not appear to have been really in dispute—that the foundations as such were perfectly properly constructed, as were the bricks and mortar of the house itself. The only defect—but, in the circumstances, a very grave matter—was the nature of the land on which the house relied for its support. It was unstable, and by its instability the house was, from the outset, doomed. Thus, for that reason, and for that reason alone, the house was unfit for human habitation: because in a foreseeable, and short, time it would collapse, through the movement of the hillside. The plaintiffs succeeded before the judge against the first defendants, though on their claim in contract only, and against the second defendants on the sole claim against them in tort. They failed against the third defendants. The judge awarded the plaintiffs, jointly against both the first and second defendants, £13,000, and he also awarded Mrs Batty £250 for the consequences of her distress—the effect upon her, physical and mental, of these events. Those sums were agreed by the parties as being the appropriate sums for damages, subject to liability.

The first and second defendants both appeal. There is a cross-notice on behalf of the plaintiffs, asserting that, as against the first defendants, the judge ought to have entered judgment on the claim in tort, as well as on the claim in contract. That would not have affected the amount to be awarded in the judgment, but it may be of practical importance to the plaintiffs none the less. There is no cross-appeal by the plaintiffs as regards the dismissal of their claim with costs against the third defendants. Counsel for all the parties concerned have, if I may say so, presented their submissions in this court with admirable clarity and conciseness, both on the issues of fact and on the issues of law which they desire to raise. I shall seek to emulate their conciseness. In particular, I do not propose to set out or summarise in any detail the evidence on the various questions of fact which were in issue before us, for it appears to me that the conclusions of the learned judge on all those issues, as set out in his judgment, are not only supported by some evidence, but they are supported by evidence which fully justifies the judge's conclusions thereon. If anyone should wish to look in more detail at the evidence, they will find it summarised, with the judge's conclusions thereon, in his judgment.

The first ground of appeal argued by Mr Brown, on behalf of the first defendants, is concerned with the terms of the contract of sale—or, rather, the contract relating to the 999-year lease—between the first defendants and the plaintiffs. That contract was made in writing on August 9 1971. It is contained in two related documents. The relevant term provides as follows, under the heading "The Vendor's Obligations": "(3) The Vendor hereby warrants that the dwelling has been built or agreed that it will be built (i) in an efficient and workmanlike manner and of proper materials and so as to be fit for habitation." That the house was not fit for habitation when the contract was made and when the plaintiffs took possession cannot, in my judgment, be disputed, on the, now unchallenged, facts which I have mentioned. How, then, do the first defendants seek to escape liability for breach of contract under this warranty? On behalf of the plaintiffs it was conceded that the warranty was not absolute: in this sense, that if there were, for example, some undetectable geological fault, at some distance away, which, after the house was built, caused an earth tremor, damaging or destroying the house, that would not be within the warranty. But if the instability of the hillside could have been detected by experts, but was not detected, and if as a result the house was, when handed over, unfit for habitation, that, say the plaintiffs, is a breach of warranty. For the first defendants it is said that the obligation imposed by the warranty "and so as to be fit for habitation," though expressed as a separate warranty, co-ordinate with the two warranties which precede it in the clause, ought to be read as though it were expressed as "and so as to be fit for habitation *so far as compliance with the two preceding warranties can achieve that result.*"

I am afraid that I cannot accept that construction. It is unnecessary, as I see it, in this appeal to seek to define whether, and, if so, to what extent, the warranty of fitness for habitation falls short of an absolute warranty. I am satisfied that on the facts of this case the warranty, on its true meaning, was broken. The warranty would be broken if the house is unfit for habitation, and is so unfit by reason of defective support from the adjoining land which a suitably qualified expert could have discovered before the building of the house was undertaken. The judge has found, and in my opinion unassailably found, that the lack of fitness for habitation—the instability of the hillside which spelled the not-far-distant doom of the house from the outset of its life—could have been discovered by expert examination. I need do no more than refer to the evidence of Mr Townsend, the first defendants' development director, the only witness to be called on behalf of the first defendants. He, in cross-examination by counsel on behalf of the plaintiffs, very fairly and very frankly made admissions as to what he should have realised as a result of his visits to the site; the need for further investigation. The judge was right to hold that the warranty was broken, and that the breach of the warranty was the cause of the house having become valueless, and the cause of Mrs Batty's personal loss.

For the first defendants it was further submitted, by reference to the evidence, that, while the judge held that on the facts known to them the first defendants should have conducted investigation into the stability of the hillside, the judge failed to give a clear answer to the question: what sort of investigation were the first defendants under a duty to carry out, having regard to whatever were the symptoms which they should have observed and which should have put them on warning. Counsel has, properly, taken us in particular to those passages in the evidence which deal with certain slips on the hillside which were regarded by some of the witnesses as being one of the factors which ought to have led to investigation. It was stressed by Mr Brown that the judge ought, on the evidence, to have regarded these factually as having been at the relevant time what he called "shallow slips," and that, being shallow slips, that meant that the investigation which ought to have been conducted, having regard to the symptoms, was not an investigation (to use the modern phrase, which may perhaps for once be appropriate here) in depth, in order to ascertain what the sub-soil contained. In my judgment there is no fault that can properly be found in respect of the learned judge's conclusions on the evidence.

It was suggested that this court should look at the evidence and form its own conclusion. If this court were to agree with the judge on the evidence that some investigation was required, in accordance with the legal duty resting on the first defendants under their contract, then this court would have to go on to consider for itself what sort of investigation was required and what such investigation would have disclosed. The right conclusion, it was submitted, having regard to the evidence in relation to the possibly shallow slips and various other matters, would be that such investigation would not have disclosed anything relevant. To that argument there appears to me to be one simple answer, offered on behalf of the plaintiffs. That was, that it was clear, on evidence accepted by the judge and which he was plainly entitled to accept, seeing and hearing the witnesses, that, if the defendants had through that investigation—any investigation—of the stability of the hillside failed to discover a cause of the symptoms which had led them to investigate they would not have allowed the building of the house to proceed. So any investigation, whatever it was and however short it properly went, would, on that basis, have resulted in the houses not being built. But no investigation was undertaken at all. I think that is right. But I am further satisfied, as was also submitted on behalf of the plaintiffs, that on the evidence the judge rightly held that the investigation which ought to have been undertaken would not, indeed, have produced merely a negative result of finding no cause for the symptoms, but would have shown, and shown clearly, that there was no safety margin from the point of view of the stability of the relevant ground; and, in the absence of a safety margin, no reasonable builder or developer would erect a building.

In this context I refer to two passages in the learned judge's judgment. The judge, at the bottom of p 31 of the transcript, had already summarised the evidence as to what would have been seen on the ground by way of warning symptoms at the time when the building of the houses was being considered. His findings as to what those symptoms would be were, in my judgment, fully justified. At the bottom of p 31/F the judge says:

> That is the state of the evidence. Where does it lead us to? I think it leads us in this direction, that these cracks

I pause there to say that the word "cracks" there may not be entirely accurate, or comprehensive, because I have no doubt that the learned judge intended to include other symptoms which he had held to exist

> on the other side of the valley and at the toe of the slope on 'our' side should have put Mr Townsend and Mr Trippier on inquiry. If they had been put on inquiry—this is on the point of the cause of this situation—the probability would be that some instability would have been found whereby houses would not have been built. That I will deal with later. But at the moment I merely find that these cracks which existed, in my judgment, in 1970 and 1971 should have put the only people who investigated that valley on their inquiry.

Then at p 33/E-G the learned judge said this:

> The evidence—I will come to the question of duty—leads me to the belief that further steps should have been taken in the sense of consulting either surveyors or soil mechanics experts and also to the view that had those persons been consulted, having regard to the situation in that valley, it would be unlikely and improbable that houses would have been built in the position that the plaintiffs' house was built.

Mr Brown in the course of his argument accepted that, in order to succeed in his appeal, he had to upset that finding. I do not think that Mr Brown's careful submissions on the evidence, with all respect to him, begin to show that that finding was wrong.

I turn to the plaintiffs' cross-notice affecting the first defendants. That is a cross-notice whereby it is asked that this court should include in the judgment against the first defendants judgment based on tort—the learned judge having refused to enter judgment for the plaintiffs other than on the basis of breach of contract. Crichton J, as I understand his judgment, thought that he was bound so to hold on his reading of a passage in the judgment of Diplock LJ (as he then was), sitting as a judge of first instance, in *Bagot v Stevens Scanlan & Co Ltd* [1966] 1 QB 197. The learned judge, at p 50/G of the transcript of his judgment, having cited that case, went on to say: "I have also had regard in that respect to the case of *Esso Petroleum Co Ltd v Mardon*, which is reported in [1976] 2 WLR 583. But I do not find that this case detracts in any degree from the finding of Diplock LJ, as he then was."

Esso Petroleum Co Ltd v Mardon is now reported in [1976] 1 QB 801. I fear that I feel bound to disagree with the learned judge's view that *Esso Petroleum Co v Mardon* does not affect the position. We in this court are bound by what was said in *Esso Petroleum Co v Mardon*, in so far as what was said was *ratio decidendi*. There can, I think, be no doubt, subject to one possible distinction which Mr Brown sought to persuade us in his reply this morning to be a relevant distinction, that the *ratio decidendi* of *Esso Petroleum Co v Mardon* necessarily requires that in a case such as the present we should hold that the mere fact that the plaintiffs have obtained judgment for breach of contract does not preclude them from the entitlement which would have existed, apart from contract, to have judgment entered in their favour also in tort. I refer to the judgment of Lord Denning MR at pp 818 to 820. I do not propose to read it. The sense of it appears to me to be entirely clear, and, incidentally, to have included the Master of the Rolls's view that *Bagot v Stevens Scanlan & Co Ltd* had been decided without the learned Lord Justice who decided it having had cited to him a number of relevant authorities.

The distinction to which I have referred, which Mr Brown seeks to make, is this: that the right of a plaintiff who sues in contract, where the facts giving rise to the breach of contract would also constitute a breach of common law duty apart from contract, to have the judgment entered on both heads is limited to cases where the common law duty is owed by one who conducts a common calling and thus is under a special type of legal liability, and to cases where the duty is owed by a professional man in respect of his professional skill. Mr Brown contends that, though there is no affirmative authority for limiting the right in that way, it ought to be treated as being so limited because there is no case in the English books, going back over many years, which shows that the right has been allowed, or possibly even claimed, in cases other than the special types of case to which he referred, and in particular the professional skill types of case. In *Esso Petroleum Co v Mardon* the right was held to arise in a case where the breach of duty was a breach by an architect involving his professional skill. I see no reason, in logic or on practical grounds, for putting any such limitation on the scope of the right. It would, I think, be an undesirable development in the law if such an artificial distinction, for which no sound reason can be put forward, were to be held to exist. In my judgment the plaintiffs were entitled here to have judgment entered in their favour on the basis of tortious liability as well as on the basis of breach of contract, assuming that the plaintiffs had established a breach by the first defendants of their common law duty of care owed to the plaintiffs. I have no doubt that it was the duty of the first defendants, in the circumstances of this case, including the fact of the joint responsibility which they undertook in arranging for erection of this house on this site, apart altogether from the contractual warranty, to examine with reasonable care the land, which in this case would

A include adjoining land, in order to see whether the site was one on which a house fit for habitation could safely be built. It was a duty owed to prospective buyers of the house. How wide or deep the examination had to be, to comply with the duty, would depend on the facts of the particular case, including the existence and nature of any symptoms which might give cause for suspecting the possibility of instability. It is clear from the facts found by the judge that, if he had thought that a finding of tort was procedurally permissible, he would have held, on his assessment of the evidence, that the first defendants were in breach of that duty. Accordingly, I would accept the cross-notice and would direct that judg-
B ment be entered for the plaintiffs against the first defendants for the tort of negligence as well as for the breach of contract.

I should add that Mr Hytner, on behalf of the plaintiffs, contended, as an alternative ground, apart from what I may call the *Esso v Mardon* ground, that he should be entitled to such a judgment in this case because on the facts of this case it would be the proper view to treat the tort and the breach of contract as being truly independent of one another. Mr Brown in his reply this morning submitted that that would not be a proper basis. He suggested that if it were to be
C upheld it would startle bankers, and also that it would be a wrong concept of time by reference to which the duty arises. In the circumstances, I do not propose to offer any view one way or the other on that dispute. It is unnecessary to do so, because of my view that apart from that alternative ground the plaintiffs are entitled to their judgment in tort.

I now turn to the appeal by the second defendants, the builders. Mr Collins very helpfully and clearly put his argument in the form of six submissions. The first submission was this: A builder should not be taken to be under any duty of care in relation to defects in or observable only upon land which is not available to him in connection with his
D operations and is not otherwise owned by him or is in his possession. Mr Collins, properly and as I think inevitably, conceded that, as he put it, "in general terms a duty situation can arise between a builder and an occupier with whom the builder is not in privity of contract." A duty of the *Donoghue v Stevenson* type can arise, it is conceded, in relation to reality. But Mr Collins contends that the duty extends only to defects —to symptoms of possible instability, for example, affecting properly-built foundations—where those defects are in, or observable upon, the actual site on which the house is to be built. If defects exist on neighbouring land which is not in his ownership or possession—or in respect of which he would
E require someone else's permission to go upon it—there is no duty, it is said, to look for or to observe or to take any action in relation to such symptoms.

For this submission, Mr Collins relies upon a passage in Lord Wilberforce's speech in *Anns v Merton London Borough Council* [1977] 2 WLR 1024. The passage on which Mr Collins relies in this context is at p 1032/A-E. It reads as follows:

> Through the trilogy of cases in this House—*Donoghue v Stevenson* [1932] AC 562, *Hedley Byrne & Co Ltd v Heller & Partners Ltd* [1964] AC 465, and *Dorset Yacht Co Ltd v Home Office* [1970] AC 1004—the position has now been reached that
> F in order to establish that a duty of care arises in a particular situation, it is not necessary to bring the facts of that situation within those of previous situations in which a duty of care has been held to exist. Rather the question has to be approached in two stages. First one has to ask whether, as between the alleged wrongdoer and the person who has suffered damage, there is a sufficient relationship of proximity or neighbourhood such that, in the reasonable contemplation of the former, carelessness on his part may be likely to cause damage to the latter—in which case a prima facie duty of care arises. Secondly, if the first question is answered affirmatively, it is necessary to consider whether there are any considerations which ought to negative, or to reduce or limit the scope of the duty or the class of person
> G to whom it is owed or the damages to which a breach of it may give rise: see *Dorset Yacht* case . . . per Lord Reid at p 1027. Examples of this are *Hedley Byrne's* case . . . where the class of potential plaintiffs was reduced to those shown to have relied upon the correctness of statements made, and *Weller & Co v Foot and Mouth Disease Research Institute* [1966] 1 QB 569; and (I cite these merely as illustrations, without discussion) cases about 'economic loss' where, a duty having been held to exist, the nature of the recoverable damages was limited: see *SCM (United Kingdom) Ltd v W J Whittall & Son Ltd* [1971] 1 QB 337 and *Spartan Steel & Alloys Ltd v Martin & Co (Contractors) Ltd* [1973] QB 27.

Mr Collins contends that there is a consideration which
H ought to reduce the scope of the builder's *Donoghue v Stevenson* duty to a potential occupier of the house which he builds. That is on the basis of Lord Wilberforce's "Secondly . . . whether there are any considerations which ought to negative, or to reduce or limit the scope of the duty. . . ." The scope, says Mr Collins, ought to be limited by reference to the builder's statutory duty under the building regulations. If he builds foundations which comply in all respects with those regulations and any other statutory provisions, and the defects are outside the area of the foundations themselves, then the builder has no further duty. With all respect to Mr Collins's argument, I do not accept it. Of
J course, the question whether or not there has been a breach of the duty will depend on all relevant considerations going to the question: Did the builder act as a competent and careful builder would have acted in what he did or did not do by way of examination and investigation? But I see no reason why, as a matter of law, or by reference to any question of policy considerations, as a matter of the existence of the duty, it should depend on whether or not the symptoms, the observable defects, are on land of which the builder has ownership or in respect of which he has a legal right of entry without requiring some other person's permission.

K I would reject that first submission.

The second submission made by Mr Collins was this: If a builder is under any such duty, it should be limited to defects which are discoverable without subsoil investigation. The way that Mr Collins put his submission was: If there is a duty on the builder, it is not a very heavy one. It would not include subsoil investigation. If a builder knows that what he sees would endanger the stability of the house, it is difficult to say there is not a duty. Mr Collins, on being asked by a member of the court why should that not apply also if he ought to have known, submitted that that would put too high a burden on a builder, and that it is undesirable
L that the law should put such a duty on him. Mr Collins accepted, as I understood it, that this submission of his must mean that he is submitting that a builder is under no duty to look at adjoining land: if, however, he does know, though he had no duty to look, of something on the adjoining land which indicates a danger without further investigation, he might be under a duty. I see no reason for limiting the duty as a matter of law in the manner in which it is suggested by that submission.

The third submission was: The second defendants were not under any higher duty than that attaching to any other
M builder who builds for a development company under a building contract. The argument as developed by Mr Collins was that the second defendants here were under no higher obligation than if they had been retained as builders by contractors at arm's-length. That is, their duty is no higher because of the particular circumstances here of the relationship between the developers and the builders, and what happened between them in relation to the inspection of the site before the building started, and their arrangement in regard to the building and what was to happen in relation to it. It may well be that on certain facts a builder would be entitled to rely upon an examination which he knew or

A reasonably supposed had been made by others on whose competence the builder could properly and reasonably rely. But in this case, on the evidence, the true view is that the decision to build on this site was a joint decision of the builders and the developers, the second defendants and the first defendants. In my judgment, if one is a party to the decision to build on the particular site in circumstances such as were shown in the evidence to exist in this case, the *Donoghue v Stevenson* duty applies. Indeed, I have difficulty in seeing how, as was a part of Mr Collins's argument at one stage, the fact that there was here an intervening contract—the contract for the 999-year lease between the first
B defendants and the plaintiffs—can affect the question whether the second defendants are under a *Donoghue v Stevenson* type of liability. For in *Donoghue v Stevenson* itself the whole question of the manufacturer's liability was considered, and decided to exist, despite the existence of an intervening contract between the retailer and the purchaser of the goods. I would reject the third submission, on the facts of this case. Again, in my view, it is not a question of the existence of the duty. It is a question whether, in a particular case, on the facts of that case, it has been broken.

The fourth submission was this: If a builder is to be taken to be under any such duty, then no cause for action
C for breach of it arises (a) in respect of damage to the house itself or (b) alternatively until the house itself is damaged or is in such a state as to present present or imminent danger to the health or safety of the occupier. This submission, essentially, as I see it, is founded on a question which was raised and the view which was expressed by Stamp LJ in *Dutton v Bognor Regis Urban District Council* [1972] 1 QB 373. At p 415/C Stamp LJ said:

> What causes the difficulty—and it is I think at this point that the court is asked to apply the law of negligence to a new situation—is that whereas the builder had, as I will assume, no
> D duty to the plaintiff not carelessly to build a house with a concealed defect, yet it is sought to impute a not dissimilar duty to the defendant council. At this point I repeat and emphasise the difference between the position of a local authority clothed with the authority of an Act of Parliament to perform the function of making sure that the foundations of a house are secure for the benefit of the subsequent owners of the house, and a builder who is concerned to make a profit. So approaching the matter, there is in my judgment nothing illogical or anomalous in fixing the former with a duty to which the latter is not subject. The former by undertaking the task is in my judgment undertaking a responsibility at least as high as that which the defendant in the *Hedley Byrne* case would, in the opinion of the majority
> E in the House of Lords, have undertaken had he not excluded responsibility.

I pause here to say that obviously this fourth proposition cannot be treated as entirely independent of the first of Mr Collins's propositions, with which I have already dealt. The doubt which was raised by that passage in that judgment of Stamp LJ was, as I see it, put at rest by passages in the speech of Lord Wilberforce in *Anns v Merton London Borough Council* [1977] 2 WLR 1024, at pp 1038 and 1039. At p 1038/F Lord Wilberforce said:

> *The position of the builder.* I agree with the majority in the
> F Court of Appeal in thinking that it would be unreasonable to impose liability in respect of defective foundations upon the council, if the builder, whose primary fault it was, should be immune from liability. So it is necessary to consider this point, although it does not directly arise in the present appeal. If there was at one time a supposed rule that the doctrine of *Donoghue v Stevenson* [1932] AC 562 did not apply to realty, there is no doubt under modern authority that a builder of defective premises may be liable in negligence to persons who thereby suffer injury. . . ."

So far, Lord Wilberforce has agreed that it would be unreasonable to impose liability for defective foundations on a council if the builder were immune. Then at the bottom of p 1038 Lord Wilberforce goes on to say this: G

> But leaving aside such cases as arise between contracting parties, when the terms of the contract have to be considered . . . . I am unable to understand why this principle or proposition should prevent recovery in a suitable case by a person, who has subsequently acquired the house, upon the principle of *Donoghue v Stevenson:* the same rules should apply to all careless acts of a builder: whether he happens also to own the land or not. I agree generally with the conclusions of Lord Denning MR on this point in *Dutton v Bognor Regis Urban District Council* [1972] 1 QB 373, 392-394. In the alternative, since it is the duty of the builder (owner or not) to comply with the by-laws, I would be of opinion that an action could be brought against him, in effect, for breach of statutory duty by any person for H whose benefit or protection the by-law was made. So I do not think that there is any basis here for arguing from a supposed immunity of the builder to immunity of the council.

The argument that the local authority should not be liable because it would be unreasonable that it should be held liable when the builder was not held liable was rejected, because Lord Wilberforce sees no reason why the builder should not be held liable. True, he specifically refers to a case where there was a breach of statutory duty—non-compliance with the building regulations. But that was the particular question arising in that case. I see no logical or practical reason for J so confining it; nor, in my view, did Lord Wilberforce so intend.

As to the question of the nature of the damage which gives rise to the cause of action, it seems to me that an answer given by Mr Hytner was a simple and full answer. If it indeed is necessary that it should be shown that there has been physical damage to the property before the action will lie against the builder, in the present case there was physical damage to the property in the landslide, or landslip, of 1974. True, the foundations of the house for the time being remained undisturbed. True, the bricks and mortar of the house, as the judge has found, remained K undamaged. But there was physical damage to the garden —a part of the property conveyed. If physical damage be necessary in order to found the action, there was physical damage. But, apart from what might be regarded as that possibly accidental element here, there is, I think, a wider reason why Mr Collins's proposition should not succeed on the facts of this case. Again I refer to the speech of Lord Wilberforce in *Anns v Merton London Borough Council.* At p 1039 Lord Wilberforce is dealing with the question "When does the cause of action arise?"; and below letter "H" he uses this sentence: "It can only arise when the L state of the building is such that there is present or imminent danger to the health or safety of persons occupying it." Was there not here imminent danger to the health or safety of persons occupying this house, at the time when the action was brought? Indeed, Mrs Batty, one of the plaintiffs, has been awarded damages for the consequences to her health and peace of mind of the foreseen disaster. Why should this not be treated as being a case of imminent danger to the safety and health of people occupying the house? No one knows, or can say with certainty—not even the greatest expert—whether the foundations of the house will move and the house perhaps suddenly tumble tomorrow, or in a year's M time, or in three years' time, or in 10 years' time. The law, in my judgment, is not so foolish as to say that a cause of action against the builder does not arise in those circumstances because there is no *imminent* danger. I would reject that submission.

Mr Collins's fifth submission is one that goes to the facts. It is this: In any event the state of the terrain was not such as to render it careless for the second defendants not to require further investigation prior to the commencement of the building. Mr Collins has properly taken us to passages in the evidence, in addition to the passages to which we had been referred by Mr Brown for the first defendants, going to this question as to what would have been seen on the site and the neighbourhood of the site at the relevant time before building started, and as to the evidence as to what that ought to have conveyed to a reasonably careful builder observing those symptoms. Once again, I do not propose to go into the evidence on those matters, helpful though counsel's references to it were. I am quite satisfied that on the evidence the judge was right in his finding that the symptoms were such that investigation was called for by a reasonably careful builder, and that if the investigation which was called for by reason of those symptoms had been made the house would not have been built.

The sixth and last of Mr Collins's submissions is this: The plaintiffs' house has not suffered damage and is not in such a state as to produce present or imminent danger to the plaintiffs' health or safety. Mr Collins made it clear that he was not challenging the judge's finding that the house was likely to be engulfed or the judge's finding that it is now valueless. This is really related to the fourth submission, with which I have already dealt, and in that I have said all that I think it is necessary to say and all that I am minded to say referring to this proposition.

I would therefore dismiss the appeals of the first and second defendants; and, by reference to the plaintiffs' cross-notice, I would direct that judgment be entered for the plaintiffs against the first defendants for liability in tort as well as for liability in contract.

Agreeing, BRIDGE LJ said: I add only a very short postscript to the passage in my Lord's judgment dealing with Mr Collins's fourth submission. Among the propositions enunciated by Mr Collins in support of his argument that the damage claimed by the plaintiffs in this case was of a nature irrecoverable against the second defendants was the contention that a party liable for negligence of the *Donoghue v Stevenson* variety is liable only for damage consequential upon having put into circulation the dangerous article which foreseeably was likely to cause injury, and not for the loss of the value of that article itself. He relied in particular on a passage from the judgment of Stamp LJ in *Dutton v Bognor Regis Urban District Council*, from which my Lord has quoted, where the point is succinctly put at an earlier page, p 414/E-F. Stamp LJ said:

> It is pointed out that in the past a distinction has been drawn between constructing a dangerous article and constructing one which is defective or of inferior quality. I may be liable to one who purchases in the market a bottle of ginger beer which I have carelessly manufactured and which is dangerous and causes injury to person or property; but it is not the law that I am liable to him for the loss he suffers because what is found inside the bottle and for which he has paid money is not ginger beer but water.

So it is argued here that whilst, if the defective house fell down and physically injured the plaintiffs, or indeed anyone else, that would impose a *Donoghue v Stevenson* liability on the second defendants, the fact that the house itself has become valueless, and indeed incapable of repair at economic cost, is not a loss which the plaintiffs can recover from the second defendants. This argument, to my mind, like the other arguments to which my Lord has adverted, is really untenable, in the light of the speech of Lord Wilberforce in *Anns v Merton London Borough Council*—with which, as I understand it, all the rest of their Lordships agreed. As my Lord has pointed out, at p 1038 Lord Wilberforce expresses his agreement with the view which had been indicated by the Court of Appeal that it would be unreasonable to impose liability upon the local authority if the builder whose primary fault had caused the defect in the building was to be immune from liability. At p 1039 he is considering what damages are recoverable against the local authority. In the context of what he had said at p 1038, in principle what he says at p 1039 must equally be applicable to the question what the damages are, in the *Donoghue v Stevenson* situation, recoverable from the builder. Lord Wilberforce says in terms, just above letter "E," "Subject always to adequate proof of causation, these damages may include damages for personal injury and damage to property. In my opinion they may also include damage to the dwelling-house itself. . . ." In my judgment that sets at rest the doubts raised by the judgment of Stamp LJ in *Dutton v Bognor Regis Urban District Council.*

I agree, for all the reasons given by my Lord, that the appeals should be dismissed, and the cross-notice allowed as he indicated.

WALLER LJ agreed with both judgments.

Appeals dismissed with costs. Cross-notice allowed, with costs against first defendants. Judgment to be entered for plaintiffs against first defendants for liability in tort as well as for liability in contract. First defendants awarded 50/50 contribution as against second defendants. Liberty to apply to Court of Appeal.

LANDS TRIBUNAL
AGRICULTURE

HULME AND ANOTHER v EARL OF AYLESFORD AND TRUSTEES OF EARL OF AYLESFORD'S SETTLEMENT (1965)

(Before the Agricultural Land Tribunal for the West Midland Area (chairman: K S W Mellor QC), on January 30 1978)

Estates Gazette March 11 1978
(1978) 245 EG 851

Applications by widow and son of deceased tenant for tribunal's direction entitling applicant to a tenancy of the holding—Applications by landlords for consent to notices to quit—Tribunal's examination of claims to eligibility and suitability—Date for determination of eligibility—Landlords' contention that eligible applicant was unsuitable—Widow held by tribunal not to be eligible, son held to be eligible and suitable—Landlords' application for consent to notices to quit refused—Direction in favour of son—Order for payment of costs

These were applications under section 20 of the Agriculture (Miscellaneous Provisions) Act 1976 to the Agricultural Land Tribunal for the West Midlands Area by Mrs W A Hulme and Peter Hulme, widow and son, respectively, of Harry Stanley Hulme for directions entitling them jointly and severally to tenancies of Patrick's Farm, Hampton-in-Arden, Solihull, part of the Earl of Aylesford's Packington Estate. The deceased tenant, who died on March 19 1977, had been the tenant from year to year of this farm, holding it under two separate tenancies. The larger part had been let to the deceased by the Earl of Aylesford. The smaller part had been let to him by the trustees of the Earl's settlement. The two parts together amounted to about 260 acres.

P G Langdon-Davies (instructed by K C K Scott, of Leamington Spa) appeared on behalf of Mrs W A Hulme and Peter Hulme. The tenant's agents were Howkins & Harrison, chartered surveyors, of Rugby, and additional expert evidence was given by Shakespear, McTurk & Graham, chartered surveyors, of Market Harborough. R V Stone, chartered surveyor, of James & Lister Lea, Birmingham, represented the Earl of Aylesford and the Earl's trustees.

The tribunal's hearing took place on December 8, 9, and 12 1977 and on January 6 1978 at the Council House, Coventry.

The following is a summary of the main points in the tribunal's decision.

Preliminary

Patrick's Farm was a difficult farm, bisected by the River Blythe and cut through by three roads. On its west side a former ammunition dump cut deeply into it, producing a generous crop of weeds and resulting in difficult-shaped fields for cultivation. On the south Siden Hill Wood thrust even more deeply into the farm. The land on each side of the River Blythe and a parallel tributary stream was level permanent pasture and suffered from drainage problems. On the north, east and west there were slopes of useful medium loam arable land. About 10 acres of the farm were waste land, 120 acres permanent pasture, and 130 acres arable. Of the arable land there were about 26 acres of leys, 26 acres of wheat and 78 acres of barley which would support a dairy herd and produce cash crops.

Procedure

By consent the applications of Mrs Hulme and Peter Hulme for directions entitling them jointly and severally to tenancies of the farm were heard together. Also by consent applications by the Earl of Aylesford and the trustees of his settlement for consent to the operation of notices to quit in respect of the two pieces of land which constituted Patrick's Farm were heard together. However, the tribunal refused applications by the Earl and the trustees for their applications to be heard separately from the applications by Mrs Hulme and Peter Hulme. Apart from the additional inconvenience and expense which such separate hearings would cause, the tribunal expressed the view that the rules contemplated that the landlords' applications should proceed in a similar manner to counterclaims in civil actions.

Eligibility

It was argued on behalf of the trustees of the Earl's settlement that the applications of both Mrs Hulme and Peter Hulme should be dismissed so far as the piece of land let by the trustees was concerned (ie the smaller part of the farm) on the ground that each of them was ineligible. The ground alleged was that each was the occupier of a commercial unit, thus being disqualified by section 18(2)(c) of the 1976 Act. The tribunal held that this argument failed at the outset in regard to Peter Hulme, since he had never been the occupier of this part. It also failed in the case of Mrs Hulme. The tribunal said:

> We read section 18 as speaking of the situation obtaining at the death of the deceased. We decline to construe section 18(2) so as to permit a person to become an eligible person at some time after such death, or to lose that status between the death and the hearing of an application. A survivor survives the deceased at the moment of death, and the use of the present tense in section 18(2)(c) seems to us to echo the use of the same tense in section 18(1) and in other subsections, in relation to that moment, the past tense being used in relation to any time prior to it. In our view a person's status as eligible or not to apply under section 20 is crystallised at the death of the deceased and the commercial unit referred to in section 18(2)(c) must be one occupied prior to, and not in consequence of, that death.

The tribunal, however, decided that Mrs Hulme was ineligible on another ground. Although she worked as hard as any farmer's wife, keeping hens and pigs, driving tractors, feeding calves, harvesting, mowing thistles, milling and mixing food, there was no evidence that her "principal source of livelihood," within the meaning of section 18(2)(b) of the 1976 Act, was derived from her agricultural work on the agricultural unit. "It was derived from her marital relationship with Harry Hulme," the deceased tenant.

The tribunal found that Peter Hulme, who had worked on the farm since leaving school in 1968, was an eligible person.

Suitability

The tribunal then considered in great detail whether Peter Hulme was a suitable person to become tenant of the holdings. They took evidence over three days and thoroughly inspected the land. They had regard to his age (nearly 26), health, experience or training, ability as a farmer, reputation in the farming community and family connections with other local farmers, and financial standing. The last factor, financial standing, was the nub of the attack made by the landlords on Peter's suitability and the matter was the subject of very close consideration by the tribunal. It appeared that by the date of the deceased tenant's death, and particularly during his lengthy illness, the farm had become run-down. He had been reluctant to relinquish control and at the date of his death his accounts showed an unsatisfactory position. However, the local bank had enough confidence in Peter to be prepared to make available reasonable overdraft facilities, a budget for improved farm profitability had been worked out, the dairy herd was to be expanded, the pasture was to be improved, investment undertaken in leys and in improving cultivations, and a programme of correcting dilapidations undertaken. The tribunal concluded that Peter would succeed in generating sufficient profits to meet the increased rent due in March 1979, in putting right the dilapidations and providing his family and his mother with a reasonable standard of life.

The tribunal considered that the landlords' attack on Peter's suitability was unjustified. Too much regard had been paid to the condition of the farm as left by a tenant who had died recently following a long and incapacitating illness. The forecasts of a deteriorating position had been confounded by an admittedly marked improvement in the condition of the holdings. A reasonable landlord, having regard to all relevant matters, would have found him to be suitable, and the tribunal were unanimously of this opinion.

Despite the careful argument of Mr Stone on behalf of the landlords, the tribunal rejected the suggestion that irremediable breaches of covenant had been committed, or, if they had, that Peter should be regarded as responsible for them. The quantum of any claim for dilapidations was a matter for an arbitrator in March 1979. The tribunal expressed confidence that the landlords would not press the claim for dilapidations so as to destroy Peter's cash resources.

Landlords' Applications for Consent to Notice to Quit

Both the Earl and the trustees applied in accordance with section 22 of the 1976 Act for consent to the operation of the notices to quit in respect of the larger and the smaller part of the farm, which, as already mentioned, were held under separate tenancies. The grounds of the applications were in each case, pursuant to section 22(2) of the 1976 Act, those set out in section 25(1) (a) and (b) of the Agricultural Holdings Act 1948. So far as the applications were based on section 25(1)(a) the purpose stated was for the landlord or a member of his family to farm the land, but it appeared that the family's intention was in fact to sell Patrick's Farm. There was no evidence as to how or by whom the land would be farmed pending sale. It seemed likely that the holdings would be sold piecemeal. The tribunal felt that it was impossible on these proposals to conclude that their implementation would serve the interests of good husbandry.

So far as the applications were based on section 25(1)(b) of the 1948 Act (sound management) the only evidence given was that, in the case of the larger holding let by the Earl, the sale of the land would assist in the payment of taxes, so as to preserve the balance of the Packington Estate generally. The tribunal held that the case made went simply to the financial position of the landlord and had nothing to do with the management of the farms on the estate. The tribunal referred to the statement of Ashworth J in *National Coal Board v Naylor* [1972] 1 WLR 908 at p 914: "It seems to me that those approaches to the problem, namely, those involving the landlord's financial position rather than the actual management of the farm, are not in accord with the intention of the section." In the case of the smaller holding the trustees had given no evidence of the purpose for which they intended to use the land if granted possession.

No reference had been made to hardship and the tribunal had no material on which to find that a fair and reasonable landlord would in either case insist on possession. The tribunal did not believe that he would do so. (This point did not really arise for decision.)

Tribunal's Order

The tribunal:

(a) Dismissed the application by Mrs Hulme.

(b) Granted the application by Peter Hulme for directions entitling him to tenancies of both the land let by the Earl and the Earl's trustees.

(c) Dismissed the application by the Earl for consent to the notice to quit served on the personal representatives of the deceased tenant in respect of the larger holding.

(d) Dismissed the application by the trustees for consent to the notice to quit served in respect of the smaller holding.

(e) Ordered the respondent landlords to pay Peter Hulme 25 per cent of his costs of the applications by and against him on scale 4 of the County Court Scale, to be taxed if not agreed. The ground for this order was that it was oppressive to introduce a large quantity of new material about dilapidations on the first day of the hearing, raising issues which the landlords in general failed to substantiate in the event.

LANDS TRIBUNAL
COMPENSATION

A **LORD BRECON'S EXORS v BRITISH WATERWAYS BOARD**
(REF/48/1975)
(SIR DOUGLAS FRANK QC PRESIDENT)
July 29 1977

Estates Gazette January 7 1978
(1977) 245 EG 57

Extinguishment of riparian owner's right to use boats on Monmouthshire and Brecon Canal—Claim under Transport Act 1968, s 105(7)—Claimants' own house and two cottages B adjoining canal—Contractual right to use one boat for 20 years—Tribunal rejects claimants' valuation based on estimated increases in licence charges resulting from inflation, applying Young v Percival principle—Board's valuation of " one bid more " at auction by estate purchaser also rejected—Award of £400 for each of three boats on basis of sale of dwellings to different occupiers all wishing to use boats

A L Price QC and G B W Nurse (instructed by Jeffreys & Powell, of Brecon) appeared for the claimants; P Weitzman QC and Miss D R Cotton (instructed by the solicitor to the British Waterways Board) for the authority.

C Giving his decision, THE PRESIDENT said: It was provided in a local Act of 1793 authorising the construction of the Monmouthshire and Brecon Canal that:

> it shall be lawful for the owners and occupiers of any land adjoining to the said canal . . . to use upon the same any pleasure boats . . . without the interruption of the said company of proprietors, or any persons acting under them, and without paying any rate for the same, so as such boats be not above 5 ft in breadth or above 12 ft in length and do not pass through any lock to be made on the said canal or any such cut. . . .

The late Lord Brecon was the owner of a house called Greenhill and two cottages used for the accommodation of D staff adjoining the canal. In 1970 he applied for a licence to put a boat on the canal without prejudice to any rights he had as a riparian owner. He was told that such rights were abolished by section 115(1)(a) of the Transport Act 1968. In fact it was section 105(5)(a) which abolished all such rights. However, subsection (7) provides thus:

> any person who suffers loss by reason of the extinguishment by virtue of subsection (5) or (6) of this section of any private right shall be entitled to be paid by the board compensation to be determined, in case of dispute, by the appropriate tribunal; and the tribunal shall, in determining the compensation, take into account any contractual right offered by the board in substitution E for the right which is extinguished.

The parties were unable to agree as to the amount of compensation, so that the matter was referred for determination by the Lands Tribunal.

Two Questions of Law

The matter first came before me in July 1975, when I determined two questions of law. The first was that the right created by the statute of 1793 was to use upon the canal a boat used for pleasure however propelled. That determination was on the board's submission that the right did not F extend to power-propelled boats. Secondly, I determined that compensation is not payable only for a financial loss. I then adjourned the matter *sine die*, and the matter did not come before me again until last month; in the meantime Lord Brecon died.

The board have offered, and the claimants have accepted, the right to use one boat on the canal for 20 years and all parties agree that that is a contractual right offered and accepted for the purpose of subsection (7) of section 105 of the Act of 1968. However, the claimants say that that alone is insufficient compensation because it does not compensate for the payments which will have to be made at the expira- G tion of 20 years, or for the use of more than one boat on the canal, nor does it take account of inflation. It was said for the claimants that although the number of boats was not restricted under the local Act, a reasonable compromise assumption to make would be that an average of two would be used, and the [claimants'] valuer [John L Harpur FSVA FIArb, of John L Harpur & Co, Brecon] made that assumption in his valuation. He then took 100 years as being in perpetuity and subdivided that into 10-year periods, applying rates of inflation at the present existing rate for the first three 10-year periods, and thereafter 5 per cent per annum. H Then for each 10-year period he applied a year's purchase of 5 per cent and by reference to valuation tables ascertained the present value of the resulting sum at 5 per cent. The resultant sum payable for the first boat, having allowed 20 years' free user, amounts to £1,677 and for the second boat, £1,951.

In support of this method of valuation Mr Price said that we are here measuring a burden, namely the capital equivalent of having to pay year by year, and likened it to the measure of damage in tort. He referred to *General Tire & Rubber Co v Firestone Tyre & Rubber Co* [1975] 1 WLR 819; *Curwen v James* [1963] 1 WLR 748; *Taylor v O'Connor* [1971] AC 115; J *Wrotham Park Estate Co v Parkside Homes Ltd* [1974] 2 All ER 321; and *S J C Construction Co Ltd's Application* (1974) 230 EG 906. As an alternative valuation the claimants' valuer assessed the depreciation in the market value of the property by reason of the loss of boating rights at £4,500, although he readily agreed that there were no transactions to which he could refer.

The main contention of the board was that rule (2) of section 5 of the Land Compensation Act 1961 applies to this case, so that the compensation has to be assessed on two assumed sales, one with the benefit of the right and the other without it. Their valuer [R Pomfret FRICS, of Digby K Turner & Co, Abergavenny] said that a purchaser would not be prepared to pay more than a nominal amount extra for the claimants' estate because of the boating rights which were enjoyed with it; and by nominal he meant one bid more at an auction, which he put at about £250. He referred to three other properties adjoining the canal which although put on the market had not been sold, but I did not find them helpful.

Mr Weitzman said that the claim here is for compensation concerning a right in land and it followed from that and from section 66 of the Act of 1961 that that Act applied and therefore the difference in market value was the right test. He conceded that a dearth of evidence made it difficult to ascertain values, but he maintained that the market test was not to be rejected unless another method was likely to be more valid. Alternatively, he argued that the claimants' method is based on pure speculation and further that, having regard to *Taylor v O'Connor* [1971] AC 115 and *Young v Percival* [1974] 3 All ER 677, no regard should be had to inflation. He submitted that if the proper basis is on the cost of a licence today then the maximum which should be paid is 10 years' purchase for one boat; that would be £165.

Mr Price in reply said that the 10 yearly increases in the estimated licence fees were not for estimated inflation as such but for the likely increase in fees resulting from inflation. Charges had risen substantially in recent years and it was to be expected that they would go on rising to keep pace with increases in prices generally; the board did not dissent from that forecast. Thus the case was distinguishable from the two cases cited by Mr Weitzman, and he further contended that the reason for the court's decision in those cases was the absence of evidence of the probable rate of inflation, whereas in this case, he argued, there was ample evidence of a continuing increase in charges.

Canal Use a Valuable Right

Having walked a considerable length of this canal, including that part of it passing the claimants' land, and seen more of it by car, I agree with Mr Price that the canal runs through an area of sensational natural scenery. The claimants' property is situated in an area of great natural beauty and has fine outlooks. The canal offers to those using it tranquillity and beauty of a very high order and I am left in little doubt that the right to use the canal was a valuable one for the adjoining owners. But how to assess that value is a more difficult matter.

I reject the claimants' valuation based as it is on estimated increases in licence charges. I think that the principle stated in *Young v Percival* (*supra*) applies and that I am bound by it. Even were it otherwise the forecasts of increases are no more than guesses, for no one can reliably forecast the inflation rate even 10 years hence. Moreover, the claimants have adopted the board's charges for a licence which entitles a boat to be used on any of the board's canals, whereas the right under the Act of 1793 is limited to the distance between the nearest locks, here 7.6 miles. I appreciate that there may not be much additional value to be attributed to the use generally of the board's canals, but the restriction between locks is a different matter, for the canal is 33 miles long and I apprehend that that is a right appreciably more valuable than for 7.6 miles.

On the other hand I do not agree that this is a case for "one bid more" nor even that that is ever a valid method of assessing compensation. Although subsection (7) of section 105 of the Act of 1968 refers to a loss suffered by a person and not to the depreciation of a value of interest in land, I think that where the compensatable interest is an interest in land, as is the case here, then compensation should be assessed by reference to depreciation in value. That view seems to be supported by cases under section 278 of the Public Health Act 1936; see, for example, *Taylor v Oldham Corporation* [1876] 4 Ch D 395. It does not follow from that that one is disbarred from using any evidence which may be available to ascertain that depreciation in value, including the value of the right in the owner's hands, and I do not reject the claimants' method of valuation merely because it is not apparently based on a market transaction, but rather because it is based on wild guesses.

I propose to assume first of all that the house and the two cottages were sold with the prospect of occupation by three different occupiers all wishing to enjoy the use of the canal, for each of whom that had been one factor in the decision to buy. The claimants' valuer said that that is the way he would market the property and I accept that that would be sensible. I think it realistic to assume that the occupiers of each of the cottages would use one boat and the occupier of the house two boats. However, having regard to the right of the claimants to use one boat for 20 years I think that that right is adequate compensation for the loss of the free use of that one boat.

Now as to the other boats. The vendor and purchaser being both willing and the latter wanting the canal use, I have to reach a view as to how far the purchaser would go and what the vendor would accept. In my judgment, the claimants' valuer substantially overvalues the benefit of the right and the board's valuer, Mr Pomfret, substantially undervalues it.

I bear in mind that the licence fee for a 12-ft craft was £33 for 1976 and that it is common ground that it will continue to increase. It is reasonable to suppose that purchasers would make that assumption, but also that the fee would be payable if they wished to use the canals beyond the two locks.

The conclusion I have reached is that a purchaser would pay an additional £400 for each of three boats. Accordingly I determine that the proper sum of compensation is £1,200 and that is the amount of my award.

The authority shall pay to the claimants their costs of this reference, such costs unless agreed to be taxed on scale 4 of the County Court scales of costs by the registrar of the Lands Tribunal.

SERVICE WELDING LTD v TYNE AND WEAR COUNTY COUNCIL

(Ref/124/1976)

(V G WELLINGS QC)

October 6 1977

Estates Gazette January 14 1978

(1977) 245 EG 143

Newcastle upon Tyne factory acquired by agreement—Question of entitlement to compensation for bank interest and charges on overdraft—No suitable alternative premises available—Capital expended on purchase of site and erection and equipping of factory equivalent in size and equipment to original—Expenditure flowed directly from reasonable mitigation of loss—Tribunal's "strong doubt" whether issue a question of law—"More properly one of fact and degree and perhaps expert evidence"—Parties ask tribunal to resolve matter inferentially from agreed facts—Decision for claimants

D G Robins (instructed by Ingledew, Mark Pybus, of Newcastle upon Tyne) appeared for the claimants; M H Spence (instructed by the solicitor to the Tyne and Wear County Council) for the acquiring authority.

Giving his decision, MR WELLINGS said: This is a preliminary issue, alleged by the parties to be one of law, as to whether the claimants are entitled under rule (6) of section 5 of the Land Compensation Act 1961 to recover compensation for certain bank interest paid by them.

The claimants, who are a company of welding engineers, owned a factory at Albert Street and Russell Place, Newcastle upon Tyne. The factory was built in 1953 and had 6,986 sq ft of gross floor area. It was equipped with a travelling crane and special high-voltage switch gear and electrical installations for arc welding.

By an agreement in writing dated July 4 1973, the claimants agreed to sell the factory to the acquiring authority at such price as might be agreed between them or in default of agreement within six months after the date of the agreement as might be determined by the Lands Tribunal as on a compulsory acquisition pursuant to the provisions of the Highways Act 1959 and other statutes mentioned in the agreement and as if notice to treat in respect of the property had been served on the claimants on July 4 1973.

Agreed Facts

The following facts and matters were agreed:

(1) After July 4 1973 the claimants purchased a piece of land at Brunswick Industrial Estate at Wideopen, Tyne and Wear, on which they erected a factory comprising 15,700 sq ft of gross floor-space and equipped the factory in manner similar to that in which the original factory was equipped.

(2) The new factory was ready for occupation and the claimants took occupation of it on July 24 1974.

(3) The acquiring authority took possession of the original factory on September 19 1974; their delay in taking possession after July 24 1974 was not due to any fault on the part of either party.

(4) At no time was there available in the open market an alternative factory suitable for the relocation of the claimants' business and the purchase of the alternative site and the erection upon it by the claimants of the new factory were necessary. Indeed in taking this course the claimants were mitigating their loss and reasonably mitigating it, because they would otherwise have been entitled to claim from the acquiring authority compensation for loss of goodwill upon the basis that their business had been extinguished and in a substantial sum.

(5) The capital necessary for the purchase of the new site and the erection and equipping of the new factory was procured by the claimants from their bank in the form of an overdraft upon and in respect of which the claimants paid interest and charges in accordance with their bank's normal terms of business.

(6) The claimants are not entitled to be compensated on the basis of equivalent reinstatement under rule (5) of section 5 of the Land Compensation Act 1961; their rights to compensation arise under rules (2) and (6) of that section.

The preliminary issue relates to the questions whether the claimants are entitled to recover the bank interest and charges paid by them on and in respect of the overdraft and if so in what amount.

This question constitutes the only item remaining in dispute between the parties. A sum of £53,464 on account of compensation has been agreed between the parties and paid by the acquiring authority to the claimants by instalments, the first instalment having been paid on December 16 1974. That sum is made up as follows:

(1) For the claimants' freehold interest in the original factory		£38,250
(2) For removal expenses and certain other items of disturbance (the details of which are not relevant for present purposes)	£14,149	
(3) By way of further compensation for disturbance: for duplication of expenditure such as double overheads and double rates attributable to the period in which both factories were available for occupation between July 24 1974 and September 19 1974	£1,065	
		£15,214
Grand total		£53,464

Three Bases of Assessment

The parties have agreed, in the event that I should hold that the claimants are entitled to compensation in respect of bank interest and charges, upon three different assessments, depending upon which of three different bases I decide to apply. The assessments and bases are as follows:

(1) If (as the claimants primarily contend) the compensation under this head ought to be assessed on the basis that the total expenditure on the new factory (excluding bank interest and charges) had never exceeded £58.509 (being a sum which it is agreed would have been the cost of building and equipping a new factory of equivalent size with equivalent equipment to those of the original factory), the agreed assessment is £4.285.

(2) If (as the claimants secondarily contend) the compensation ought to be assessed on the basis that the total expenditure on the new factory (excluding bank interest and charges) had never exceeded the sum of £52.399 (that is to say the agreed compensation of £53.464 less £1,065 for duplicated expenditure), the agreed assessment is £3,985.

(3) If (as the acquiring authority would contend if the claimants were entitled to any compensation under this head) the compensation ought to be assessed on the basis that the total expenditure on the new factory (excluding bank interest and charges) had never exceeded £38.250 (the aggregate of compensation for disturbance so far agreed less £1,065 for duplicated expenditure), the agreed assessment is £2.235.

Mr Robins, for the claimants, said that the bank interest and charges were expended in order to service the necessary capital and this expenditure was directly consequent upon the taking: see *Harvey v Crawley Development Corporation* [1957] 1 QB 485 at 492 per Denning LJ. The expenditure, he said, was within the same category as double overheads or bridging loans which were accepted as attracting compensation for disturbance: see *B & T (Essex) v Shoreditch Corporation* (1958) 9 P & CR 471 in which an award of compensation for double overheads and bank interest was made by the tribunal); *Coulson v Borough of Bury St Edmunds* (1969) 210 EG 1245 (interest on bridging loan). Mr Robins submitted that although the expenditure in question was anterior in time to the date of the taking, the expenditure nevertheless flowed directly from the taking. In the alternative, the claimants had reasonably mitigated their loss, the expenditure had flowed directly from the mitigation and, upon principle, the claimants were entitled to compensation for it for that reason.

Mr Spence for the acquiring authority said that the claimants were not entitled to compensation under this head because it was to be presumed that in the price which they paid for the land and the erection and equipping of the new factory they had got value for money in that when they came to sell the new factory they would receive a value which would include or reflect the interest paid. The expenditure was in the same category as architects' fees: see the decision of the tribunal in *W Rought v West Suffolk County Council* (1954) 164 EG 101, where it was held that architects' design drawings for a building alternative to the premises acquired but never put into effect were of value to the claimants and should not be the subject of compensation.

Threat to Adjourn

During the argument I expressed strong doubt as to whether the preliminary issue was a question of law at all and threatened to adjourn this matter and fix a date for the hearing of the reference in order to enable the parties to call evidence including, if they so desired, expert evidence. I was asked by counsel for the parties, in order to save costs, not to take that course but to resolve the matter as best I could. In order to assist me in that course they came to certain further agreements as follows:

(1) That in any building project there are four elements in the cost of the completed building:

(a) the cost of the site;

(b) payments to the building contractor;

(c) architect's and other fees;

(d) interest charges or loss of interest on all the first three elements.

(2) That the tribunal (that is to say I), as arbitrator, should be at liberty to draw any inferences it might think necessary from the agreed facts or any of them, including the four elements stated above, and should be treated as empowered by the agree-

A ment dated July 4 1973 to use my own expertise and draw such conclusions as to what a purchaser might take into account with respect to the value of the new factory as the tribunal might think fit and to use judicial knowledge in the widest possible manner.

I heard very little argument as to which of the three bases of assessment would be correct if I came to the conclusion that the claimants were entitled to compensation under this head. I heard no argument in respect to the third basis except that Mr Robins said that it was an impossible basis.

There is no doubt that in purchasing the new site and erecting and equipping the new factory the claimants were B mitigating their loss and taking reasonable steps to do so. In my opinion, whether or not the expenditure in question flowed directly from the taking, it, in so far as it was reasonable in amount, flowed directly from the mitigation: see *Halsbury's, Laws of England,* 4th ed, vol 12, para 1193. In the circumstances the claimants are, in my judgment, *prima facie* entitled to compensation for at least some of the expenditure. There is, I believe, no presumption, either of law or fact, that the claimants have received value for money expended. I am unwilling to make any assumption as to what a prospective purchaser or vendor would take into account in the matter of bank interest and charges except that I think it is likely that C the value of the new factory would be precisely the same whether or not any bank interest and charges had been paid. It appears to me that the preliminary issue is more properly one of fact and degree and perhaps expert evidence rather than of law. I have not been satisfied that the claimants received value for money for the interest and charges paid away. Accordingly I shall decide the issue in favour of the claimants.

The question which is the proper basis is, in my view, one of reasonableness. In my opinion, it was reasonable of the claimants to expend capital upon the purchase of the site and D upon the erection and equipping of a factory equivalent in size to that of the original factory. I thus find in favour of the first basis with the result that the compensation to which the claimants are entitled for bank interest and charges amounts to £4,285. The acquiring authority will pay to the claimants their costs of this preliminary issue such costs if not agreed to be taxed by the Registrar of the Lands Tribunal on the High Court Scale.

E

J D BRITTON & SON LTD v LONDON BOROUGH OF HILLINGDON
(REF/202/1976)

(W H REES FRICS)

September 7 1977

Estates Gazette January 28 1978

(1977) 245 EG 317

Small area of land at Ickenham, Middx, separating development site from road—Planning consent for development site provides for access from another road, but line of access passes through house within conservation area— F **Access across subject land substantially shorter—Cost saving of £10,000—Claimants' valuation approach of proportion of cost saving upheld, but at 25 per cent instead of 75 per cent**

M H Spence (instructed by H B Wedlake Saint & Co) appeared for the claimant; J A R Grove (instructed by the solicitor to Hillingdon Borough Council) for the acquiring authority.

Giving his decision, MR REES said: This is a reference to determine the compensation to be paid for the acquisition of a small piece of land at the end of Boniface Road, Ickenham, Middlesex. The subject land is irregular in shape and the agreed area is 0.0066 acre. The claimants alleged that its G importance lies in the fact that it separates a 2-acre site known as The Buntings, Swakeleys Road, Ickenham, from Boniface Road. The claimants at one time considered that the subject land was "a golden strip," albeit of nine carats, but by the time the matter came to a hearing before me the claimants were thinking in terms of a baser metal because they claimed £7,500 for the value of their freehold interest. The acquiring authority considered that compensation should be on the basis of existing use value, £200, an agreed figure between the parties on this footing.

The subject land is being acquired under the London Borough of Hillingdon (Swakeleys Road, Ickenham) (no 2) H Compulsory Purchase Order 1974. This order includes the 2-acre site adjoining the subject land, and other land adjoining the 2-acre site towards its southern end was acquired under another compulsory purchase order at the same time. After a joint inquiry into these two orders, that referring to the subject land was confirmed by the Secretary of State with modification on February 18 1975. Notice to treat was served and entry was made on the land on February 2 1976, which thus becomes the date at which the valuation for compensation is to be made.

Towards the southern end of the 2-acre site is a detached house about 50 years old, the accommodation of which is J on two floors and comprises four bedrooms, a bathroom, three reception rooms, kitchen etc. All main services were connected to this property at the relevant date, but the property was then in a derelict condition and unsuitable for occupation.

Planning History

The planning history, which mainly relates to the 2-acre site and not to the subject land, is a complicated one. At the relevant date both the subject land and the 2-acre site were allocated for residential use in the initial development K plan for Greater London at a density of 60-70 habitable rooms per acre. The 2-acre site together with land adjoining is in a conservation area and the 2-acre site is the subject of a tree preservation order. Between 1964 and 1966 the claimants extended Boniface Road eastwards and erected about 22 homes on each side of the extension on an area of land (known as the Mallows Estate) adjoining the western boundary of the 2-acre site. I assume that they sold these houses, but it is certain that they retained the subject land.

At the end of 1972 a firm of developers, Royco Homes Ltd ("Royco") contracted to purchase land which comprised, in the main, the 2-acre site but only part of the L house itself, the boundary going through the building. This contract was subject to two conditions: first, that Royco had the right to enter on to the land retained by the vendor to demolish that part of the house which was excluded from the sale and, second, Royco were under an obligation to construct a vehicular access to the land retained by the vendor at Royco's expense. This purchase was completed in October 1973.

In May 1973, Royco applied for planning permission for residential development on the land they had purchased, but this was refused in August 1973. However, the planning authority indicated that permission might be forthcoming for M a rather similar form of development and so in October 1973 Royco applied again. This application was granted subject to conditions in December 1973, but Royco were successful in obtaining an order of certiorari from the High Court in March 1974 quashing this grant of planning permission and an order for mandamus was adjourned *sine die* with liberty to apply in case it subsequently proved necessary (see *R v Hillingdon London Borough Council ex parte Royco Homes Ltd* [1974] 2 WLR 805). These orders related to the conditions attached to the planning permission granted in December 1973, which Royco considered onerous. As a result the application made in October 1973 was granted planning consent in April 1974, but the onerous conditions were excluded, and this planning permission was still in existence at the relevant valuation date. In May 1974 the acquiring authority contracted to purchase the 2-acre site from Royco for £200,000 and this purchase was completed in June 1974. The planning consent granted in April 1974 provides for access to be off Swakeleys Road, not from Boniface Road. The conservation area was designated on January 29 1970, the house, The Buntings, being within it.

Under the Town and Country Planning Amenities Act 1974 (now section 277A of the Town and Country Planning Act 1971) it is necessary to have planning permission to demolish buildings in such an area. One of the plans submitted by Royco in connection with their planning applications showed the line of the access road from Swakeleys Road to the 2-acre site as passing through the house, The Buntings, itself, thus necessitating its demolition in whole or in part. So, after the Town and Country Planning Amenities Act 1974 came into effect, either the local planning authority would have to grant permission for the house to be demolished in whole or in part if the road were to follow the line indicated or the line of the road would have to be altered so as to squeeze between the house and the nearest boundary of the 2-acre site.

Council's Scheme

In December 1975 the council was granted planning consent for the erection of 36 aged persons dwellings together with a warden's flat and guest room on land comprising the 2-acre site and some adjoining land. The plan accompanying this application showed an access for the 2-acre site being via Boniface Road, with another access for a proposed hostel to be built on adjoining land, access being via the 2-acre site.

The parties agreed a number of values for the 2-acre site on various bases, but I do not need to set out all of these; by the time of the hearing before me some were of no significance. Both parties asked me to ignore the council's scheme and thus I have to consider what would have happened if the council had not come on to the scene in 1975-76.

I note that if, apart from the council's scheme, access were to be obtained from Boniface Road the parties agreed that the length of road required to be constructed within the 2-acre site for its development would be 200 ft less than if access were to be from Swakeleys Road and that, mainly on this account, but also because of other factors, the cost saving would be £10,000.

Counsel for the claimants called M T Godfrey FRICS [partner of Biscoe & Stanton, of London WC1]; counsel for the authority called R Dickins RIBA MRTPI, Assistant Director of Planning (Implementation), London Borough of Hillingdon, and K F Salmon ARICS, Assistant Borough Valuer, London Borough of Hillingdon.

Mr Godfrey said that in his opinion the owner of the 2-acre site would pay 75 per cent of the cost saving which was agreed at £10,000 for the subject land (ie he would pay £7,500). He considered that physically it would be possible to squeeze an access road from Swakeleys Road to the 2-acre site past the house, The Buntings, but this would take no account of any possible building line and no developer of the 2-acre site would prefer to deal with the matter in this way; he would prefer to demolish the house in whole or in part. In his opinion an access from Boniface Road would give a slight increase in value to the 2-acre site because it was a safer access for children than one from Swakeleys Road. Because of the shops in Swakeleys Road opposite the 2-acre site cars would be likely to be parked in any access road to the 2-acre site from Swakeleys Road which was a dual carriageway. He saw no reason why planning permission would not be granted to change the access from Swakeleys Road to one from Boniface Road and in this connection he pointed to the planning permission granted in December 1975 for the council's development which showed an access from Boniface Road.

Under cross-examination, he agreed that the three factors which determined the percentage of 75 which he had chosen, were:

(1) Section 277A of the Town and Country Planning Act 1971;

(2) The 2-acre site would be enhanced in value because of the better access; and

(3) The flats, when built, would fetch £250 more each when sold because the site would be more attractive with an access from Boniface Road.

He did not agree that if an application were made for development of the 2-acre site with access from Boniface Road, there would be more delay than there would in dealing with the matters reserved, which included access, when the consent was granted in April 1974: both might have taken a long time to be dealt with. He did not consider that the Community Land Act 1975 would cause difficulties. Because of the difficulties of section 277A, in his view the local planning authority would not hesitate to grant consent if an application were made showing the main access to the 2-acre site from Boniface Road. He referred to the report of the local inquiry in this connection and he said that he knew of no indication that the London Borough of Hillingdon thought that an access from Boniface Road would be a poor one. If the main access to the 2-acre site were to be from Boniface Road he would envisage a pedestrian access to Swakeleys Road.

Mr Dickins said that, in his opinion, if the sole access to the 2-acre site were to be from Boniface Road this would be a substantial change and would require an entirely new planning application for development of the 2-acre site. The matter could not be dealt with by variation or by an application relating solely to the access and to small pieces of land required to bring about the alteration. To have the main access through Boniface Road would result in generation of traffic through a residential area. He thought that a fresh planning application in respect of the 2 acres of land with access off Boniface Road would have to go to appeal to the Secretary of State and that it would take nine to 12 months to get the result. There was a great public interest in the development of the 2-acre site. He doubted if planning permission for the demolition of the house, The Buntings, would be granted but thought that it would be acceptable for the access road to be squeezed between the house and the nearest boundary.

Delay in Getting Planning Permission

Mr Salmon said that he did not consider that a developer of the 2-acre site would be prepared to pay more than the existing use value of the subject land (agreed at £200). A developer would prefer the access to be from Swakeleys Road and such a developer would prefer to rely on the existing planning permission (ie that granted in April 1974, which showed access off Swakeleys Road) in this sensitive area. If a developer attempted to vary the planning permission there would be a long delay which would be at the developer's expense: such a delay might be up to 12 months. If the approach used by the claimants, a proportion of the cost saving, were correct, he would adopt only 25 per cent of £10,000. He would do so because there would be a delay in getting planning permission for an access from Boniface Road, and if an entirely new application were to be required the result might be a lower, not a higher, density than that envisaged in the consent of April 1974.

The owner of the subject land negotiating with the owner of the 2-acre site would keep in mind that anyone other than the owner of the 2-acre site would only give a nominal sum for it and thus the owner of the subject land would not wish finally to break off negotiations with the owner of the 2-acre site. He had never heard of a percentage as high as 75 in these circumstances. He considered that the owner of the 2-acre site was in a strong negotiating position, *vis-à-vis* the owner of the subject land. On this approach he preferred a lump sum rather than a percentage and he put this at £2,000. He did not consider that access from Boniface Road was to be preferred to access from Swakeleys Road. He thought that the local planning authority would permit the access road from Swakeleys Road to be squeezed between the house, The Buntings, and the nearest boundary. Under cross-examination, he agreed that access from Boniface Road had the advantage of no car parking for shopping purposes and that it would be safer for children, but he thought that there would be both advantages and disadvantages to prospective purchasers of any flats built on the 2-acre site.

I heard evidence as to possible accesses from Boniface Road to the 2-acre site over the gardens of numbers 2 and 28 Boniface Road, which adjoin the subject land, but counsel for the authority conceded that neither of these was likely to be approved.

I viewed the subject land, the surrounding area and most of the 2-acre site.

I accept the submission made by counsel for the claimants that the owner of the 2-acre site would be willing to pay more than £200 for the subject land at the relevant date: that figure would be a very attractive one to the owner of the 2-acre site. I accept the submission made by counsel for the authority that 75 per cent of £10,000 is too high, particularly because the owner of the 2-acre site at the relevant date did not have planning permission for an access road from Boniface Road and without such permission there would be no cost saving. There might be a long delay in obtaining such a consent: it might never be granted.

The advantage which the owner of the 2-acre site would obtain by purchasing the subject land would be that he could offer an alternative course to the local planning authority to those which they already had, viz: either to give permission for the demolition in whole or in part of the house, The Buntings, or to agree that the access road from Swakeleys Road would have to be squeezed between the house and the nearest boundary. Once the 2-acre site and the subject land were in the same ownership, the planning authority could (on application being made) grant access from Boniface Road and leave the house, The Buntings, standing in its entirety, unaffected by any road close to it.

I think that the owner of the 2-acre site would relate his bid to a percentage of the possible cost saving of £10,000 rather than pick a lump sum out of the air, and for this reason I prefer Mr Salmon's figure of £2,500 to his £2,000. £2,500 is a small proportion of £200,000 paid for the 2-acre site and an even smaller proportion of the total costs of its development. The compensation to be paid for the subject land will therefore be £2,500, in addition to which the acquiring authority will pay the claimants a surveyor's fee based on scale 5(a) of the scale of charges of the Royal Institution of Chartered Surveyors.

Having opened the sealed offer dated June 21 1977 lodged by the respondent authority and addressed to the claimants' surveyors, I find that the amount thereof is less than the amount of my award. Having been addressed by the parties, I order the acquiring authority to pay the claimants' costs of this reference which, if not agreed, shall be taxed by the Registrar of the Lands Tribunal on the High Court Scale.

SUCCAMORE v LONDON BOROUGH OF NEWHAM

(REF/216/1977)

(R C WALMSLEY FRICS)

December 6 1977

Estates Gazette February 4 1978

(1977) 245 EG 404

Disturbance claim—Acquisition of Silvertown, East London, terrace house—Claimant moves to "comparatively modern" bungalow 31 miles away—"Crawley" costs—Tribunal allows in full (1) solicitor's fees for both parties to mortgage advance, (2) removal expenses as "reasonable consequence," (3) survey fees and search fee on abortive purchase although incurred before building society advance secured, (4) survey fee on property purchased, and (5) travelling expenses

H M Boggis-Rolfe (instructed by Crane & Hawkins) appeared for the claimant; A B Kitson for the acquiring authority.

Giving his decision, MR WALMSLEY said: This reference arises from the compulsory acquisition of a house, 5 Wythes Road, E16, by the London Borough of Newham from the owner-occupier Sydney Oliver Thomas Succamore. The compensation payable for the claimant's freehold interest has been settled, at £7,500. The notice of reference to the tribunal describes the nature of the dispute as "the level of disturbances suffered by the claimant and the amount of compensation properly payable therefor in consequence."

The acquisition is by virtue of the Newham (Drew Road No 1 Area) Compulsory Purchase Order 1974. Notice to treat was served on October 17 1975 and possession was taken on October 28 1976. No 5 Wythes Road was a two-storey mid-terrace three-bedroom house. The claimant first sought to relocate himself in a house about 10 miles away, 10 Bruce Avenue, Hornchurch, but that purchase was abortive. Later he purchased, for £13,250, a two-bedroom bungalow about 31 miles away, 30 Ethelred Gardens, Wickford, Essex.

The claim presented before me was in the sum of £588.66 made up as follows:

1.	Solicitor's costs and disbursements on purchase of 30 Ethelred Gardens, Wickford	£199.05
2.	Removal expenses	53.46
3.	Surveyor's structural survey fees on abortive purchase of 10 Bruce Avenue, Hornchurch	56.16
4.	Local search fee re 10 Bruce Avenue	5.25
5.	Surveyor's structural survey fees on purchase of 30 Ethelred Gardens, Wickford	65.34
6.	Purchase of "essential carpets"	140.00
7.	Purchase of "essential curtains"	46.00
8.	Expenses incurred travelling to view properties	13.00
9.	Fitting of gas stove	5.00
10.	Telephone connection charge	5.40
		£588.66

H M Boggis-Rolfe, appearing for the claimant, called evidence from M A Ashton, an associate solicitor with the claimant's solicitors, Crane & Hawkins. A B Kitson, appearing for the authority, called two witnesses from the valuation division of the Department of Engineering and Surveying of the London Borough of Newham—D R M Bennett BSc (Est Man) and A G Helps FSVA. The claimant himself was not present at the hearing, being apparently on an oil rig in the North Sea. I will take each item of claim in turn.

Item 1. Legal Costs

Mr Ashton told me that the item of £199.05 represented a charge of nine hours' time spent at £20 per hour rounded down to £175, plus VAT £14, and with disbursements added of £10.05. The account rendered to the claimant had shown the £175 as being apportioned ("on a guesstimate basis"), as follows:

To our professional charges in connection with the purchase of this property on your behalf at a price of £13,250 .... £125

A To our further professional charges in the grant to you of a mortgage advance of £5,750 by the Woolwich Equitable Building Society £50

The authority challenge only the £50 relating to the mortgage advance. They are prepared to allow the amount which is shown as being the scale fee payable to the solicitor acting for a building society, on a scale published by the Building Societies Association.

I was informed that such scale fee in the present case would have been £35. Technically the building society discharge this fee to the solicitor acting for them, and then recover it from the mortgagor. I gather that, in practice, if the solicitor con-
B cerned is acting for the purchaser-mortgagor as well as for the building society, this technicality is omitted and the solicitor charges the mortgagor direct.

On studying an extract from the *Law Society's Gazette* for March/April 1971, being a council statement dealing with "Building Societies—Mortgage Charges," I observe a note therein which reads "(ii) It is also emphasised that these scales only apply to the charges of the solicitor acting for the building society as mortgagee and *not* to the mortgagor's solicitor's charges." In the present case Mr Ashton's firm were acting for both parties to the mortgage advance. As I understand the position, their charge of £50 covers not only
C the work done for the building society, meriting a scale fee of £35, but also their work acting for the claimant as mortgagor.

I allow this item as claimed.

Item 2. Removal Expenses

In respect of the removal costs incurred, of £53.46, Mr Bennett is prepared to allow only £40. He reduces the amount claimed on the grounds that at the relevant time there were many properties for sale in Newham and adjoining areas, and that it was therefore unnecessary for the claimant to have relocated himself 31 miles away. If, said Mr Bennett, the
D claimant had moved only a reasonable distance, say within a radius of 10 miles, the cost of the removal would have been not more than £40.

The authority's valuer has, I think, applied the wrong test. The appropriate test is that defined by Romer LJ in *Harvey v Crawley Development Corporation* [1957] 1 QB 485, as follows:

> It seems to me that the authorities to which our attention was drawn do establish that any loss sustained by a dispossessed owner (at all events one who occupies his house) which flows from a compulsory acquisition may properly be regarded as the subject of compensation for disturbance, provided, first, that it is not too remote and, secondly, that it is the natural and reason-
> E able consequence of the dispossession of the owner.

Applying the above test I find that the removal expenditure of £53.46 (a) was a loss sustained by the claimant as a dispossessed owner, (b) was a loss flowing from a compulsory acquisition, (c) was not a remote loss, and (d) was a natural consequence of his dispossession. I accept that the claimant's removal to a distance of 31 miles was not a *necessary* consequence of the dispossession, but the authority have failed to persuade me that the cost of that removal was not a *reasonable* consequence of the dispossession.

I allow this item as claimed.

F *Items 3 and 4. Survey fees and search fee on abortive purchase*

At the stage when the claimant was proposing to purchase 10 Bruce Avenue, Hornchurch, he instructed a firm of surveyors to carry out a structural survey of that house, and his solicitors instituted a local search. The acquiring authority do not concede either of these items. In Mr Bennett's opinion a prudent purchaser would have secured his building society advance before incurring expenditure on either a structural survey or a local search. I do not accept this. I prefer the evidence given by Mr Ashton that, except perhaps in a problem area where a building society might reject any advance, it is both reasonable and prudent for a prospective purchaser to obtain a structural survey report, and have a G
local seach made, at the same time as, or before, making an approach to a building society. The surveyor's report on 10 Bruce Avenue, I was told, had proved adverse; and the building society had attached conditions to the making of any advance.

I allow these items as claimed.

Item 5. Survey fees on the property purchased

In respect of the structural survey fees, of £65.34, incurred on the property purchased at Wickford, Mr Bennett is prepared to allow only £50. He reduces the amount claimed on the grounds that these fees relate to a property worth £13,250, H
and that the claimant could have found a property to buy at a price more comparable with his own. "That," said Mr Bennett, "would have cost him only £50." With respect, this latter statement seems to me a *non sequitur*. The property purchased by the claimant is, from its description, a comparatively modern bungalow. The work involved in the structural survey that was carried out may well have been no more than would have been involved in a survey of some older and less valuable property.

I allow this item as claimed.

Items 6 and 7. Carpets and curtains

The claimant purchased two fitted carpets in his bungalow J
at Wickford and he also bought curtains. In the absence of the claimant himself, no explanation was forthcoming as to why the amounts claimed, of £140 and £46, had been upped from the figures shown in a disturbance allowance claim lodged by the claimant's solicitors on November 19 1976, which were:

Purchase of fitted carpets from vendor of 30 Ethelred Gardens (bill can be supplied if required)	£120.00
Curtains for 30 Ethelred Gardens (bill mislaid)	£17.25

Mr Bennett is prepared to concede only £30 to cover both these items. I have studied a "Valuer's survey in anticipation K
of disturbance claim" taken by Mr Helps at 5 Wythes Road on December 12 1975. This gives details of the furnishings in that house, room by room. In respect of the curtains I allow £17.25. As regards carpets, it was doubtless sensible for the claimant to take over the carpets as fitted in two rooms in his new bungalow, and to pay £120 for these. I doubt not, however, that he both got good value for his money and also is now much better carpeted than he was before. I allow £30.

Item 8. Travelling

In the detailed claim just referred to, dated November 19 1976, this item is shown as: L

Expenses incurred in travelling to view properties:

Petrol	£7.50
Rail fares	£3.20
	£10.70

Mr Bennett is prepared to concede only £5, on the grounds that there was no necessity for the claimant to search as far out as Wickford.

I allow this item at £10.70.

Items 9 and 10. Fitting of gas stove and telephone connection

The claim for these items is not contested. M

Summarising my findings:

1.	Legal costs	£199.05
2.	Removal expenses	53.46
3.	Survey fees on abortive purchase	56.16
4.	Search fee on abortive purchase	5.25
5.	Survey fees on property purchased	65.34
6.	Carpets	30.00
7.	Curtains	17.25
8.	Travelling expenses	10.70
9.	Fitting of gas stove	5.00
10.	Telephone connection	5.40
		£447.61

A I determine the compensation payable by the acquiring authority to the claimant in respect of disturbance in the sum of £447.61.

Having opened the sealed offer of compensation lodged by the acquiring authority I find that the amount thereof is slightly less than the amount of my award. The acquiring authority will pay the claimant his costs of this reference, such costs unless agreed to be taxed by the Registrar of the Lands Tribunal on Scale 2 of the County Court Scales of Costs.

B

IND COOPE (LONDON) LTD v ENFIELD LONDON BOROUGH

(REF/182/1976)

(J D RUSSELL-DAVIS FRICS)

December 21 1977

Estates Gazette March 18 1978

(1977) 245 EG 942

Off-licence premises with tied tenancy at Enfield—Refusal of alternative accommodation in new unit ¼-mile away held C not unreasonable—Commercial decision which claimants entitled to take—Off-licences to be treated no differently from other retail premises in valuation approach—Average trade basis rejected—Tribunal accepts that tied trade has more security than free one or than goodwill of any other retail business

Charles Fay (instructed by Routh Stacey Pengelly & Boulton) appeared for the claimants; Michael Fitzgerald (instructed by solicitor to London Borough of Enfield) for the acquiring authority.

Giving his decision, MR RUSSELL-DAVIS said: This is a D reference to determine the amount of compensation to be paid to the claimants, Ind Coope (London) Ltd, by the London Borough of Enfield in respect of the compulsory acquisition of the freehold interest in the Bush Hill Stores, 24-26 Fourth Avenue, Bush Hill Park, Enfield.

The subject land comprised two adjoining terraced properties on the east side of Fourth Avenue situated in a densely populated residential area between Main Avenue, the local shopping thoroughfare, and Lincoln Road, a short distance west of the Great Cambridge Road, A10, and about half a mile from Enfield Town. The premises were once dwelling-houses, but had been adapted to form an off-licensed E shop, stockroom, stores and dwelling-house all on two floors, and a small cellar, the two premises being intercommunicating, and there was a garage in rear.

Small Net Population Gain

The subject land formed part of the Enfield (The Avenues) Compulsory Purchase Order 1970, made under Part V of the Housing Act 1957, which was duly confirmed by the Secretary of State for the Environment. Notice to treat was dated December 10 1971, to which a formal statement of claim was returned on January 26 1972 in the sum of £21,050, plus fees, and for the loss on forced sale of the inventory. After formal F notice, entry was taken on November 22 1974, which is the date of valuation for this purpose. The premises have now been demolished and the area is in course of redevelopment for residential purposes: 276 dwellings including 22 shops are being replaced with 300 dwellings and 9 shops, so there is a small net population gain.

The premises were let on December 16 1968 by the claimants to W J Warby on the usual form of agreement for such premises on yearly tenancy from November 12 1968, subject to six months' notice on either side, at a rent of £377 per annum, tenant paying rates and being responsible for internal repairs. The tenant accepted a tie to purchase beer, cider, wines and spirits etc from the landlords and this tenancy G was subsisting at the date of entry.

The claim for the inventory has been settled and paid to the tenant and is not the subject of this reference.

For the claimants Mr Fay called J H Hayns FRICS, a partner in Harold Williams & Partners, of London, who has experience in valuation of licensed properties. Mr Fitzgerald called S P Mason FRICS, regional licensed property valuer, Inland Revenue, for London, on the approach generally to valuation of licensed property, P G Elliott ARICS, assistant valuer, Enfield Council, who gave factual evidence on structural condition and repair, and J E Hartley ARICS, principal valuer in the Borough Valuer's Department. H

The valuations as at November 22 1974 are as set out opposite (abridged).

A. *Mr Hayns' Primary Valuation* (*C*2) as revised at the hearing:

Average trade 1970–74 incl adjusted for rising trend			
Bottled beer 150 barrels at £7.50 pb	£1,125		
1,525 gals wines and spirits at 90ppg	£1,373	£2,498	
Add tied rent as 658 converted barrels @ £1.75		£1,152	£3,650
(Note 3 gals of wines and spirits = 1 barrel)			
Less repairs			250 J
			£3,400
Years' purchase			10
		Valuation	£34,000

B. *Alternative Valuation* (*C*12)

Property				
Shop (no 26) 198 sq ft @ £2	£396			
Rear office 142 sq ft @ 75p	106			
Stock 152 sq ft @ 60p	91			
Store (no 24) 238 sq ft @ £2	476			K
Cellar	25			
Forecourt	25	£1,119		
Residential 762 sq ft @ 75p		571	£1,690	
Garage			50	
			£1,740	
Years' purchase			11	
			£19,140 say	£19,150
Loss of brewers' wholesale profits				
Trade as ascertained above	£2,498			L
Years' purchase	5		£12,490 say	£12,500
			Valuation	£31,650

C. *Mr Hartley's Primary Valuation* (A2(a)) (compared with B)

Freehold premises with vacant possession			
Shop (no 26) 198 sq ft @ £2	£396		
Rear office and stock room 142 sq ft @ 75p	106		
Stock/kitchen 152 sq ft @ 60p	91		
Store (no 24) 238 sq ft @ £1	238		M
Cellar	25		
Forecourt	25		
Residential 762 sq ft @ 75p	571		
Garage in yard	50	£1,502 say	£1,500
Years' purchase @ 10% deferred 1 year			9.09
			£13,635
Term tied rent	£377		
Less external repairs	200	£177	
Years' purchase 1 year @ 7%		0.9	say 150

A

*Going concern—goodwill of brewers**			
Estimated trade disturbance			
Average trade 1971–73	£2,400		
Years' purchase	2		
	£4,800	say	5,000
			£18,785
		say	£19,000
Less compensation paid to tenant under section 20 of the Compulsory Purchase Act 1965			
Business claim £365 × 1 YP			365
		Valuation	£18,635

B

D. *Check Valuation* (compared with A)

Tied rent fixed in 1968	£377			
Estimated for 11/1974	2¼	£848		
Less repairs 10%		84	£764	
Brewers' profits (average 1971-73)			2,400	3,164
Years' purchase				6
				£18,984
			Valuation, say	£19,000

C * Since alternative premises were refused it is in doubt whether this should be on a going concern basis.

The fundamental differences between the parties are:

(1) Whether the claimants can be held to have mitigated their loss by their refusal to accept alternative premises offered to them by the authority in Main Avenue close by;

(2) Valuation approach, whether valuations A or D, or B or C should be adopted; and on that being decided;

(3) Differences in make-up of comparative valuations.

(1) *Mitigation of Loss*

D As part of the redevelopment scheme the authority redeveloped the north side of Main Avenue, eastwards from the junction of Fourth Avenue as far as the schools. They had available a lock-up shop (unit 5) of 360 sq ft or a larger unit of 750 sq ft. On inquiry from the claimants in October 1973 the premises were offered in February 1974 on 20-year lease at £700 per annum exclusive for the smaller and £1,300 for the larger, subject to five-year reviews. The premises were to be fitted out (including shop fronts) by the lessee. The rent for the smaller unit was reduced on June 7 1974 to £650, subject to a trade restriction as an off-licence only. A self-contained flat, above, in the same parade, but not intercom-
E municating, was available at £7.07 per week for three bedrooms (comparable with the subject property). On July 1 1974 the claimants declined without giving reasons, other than —"After considering the matter in great detail, I regret to inform you that we do not feel able to proceed with taking a unit on the terms outlined." In the outcome the smaller unit was taken by a greengrocer, who assigned to an independent off-licensee in January 1977.

Reasons for Refusal of Alternative Premises

No representative of the claimants gave evidence, but Mr
F Hayns enumerated the reasons for refusal as follows:

(a) Loss of population in the locality. In the event there is a small gain.

(b) One unit would be too small. Although the sales area would be about the same, stock storage space would be substantially less.

(c) Substitution of freehold by tenancy. The outgoings in the way of rent for shop, flat and garage coupled with interest on, and a sinking fund to replace, capital expended in fitting out especially of the larger unit, would make greater inroads into the net trade of the business than could be afforded. The rent would be subject to review.

(d) Main Avenue already had an established grocer's off-licence with cut-prices, opposite and in competition with which the catchment would have to be shared. The claimants considered that they had the exclusive business in their existing catchment area around Fourth Avenue, which but for the scheme would have endured; it was quite different from Main Avenue, despite changes in shopping trends, in that liquor bought and carried away is said to form part of the housewife's daily shopping. G

(e) No integral living accommodation and hence loss of security of stock and of the image of the resident "your friendly off-licensee."

I have viewed the locality including the local shopping centre in Main Avenue. Having heard the arguments I think the refusal of the alternative accommodation was not unreasonable and was valid. There was no compulsion on the claimants to continue to trade a quarter of a mile away with the interruption of business as a result. It was a commercial decision they were in their discretion entitled to take. Thus I find that the claimants cannot be held to have failed to mitigate their loss. H

(2) *Valuation Approach*

Mr Hayns asserted that the only correct approach is to ascertain the loss to the brewer claimants, and that this can only be done in the traditional way by valuing the trade on the average over three or five years up to the acquisition with adjustment for rising trends in the latter years, as shown in valuation A. J

The respondent authority say that as at 1974 the correct approach should be the market value as a freehold shop plus compensation for disturbance, or extinguishment of the business, as shown in valuation C.

Errol Committee's Recognition of Changes

Cogent evidence for the authority was given by Mr Mason, who is an experienced licensed property valuer, indicating that since monopoly value was abolished by the Licensing Act 1961 and resale price maintenance by 1964, off-licensed premises were no different from any other sort of retail shop. Off-licences were more freely granted, and the modern trend was that supermarkets, grocers and other stores, unlike public houses, sold liquor outside licensing hours. My attention was drawn to the report of the Departmental Committee on Liquor Licensing (the Errol Committee) of December 1972, which recognised the changes, and said at paragraph 2.21: K

> After a number of cases in which the decisions of the justices had been overruled, it became increasingly rare for applications to be refused simply on the grounds that a "need" did not exist. This change of practice was not immediately universal, and there are still occasionally refusals on the grounds of lack of need; but, generally speaking, it is now easier to obtain an off-licence. . . . Second, the 1961 Act changed the permitted hours of off-licensing hours of the district, and multiple stores and supermarkets appeared to take the view that it would be more trouble than it was worth to have a separate off-licence section, which had to conform to licensing hours, when the rest of the shop was open for business. The 1961 Act removed this obstacle by enabling off-licences to remain open during normal shop hours. The last major development was the abolition of resale price maintenance in 1964. This tempted more supermarkets and discount houses into the field. Off-licence sections in supermarkets are now very common and it is estimated that their numbers will grow. L M

In 1968, said Mr Mason, his department had come to the conclusion that, unlike public houses, there was no longer any essential difference between off-licences and other retail premises, and in preparation for the 1973 rating revaluation responsibility for such valuations had been transferred to district valuers' offices for all purposes.

He had not seen or valued the subject property, but he considered that Mr Hartley's Valuation C was the right

A approach, and that it recognised loss to the claimants by valuing the premises, and then the trade under rule (6) of section 5 of the Land Compensation Act 1961.

Comparables

Mr Hayns produced details of two settlements under compulsory acquisition, both for normal off-licences. One, the Star at Canning Town (valuation date April 1975), was, he said, agreed with the district valuer's office in 1975 on his method; and the other, the Cornwall Stores, Westbourne Park (valuation date 1973), with the valuer's department of West-
B minster City Council in August 1976 likewise. But I am not convinced that the latter at least was not his analysis of an agreement which had been reached by valuers by differing approaches. Two other off-licences belonging to the same claimants, the Merton at Watford (June 1973) and the Napier at Ashford, Middlesex (June 1976), has been sold to off-licensees free of tie; I understand, however, that Mr Hayns was not concerned personally, but with knowledge of the trade supplied to him he was able to analyse the prices by his method.

Mr Hartley produced three relevant comparables, which his office had handled for acquisition, and one with which he had
C been personally concerned; all had been valued by his method and so agreed with claimants' valuers.

I think that Mr Mason's evidence, which was given with authority, has much force, and in the light of modern trends, having heard the arguments, I cannot see any real reason to treat off-licences as at November 1974 any differently from other retail shops. I accept that the trade loss, including the value of the tie, can be fully reflected in the disturbance claim on value-to-claimant basis, as is implicit in any claim which comes under rule (6). Accordingly I have to consider the merits of Mr Hayns' valuation B, which he helpfully produced during the hearing at the request of the tribunal, alongside
D Mr Hartley's valuation C.

(3) *Differences between the two valuers shown in valuations B and C*

The freehold premises

(a) At the outset I should say that, having heard the evidence of Mr Hayns and Mr Elliott, I accept that the subject premises were in reasonably good repair, and, while under modern conditions not ideal in layout, they did give ample space for bulky stock storage, which I judge to be essential at busy seasons. I consider that the premises could, but for the scheme, have had a useful life for many years.

E The estimated rental value to be capitalised is: Mr Hayns £1,740; Mr Hartley £1,500. The only difference between the two valuers is in respect of the store (no 24) estimated at £2 and £1 per sq ft respectively. This was in the front part of the house facing on to the forecourt with an opening cut in the party wall to the sales area in the front of no 26, where the street entrance was. The passage to the dwelling-house was partitioned off on the other side. Unlike no 26, no 24 had no shop front other than a facia and stall riser, between which was fitted a wall of translucent glass bricks.

Mr Hayns said it could be used equally as a shop or shop extension to that existing and was worth the same; he had
F put 75p on the office-store in rear of the shop, which represented the less valuable storage element in the premises. Mr Hartley did not deny that it could be used as selling space and that no planning permission would have been required for its use or adaptation; it had been used simply as a store by the claimants and there was no need for more selling space, and in that situation there would have been no demand for a separate underletting to another trade.

I take the view that this part served as a useful and valuable adjunct to the shop, convenient, and in which some display could have been set out. For an alternative trade, for example a small local draper, customers could move from one
part to another on the same level without inhibition. Nevertheless, I come nearer to Mr Hartley's opinion and find for G
£1.25, which brings this figure to £297 instead of £238, and the estimated rental of this whole to £1,561, say £1,560.

(b) *Multiplier*. Mr Hayns used 11 years' purchase without deferment. Mr Hartley deferred for one year at 7 per cent and then applied the 10 per cent tables.

The tied tenancy was from year to year at a low rent of £377, fixed in 1968, but subject to three months' notice on either side. It seems that the claimants' policy is to review tied rents every five years, but in view of the compulsory purchase order did not do so here in November 1973. Such review
would certainly have resulted in some increase in the rent, H
and I cannot in the circumstances find that there is any case for a deferrer.

On the principal multiplier, that for Mr Hartley's five other acquisitions, adduced, varied from 9 to 10¼ YP. That for 68 Main Avenue, on the corner of Fourth Avenue, a shop and house (not an off-licence) was 10 and I think offers the best comparison. Mr Hayns did not support his figure of 11 by any comparable. Thus I find for 10 YP, giving a capital value of the freehold of £15,600.

(c) *Deduction for compensation to tenant*. In the no-scheme world there would have been no question of deter-
mining the tied tenancy. It might have been renewed, but J
there would be no compensation payable to the tenant. I can find no logical reason for the deduction of £365.

Loss of Trade to the Brewers

(d) *Annual income*. Mr Hayns used £2,498, say £2,500, derived from valuation A. Mr Hartley adopted £2,400, derived from the average for 1971-72-73, because he had not been supplied with the figures for 1974.

But an examination of the three years 1972-73-74 shows an average, actually, of 148 barrels of beer and 1,509 gallons of wines and spirits. The prices of £7.50 per barrel and 90p
per gallon are not in dispute. There is no proof that in K
1975-76 conditions the volume of trade would necessarily have expanded, although a rising trend might have been there: this could be reflected in the multiplier to be applied. I would not, therefore, have used Mr Hayns' adjusted figures in the uncertain prospect. The averages come to £1,110 and £1,458 respectively, making £2,468, which I adopt.

(e) *Years' Purchase*. This is the major difference between the valuers: Mr Hayns, 5 YP: Mr Hartley a spot figure of £5,000, which he later quantified as 2 YP of £2,400.

Mr Hayns thought that the security coming from a tied trade was exceptionally good and quite different from that
in a free trade. The premises enjoyed a virtual monopoly L
position in the centre of a dense residential area with no other comparable off licences within a quarter of a mile. It opened long hours and was handy for evening and Sunday visits without the need to have to carry heavy goods very far. The tenant resided on the premises and would be known personally to many of his customers. There was no likelihood of competition setting up in the immediate vicinity. The volume of trade had risen up to the inception of the scheme and continued to rise until the start of the population removal. Other brewers or off-licence chains would have been interested had the property come into the market. There was little to be
feared from the grocer's off-licence in Main Avenue, which, M
even at cut prices, had very little impact, because it really catered for a different catchment.

Value in Tie

Mr Hayns was not impressed with Mr Hartley's comparables, which he said were all at the lower end of the trade, which Bush Hill Stores by its own showing was not. He thought that 5 YP bore a proper relativity with 10, which he had used in his direct method valuation A. In such circumstances he considered 5 YP not to be too high; it was the tie wherein the value lay, unlike the retail trade of another kind of shop.

Mr Hartley said that £5,000 stood as his estimate of the loss of trade, because the claimants could have re-established themselves elsewhere. His information as to the amount of the trade was sketchy. "I had very little evidence," he said, and "the figure was somewhat plucked out of the air: it happens to be 2 years' purchase of the trade figure of £2,400." Under questioning later he said, as I understood him, that it represented total extinguishment. His statements were not very consistent, and I gained the impression that he had little experience in valuing off-licences: I must however not be thought to be criticising him for this.

An analysis of his previous dealings with off-licences in his district indicated:

(1) 13 Station Road, N11. No value attributed to the licence, because the claimants had ceased to trade.

(2) 215 Chase Road, N14. Trade not valued. Premises valued as a shop.

(3) 79 Swinnerton Street, E9. As at June 30 1972, small trade valued at 7 YP on a similar approach of "shop plus." This was an Ind Coope off-licence.

(4) 170 Town Road, N9. Valued as ordinary trade disturbance at 1½ YP of net profit.

I understand that all these were at "the lower end of the trade," and I must observe that they really afford no helpful evidence for the present case. Transaction (3) above was dealt with in Mr Hartley's previous appointment at Hackney, and there really seems no consistent explanation for 7 YP, which seems to stand out like "a sore thumb."

Mr Mason suggested that in good cases 2-3 YP of the tied trade was generally to be expected as a true reflection of a brewer's loss. But he conceded that compensation for goodwill for a retail business (he mentioned, as an example, a butcher) might realise 4 YP. But, as has already been said, he had no acquaintance with the subject property; he was not giving expert valuation evidence, but attended more to instruct on the skill of licensed valuation in support of Mr Hartley. I judge that he had no particular knowledge of off-licences in the Enfield area; nor is he to be regarded as having any expert or detailed knowledge of the valuation of goodwill of retail businesses.

Mr Hayns has much experience in this type of valuation work, and I prefer his evidence to that called for the authority. On his own showing he did not accept the "shop plus" method, and on previous occasions he had always applied the direct method, capitalising the trade and ignoring the bricks-and-mortar element, whether or not that method had also been used by those with whom he had been in negotiation. He had not been prepared to deal with this case on any other basis until he was asked by the tribunal to submit the alternative valuation, although, without disrespect, I suspect that his valuation, B(12) was drawn so as to support his primary valuation, although the total fell short by some £2,300.

Having heard the evidence, I accept that a tied trade has more security than a free one or than the goodwill of any other retail business. For the latter, in the better cases 3 YP or even more is frequently adopted.

The subject property was certainly well above "the lower end of the trade," and in the exercise of my judgment I find in this case for a multiplier of 4½, which applied to the trade, as found, of £2,468 comes to £11,106. The total valuation is thus £26,706, which I round down to £26,700.

There is one matter on which I should comment. Under cross-examination Mr Hayns was asked, if his opinion of value as at January 1972 was £21,050, the figure returned, how he now justified £34,000. His reply was that the valuation dates were different by more than 2½ years, he had access to up-to-date figures of trade, and he had gained more experience in the meanwhile from other negotiations. I accept that answer.

Accordingly, the acquiring authority will pay the claimants the sum of £26,700 in full settlement under all heads of compensation in respect of the compulsory acquisition of the Bush Hill Stores, 24-26 Fourth Avenue, Enfield, together with a surveyor's fee calculated in accordance with scale 5A of the scales of charges of the Royal Institution of Chartered Surveyors, and any legal costs properly incurred since the date of the notice to treat.

Having read this decision and opened the copy of a sealed offer made by the respondent authority to the claimants, dated November 29 1976, I find that the amount thereof is less than the award. Accordingly the respondents will pay the claimants' costs of this reference which, if not agreed, shall be taxed by the Registrar of the Lands Tribunal on the High Court scale.

REED EMPLOYMENT LTD v LONDON TRANSPORT EXECUTIVE

(Ref/59/1976)

(V G WELLINGS QC and R C WALMSLEY FRICS)

November 25 1977

Estates Gazette April 15 1978 and April 22 1978

(1977) 246 EG 233; (1977) 264 EG 323

Oxford Street, London, leasehold premises used as branch of large employment agency group with four other branches in same street—lease 5¾ years unexpired—Claim for loss of profits pre-entry and anticipated profits post-entry—Acquiring authority's view that entire profitability lay in site rejected—Pre-entry loss allowed on basis of budget "with-scheme" forecast—Claimants' figures of anticipated future profit based on "sideways projection" of average performance of Oxford Street branches held not sufficiently dependable—Tribunal bases award on forward projection of average annual level of branch contribution to company profit—Deductions made for profit rental, head office expenses, interest on working capital and "transference factor"—Claimants' valuer's year-by-year discounting method of evaluation of loss adopted—Reserve assessment on "classical approach" of applying YP multiplier to historic net profit—Inherent difficulty of this approach—Tribunal's view that "loss of goodwill" may be misleading description for business loss suffered by trader on dispossession—"Wood Mitchell" decision followed for taxation purposes—Award of £64,700

M H Spence and N C Nardecchia (instructed by Lovell, White & King) appeared for the claimants; K R Bagnall QC and D E Neuberger (instructed by the Chief Solicitor to the London Transport Executive) for the acquiring authority.

Giving their decision, MR WELLINGS and MR WALMSLEY said: This is a reference of the compensation payable to Reed Employment Ltd by London Transport Executive on the acquisition of the claimants' leasehold interest in premises 377 Oxford Street, London W1. The issue for the tribunal to resolve is that of trade disturbance. Reed assess their compensatable loss under this head at over £180,000. The authority contend that the compensatable loss is nil.

The enabling statute is the London Transport Act 1969, which empowers the London Transport Board to construct works and acquire lands. The works relevant to the present reference are part of the new underground railway currently under construction and which was formerly referred to as the Fleet Line but has been recently rechristened the Jubilee Line. The site of the subject premises is being incorporated in the rebuilt Bond Street underground station, which will serve that new line.

The premises lie to the south side of Oxford Street, within the Grosvenor Court Hotel development, which forms part of the island site bounded by Oxford Street to the north, Davies Street

A to the east, Weighhouse Street to the south, and Gilbert Street to the west. The premises immediately adjoin the entrance to Bond Street station on the existing Central Line. The accommodation, prior to the demolition of the property, was on three floors, ground, mezzanine and basement, each of about 200 sq ft in area. The frontage to Oxford Street is about 8 ft and the depth about 25 ft. The use was for an employment agency office. The claimant company is one of a group of wholly-owned subsidiaries of Reed Executive Ltd.

Notice to treat was served on the claimants on December 20 1971. A three-months' notice of entry was served on February 1 1972, but by agreement, the date for vacation was extended from B May 1 1972 to August 1 1972. The date of entry and the valuation date for our purpose would ordinarily therefore be August 1 1972. However, the claimants had in fact ceased trading nearly a month earlier and had vacated on July 7 1972. For claim purposes, it is common ground that figures of profit are sufficiently accurately taken if the valuation date be treated as having been June 30 1972.

Reed had opened this branch of theirs at 377 Oxford Street in September 1969. At the date of entry they were holding under a lease dated September 11 1969 for the unexpired term of seven years from March 25 1971, and at a rent of £3,500 per annum.

C Compensation for that interest in land has been agreed with London Transport Executive in the sum of £17,000, representing the value of a £6,500 per annum profit rental for five-and-three-quarter years. The claimants, by a deed of assignment dated March 16 1977, have indeed already assigned the residue of their leasehold interest to the authority in consideration of the principal sum of £17,000 and interest thereon of £9,840.85.

In skeleton form, the claim for compensation put forward may be expressed thus:

Interest in land (agreed and already paid)		£17,000
Trade disturbance		
Loss of profit pre-entry	£5,678	
Loss of anticipated profits post-entry	176,000	181,678
Fixtures and fittings		1,060
		£199,738

D

Also Professional fees.

As indicated, the figure for interest in land is not in issue. And by the end of the hearing the amount payable for fixtures and fittings had also been agreed, as above. The question of professional fees will be referred to at the end of this decision. We E confine our attention to compensation for trade disturbance.

At the hearing, Malcolm Spence and Nicholas Nardecchia appeared for the claimants; Kenneth Bagnall QC and David Neuberger appeared for the authority. Witnesses for the claimants included (1) J M P Cooper FCA, group finance director of Reed Employment Ltd (2) T N Doran MSc, marketing director of the Reed Employment Ltd (3) M J Whittaker, property director for Reed Executive Ltd and (4) A P Musto FRICS, one of the senior partners in the firm of Montagu Evans & Son. Witnesses for the authority included (1) J D Orme MA ACA, a partner in the firm of Robson Rhodes and (2) G A Sullivan FRICS, deputy estate manager of London Transport F Executive.

For the sake of brevity and convenience, but with no apology for errors of syntax, we shall from time to time in this decision use the following expressions:

"perm" and "temp" (for "permanent" and "temporary").

"permanent placement", "temporary placement" (the placing of an applicant in, as the case may be, a permanent or temporary situation)

"permanent staff", "temporary staff" (staff dealing with applicants for, respectively, permanent and temporary posts).

"permanent income", "temporary income" (income derived from, respectively, permanent and temporary placements).

The Claimants' Business G

We start by describing the nature of the claimants' business, the way in which that business operates and the accounting system of the company.

Reed Employment Ltd was founded in 1961. Its principal trade is the supply of permanent and temporary staff to clients, mainly for secretarial, clerk, typing and telephonist work. Reed are one of the three leading agencies in this field, the other two being Brook Street Bureau and Alfred Marks Bureau. The company operates in some 120 agency branches in England and Scotland. Branches are geographically grouped into divisions. No 377 Oxford Street was a branch within the Central 1 London Division. Each branch is the responsibility of a manager, assisted H by a temporary supervisor and one or more interviewers.

The provision of permanent staff to a client involves the payment to the agency, by the client, of a fee based as a percentage of the applicant's starting salary. The higher the salary, the greater the percentage charged. If the applicant should leave the position within a short period (formerly four weeks, but now five weeks), a pro-rata credit is issued. The income from the perm business (the "permanent income") is expressed in terms of "net invoice value" per placement, being the actual fees invoiced, less any rebates given. In respect of perms the business is founded on an excess of demand for specified skilled workers over supply. J It is purely an introduction service. It is attractive to applicants, because it offers a convenient office location for the discussion of opportunities with an interviewer. It is less expensive to clients than advertising, which is an open-ended commitment whereas agency fees are incurred only when an applicant has started work.

The provision of temporary staff to a client involves the payment, by the agency, of an hourly rate to the temp, and the charge of an hourly rate to the client. The difference in rates (the "gross margin") represents the agency profit out of which must be borne national insurance and pension contributions, leaving a net margin. The income from the temp business (the "temporary income") is expressed in terms of "net margin" per K hour sold. In respect of temps, the business is dependent on the supply of applicants, wishing to work on a short-term basis, being matched with a demand for such assistance from clients.

Reed's accounting system is highly sophisticated. Each autumn a budget forecast is made and this is done in three stages: *Stage* 1 Macro Economics. *Stage* 2 Total company figures, or "the cake". *Stage* 3 A budget for each division (and therefrom for each branch), or "the cake slice".

Stage 1 is a correlation of the company's trends with statistics of "numbers employed" as shown in the regular CBI Industrial Trends Survey. CBI statistics apparently disclosed a pattern of L peaks and troughs of numbers employed, having a cycle of just under four-and-a-half years. Reed have found that their own moving annual total ("MAT") of perm placements follows the CBI Index, but is about nine to 12 months later, and that their economic cycle of temp hours takes place about three months ahead of the placement cycle.

Stage 2 of the annual budget extrapolates the graphs of sales units in line with the above indices. Mr Doran explained one example of extrapolating the temps graph in this way:

In July 1976 the MAT was 450 hours per branch per week. We extended the graph to reach about 1,150 hours per branch per week by January 1978 before reaching a peak in 1978. We then converted M the graph into a monthly seasonal one to arrive at a budget for each month. This was done by adjusting our normal seasonal pattern with a cyclical factor.

Stage 3 Having forecast the total placements and hours for the company, the "cake" is then divided into "cake slices", according to how each division has historically contributed. Finally, each cake slice is divided into branch contributions, again based on their history. The following example was given by Mr Doran of how, in respect of their branch at 70 Oxford Street, a budget of 2,627 temp hours for the month of July 1976 came to be set:

Out of the total cake budget of 304,148 hours for July 1976 Central 1 Division was asked to contribute 39,204 hours or 12.9 per cent.

This percentage represents the division's contribution to the company hours over the previous twelve months, with a weighting towards recent months. During March, April, May 1976, 70 Oxford Street contributed 6.7 per cent of hours in Central 1 Division. 6.7 per cent of 39,204 = 2,627 hours.

In due course the performance of each branch is matched against its budget.

Locational Factors

The policy of Reed regarding the location of the company's branches was described to us in this way. The whole emphasis of the agency is centred around applicants. Success depends on an ability to appeal to those seeking employment. In Central London it has not been Reed's policy to advertise, but to concentrate instead on the quality of branch sitings, coupled with a good window "image", aimed exclusively at passing trade. For this purpose only prime sites at ground-floor level and in busy thoroughfares would be considered. In highly active locations, branches could be closely located and yet profitable; for instance, the Reed branch at 71 Oxford Street is within 50 yds of their branch at 115 Oxford Street and almost opposite their branch at 70 Oxford Street.

In 1969, Reed had the above three branches in Oxford Street, all east of Oxford Circus. They had none in the length of Oxford Street west of Oxford Circus—a length that was referred to as "the best part of the busiest pedestrian thoroughfare in Europe". The evidence was that it is difficult to set up an employment agency in this part of the "West End" zone, because neither the planning authority nor landlords favour a change of use from "shop" to "office". Reed's view was (and is) that they would be able to trade successfully in at least six sites in Oxford Street between Oxford Circus and Marble Arch. They were able, in the autumn of 1969, to secure and start trading from one such site, namely the subject property, which lies about a quarter of a mile west of Oxford Circus, in what is agreed to be one of the principal locations in Oxford Street. Then in May 1971, Reed opened a second branch in this prime length, at 271 Oxford Street, very near to Oxford Circus.

Main Claim

The main claim is in the sum of £176,000. It relates to lost anticipated profits for the unexpired period of the lease for the subject premises, viz for the five-and-three-quarter years to March 1978. Mr Spence explained that the reason Reed were not seeking to go beyond the five-and-three-quarter years was in order to avoid argument as to the projection of profit into a period contingent on a renewal of the lease being secured under the provisions of the Landlord and Tenant Act 1954.

By the date when Rule 42 documents had to be lodged in this reference Reed had in their possession full information about the performance of all their branches up to June 30 1976. They were able therefrom to analyse, in particular, the performance of their other four Oxford Street branches in respect of each of the four years from June 30 1972 to June 30 1976. And they were able to prepare budgets for each of these branches forecasting performance for the ensuing one-and-three-quarter years to March 25 1978. By the date of the hearing in July 1977, the factual information about actual performance was at hand for a further year, to June 30 1977, so that the period to be budgeted had become reduced to only nine months. We are satisfied that the statistical evidence presented by the claimants in respect of the performance by the Reed branches at 271, 71, 115 and 70 Oxford Street over the five-and-three-quarter-year period from the valuation date, can be accepted as correct.

In respect of the subject branch, Reed knew its performance up to the date of its closure, which was (for practical purposes) at the end of June 1972. That performance could be compared with the average performance, over a matching period, for the other four Oxford Street branches (an average which we shall now call "the average"). No 377 Oxford Street had apparently outperformed the average, both in its placements of perms and in its number of hours sold of temps. The calculated ratio for the perm income by reference to the average was 1.4035398:1 and for the temp income it was 1.6187708:1. On the assumption that no 377 would have continued at this same level of outperformance over the next five-and-three-quarter years, it became, for Reed, no more than a matter of intricate arithmetic to calculate, by a sideways projection, the net profit which 377 Oxford Street would have contributed to the company's total income over that period had it not been closed. The outcome of this arithmetic was presented to us by Mr Cooper, supported by a mass of statistics, in the form shown below.

At this stage Mr Musto took over the assessment of compensatable loss. He adopted the figures of projected net profit as calculated by Mr Cooper and therefrom he arrived at the claim figure of £176,000 in the following way [see opposite]:

Legal Argument in Rebuttal

The primary case for the authority is that the claimants are, as a matter of law, not entitled to any compensation for trade disturbance. As part of this submission, Mr Bagnall makes two points:

(1) relying on admissions, elicited by him in cross-examination of Mr Cooper and Mr Whittaker, that the whole of the profitability of the claimants' business arises from the position of their site, Mr Bagnall says that all the profitability has already been compensated for in the compensation agreed for the interest acquired and that any further claim to profitability under rule (6) fails for duality.

(2) anticipated future profit cannot in law be the subject of a claim under rule (6): it is not capable of being forecast and is too remote.

	Actual		Projected					
	July 1970 to June 1971	July 1971 to June 1972	July 1972 to June 1973	July 1973 to June 1974	July 1974 to June 1975	July 1975 to June 1976	July 1976 to June 1977	July 1977 to June 1978
Total turnover	45,189	61,100	171,378	241,499	204,566	126,167	153,818	210,755
Permanent income	18,822	20,727	48,398	63,263	68,914	47,052	50,026	51,717
Temporary income	7,574	15,048	22,626	34,009	35,640	24,494	31,916	47,909
Total gross income	26,396	35,775	71,024	97,272	104,554	71,546	81,942	99,626
Expenses	13,772	13,323	15,557	17,853	20,656	25,961	28,123	22,472
Profit rental	5,937	6,500	6,500	6,500	6,500	6,500	6,500	4,875
Total costs	19,709	19,823	22,057	24,353	27,156	32,461	34,623	27,347
Net profit	6,687	15,952	48,967	72,919	77,398	39,085	47,319	72,279

In support of his submissions Mr Bagnall referred *inter alia* to *Collins v Feltham Urban District Council* [1937] 4 All ER 189; *George Wimpey & Co v Middlesex County Council* [1938] 3 All ER 781; *Watson v Secretary of State for Air* (1954) 5 P & CR 13; *D McEwing & Sons v Renfrew County Council* (1959) 11 P & CR 306; *Bailey v Derby Corporation* [1965] 1 WLR 213; *Hoey v Glasgow Corporation* 1972 SC 200.

We do not accept that the entire profitability of the claimants' business lay in the site. We prefer the view (expressed by Mr Musto and confirmed by Mr Orme) that, while the location of the business was important because of the steady stream of potential applicants for employment passing the doors of the shop, it was the window display and presentation which attracted them into the subject premises and it was the skill of the claimants' management and staff, in matching applicants to jobs, which played the crucial part in earning profits for the claimants.

The cases cited by Mr Bagnall in our view do not support his submissions, even by analogy. For the most part they are cases in which the *ratio decidendi* was in effect the same as that in *Horn v Sunderland Corporation* [1941] 2 KB 26, that is to say the claimant received compensation for his interest acquired, not on the basis of its existing use value (as in the present case) but having regard to the development potential of the land, with the result that he could not claim trade disturbance because he would have to disturb himself in order to obtain the higher value: see *Horn v Sunderland Corporation supra*, at p 35 *per* Sir Wilfrid Greene MR. *McEwing's* case, which was strongly relied upon by Mr Bagnall, is in that category and expressly stated to be so by the Court of Session in *Hoey v Glasgow Corporation* 1972 SC 200. Looked at in another way, the cases cited by Mr Bagnall are examples of cases in which the land acquired was in the nature of the claimant's stock-in-trade or raw material of his business. Thus Lord Clyde (Lord President) in *McEwing's* case said, at p 311: "By their processing of this raw material and selling the result they anticipate making a profit. They are entitled to the market value of their raw material, so that they may use the surrogatum for making profits in other ways, but if they get the price of their raw material they cannot also get something in respect of the profit they hoped to make upon it. For, if so, they would have the means of securing that profit twice over. Prospective future profits on *future prospective developments* cannot therefore be claimed in addition to the market value of the land" (our italics). The authorities cited by Mr Bagnall do not establish that anticipated future profits of

Year July 1972 *to June* 1973		
Net profit	£48,967	
Add back depreciation (note A)	460	
	49,427	
Deduct Corporation Tax 43% (note B)	21,253	
	28,174	
Discounted 1 year 9¼% (note C)	0.913	£25,722
Year July 1973 *to June* 1974		
Net profit	£72,919	
Add back depreciation	460	
	73,379	
Deduct Corporation Tax 52%	38,157	
	35,222	
Discounted 2 years at 13¾%	0.803	28,283
Year July 1974 *to June* 1975		
Net profit	£77,398	
Add back depreciation	460	
	77,858	
Deduct Corporation Tax 52%	40,486	
	37,372	
Discounted 3 years at 12 %	0.717	26,796
Year July 1975 *to June* 1976		
Net profit	£39,085	
Add back depreciation	460	
	39,545	
Deduct Corporation Tax 52%	20,563	
	18,982	
Discounted 4 years at 13½%	0.632	11,997
Year July 1976 *to June* 1977		
Net profit	£47,319	
Add back depreciation	460	
	47,779	
Deduct Corporation Tax 52%	24,845	
	22,934	
Discounted 5 years at 13½%	0.556	12,751
Nine months July 1977 *to March* 1978		
Net profit	£72,279	
Add back depreciation (9 months)	343	
	72,622	
Deduct Corporation Tax 52%	37,763	
	34,859	
Discounted 5¾ years at 13½%	0.504	17,569
Total net deferred lost taxed profits in 5¾ years expressed as at the valuation date		£123,118
Adjustment to obviate duplication of Corporation Tax and Capital Gains Tax. Gross-up to allow for 30% tax on award		52,764
		£175,882
	say	£176,000

Note A. The depreciation allowed in the accounts of the company makes a deduction for depreciation in the original price for the lease As a deduction has already been made in arriving at the net profit figures above for profit rental, the original depreciation allowance requires to be added back in order to avoid duplication.

Note B. The 43% rate of Corporation Tax for the year to June 1973 has been calculated on the basis of nine months at 40% and the remaining three months at 52%.

Note C. The interest rate on the discounting factor has been based, *for each year*, at a level 2% above the prevailing Bank Rate.

an existing business on land paid for at existing use value cannot be the subject of separate compensation. Whether or not they can be forecast is a question of evidence and inference from evidence. We do not agree that they are too remote. We think, with respect, that Mr Bagnall confuses remoteness with proof.

For these reasons we hold that the claimants are not disentitled in law from claiming compensation for trade disturbance and we turn to consider the evidence presented on behalf of the authority.

Evidence in Rebuttal

The secondary case for the authority is that, even assuming entitlement, the claimants have not satisfactorily established any loss at all. This, as will be seen, is the effect of the evidence given by Mr Orme.

The basis of assessment adopted by Reed, said Mr Orme, was that which would be applicable to the total extinguishment of an isolated business. But the business carried on at 377 Oxford Street was not isolated, it formed part of a totality. And the branch business itself was not wholly lost. Reed were conducting a business of employment agency work through many branches: the maintenance and growth of this business in terms of volume and profit had continued unchecked by the closure of their branch at 377 Oxford Street; turnover, for instance, had increased from £3.3m in 1971 to £4.7m in 1972 and to £9.7m in 1973. The claimants were not to be compensated for the loss of some whole business, but for the element of disturbance caused to an otherwise continuing business. In any event what is actually lost on the acquisition of any business was not future profits (which do not exist) but the opportunity of carrying on the business at that location for such length of time as the occupant might reasonably contemplate.

While accepting that Reed's system of budget forecasting was well founded, Mr Orme emphasised that his acceptance of its general accuracy related solely to the company results as a whole and did not include the dividing up of the company "cake" into divisional "cake slices" and the sub-dividing of those slices into branch contributions. Nor did Mr Orme accept that it was practical to make a sideways projection based on the assumption that the anticipated future performance of any one branch would have been the same as the average actual performance for any group of other branches; whereas the general performance of a group of branches might, on average, behave on a particular pattern, the results of individual branches could well be erratic. Mr Orme's reservations were so great that he had found himself unable to offer any alternative projection derived from the statistical data underlying Mr Cooper's figures of projected net profit.

What Mr Orme had done instead was to look primarily at the contribution made by 377 Oxford Street to the company's profit in 1970 and in 1971, and to a document dated July 19 1972 prepared by Reed themselves in connection with their original compensation claim (which had been lodged on January 7 1972 in the total sum of £110,000). The relevance of this particular document ("document 22") became increasingly apparent during the course of the hearing. It shows the results (actual and budgeted) of the subject branch for 1972, as follows:

Half year	*Budgeted*	*Actual* (*see Note*)
January to June 1972		
Temp income (net)	£7,072	£8,333
Perm income	10,433	8,260
Total income	£17,505	£16,593
Less Branch expenses	6,612	6,675
Branch contribution	£10,893	£9,918
July to December 1972		
Temp income (net)	£10,319	N/A
Perm income	7,616	N/A
Total income	£17,935	N/A
Less Branch expenses	6,648	N/A
Branch contribution	£11,287	N/A

Note (*by the tribunal*). The actual figures shown above are as now agreed. They differ only slightly from the actual figures as shown in document 22.

The historic record of the business of the branch over the two previous *calendar* years had been:

	1970	*1971*
Temp income (net)	£5,855	£10,353
Perm income	21,150	20,723
Total income	£27,005	£31,076
Less Branch expenses	14,584	12,875
Branch contribution	£12,421	£18,201

Mr Orme considered that the appropriate level of future profits, estimated as at July 1972, was to be found from the past record of the branch and what might have then been reasonably contemplated for the future. Having looked at (a) the actual results of the branch business prior to closure (b) the budgeted results for January to June 1972 and for July to December 1972, and (c) the actual results for the first half-year of 1972 until closure, Mr Orme expressed the opinion that "both in historic terms and in terms of what might reasonably be contemplated at the date of closure, the appropriate level of branch profits, before any further adjustments, is £20,000 per annum."

Mr Orme turned his attention next to the question of adjustments. He first deducted the profit rental of £6,500, thereby reducing the level of branch profits to £13,500 per annum. Then he deducted Corporation Tax at 40 per cent leaving post-tax profits of £8,100 per annum. He further considered that head office expenses and working capital should be brought into the calculation. Finally, there was transference of business to be considered: he had taken note that on the closure of 377 Oxford Street, there had been an increase in the perm business at 271 Oxford Street and also an increase in the temp business at the Reed branch in Queensway. Bringing the increased business at these two branches into account, Mr Orme came to the conclusion that "the total transfer was sufficient for no compensation to be payable for loss of profits."

Loss of Profit Pre-entry

Before dealing with the main claim, which relates to loss of anticipated profits post-entry, we propose to consider the smaller claim of £5,678, which relates to loss of profit (as alleged) pre-entry. Reed say that, as a result of disturbance caused by the authority, their business fell away during the two or three months before the branch closed.

As to what had been the ambient circumstances between April and June 1972, we had evidence for the claimants from Mr Whittaker and from one Susan Ormsby. On this same matter M P O'Connell gave evidence for the authority.

Mr Whittaker's evidence was that demolition of nearby premises by the authority began in May 1972, creating a dusty environment which caused pedestrians to avoid the length of pavement in front of no 377. During the last weeks in June, the staff there were complaining that the noise outside was interfering with telephone calls, and that the permeation of dust was discouraging applicants. In view of the worsening conditions, two staff were considered sufficient; and after both these girls became ill, it was decided to close down the branch on July 7 1972, three weeks ahead of the agreed date for vacation.

Susan Ormsby (we do not know how she should be described) is a senior group manager for the South Division of Reed who was posted to 377 Oxford Street in the capacity of a floating manager to be responsible for the closure of the branch. Her evidence, given on affidavit, was that:

> During the three weeks that I worked there I was responsible for the continuing day to day business of the branch and I was given one person to help me carry out those duties. The branch had a large turnover on both temporary and permanent business and normally operated with a staff of four people, namely a Branch Manager, a Temporary Supervisor and two Permanent Interviewers. The branch premises were divided between three levels, ground floor being manned by the Manager and one Interviewer, the first floor manned by the Temporary Supervisor and dealing with temporary business, and the basement manned by the remaining Interviewer. It was hoped that both myself and the Interviewer would maintain the business despite the fact that it would not be possible to bring relief staff in from other branches.
>
> During the three weeks we worked under appalling conditions with constant pressure. From 9 am to 5.30 pm we suffered the constant noise from pneumatic drills working directly outside the entrance on the ground floor and the dust that resulted from this, along with the high July temperature, resulted in both of us suffering from exhaustion. In fact there were two occasions when my colleague actually fainted and this was directly related to the conditions under which we worked. As we could only work on the ground, ie street level, there was no relief in any way.

We registered approximately 15 applicants a day though a high proportion were catering staff which are not placeable. At times, we had perhaps five people actually waiting to be interviewed—in some cases they had to be left sitting on the stairs because of the shortage of space. Those applicants that we could work on were sent out on interviews but as for the remainder they were lost from us. Ideally we should have walked them to our nearest branch at 271 Oxford Street but practically this was out of the question.

We told them to register at the aforementioned branch and gave them directions but many of them did not arrive and I believe this was due to the competition from other agencies in between. Eventually we were obliged to close because conditions became too bad to continue.

Mr O'Connell is employed by the consulting engineers responsible for the design and supervision of the Fleet (Jubilee) Line. From February 22 1972, Mr O'Connell had been deputy resident engineer on the site and he was able to give factual evidence from his written records. Outside the subject premises areas of pavement were apparently excavated over a period from September 27 1971 to April 22 1972, and again for a further period from May 15 1972 to July 9 1972. All this work, however, which included the use of pneumatic drills, had been carried out at night; and during the daytime the pavement excavations had been covered with iron plates. During the period May 3 1972 to July 19 1972, garage premises at the rear of 377 Oxford Street and fronting to Weighhouse Street were being demolished: this work was carried out in the daytime between 7 am and 7 pm; the outside walls of the garage were left standing during that period and precautions, including the use of sprays, had been taken against dust nuisance. In June 1972 there had been the daytime use of pneumatic drills under the road surface of Oxford Street, in the basement of Lilley & Skinner opposite to no 377, in connection with work for a necessary sewer diversion.

Having heard the evidence we are satisfied that there was some considerable disturbance of quiet enjoyment during several weeks preceding the date of closure. The surroundings of no 377 had already become less attractive to pedestrians, in that adjoining properties were either already closed or were in the process of their business being run down prior to closure: no 379 on one side had been closed since March 25 1972 and no 375 on the other side was running down for closure on August 15 1972. Discomfort and loss of staff morale are, of course, not compensatable items, but the conditions described lend colour and background to the assertion by Reed that there was a loss of profit also.

The claim figure put forward is made up in this way. During the six months to June 1972, no 377 effected 123 placements of perms whereas, on a sideways projection from the performance of the other four Oxford Street branches (as explained earlier), the expectation would have been for 211 placements over that period. Each placement was at that time bringing in a net income of £64.52, and the shortfall of 88 placements represented, therefore, a loss of profit of £5,678. We note, however, as to this, that some of the apparent shortfall in the perm business was offset by an increase in the temp business: the temp supervisor at no 377 was, on the evidence, exceptionally capable and in consequence the branch had become temp orientated.

A more dependable indicator of any loss of profit is, we think, provided by document 22, the budget figures in which had been prepared in or about October 1971. Reed had been aware since August of that year that the subject premises were to be compulsorily acquired, and their budget of branch contribution made some two months later would therefore be a "with scheme" forecast. As such it was, inferentially, lower if anything than it would have been in the deemed "no scheme world." Although we shall have observations to make later about the reliability of Mr Cooper's sideways projection of anticipated profits for no 377, we see no cause to doubt the general dependability of the forecast made by Reed in October 1971 in respect of this branch (or of any other of their branches) for the ensuing half year and full year. At that time, Reed were not using the sophisticated system of accounting that we have described: prior to 1973, Mr Doran told us, the company's practice had been for the central controller to make an estimation of branch performance and then to sit down with the branch manager and decide what the branch contribution should be.

For the half-year to June 1972, Reed had forecast a branch contribution for no 377 of £10,893. The actual contribution that the branch had made was £9,918. Mr Orme had regarded this shortfall as an indication of a less-than-successful branch. We regard the shortfall as an indication of loss sustained by the claimants as a direct consequence of the authority's scheme. We allow £1,000 as loss of profit pre-entry.

Approach to Main Claim

Returning now to the main claim, this relates to the value to the claimants of the loss of their ability to derive a future profit out of the premises from which they had been dispossessed. The measure of compensation payable was described in *W Rought Ltd v West Suffolk County Council* [1955] 2 QB 338 at p 348, by Lord Evershed MR in this way:

> What the payer has to pay by way of compensation is, as stated by Scott LJ in the passage I have already twice read, a sum so as to put, so far as money can do it, the owner in the same position as if his land had not been taken from him; and this, I observe, is exactly the same measure as the measure of damages applied to the case of a man liable to pay compensation for breach of contract or, for that matter (where there is no question of punitive damages), in tort.

In considering what has been the claimants' loss, we have in mind also Lord Macnaghten's words in the *Bwllfa* case (*Bwllfa & Merthyr Dare Steam Collieries* (1891) *Ltd v Pontypridd Waterworks Co* [1903] AC 426 at p 431):

> If the question goes to arbitration, the arbitrator's duty is to determine the amount of compensation payable. In order to enable him to come to a just and true decision it is his duty, I think, to avail himself of all information at hand at the time of making his award which may be laid before him. Why should he listen to conjecture on a matter which has become an accomplished fact? Why should he guess when he can calculate? With the light before him, why should he shut his eyes and grope in the dark?

We deem it proper, therefore, while putting ourselves at the date of valuation in 1972 and looking ahead, to assess the future in the light of all the facts that have become known by the date of the hearing. Those known facts include, of course, the performance figures for the claimants' other branches.

It is well established that loss of profits by reason of expulsion from premises depends on the profits being earnable, although not earned. Mr Bagnall submits that there is a fundamental difference between the right to earn a profit and the right to recover a known loss. We recognise that there is a difference but it does not seem to us a fundamental one; a known loss can be quantified with certainty whereas any quantification of a loss *in futuro* can be no more than an estimate. We are conscious that any assessment of anticipated profits can be no more than the estimation of a likelihood. In the present case, for instance, supervening factors could have arisen to render invalid any particular projection over the period to March 1978, as for instance:

(a) Reed might conceivably have decided to sell their interest before the expiration of their lease. In the Report and Accounts for Reed Executive Ltd for 1971, we note a sentence in the chairman's statement that reads "The group's activities are such that our property portfolio now includes many shops in the leading retailing centres in the country, and it should from time to time present the possibility of realising profits as a result of development."

(b) Reed might have found themselves unable to continue recruiting personnel of the right quality for their staff.

(c) There could have been a falling-off of applicants. This contingency casts a sidelight on one of the agreed facts:

> "In the employment agency business the agency has more jobs available than applicants to fill the jobs. The agency's clients are the businesses who have jobs available which they wish the agency to fill for them. The number of these jobs exceeds the number of applicants

A an agency can expect to find. Accordingly it is the number of applicants which is critical to an agency's business, Further, the number of applicants placed varies, not so much with the level of unemployment at any given time, but with the level of "employment activity", that is to say, the extent to which people are changing their jobs."

Contingencies similar to the above were reviewed by the Court of Appeal in *Phillips v The London and South-Western Railway Co* [1879–1880] 5 QB 78. This was an early case concerning damages to be awarded a claimant for loss of professional income arising from personal injuries. It was held, contrary to the submission for the plaintiff, that there had been no misdirection of the jury by the trial judge, Field J. In the judg-
B ment of James LJ there is this passage at p 86:

> Now, on the first point, taking the whole of the summing up together, it seems to me that the case was put to the jury in the way in which the plaintiff's counsel contend that it ought to be put. Mr Justice Field says, "An accident might have taken the plaintiff off within a year. He might have lived, on the other hand, for the next twenty years, and yet many things might have happened to prevent his continuing his practice. If it had been a question of trade or business, bankruptcy might have supervened. That does not come into account here, and I only give it by way of illustration of what must pass through your minds for the purpose of seeing what sum is to be given. It is given, recollect, once for all, and once only, you must not forget that, and
> C it must be given on the fairest estimate you can make of what the probable continuance of the plaintiff's professional income would have been." That comes to this, you are to consider what his income would probably have been, how long that income would probably have lasted, and you are to take into consideration all the other contingencies to which a practice is liable. I do not know how otherwise the case could be put. Again he says: "The damages to which a man is entitled are the consequences of the wrongful act by which he suffers. The consequences of the wrongful act here are undoubtedly that Dr Phillips has been and is prevented from earning such a sum of money as you think he would have been likely to earn if this accident had not happened." The case could not have been put more favourably for the plaintiff than it was thus put by Mr Justice Field.

D

Assuming that sufficiently dependable figures of anticipated future profit can be arrived at, we are attracted by Mr Musto's method of evaluation of the claimants' loss. The approach is one that is already familiar to the tribunal. It was used, and approved, for instance, in *Festiniog Railway Co v Central Electricity Generating Board* (1971) 220 EG 1699, the hearing of that case being in 1971, although entry had been taken some 15 years earlier in 1956. In *Festiniog* 15 years' loss of estimated profit (covering the years 1962–1976) were brought into calculation, discounted year by year to the date of valuation. The same approach was used in the more recent case of *Wakerley and*
E *Another v St Edmundsbury Borough Council* (1977) 241 EG 921 where both parties brought into account an estimated loss of profit for the lifetime of the claimant. The actual calculation by the valuers in *Wakerley* was by applying a multiplier in order to capitalise an ascertained (and assumed level) annual net profit, but capitalisation in this way is, of course, no more than a telescoped form of addition.

Reverting now to the present case, the first question to be looked at is whether or not Mr Cooper's sideways projection of figures from other branches of Reed produces sufficiently dependable figures of anticipated profit in respect of the subject branch. The method of calculation adopted depends for its
F dependability on various assumptions, all of which must be sound. We select just three:

1. That the trading periods adopted for the computation of the ratios of 1:4035398 and 1:6187708 provide a sufficiently sound foundation for the intricate superstructure of calculation based thereon.

2. That there are adequate grounds for expecting the future performance of the branch at 377 Oxford Street to follow the average pattern of the four other Reed branches in Oxford Street.

3. That the trading performance of Reed's other branches after the closure of 377 Oxford Street was unaffected by that closure.

Assumption 1. In respect of the temp income, the base period used by Mr Cooper for comparison was the 12 months prior to closure, July 1971 to June 1972. But in respect of the perm income a six-month period was used, viz July to December 1971, because (a) the perm income for the *ensuing* six months, January to June 1972, had been affected by disturbance, as described earlier and (b) during the *previous* six months, January to June 1971, the Reed branch at 271 Oxford Street had been opened, and this would have confused the arithmetic. We accept that the base periods adopted were the best available, but nevertheless, the trading results for only six months provide a narrow base from which to project future profitability over a period of five- G
and-three-quarter years. The use by Mr Cooper of multipliers carried to the seventh place of decimals conveys a misleading impression of dependable accuracy. H

Assumption 2. Is it a dependable assumption that 377 Oxford Street would have performed over the ensuing five-and-three-quarter years in ratio with the average of Reed's other four Oxford Street branches? We do not think it is. It is true that the three branches at the easterly end of Oxford Street, near to Tottenham Court Road, have performances which follow a conformable pattern one with another, but the performance of 271 Oxford Street from its opening in June 1971 was dramatically
good—Mr Cooper described its early growth as having been "euphoric". The bringing into account of the exceptionally high figures for no 271 leads, we think, to a projection for no 377 which is unwarrantably high. J

Assumption 3. Mr Cooper's calculations require the fundamental assumption that the performance of the other branches of Reed subsequent to June 1972 were unaffected by the closure of the subject branch. This is a matter which we shall be considering in some detail later. For our present purpose, we state simply that we do not accept that assumption.

The combined effect of the doubts expressed above, coupled with a general reservation we have as to the proposition that a
figure of profit can be dependably projected sideways from one trading position (or a group of trading positions) to another trading position, even in the same street, causes us to look askance at the figures on which Mr Musto bases his assessment of compensation payable. Those figures, it will be recollected, are: K

Anticipated profit for year to June 1973	£48,967
" " " " " " 1974	72,919
" " " " " " 1975	77,398
" " " " " " 1976	39,085
" " " " " " 1977	47,319
" " " nine months to March 1978	72,279
The historic figures, for comparison, had been: L	
Actual profit for year to June 1970	£6,687
Actual profit for year to June 1971	15,952

All the above figures are net, after deduction of profit rental. Our reservations, like Mr Orme's, are so great that we do not propose to seek to derive an alternative projection derived from Mr Cooper's statistics.

This finding does not detract, in any way, from the merit of Mr Musto's method of evaluation of compensatable loss, but it does mean that we must look elsewhere for a sufficiently dependable figure of anticipated future profit. Happily such a
figure is, we find, derivable by the same approach as was adopted by Mr Orme. We accept the correctness of Mr Orme's first step, which was to make a forward projection, as at the date of valuation, of the average annual level of branch contribution that might reasonably have been anticipated from 377 Oxford Street over the claim period to March 1978. From that projected figure must be deducted the profit rental, and then there falls to be considered (a) the question of head office expenses (b) the question of working capital and (c) the question of transference of business. We will take these matters one at a time, viz anticipated level of contribution, head office expenses, working capital and the transference factor. M

Anticipated Level of Contribution

Mr Orme's conclusion was that the appropriate level of branch contribution from 377 Oxford Street would be £20,000 per annum. This opinion of his was based primarily on (a) the actual results of the branch for the two-and-a-half years up to its closure (January 1970 to June 1972) and (b) the budgeted results for the half-year after its closure (July to December 1972), as shown in document 22.

We think Mr Orme's estimate is too low, on three counts.

First. As regards the shortfall of about £1,000 in the actual results for the half-year to June 1972, as compared with the budgeted results, Mr Orme regarded this result as evidence of a less-than-successful branch, whereas (as has been seen) we regard that shortfall as attributable to the authority's own scheme.

Secondly. As regards the budget of branch contribution, £11,287, for the half-year to December 1972, Mr Orme told us that he had looked on this as "a genuine forward-looking figure".

We also view that forecast as a genuine forward-looking figure, but we are in no doubt, on the evidence, that it was a "with scheme" forecast and was lower than would have been budgeted in the deemed "no scheme world".

Thirdly. Mr Orme gave no indication of the weight, if any, that he attached to the actual figures of contribution to the company's profit achieved by Reed's other branches in Oxford Street over the ensuing five years from July 1972 to June 1977. The demonstrated performance of these other branches over that period, even allowing for some transference of business, points to a level of anticipated profitability for the subject branch much higher than had been historically achieved over the two-and-a-half years to June 1972.

We adopt, as the average annual level of anticipated branch contribution, the figure of £30,000. After deduction of £6,500 profit rental, this becomes a contribution of £23,500.

Head Office Expenses

In Mr Orme's opinion it is necessary to bring into the calculation of profitability an amount for central expenses such as remuneration of staff at levels above that of branch manager, accounting costs, marketing costs, personnel and training costs and bad debts.

We understand that it is the practice of Reed not to allocate a figure of central costs to their branches, when budgeting for branch contributions to central funds. Mr Cooper told us that the closure of a single branch would not in any way reduce the central overhead costs of administration; the true loss following from the closure of 377 Oxford Street was, he said, the "slice of profit", viz the net income from perms and temps that the branch would have contributed to central funds.

We accept Mr Orme's view as in principle correct, and we adopt as the proper allowance for head office expenses a figure put forward by Mr Sullivan, namely £5,000.

This deduction reduces the anticipated branch contribution from £23,500 to £18,500.

Working Capital

In Mr Orme's opinion, it is correct accountancy practice to bring into the calculation of profitability an allowance in respect of working capital. At the relevant period, said Mr Orme, Reed were constantly expanding, and all sources of money could be regarded as available to sustain their increased investment; the cash that had been required by 377 Oxford Street to carry on business would be released and become available for use elsewhere.

Mr Cooper told us that Reed derived no benefit from a release of capital on the closure of a single branch, because the company operated on borrowed, interest-free, funds and any moneys released would, in effect, merely reduce the amount owing to creditors.

We accept Mr Orme's view as in principle correct, and we adopt as the proper allowance for interest on working capital a figure put forward by Mr Sullivan, namely £800, representing working capital of £8,000 at 10 per cent.

This deduction reduces the anticipated branch contribution from £18,500 to a net figure of £17,700.

The Transference Factor

We come now to a matter on which a great deal of time was spent during the 11-day hearing. For convenience, we have adopted the expression "transference factor", but this is used with a particular connotation. The question involved is whether or not the net income that, but for its closure, 377 Oxford Street would have contributed to company funds has been totally lost.

Some description of the company's business practice is necessary. When a branch of Reed is opened it receives a file of all the job vacancies held by all the other branches. When a branch is closed, Reed do not inform their clients, and do not need to. There is no exclusivity of clientele. The client will register job vacancies with many agencies, to spread recruitment. Reed do not undertake directly or continuously job advertising, or the advertising of a corporate image. "We have no identity", said Mr Cooper "we adopt a purely passive role, relying on the strength of branch location to attract applicants, and we neither expect nor receive any applicant loyalty"; although it was clearly necessary to have clients and to have vacancies, the limiting factor, and the scarce resource, was not vacancies but applicants.

A simile, used by Mr Spence, was that of a magnet set to attract iron filings. The presence of an employment agency branch in Oxford Street, said Mr Spence, was like a magnet drawing in any potential applicants ("the iron filings") who might come within its field of attraction. The closure of no 377 left the iron filings exposed to the magnetic fields of competing agencies. In the length of Oxford Street between no 377 and no 271, there were no fewer than eight branches of other agencies, three of these with ground-floor displays including two branches of Brook Street Bureau. We are invited by the claimants to accept that these other magnets would between them have intercepted all potential applicants passing that way.

The authority do not allege that there was any transfer of temps from no 377 to no 271, but they contend that there was a transfer of perms. The claimants say there was no transfer to no 271 at all.

Temps are different from perms in being based on a particular branch and returning each week to that branch in order to draw the payment due for hours worked. On the closure of no 377, the capable temp supervisor there was transferred to the Reed branch at Queensway and was able to persuade some of the temps to transfer with her.

Mr Orme produced coloured graphs which, in his opinion, indicated that the coincidental improvement in the growth at Queensway could be attributed, at any rate in part, to the closure of 377 Oxford Street.

We are not satisfied that any transference of business to no 271 can be demonstrated either graphically or statistically. As regards Queensway, we accept the evidence of Mr Cooper that the improvement at that branch was both of short duration and minimal: "Temps" said Mr Cooper "were highly transient; the average 'life' of a temp was something short of six weeks, and in order to maintain its temp business, each branch had to cope with high turnover by perpetual recruitment."

Yet, on the other hand, the claimants' contention that the whole of the future profitability of the closed branch was a total loss, requires our acceptance of the proposition that all the applicants who would have gone into no 377 went into some agency but never entered a Reed branch—always the branch of a competitor. We find this proposition unacceptable. Reed are, we think, too modest in speaking of their lack of identity and their passive role. As one of the three leaders in the employment agency field, we would expect their name to be sufficiently

A magnetic for their outstandingly well-placed branch at 271 Oxford Street to pull in potential applicants from quite a wide field. We do not think that all the "iron filings" that would have been attracted into 377 Oxford Street were lost to Reed.

From the ascertained figure on a total loss basis, there should, we think, be a token 10 per cent deduction to bring into account the transference factor.

The foregoing findings will enable us to arrive at a figure of compensatable loss, because no challenge was raised to the other factor in Mr Musto's assessment, namely the risk rates adopted by him for the discounting back of future profits, year by year, to the valuation date. First, however, we should refer to a B reserve basis of assessment which was put in for the claimants.

Reserve Basis of Assessment

The claimants' reserve assessment prepared by Mr Musto in case we had decided, for one reason or another, to reject his substantive method of calculation, was on the basis of what the parties referred to as "the classical approach" by which was meant the application of a years' purchase multiplier, to a figure of historic net profit. On this approach Mr Musto's assessment was in the sum of £77,315 as follows:

Net profit for the year to June 1972 after deduction	
C of £6,500 profit rental	£15,952
Add Loss of profit during that year due to disturbance from the scheme	5,678
	21,630
Add back depreciation, as in primary assessment	460
	22,090
YP	3.5
	£77,315

Against the eventuality of our deciding to arrive at a compen- D sation figure by reference to the classical approach, Mr Sullivan produced the following calculation:

Net profit for the year to June 1972, after deduction of £6,500 profit rental		£15,952
Deduct Head office expenses	£5,000	
Interest on working capital	800	5,800
		10,152
YP		2.25
		£22,842

E The above figure of £22,842 was put forward by Mr Sullivan contingent on the assumption (not accepted by him) that a total extinguishment basis was appropriate, because no mitigation of loss was possible by transfer of business or by retention of business name, managerial or personnel skills, business clients and contacts.

Both valuers were cross-examined in particular as to their choice of a multiplier of net profit.

Mr Sullivan's evidence in support of his 2.25 YP was that on the classical approach the usual multiplier adopted for a leasehold interest was 2 YP or possibly 2.25 YP and for a freehold interest the usual multiplier was 3 YP. Mr Musto relied in part F for his 3.5 YP on previous decisions of the tribunal, as for instance *G E Widden & Co Ltd v Royal Borough of Kensington and Chelsea* (1970) 213 EG 1442, but also in part on an open-market transaction in 1968 by which Brook Street Bureau had acquired the whole of the issued share capital of Phoenix Employment Agency, which at that time had some 29 branches. The purchase price in that transaction was £183,350, representing apparently about 7 YP of a net profit of £26,114 after charging Corporation Tax, and the price was paid in the form of shares. In addition there was to be a payment of (a) seven times the amount by which the taxed profits for the year to June 30 1969 exceeded £26,114 and (b) seven times the amount by which the taxed profits for the year to June 30 1970 exceeded the taxed G profits for the year to June 30 1969. Mr Musto relied on the Phoenix transaction as an indication that in the employment agency business 7 YP of taxed profits (equivalent to about 3.5 YP of pre-tax profit) was a multiplier in actual use in the market.

The nature of the above evidence illustrates the difficulties which confront parties who seek to use the classical approach for evaluation of a loss which is described in *Cripps* (11th edition pp 925–927) as "loss of goodwill". The use of this classical approach can occasion no less difficulty to the tribunal than to the parties. In *W Clibbett Ltd v Avon County Council* (1976) 237 EG 271 the learned President observed:

> H It is common ground that the amount of compensation should equal the monetary loss to the claimants and that I am not concerned with market value, that is to say what amount the goodwill if offered in the market would have realised; indeed that would have been an impossible exercise. . . . I was asked to look at certain previous decisions of the tribunal, but as I have said and emphasised before, decisions are relevant only to arguments on law or procedure. The assessment of compensation must be decided on, and only on, the evidence. Useful though the respective valuers' calculations are in assisting me to reach a decision in this matter, I do not propose to determine each item in dispute. How can I, for example, say what is the appropriate years' purchase when on the evidence I can choose any number between 1.5 and 10 without any evidence which I should J choose other than the *ipse dixit* of each of the valuers? I propose, therefore, to adopt a robust approach similar to that used by the courts in assessing general damages and to award a sum which, in my judgment, in all the circumstances is reasonable. It seems to me that my function is comparable to that of the courts in assessing damages for loss of future earnings.

The inherent difficulty in the classical approach is that it seeks to express a loss of future earnings in terms of a loss of "goodwill." While it is no doubt a matter of convenience to describe the loss which a trader suffers in his business on dispossession by an acquiring authority as loss of goodwill or diminution in goodwill, we think that the description may be misleading; the goodwill is K in no sense purchased by the authority, but remains the property of the trader: *Bailey v Derby Corporation* [1965] 1 WLR 213 at p 220, per Lord Denning MR. The true loss suffered by the dispossessed trader is the ability to derive a future profit out of the premises from which he has been dispossessed. The measure of that loss is a matter for evidence.

Incidence of Taxation

Should tax be brought into account? As has been seen, Mr Musto's assessment of loss has incorporated a deduction for Corporation Tax and also an end-addition to obviate duplication L of Corporation Tax and Capital Gains Tax. Mr Musto's calculation is related to an assumed Capital Gains Tax liability at 30 per cent on a chargeable gain representing the difference between (a) the total adjusted net taxed profit (taken at £123,118) and (b) the cost of acquisition of goodwill (assumed to have been nil). This basis follows the tribunal's decision in *Hobbs (Quarries) Ltd v Somerset County Council* (1975) 234 EG 829. The calculation appears, incidentally, to assume that the asset for which the compensation is being paid is in fact goodwill.

Since the *Hobbs Quarries* decision was given, however, the tribunal has given its decision in *Wood Mitchell & Co Ltd v Stoke on Trent City Council* (1977) 241 EG 857, where the tribunal M distinguished the *Hobbs Quarries* case on the basis that the compensation payment in that case arose from revocation of a planning consent, whereas in the *Wood Mitchell* case the compensation payable was in part for disturbance where land was taken, and the provisions of paragraph 11 of Schedule 19 to the Finance Act 1969 applied. In the light of this later decision, Mr Musto had helpfully prepared, and produced, a reworking of his figures on the basis of the *Wood Mitchell* decision and this showed a figure of claim for loss of anticipated profits post-entry of £248,031 in place of £176,000. We prefer, and adopt, this latter method of arriving at our award.

Award

Our finding of compensation is the total sum of £82,760 representing:

Interest in land (agreed, and already paid)			£17,000
Trade disturbance			
Loss of profit pre-entry		£1,000	
Loss of anticipated profits post-entry calculated as hereunder		63,700	64,700
Fixtures and fittings (agreed)			1,060
			£82,760

The figure above, of £63,700, for anticipated profits post-entry represents:

Year to June 1973			
Anticipated contribution to company profit		£30,000	
Deduct Profit rental		6,500	
		£23,500	
Less Head office expenses	£5,000		
Interest on capital	800	5,800	
Net contribution		17,700	
Discounted 1 year at 9½%		0.913	£16,160
Year to June 1974			
Net contribution, as above		17,700	
Discounted 2 years at 13¾%		0.803	14,213
Year to June 1975			
Net contribution, as above		17,700	
Discounted 3 years at 12%		0.717	12,691
Year to June 1976			
Net contribution, as above		17,700	
Discounted 4 years at 13½%		0.632	11,186
Year to June 1977			
Net contribution, as above		17,700	
Discounted 5 years at 13½%		0.556	9,841
Nine months to March 1978			
Net contribution, three-quarters of above		13,275	
Discounted 5¾ years at 13½%		0.504	6,691
			£70,782
Less for element of branch contribution not lost to the company 10%			7,078
			£63,704
		say	£63,700

As has been noted, the claim includes a heading for professional fees. We were told that the parties hoped to be able to reach agreement on the total figure to be added to the amount of our award to cover legal costs, surveyor's fees and accountants' costs and fees. Accordingly we are making this an interim award and there will be leave to apply.

In conclusion we would express our indebtedness to learned counsel and instructing solicitors and to the expert witnesses for the care and facility with which the complicated evidence in this case was presented.

Having opened the sealed offer of compensation lodged by the acquiring authority the amount thereof is found to be less than the amount of our award. The acquiring authority will pay to the claimants their costs of this reference, such costs unless agreed to be taxed by the Registrar of the Lands Tribunal on the High Court Scale.

TAYLOR v CHELTENHAM BOROUGH COUNCIL
(REF/3/1977)
(R C WALMSLEY FRICS)
January 30 1978

Estates Gazette June 10 1978
(1978) 246 EG 923

House in terrace of listed buildings in Cheltenham acquired as "in need of repair"—Evidence of advanced decay resulting from long-standing neglect—Tribunal's finding that "mathematically the house had a minus value"—Logic of approach of council's valuer accepted, that purchaser almost certainly not householder for own occupation but speculator or builder at "knockdown price"—Restoration a loss-making enterprise for council—Award of £2,000

N Walker (solicitor of John Duncombe & Co, of Cheltenham) appeared for the claimants; J M Sullivan (instructed by the Solicitor to Cheltenham Borough Council) for the acquiring authority.

Giving his decision, MR WALMSLEY said: This is a reference of the compensation payable to Morley Gordon Taylor by Cheltenham Borough Council on the acquisition of a property, 23 Lansdown Parade, Cheltenham. No 23 is the end house of a terrace of listed buildings. The enabling statute is the Town and Country Planning Act 1968, section 50 of which provides for "compulsory acquisition of listed buildings in need of repair," where it appears to the Secretary of State for the Environment that reasonable steps are not being taken for preserving such a building. Under section 51 a prerequisite to the start of compulsory acquisition is the service, on the owner, of a "repairs notice" at least two months previously. These provisions are now to be found in the consolidating Act of 1971 as sections 114 and 115.

Origin of CPO

The origin of the present acquisition is sufficiently shown in the following statement of reasons lodged by the council with the Secretary of State on the making of the Borough of Cheltenham (23 Lansdown Parade) Compulsory Purchase Order 1970 on December 9 1970:

1. Number 23 Lansdown Parade, Cheltenham, is the end of a terrace of listed buildings all, bar number 23, being in first-class repair. The council, at the suggestion of the other owners of Lansdown Parade, put forward a scheme for the restoration of the parade, which has been done successfully. The owner of number 23 did not wish to join in.

2. As can be seen from the repairs notice, number 23 is in extremely poor condition and the owner was asked whether he would be prepared to sell. He is under an illusion that he is entitled to war damage in respect of damage to the property when a stick of bombs fell during the war, and it is therefore not possible to negotiate reasonable terms with him, as he considers the deterioration of the building is due to the war damage. Inquiries of a school teacher who resided in number 22 during the war reveal that this is not the case, and the owner has not spent any money on the property for many many years.

3. Accordingly, the council had no option but to serve a repairs notice under the Town and Country Planning Act 1960 [*sic*], section 50, with which the owner has failed to comply.

4. The owner is in an old people's home and was originally moved there because the medical officer of health considered that he was not capable of looking after himself.

5. The property is full of furniture and it is only just possible to walk up the stairs. Most of the furniture is valueless.

6. Unless steps are taken to repair this property the whole terrace will be affected and it is likely that if the property is left in its present derelict condition it may eventually collapse.

The compulsory purchase order having been confirmed on October 29 1971, a notice to treat was duly served, followed by a notice of entry, and possession was taken on December 3 1971. At some date in 1972 the council spent about £300 on sufficient repairs to make the house watertight. They retained a firm of architects to draw up a scheme for rehabilitating the property as two dwellings, making the ground floor and the first floor into a maisonette and the basement into a flat. At the rear of the house was a wing in derelict condition and requiring demolition: this first scheme provided for the replacement of this wing with a new extension. The proposed work was put out to tender, but in the event the refurbishment of 23 Lansdown Parade was carried out by the direct works department of the council pursuant to the following minute of the Finance and General Purposes Committee dated August 29 1973:

23 Lansdown Parade (Min 145—July 10, 1973).

With the consent of the council, this minute had been brought back by the chairman for further consideration.

Three tenders for the restoration, reconditioning and conversion of this property into two units of accommodation had been received. The lowest tender price of £29,284 was considered to be too costly in relation to the accommodation to be provided. It was proposed that none of the tenders submitted be accepted and the borough engineer be asked to carry out a modified and less ambitious scheme by direct labour and that, to facilitate the raising of the necessary loan, the property be appropriated to the housing committee.

The borough engineer reported that a revised bill of quantities had been priced by his department, at £20,500. Savings had been made by deleting the extension at the rear of the property, and by a reduction of the specification. Further savings would probably be made by slight amendments to the plans.

The revised scheme would still provide two separate units of accommodation and improvement and restoration grants would be available.

The building was in extremely poor condition and the borough engineer stressed that it was most important that work be commenced without further delay. The original three tenderers had not been invited to submit new prices (based on the reduced scheme) because of the time factor involved.

Resolved, (1) that the previous decision not to accept any of the tenders submitted by private contractors be confirmed.

(ii) that the report and price submitted by the borough engineer be accepted.

(iii) that the borough engineer be authorised to proceed with the work by the direct works department forthwith.

(iv) that the property be appropriated to the housing committee to facilitate loan sanction.

Work was begun immediately and was completed some nine months later. On September 17 1974 a minute of the Policy and Finance Committee records:

23 Lansdown Parade. The town clerk and chief executive reported that the conversion and restoration of this property had been completed at a cost of £21,370 (including extra works) as against the borough engineer's price of £20,500. Architect's and quantity surveyor's fees amounted to £3,460.

The Historic Buildings Council had made a grant of £1,760 and the Department of the Environment had approved an improvement contribution of £316.50 per annum for 20 years, making a total for the scheme of £16,740 plus the cost of acquiring the property, which had not yet been determined.

At the hearing before me N Walker appeared for the claimant and called evidence from D N Perry FRICS. J M Sullivan of counsel appeared for the council and called evidence from J C A H Hancock FSVA. Mr Perry has been in private practice for some 17 years and was formerly deputy county valuer to the Gloucestershire Council for 13 years. Mr Hancock is the borough valuer, Cheltenham, and has had some 23 years' professional experience.

Facts about other Houses in Terrace

Mr Perry's valuation was in the sum of £8,000. Mr Hancock, on the basis that the subject property was little more than a ruin, proposed a purchase price of £2,000. Before summarising the valuation evidence I propose to set out certain matters of a factual nature relating to various of the houses in the terrace.

2 Lansdown Parade. Evidence was given on affidavit that in November 1971 no 2 had been sold, in reasonable condition, for £7,650.

17 Lansdown Parade. Evidence was given on affidavit that in August 1971 no 17 had been sold for £6,600. This house is somewhat narrower than no 2 with a frontage of about 15 ft as compared with about 20 ft.

22 Lansdown Parade. In March 1970 no 22 had been on offer for sale at £7,650. In May 1972 it was valued by Mr Perry for an intending purchaser at £12,000 subject to certain defects which were later put right at a cost of £1,000. Mr Hancock accepts that the value of no 22 in good condition in May 1972 may be taken to be £13,000. The client for whom Mr Perry had prepared his report did purchase no 22 but not until August 1972 and he was then "gazumped" into paying £14,800. That purchaser carried out some modifications, and some three years later, in November 1975, Mr Perry resold the house for him for £20,500.

The subject house. After three years of negotiation Mr Perry and Mr Hancock had reached the stage, by December 1975, of agreeing that the value of no 23, in its then condition reconstructed as two dwellings, could be taken to be £23,000.

For comparison purposes it may be mentioned that nos 2, 17 and 22 all have rear wings. Although no 23 had also been built with a rear wing, in December 1971 that extension was in a derelict condition, as has been mentioned, and in 1973-74 it was demolished as part of the reconstruction scheme.

Valuation Evidence

Mr Perry's valuation may be summarised in this form:

Value of the subject house at the relevant date but in its 1975 condition		£12,290
Deduct Cost of putting it in that condition	£7,523	
Less Grants obtainable	3,177	4,346
		£7,944
	Value, say	£8,000

I will take each element of the above valuation in turn:

Value of the subject house at the relevant date, but in its 1975 condition.

Mr Perry arrived at his figure of £12,290 in the following way. His starting point was the agreed value of no 22 in May 1972 of £13,000. Using the House Price Indices published by the Royal Institution of Chartered Surveyors, and applying the figure (166) for the fourth quarter of 1971 as compared with the corresponding figure (197) for the second quarter of 1972 (December 1965 = 100), Mr Perry calculated the value of no 22 in December 1971 to be £10,955. He then assumed the same relationship in value between no 22 and the subject house at the end of 1971 as was shown by their relative values at the end of 1975 (20,500:23,000), and thereby arrived at £12,290.

Mr Hancock took the view that, helpful as the RICS indices might be as a general guide to the movement of house prices on a national basis, as a guide to the value of any particular house at a particular time these statistics were less dependable than the evidence provided by the sale prices or agreed values for similar houses in the same locality. By reference in particular to the sale price of no 2 in November 1971 (£7,650), but having regard also to the asking price for no 22 in March 1970 (£7,650), the sale price of no 17 in August 1971 (£6,600) and the agreed value of no 22 in May 1972 (£13,000), Mr Hancock put the value of the subject house, in December 1971 but in its 1975 condition, at no more than £9,000. This figure, Mr Hancock emphasised, related to no 23 as reconstructed to form two dwellings. Its value as a single dwelling in December 1971, in good condition, was put by Mr Hancock at not more than £8,000.

Having heard the evidence and having inspected, my finding of value of no 23 in December 1971, in its 1975 condition as two dwellings, is in the sum of £11,000. And as a single dwelling, £10,000.

A *Cost in December 1971 of putting the subject house in good condition.*

Mr Perry arrived at his figure of £7,523 in this way. He estimated at £7,233 the cost of remedying dilapidations, and then added £290 for the cost of finance, viz £7,233 at 12 per cent for four months. The figure of £7,233 represents the total of nearly 300 items comprised in a detailed schedule of costings. This schedule is itself based on a seven-page "Schedule of State and Condition" made on July 24 1972. The schedule of condition had been taken by Mr Hancock himself and it is an agreed document. But there is the sharpest of conflicts as to what that document represents.

B

Ill-founded Assumption

Mr Perry had assumed that he need do no more than put a price on each of the defects mentioned in the schedule of condition, and by then allowing for the cost of remedying these, the house could be valued as having been put in good condition. This assumption, I am satisfied, is wholly ill-founded. It is clear on the face of the document that the record taken was of *visible* defects.

Mr Hancock said that, if the intention had been to prepare C a schedule of required works, a structural survey would have been undertaken to ascertain the existence and extent of defects behind plasterwork and/or beneath floorboards. He was in no doubt that there were serious hidden defects. Taking one example, the first item in the schedule of condition records the floor of the front bedroom as:

> Holed and generally worn in places. In one corner there is extreme rot which has caused the whole floor to sink about 3 in. Two lengths of floorboard are missing, having rotted away completely exposing rotten end of three joists.

Mr Perry allows £50.60 for works to make good, but Mr Hancock's evidence was that this particular rot was on a D line which went from the roof right down to the basement, indicating the necessity of extensive work to the timbering. In addition to Mr Perry's basic misapprehension as to the nature of the schedule of condition, his prices make no allowance for various specific matters, including preliminaries, scaffolding, demolition and clearance of the rear wing, contingencies and supervision.

I do not propose to include in this decision a record of the lengthy evidence given on these and other matters. A comprehensive set of photographs, taken in July 1972, was put in and these provide a vivid record of advanced decay resulting from long-standing neglect. My finding on the E evidence is that Mr Hancock cannot be far out in his opinion that to put the subject house in good condition in December 1971 would have cost "twice the amount allowed by Mr Perry." I adopt a figure of £15,000.

Grants obtainable in December 1971 towards the cost of putting the subject house in good condition.

Mr Perry's figure of £3,177 represents:

Two improvement grants at £1,200	£2,400	
Grant for renovation of listed building 50 per cent of £1,554	777	£3,177

The bringing into credit of *two* improvement grants is F clearly an error, because the remedying of the defects existing at December 1971, estimated by Mr Perry to cost £7,523 and found by myself to be £15,000, would do no more than restore the subject house as a single dwelling. Mr Hancock very fairly told me, however, that a second improvement grant would probably have been obtainable by the expenditure of an additional £1,000 on a conversion to two dwellings.

The 50 per cent renovation grant brought into credit by Mr Perry relates to the restoration of the facade of a listed building and represents grant-aid from two separate sources. Provided that such work is carried out to a proper standard and with approved materials, an owner may apparently look to the Historic Buildings Council for a grant of 25 per cent G of the cost, and to the local authority for a similar amount. Mr Hancock's comment on this credit item of £777 was that Mr Perry's pricing had only allowed for the use of "plastic stone," and this standard of renovation did not rank for grant-aid.

Assumption in Claimants Favour

Although given no specific evidence on the point, I propose to assume, in the claimant's favour, that if the house had been notionally put in good condition at a cost of £15,000, this refurbishment would have included work of a standard quali- H fying for grant-aid on a figure of, say, £3,000 (viz "twice the amount allowed by Mr Perry").

Value of house.

Using Mr Perry's approach, but applying the revised figures as found by myself, it will be seen that the value of 23 Lansdown Parade at December 1971 comes out, not at £8,000 as calculated by the claimant's valuer, but at a minus quantity, thus:

Value of the subject house at the relevant date, but in its 1975 condition			£11,000 J
Deduct Cost of putting it in that condition, including cost of conversion into two dwellings		£16,000	
Less Grants obtainable			
Two improvement grants at £1,200	£2,400		
Historic Buildings Council, 25% of £3,000	750		
Local authority, 25% of £3,000	750	3,900	12,100
		Value (Minus)	£1,100

K

This finding of mine, that mathematically the subject property had a minus value at the relevant date, was the basic assumption which had been adopted by Mr Hancock, whose approach to the compensation properly payable then ran on the following lines.

Valuation, says Mr Hancock, is not mathematics. No one would expect actually to be paid money for buying the house. Any purchase would be as a speculation. Almost certainly the purchase would not be by a householder for his own occupation, because, on seeing the condition of the house and/or getting advice, such a buyer would leave it alone. The purchaser might be a speculator or possibly a builder wishing to L keep his men employed. But whoever the purchaser was, the temptation would have to be a knockdown price. As a Regency building there was little doubt that, at a price as low as £1,000, the house would sell. It was possible that, at £1,500, it might also attract a buyer. But to expect a sale at more than, say, £2,000 would be to stretch credibility to breaking point.

I accept the logic of this approach, and I adopt the end-figure.

Suggestion of "Substantial Profit" by Council M

The acquiring authority will pay the claimant the sum of £2,000, to which will be added a surveyor's fee based on scale 5(A) of the scales of professional charges of the Royal Institution of Chartered Surveyors, plus VAT. I was told that there was a claim for disturbance outstanding, relating to costs of removal, but that it was hoped to dispose of this by agreement. Failing such agreement there will be leave to apply.

One further matter should be referred to. At the hearing, evidence was given by Mr Perry to the effect that, if the council succeeded in purchasing 23 Lansdown Parade at a

A price of only £2,000, they would be making a substantial profit out of the deal at the expense of the claimant who (as has been seen) is in an old people's home. This evidence was related to the figures shown in the committee minute of September 17 1974 as set out earlier. Mr Perry's calculation may be expressed thus:

Value of the subject house at September 1974			£23,000
Deduct Cost of conversion and restoration		£21,370	
Professional fees		3,460	
		£24,830	
Less Historic Buildings Council grant	£1,760		
Local authority grant	1,760		
Grant from the Department of the Environment	6,330	9,850	
		£14,980	
Net cost of works, say		£15,000	
Add Purchase price		2,000	17,000
		Profit	£6,000

B

C The two major flaws in the above calculation, as pointed out by Mr Hancock, are:

(a) that the improvement contribution of £316.50 per annum for 20 years from the Department of the Environment was not in the nature of a grant but was by way of a reduction in loan charges.

(b) that although a figure of £23,000 had been agreed between valuers this related to values at December 1975 and could not be set directly alongside figures of cost incurred over two years earlier.

D Mr Hancock presented a reworking of Mr Perry's calculation. This shows, instead of a *profit* of £6,000, a *loss* of over £12,000 as follows:

E	Agreed value of the subject house as at December 1975			£23,000
	Deduct Cost of conversion and restoration		£21,370	
	Professional fees		3,460	
	Cost of finance £21,370 at 12% for 4 months		855	
			£25,685	
	Less Historic Buildings Council grant	£1,760		
	Local authority grant	1,760	3,520	
	Net cost of works at August 1973		£22,165	
F	Conversion factor (per *Spon* 218:143)		1.524	
	Equivalent net cost of works at December 1975		£33,779	
	Add Purchase price		2,000	35,779
			Loss	£12,779

"This was", said Mr Hancock, "a loss-making enterprise from the start. But the council had to do something. They were under a statutory duty to preserve a listed building. We were about the only people who would have bought. Mathematically the house had no value at all." This statement seems to me a very fair summary of the background to the case. G

Having opened the sealed offer of compensation lodged by the acquiring authority, I find that the amount thereof exceeds the amount of my award. The acquiring authority will pay the claimant his costs of this reference up to the date of the sealed offer, December 16 1977, and the claimant will pay the costs of the acquiring authority as from that date, such costs unless agreed to be taxed by the registrar of the Lands Tribunal. Mr Sullivan told me that, having regard to the personal circumstances of the claimant, he had been instructed not to press for costs on the High Court scale. Having received further submissions. I direct that taxation shall be on scale 3 of the County Court Scales of Costs. H

LANDS TRIBUNAL

RATING

A **CENTRE HOTELS (CRANSTON) LTD v RIDGEON (VO)**
(LVC/671/1975)

(J H EMLYN JONES FRICS)

November 18 1977

Estates Gazette February 18 1978
(1977) 245 EG 577

Hotel on perimeter of Heathrow Airport—Valuation on comparables basis—Ratepayers claim 50 per cent end-allowance for location, standard of accommodation and general disabilities—Such end-allowance not to be rejected
B **merely because of size or of disparity it suggests between appeal property and comparables—Tribunal reduces end-allowance to 37½ per cent**

A J Anderson (instructed by Freshfields) appeared for the appellants; Solicitor of Inland Revenue for the respondent valuation officer.

Giving his decision, MR EMLYN JONES said: This appeal concerns the assessment for rating purposes of the Centre Airport Hotel situated on the perimeter of London Heathrow Airport. The issue is entirely one of valuation.

The hotel was entered in the valuation list for the rating area of the London Borough of Hillingdon and assessed at
C gross value £186,000, rateable value £154,972. Following a proposal made by the appellants on September 14 1973 the matter was referred to the local valuation court, who determined the assessment at gross value £145,000, rateable value £120,805. The appellants being dissatisfied with this decision appeal to the Lands Tribunal and allege that the correct assessment of the hereditament does not exceed gross value £109,000, rateable value £90,805. The respondent valuation officer, Mr Ridgeon, claims that the assessment determined by the local valuation court is not excessive.

The hotel lies some 1½ miles to the west of the tunnel entrance leading to the passenger terminal buildings of the
D airport. It is situated on the south side of the old Bath Road in the village of Longford. The hotel was built in 1968 and extended in 1971. At that time it was possible to proceed along the old Bath Road past the entrance to the hotel, although most traffic proceeded along the A4 and bypassed the village of Longford. Similarly all long-distance traffic to and from the west tended to use the M4 motorway rather than the A4 trunk road.

The hotel is on two storeys contained in a number of linked blocks, the height of the building being restricted by reason of its proximity to the airport, adjoining as it does the northern perimeter fence. The hotel is fully air-conditioned,
E largely double-glazed, most of the bedrooms being in fact triple-glazed, and within the bedrooms there is remarkably little disturbance caused by the noise of the aircraft. All the bedrooms contain their own private bathroom, and the bedrooms forming part of the 1971 extension also include a small kitchenette. There are 171 twin and double rooms and 111 single rooms in the 1968 original hotel and 66 twin and double rooms in the 1971 extension.

For the purpose of comparison the parties adopt a factor of 1.5 bedrooms per double room and arrive at a total for the hotel as it exists today of 466 equivalent single bedrooms. F
The hotel also contains a restaurant, coffee shop, bars and a small lounge; and extensive banqueting facilities also capable of being used for conferences and other gatherings. There is adequate car-parking on a site extending in all to some 9 acres.

For the appellants Mr Anderson called the hotel manager, M A Richards, and R M Williams ARICS, a partner in Leavers and an experienced rating surveyor. Mr Brown appeared for the valuation officer, D W Ridgeon ARICS, who gave evidence on his own behalf.

Comparison with Nine Other Hotels

G
The appellants occupy the hotel under a sale and leaseback arrangement, but both valuers agree that the so-called rent provides no reliable guide as to the value of the appeal hereditament at the relevant date. Instead they base their valuations on a comparison with the assessments of nine other hotels in the immediate locality where the assessments have either been agreed or have been the subject of the local valuation court's decision. The two valuers accept the correctness of the assessments of these other hotels and the relatively of the values applied to them. Mr Williams devalues the assessments applying constant figures to them all, in so far as that is possible, by applying prices per square metre to the H
public parts and the service areas and a fixed unit price of £200 per equivalent single bedroom. Some of the hotels have swimming pools for which Mr Williams adds a lump sum. He arrives at the relative value of the different hotels by applying end-allowances under three headings: (1) location; (2) standard; and (3) disabilities. In the case of the appeal hereditament Mr Williams makes an adjustment for location of 10 per cent, 17 per cent for standard and 23 per cent for disabilities, making a total allowance of 50 per cent. The application of the constant rates to the different parts of his valuation provides a total of £217,896. After making the 50 per cent deduction, Mr Williams arrives at a total of £108,948, J
which he rounds up to £109,000.

Mr Ridgeon first of all values the different hotels by applying prices per square metre and per bedroom and takes into account the varying factors which differentiate them and makes no end-allowances in any case, except one where there is a specific adjustment made in respect of structural defects. For the purpose of comparison Mr Ridgeon also provides a schedule in which he follows the method adopted by Mr Williams, from which it is clear that apart from the question of swimming pools there is no disagreement between the valuers as to the relative values to be attributed to the parts of the various hotels. From the same total of £217,896 Mr K
Ridgeon makes an end-allowance for disabilities of 33.4 per cent, which provides in round terms a figure of £145,000.

Fitting into the Proper "Slot"

The issue can therefore be simply stated. In order to fit the assessment of the appeal hereditament into the proper "slot" in relation to the nine other hotels in the vicinity of London Airport, what is the appropriate allowance to be made as a deduction from the accepted basis in order to reflect the pecu-

A liar characteristics of the appeal hereditament? Mr Williams says 50 per cent; Mr Ridgeon 33⅓ per cent.

I will deal first with one objection to the higher percentage which was submitted on behalf of the valuation officer. Where an allowance of the order of 50 per cent is put forward, it is said, then it can no longer be true to say that the hereditament with which a comparison is being made can be regarded as comparable. This, of course, is mainly a question of degree, and it might be said that if that were the reason for rejecting such an approach the same is only slightly less true where an allowance of 33.4 per cent is being suggested. In the circumstances of this case it is quite clear that the assessments of B the other hotels in the immediate locality of broadly similar character provide the most reliable guide. The 50 per cent suggested by Mr Williams emerges only because he is adopting the same rate per square metre for the different parts of the hotel as those applied to the other hotels.

Some of the factors giving rise to the end-allowance—for example the question of standard—can also be expressed by varying rates per square metre and the price per bedroom in order to reflect variations in quality. This is what the valuation officer has done on his first approach and if that method were to be applied then Mr Williams' final figure would incorporate an end-allowance of 25 per cent. I there- C fore do not consider in the light of the facts as presented in this case that an end-allowance of 50 per cent is to be rejected merely because of its size or because of the disparity which it suggests between the appeal hereditament and the comparables.

Having said that, it is necessary to look more closely at the grounds on which Mr Williams seeks to justify his allowances.

He says that the appeal hereditament is further away from the tunnel entrance to the airport than most of the other hotels. I do not think that these differences justify any such allowance. All the hotels save one are within the same sort of distance of the tunnel entrance and they all provide courtesy D buses to carry guests to and from the passenger terminal. No one, as I understand it, arrives at these hotels on foot, and the comparatively small differences in distance do not, I think, justify any significant difference in value. At the same time it is quite clear that although most of the residential business is linked to the airport, the conference and banqueting facilities are on the whole utilised by businessmen in the locality not necessarily connected with the airport. The location of the hotel for this latter purpose is not quite as convenient as most of the others and this factor represents a disability which in my judgment does justify some allowance.

E As to the general standard of the accommodation, both valuers recognise that this is inferior to most of the other hotels. Mr Ridgeon suggests an allowance of 18.4 per cent on this account and Mr Williams 17 per cent, so there is no significant difference between them.

As to general disabilities, Mr Williams itemises them as follows: (1) spread; (2) separate kitchens; (3) Belvedere suite detached from kitchens; (4) entrance foyer small; (5) blocks not connected at ground and first floors; (6) lounge—size and position; (7) management problems; (8) TV reception and function of address system.

The last item refers to interference from certain radar F equipment which affects some of the bedrooms on that side which immediately fronts on to the airport. I am left with the firm impression that some of the factors are more properly to be reflected in the general standard of the hotel and by including them under this heading Mr Williams is in effect taking them into account twice.

It will be apparent therefore that I consider the allowances made by Mr Williams to be too great. At the same time I do not think Mr Ridgeon has adequately taken into account the difficulties of management which, as Mr Williams claims, stem from the spread of the buildings. On the other hand the local valuation court carried out an inspection not only of the appeal hereditament but of all the other hotels and they G came to the conclusion that the allowances made by the valuation officer were sufficient.

I have inspected the appeal hereditament in some detail and I have also been into all the comparable hotels, both into the public parts and into the bedrooms, kitchen and service areas. At the end of the day the valuation of the appeal hereditament must remain a matter of impression, but having heard the evidence and in the light of my inspections I have come to the conclusion that the valuation officer has underestimated the difficulties of access from the reception area to the different parts of the hotel, and the effect on security of the large number of exit doors, 25 in all, which are to be found H throughout the hotel. I think, therefore, that the appropriate end-allowance to be made in the valuation where no adjustment has been taken into account in prices per square metre and in unit prices is of the order of 37½ per cent. I therefore allow the appeal and determine the assessment at gross value £136,000, rateable value £113,305.

The respondent valuation officer will pay the costs of the appellants, such costs failing agreement to be taxed by the Registrar of the Lands Tribunal on the High Court Scale.

J

SHEPPARD (VO)'S APPEAL
(LVC/192-193/1977)
(R C WALMSLEY FRICS)
July 18 1977

Estates Gazette February 25 1978
(1977) 245 EG 665

Temporary nuisance or disability—Two adjoining houses at Blackburn affected by noise and dirt from sewer-laying work during 6-months' period—Specification of dates by LVC bad in law—LVC not empowered to make two reductions on one proposal or a reduction and then an increase—Review of earlier decisions of tribunal on temporary nuisance or disability—Observations in "Dick Hampton" case on "transience" and "12-months' working rule" applied—Held that hypothetical tenant would depress his rental bid—Assessments amended from proposal date K

R H Sheppard, the appellant valuation officer, appeared in person.

Giving his decision MR WALMSLEY said: These are appeals by the valuation officer from decisions of the Lancashire Local Valuation Court in respect of two houses at Blackburn. The houses adjoin each other in Selous Road. One is an L inner-terrace two-bedroom house, 57 Selous Road. The other is an end-terrace three-bedroom house on the corner of Selous Road and Cartmel Road, its address being 3a Cartmel Road.

The County Borough of Blackburn have for some time been laying a very large and deep sewer through the town. The scheme involves sinking of successive shafts, 13 ft 6 in in diameter and up to 40 ft in depth, these shafts being connected one to the other with tunnels 6 ft in diameter. On December 16 1974 work on the scheme started in Selous Road. A rectangular compound with corrugated sides 6 ft high and measuring about 75 ft square was constructed within about 25 ft of the appeal houses. Within that com- M pound and about 30 ft from the houses a shaft was then sunk. This was done by first breaking up the pavement with pneumatic drills, then digging down mechanically for the first 20 ft. Below the 20-ft level, to a finished depth of 30 ft, the shaft was dug with the aid of compressed-air clay spades. Thereafter the shaft was lined with pre-cast concrete segments. Work continued from 7 am to 7 pm on Mondays to Fridays each week for a period of six months. The air compressor (housed in a cabin to muffle the noise) was running for eight hours out of the 12. A crane was used to take the material out of the shaft and then to lower the segments. There were lorries coming and going. There was noise and dust and general disturbance. By June 24 1975 all work near the appeal houses had been completed and the operations were resumed elsewhere.

The above matters were described to me by Keith Shufflebottom BEng, an engineer employed in the Department of the Director of Public Services with Blackburn Corporation, who was in charge of the work. Mr Shufflebottom was called as a witness by the appellant valuation officer, R H Sheppard ARICS.

For convenience I will deal with the two appeals separately.

3a Cartmel Road

The owner-occupier of the end-terrace house 3a Cartmel Road, P J Boyle, lodged a proposal on February 1 1975 for "temporary rate reduction" on the grounds that:

> New sewage pipes are being laid past the above property and a 30-ft shaft has been made outside the lounge window. The constant noise, which we are told will increase as the work progresses, and the dirt from all this makes us feel that we should be entitled to some compensation. The work is going to take at least six months to complete and as it is being undertaken on behalf of the local authority, a rate reduction is the obvious answer.

The appeal resulting from the above proposal came before the local valuation court on December 13 1976. The court decided that the assessment of this hereditament, then standing in the valuation list at gross value £195, rateable value £136, should be as follows:

Gross Value	*Rateable Value*
£181	£124 to 15.12.74
£157	£104 from 16.12.74 to 31.5.75
£181	£124 from 1.6.75

From the above decision the valuation officer now appeals to the tribunal on the stated grounds that "the decision of the local valuation court is considered to be incorrect, and the assessment should be gross value £181, rateable value £124."

(It may be noted here that 3a Cartmel Road has, in fact, been standing in the valuation list at the above figures of gross value £181, rateable value £124, since June 1 1975 in pursuance of the local valuation court's own decision.)

At the hearing before me the valuation officer amplified his grounds of appeal, putting these forward under five heads which I propose to deal with in the following order:

(a) that the decision of the local valuation court was bad in law in that it specified dates, usurping in that respect the prerogative of the rating authority.

(b) that the local valuation court was wrong to have made two reductions on one proposal, first from £195 gross value to £181 gross value and then from £181 gross value to £157 gross value.

(c) that the local valuation court, having reduced the assessment to £157 gross value had, in any event, no power to increase that assessment to £181 gross value in the absence of any proposal for an increase.

(d) that, leaving out of account the effect of the sewerage works, the assessment of 3a Cartmel Road should be gross value £173, rateable value £117, and not gross value £181, rateable value £124.

(e) that, even bringing into account the effect of the sewerage works, the assessment should still be gross value £173, rateable value £117, because the nuisance suffered was for too transient a period to affect rental value on the rating hypothesis.

Specification of dates

I accept the contention that the decision of the local valuation court was bad in law, to the extent that it specified dates. This question was fully argued before the tribunal in the case of *Sheffield Corporation v TB & W Cockayne Ltd and another* (1958) 172 EG 711, where it was held that the valuation officer and the ratepayers had no power to agree, and neither the local valuation court nor the Lands Tribunal had power to determine, a date as from which an alteration of the valuation list resulting from a proposal should take effect.

Making of two reductions in one proposal

I accept that the local valuation court had no power to make two reductions on a single proposal. The duty of the court, and of the tribunal, is "to determine what rent a hypothetical tenant would pay on the date of the proposal, for a yearly letting" (*Barrett and another v Gravesend Assessment Committee* [1941] 2 All ER 308), and there cannot be two values at the one date.

Making a reduction and then making an increase

I accept the contention that the local valuation court, having made a reduction on a proposal, had no power to increase that reduced assessment. Here there was a proposal for a reduction but no proposal for a subsequent increase.

Gross value, leaving out of account the sewerage works

I accept the opinion of the valuation officer that the proper gross value of 3a Cartmel Road, leaving out of account the effect of the sewerage works, is £173. The figure of £181 gross value which appears in the local valuation court's decision was one which had been recommended by the valuation officer himself, but Mr Sheppard told me that as the outcome of a subsequent further study of the assessments of similar-sized houses in the locality, he had decided that a somewhat lower figure, of £173 gross value, would be more correct.

Gross value, bringing into account the effect of the sewerage works

In support of his submission that the nuisance suffered in the present case was for too transient a period—six months in all, of which less than five months were after the proposal date—for the rental value on the rating hypothesis to be affected, the valuation officer drew my attention to a number of cases decided by the tribunal. I am attaching, as an annex to this decision, a summary of six earlier decisions in cases relating to a temporary nuisance or disability: *Lewis v Holman (VO)* (1961) 179 EG 371, *Heath v Holman (VO)* (1961) 179 EG 41, *King v Dyer (VO)* (1964) 189 EG 1095, *Hart v Gregory (VO)* (1964) 191 EG 855, *Dennis and another v Gilmore (VO)* LVC/1078, 1101/1965, and *Nightingale v Harrap (VO)*. LVC/87/1971.

I accept that these decisions support the valuation officer's submission in the present case. The decisions indicate, in effect, the observance by the tribunal of a working rule that even if there is at the date of the proposal some nuisance sufficiently severe as to affect rental value, this circumstance would only result in a reduction of assessment if the conditions complained of lasted for a substantial period, say of the order of 12 months and mainly from the date of the proposal. Where the nuisance has extended over some lesser period, then the tribunal has hitherto held that the disability would not affect the rent offered for a year-to-year tenancy because the conditions complained of were only "transitory" or "transient."

Rateable Occupation Parallel

The above line of thought ran closely parallel to the tribunal's previous thinking about "transience" as an ingredient of rateable occupation. That thinking has now come into question since the Court of Appeal, in June 1975, allowed an appeal by the valuation officer against a decision of the tribunal (Sir Michael Rowe QC and R C G Fennell FRICS) holding that certain limestone workings were not in rateable occupation because the periods of occupation were too transient (*Dick Hampton (Earth Moving) Ltd v Lewis (VO)* (1975) 235 EG 619). The Lands Tribunal had said [1973] RA 227:

> There is no doubt that one of the ingredients of rateable occupation is that the occupation shall not be too transient. . . . In [cases] which related to temporary structures the courts seem to have accepted as a general working rule that occupation for a year or more is not too transient and the implication has been

> that if it has been for less than a year, it is too transient. . . . These borrow pits are closely comparable to temporary structures like builders' huts. Applying the working rule established for such structures it appears to us that the occupation was too transient to constitute rateable occupation and we hold this as a matter of fact.

Referring to the above Lord Denning MR, observed at p 619:

> It has often been stated that one of the necessary ingredients of rateable occupation is that "possession must not be for too transient a period". . . ; and in relation to builders' huts, that 12 months is the working rule. More than 12 months, there is rateable occupation: less than 12 months, no rateable occupation. . . . It is, I think, a mistake to elevate this ingredient into a principle of law or to construct out of it any working rule. . . . I cannot accept the supposed "working rule" of 12 months.

And Roskill LJ observed at p 623:

> We were told that in the "building hut" cases 12 months has been taken as the working rule. I say nothing to disturb a practice which no doubt has been found administratively convenient in a particular class of case. But the existence of an administratively convenient practice must not be allowed to obscure what is the real question of principle involved, the determination whether the occupation is of sufficient permanence as properly to amount to beneficial occupation. It cannot be right in every case of every kind where this question arises to answer it by reference solely to the question whether the period of occupation is a few months, weeks or days more or less than some arbitrary period such as the 12 months' period which I have mentioned . . . [Transience] will always be a factor in determining whether or not the requisite element of permanence is present, but the degree of importance attaching to this factor must vary with the circumstances of the particular case. In the present class of case I regard the period of occupation as of vastly less importance than its quality and its consequences.

It is true that the above considerations relate to rateable occupation, which is a different question from that with which I am concerned here. Nevertheless, it appears to me that the above observations, *mutatis mutandis*, are equally relevant to the ingredient of "transience" when considering the effect of a nuisance or a disability on the rental value of a hereditament in gross value terms.

The degree of importance attaching to the factor of transience must vary with the circumstances of the particular case. The adoption of a 12-month minimum period is recognisably a convenient practice where (as for instance with an unfinished estate road) the rental value is not being greatly affected. But where the nuisance or disability, while it lasts, is a severe one and is reducing the rental value substantially, then considerations of administrative convenience must clearly take second place.

The tribunal should, it seems to me, put itself in the situation of a reasonable man negotiating, at the date of the proposal, a tenancy for a year with reasonable expectation of its continuance. If the hypothetical tenant in these circumstances may reasonably be expected to offer the same rent as if the temporary nuisance or disability did not exist then the gross value will stay unreduced. But if, on the other hand, the hypothetical tenant would say: "But for this nuisance or disability the property is worth so much rental value, and when the nuisance or disability passes I shall be able to enjoy the full rental value; but meanwhile the nuisance or disability is sufficiently severe and/or is likely to continue sufficiently long as to reduce my rental bid for a year to year tenancy of the property," then the assessment will fall to be reduced accordingly.

It is against the above background that I turn to a consideration of the circumstances in the present case. At the date of the proposal the ratepayer had been suffering, for about six weeks, what was unquestionably a severe degree of nuisance in his occupation of 3a Cartmel Road, and (incidentally) without there being open to him any claim for statutory compensation. He apprehended, correctly, that this nuisance was likely to continue for a total period of at least six months. In financial terms, and adopting for the moment the tone of value as found by the local valuation court, the rental value of his house had become reduced from £181 to £157, and it was likely to continue at that reduced level for about a further five months before quiet occupation and a full rental value were to be again enjoyed.

Would a reasonable man, coming fresh to the scene and finding this situation, put in a reduced rental bid? or would he shrug off the nuisance? In my judgment he would not reduce his bid by as much as £24 because that would presume the nuisance to be continuing indefinitely, or at any rate for, say, a further 12 months or so from the proposal date. But neither would he, I think, shrug off the nuisance and offer the full rental value. In my opinion he would have depressed his bid by, say, £10, and that reduced bid would represent the reasonably expected rent on the rating hypothesis.

It is as well that I should add at this point that, in bringing into account the probability of the completion of the operations complained of, I have had regard to the judgments given in the House of Lords in *Dawkins (Almond) v Ash Brothers & Heaton Ltd* [1969] 2 All ER 246, and also to the decision in the later Lands Tribunal case of *Hounslow London Borough Council v Rank Audio Visual Ltd and Bryant (VO)* (1971) 217 EG 1225, 1421.

The sewerage work here (including the probability of its being completed within an estimated period of months) is a matter which (a) results from statutory powers, (b) impends from without and (c) represents a fact which is *essential to* the hereditament itself. In this respect the circumstances are distinguishable from those in the *Rank Audio Visual* case, where the continuance and completion of certain factory alterations was found to be a matter personal to the parties and therefore to be disregarded as a fact which was *accidental to* the hereditament itself.

The appeal, on the ground of incorrectness, is allowed. Adopting the tone of full value as now put forward by the valuation officer, of £173 in place of £181, the assessment is determined at gross value £163, rateable value £109. This determination represents what in my judgment is the correct gross value at the date of the proposal in February 1975. The necessary order will be issued directing that the valuation list be amended accordingly.

57 Selous Road

The other appeal may be dealt with more briefly. The owner-occupier of the adjoining house, 57 Selous Road, W H Brown, lodged a proposal on the same date as his neighbour and on similar grounds. The resulting appeal was heard by the local valuation court together with that for 3a Cartmel Road. The court decided that the assessment, then standing in the list at gross value £133, rateable value £84, should be:

Gross Value	*Rateable Value*
£133	£84 to 15.12.74
£113	£70 from 16.12.74 to 31.5.75
£133	£84 from 1.6.75

The valuation officer's grounds of appeal are that "the decision of the local valuation court is considered to be incorrect, and the assessment should be gross value £133, rateable value £84."

(This hereditament, since June 1 1975, has in fact been standing in the valuation list at the valuation officer's figures, in pursuance of the local valuation court's own decision.)

At the hearing before me the valuation officer put forward submissions similar to those for the other appeal houses, viz:

(a) That the decision of the local valuation court was bad in law in that it specified dates, usurping in that respect the prerogative of the rating authority;

(b) that the local valuation court, having reduced the assessment to £113 gross value, had no power to increase it to £133 gross value in the absence of any proposal for an increase;

(c) that the assessment should stay unreduced at £133 gross value because the nuisance suffered was for too transient a period to affect rental value on the rating hypothesis.

I accept the first two submissions. As to the third submission, and for the same reasons as set out for the other appeal house, I think a hypothetical tenant would have reduced his rental bid for 57 Selous Road by, say, £8 to £125.

The appeal, on the ground of incorrectness, is allowed. The assessment is determined at gross value £125, rateable value £78. The necessary order will be issued accordingly.

There being no respondent in either appeal, there can be no order as to costs.

I took note of an intimation by the valuation officer that a number of other proposals, similar to those lodged by Mr Boyle and Mr Brown and on similar grounds, have been received from ratepayers in other parts of Blackburn, and that these are being held up pending the outcome of the present appeals.

Annex

Summary of certain earlier decisions of the tribunal in cases relating to a temporary nuisance or disability

Lewis v Holman (VO) (1961) 179 EG 371 (John Watson FRICS)

The value of an attractive three-bedroom house in a high class residential area was held to be seriously affected by the use of adjoining land, ultimately lasting for nearly 18 months (a year after the proposal date), this being for huts, shelters, latrines, bulldozers etc.

"I must put myself, it seems to me, in the situation of a reasonable man negotiating a rent for the house on a yearly tenancy, with a prospect of the tenancy continuing. . . . He would realise that the nuisance would not go on for ever but would anticipate having to put up with it for a considerable time. I think he would say to himself: 'But for this nuisance I would offer so much for the property. While the nuisance persists I will reduce that amount by so much. When it ends I will make a new bargain."

Heath v Holman (VO) (1961) 179 EG 41 (R C G Fennell FRICS)

A 12-in main sewer, 18 ft deep, was laid across the garden of a bungalow. The work took from June to December 1960. The ratepayer's proposal was lodged in July 1960. The local valuation court reduced the assessment. The valuation officer did not appeal ("he regarded the reductions made by the local court, whose decisions he was not disputing, as of a temporary nature, the removal of which he would propose in due course"). The ratepayer's appeal, seeking a reduction to a nil assessment, was dismissed.

"Mr Heath claims that a tenant would pay nil rent, and presumably, therefore, that the landlord would be prepared to accept nil, but only, of course, for the period of disturbance which would certainly be estimated in this case at less than a year. It seems to me, however, that in such circumstances the practical approach would be something like this: The landlord and tenant would agree the annual rent as if the tenancy were to go on for an indefinite period, and either the landlord would then offer, and a bargain would be struck as to lump sum compensation for the temporary loss of quiet enjoyment, which loss the landlord could presumably claim as a damage to his estate interest, or, alternatively, the tenant would claim directly for the damage and nuisance suffered. The hypothetical rent would not therefore be affected. Statutory compensation under the Public Health Act does not, however, extend to temporary annoyance, but in this case the source of annoyance would run with the nuisance, and the limited duration would dispel any recognition within the rating hypothesis. . . . I have come to the conclusion that there is no cause to reduce the hypothetical rent for a short-term disturbance of this nature."

King v Dyer (VO) (1964) 189 EG 1095 (Sir William FitzGerald QC)

Building operations likely to last at least two years, causing noise, mud, dust and disturbance, were held not too transitory to affect the value of a finely-situated house across the road, the assessment of which was reduced by about 20 per cent.

"Now it is quite obvious that the owner of property must expect to put up with a certain amount of inconvenience from building operations, and repairs carried out by his neighbour, and those would not affect the rent provided they are transitory. That is the crux of the matter. What period of time can be regarded as transitory? I will not attempt to give a general answer to this question, but in this particular case I do not think the operations are so transitory as not to affect the rent. [The ratepayer] says they will continue for at least two years and I see no reason to quarrel with this estimate. My inspection endorses [the ratepayer's] view that because of these operations his house is suffering such inconvenience as would lower its rental value temporarily."

Hart v Gregory (VO) (1964) 191 EG 855 (R C Walmsley FRICS)

No reduction was made in the assessment of a converted flat in a residential area for demolition of six houses opposite, extending over a period of more than a year, with incessant noise for a particular period of three months from a pile-driver.

"[The valuation officer] regarded these demolitions as a transient nuisance, not affecting the gross value of the appeal flat. I think the valuation officer is right about this; having seen the work in progress. I recognise that the demolitions must be an annoyance in this quiet residential locality, but I do not think they would affect the rental bid of an intending tenant on a year-to-year tenancy."

Dennis and another v Gilmore (VO) LVC/1078, 1101/1965 (Sir Michael Rowe QC)

No reduction was made in the assessments of two pleasant modern suburban houses for nuisance caused by the tipping of a large quantity of domestic refuse on the field immediately beyond their back gardens.

"The appellants had to suffer the noise, smell, dust and flies inseparable from refuse tipping, however well it is done, in their immediate vicinity for two to three months and altogether for about eight months. If the hypothetical tenancy for rating was for a year only, it would have reduced the rent for both properties very considerably, but it is not and I feel I must accept the valuation officer's contention that the hypothetical tenant from year to year would not reduce the rent he would pay because of the tipping, because it was a transient matter."

Nightingale v Harrap (VO) LVC/87/1971 (R C G Fennell FRICS)

No reduction was made in the assessment of a house for disturbance by road and building developments just beyond the rear boundary of its garden, involving the noise of heavy traffic for not much more than a year and the working of heavy earth-moving and constructional equipment in the immediate neighbourhood for a much shorter time.

"While I sympathise with the ratepayer and his family over the discomforts of the 1970 summer, he has not been able to convince me that the exceptionally noisy and dusty conditions lasted for a substantial period, say of the order of 12 months and mainly from the date of the proposal; nor that the general noise of other development traffic was so unusual and unexpected that in June 1970 [the proposal date] the average prospective tenant would have anticipated such prolonged disturbance that he would have demanded a reduction in the normal hypothetical rent, that is the gross value, the subject of this appeal."

HAUPTFUHRER v THOMPSON (VO)
(LVC/623/1977)

(J D RUSSELL-DAVIS FRICS)

December 20 1977

Estates Gazette April 29 1978

(1977) 246 EG 407

Grade I listed 18th-century Palladian villa at Richmond, Surrey—Condition of restoration grant that certain rooms be open to public view at reasonable times by written appointment—Other disabilities, including vandalism—Public-viewing obligation does not exclude full private occupation—Basic gross value not excessive to any measurable degree against comparables—LVC's £200 reduction upheld—Treated as end-allowance

Both parties appeared in person.

A Giving his decision, MR RUSSELL-DAVIS said: The appellant Fred Hauptfuhrer, an American journalist, is the Crown lessee of Asgill House, Old Palace Lane, Richmond, Surrey. In March 1969 he acquired the lease for 83½ years at a ground rent of £100 per annum for a premium of £8,500 in consideration of the expenditure by him of the sum of £35,000 on its restoration and improvement; this work was effectively completed by November 1970.

Asgill House stands in its own grounds of about 1.29 acres at the foot of Old Palace Lane, which leads out of The Green to the River Thames public towpath known as Cholmondeley Walk to which the garden has a frontage, the boundary B being a low brick wall with parapet surmounted by an ornamental iron railing. It has a fine vista up-river to Richmond Bridge 600 yds away, but down-river the view is impeded by the iron railway bridge a few yards away and then Twickenham Bridge. The house itself stands some 15 ft above river level.

It was built about 1760 on the site of Henry VII's old palace for Sir Charles Asgill, Lord Mayor of London 1757-58, and was one of the first major projects of Sir Robert Taylor, later Architect of the King's Works. It is a Palladian villa having Oxford stone elevations and slated roofs in the Italian renaissance style of three storeys and basement, the principal C rooms being grouped around three sides of a vaulted hall and staircase.

"Beautifully Restored" Interior

It is one of the most notable houses in Richmond, in which there is an abundance of historic buildings. It is listed as a building of historic and architectural interest, Grade I, and particularly recorded are its entrance portico constructed in the later Ionic style, its river elevation with semi-octagonal central bay, flanked on either side by single windows, and the round-headed windows, false parapet and balustrade at ground D level. The interior is beautifully restored and has original wall paintings.

Mr Hauptfuhrer occupies as his own household the ground and first floors comprising entrance hall, cloaks and wc, octagonal dining-room and sitting-rooms, a further rectangular sitting-room, two rectangular bedrooms, bathroom and wc and well-modernised-and-equipped kitchen. The second floor is occupied by a caretaker and comprises three rooms, dressing and utility rooms and wc, and the basement provides storage and houses the central-heating boiler. It has a total area of some 6,207 sq ft. There is a separate garage and wc and fuel store in a detached building and a brick garden E shed.

It is an exceptional house in every way and the restoration was carried out with the aid of a grant from the Historic Buildings Council, subject to two important conditions, namely, that the house should be resided in at all times as a private dwelling and that the rooms on the ground and first floors should be open to the public for view at reasonable times by appointment. It is accepted that Mr Hauptfuhrer observes these conditions further than the written letter.

The Richmond Council accepts that the building is one of its architectural treasures and it is included in an illustrated brochure and in a handout called "Notes on Local History," F and drawings were included in an exhibition in 1972-73 in the Orleans House Gallery, all of which encourages public interest and requests to view the house from visitors from well outside the district.

Asgill House was entered in the 1973 valuation list as house and garden at gross value £1,600.

On June 21 1975 Mr Hauptfuhrer made a proposal seeking a reduction in assessment on the grounds, summarised, (a) that the reduction given five years ago (ie following the restoration) was not commensurate with the reduction in the size of the house, (b) the hereditament was not only a private house but also a Grade I building of public interest, accessible to the public and experts, and as one of the most important G single buildings in the borough, which public amenity imposes on him a heavy outlay for maintenance etc, and (c) that the rates payable should only be to the extent of the domestic occupation, excluding those parts which are open to the public.

On April 19 1977 the resulting appeal was heard by the Central London (South Western) Valuation Court, which awarded a reduction to gross value £1,400, recognising its accessibility to the public, the borough's exploitation of it without providing adequate protection, and the precedent of *Castle Howard Estate Ltd v Phillipson (VO)* (1959) 174 EG 1011. H

Mr Hauptfuhrer was aggrieved by this decision and appealed to the Lands Tribunal on May 27 1977 on the grounds that the court had recognised the disability but had failed to make commensurate reduction in assessment; he spoke to these grounds at the hearing before me.

The valuation officer, N A Thompson, responded and defended a gross value of £1,400, rateable value £1,138.

Ratepayer's Case

The appellant said that Asgill House was unique; there was no other building in Richmond to match it. Therefore, com- J parison on the basis of size with other houses of historic or architectural distinction in full private occupation was only of marginal relevance. Apart from having to be open to the public, Asgill House suffered disabilities which the others did not. The lack of view down-river was insignificant compared with the nuisance of noise and vibration of electric trains crossing an iron bridge a few yards away. Old Palace Lane was congested at weekends; his garage entrance was often obstructed and had been more than once damaged by clumsy drivers. Richmond attracted more than its share of disreputable crowds, especially in the summer, and Asgill House had suffered from vandalism from hooligans using the K towpath, even to having windows broken, and the throwing of litter into the garden was a common occurrence. Although the house itself stood above the level, flooding of the towpath happened perhaps 30 times a year and sometimes trapped unwary passers-by, who took refuge by climbing the wall into the garden and damaged shrubs.

He said that Richmond Council exploited to their own advantage the beauties of Asgill House by advertising it as an attraction to visitors to the town, but would or could do nothing to afford protection to it. He chose to make no charge for visitors, but the council made no contribution in any way at all. L From a residential point of view the house was a white elephant and he could not conceive of any tenant prepared to rent it at anything like £1,400 per annum. He thought a fair contribution to the rates would be £200 per annum which, translated to gross value, would mean an assessment of about £420.

In the *Castle Howard Estate* (first) appeal the valuation officer accepted a nominal sum of £10 gross value (the Lands Tribunal decided nil) in respect of those parts of the house open to the public, leaving a fair assessment only on the inhabited private apartments. Mr Hauptfuhrer asked for the same treatment. M

VO's Submissions

Mr Thompson said that the basic gross value of the house devalued at 20½p pe rsq ft on 6,207 sq ft to give £1,275, to which £125 was added for central heating, garage and garden shed, making £1,400. He compared this with from 25½p to 23p on the somewhat smaller houses 1, 2, 3 and 4 Maids of Honour Row, having from 4,463 to 5,240 sq ft, all terraced and having three storeys and semi-basement, and with 23.3p on 4,889 sq ft on the Old Palace, semi-detached of three

A storeys and no basement; all were close by in Richmond Green. Asgill House was also to be related with 30p on 6,645 sq ft for The Wick, enjoying an unrivalled situation on Richmond Hill, detached, of semi-basement and three upper floors. Maids of Honour Row had only small back gardens and that of the Old Palace was all in the front. He conceded that there was limited comparability with the six, but all are privately-occupied and listed houses. None of the assessments had been challenged. Only Asgill House had the restriction on the enjoyment of occupation by the obligation to make the house available to view by the public.

He appreciated the disadvantages of the relatively isolated B position facing the towpath, and of living at the bottom end of Old Palace Lane. The house itself was not prone to flooding and he dismissed flooding as an insignificant disability. Asgill House was a well-modernised, extremely habitable house of considerable local but not national importance, and despite the unwelcome actions of a few badly-behaved individuals, such a situation in Richmond would readily find a tenant at the amount of the gross value; the prestige and attraction of living there outweighed its drawbacks. Such a tenant would accept the necessity to show the house as a small price to be paid. The fact that the local authority gave no C protection or financial aid was not relevant.

The valuation officer said that Castle Howard in North Yorkshire was quite dissimilar. It was a national monument and a vast edifice of 52,500 sq ft and 130 rooms. It was quite unlettable except for a small part in private occupation. It attracted 70,000 visitors in 1961 and by 1963 was making a substantial deficit. Neither the first decision in 1958 nor the second appeal, *Mullineaux (VO) v Castle Howard Estate Ltd* (1968) 205 EG 140, had any relevance to Asgill House, which had only 18 rooms and visitors running to perhaps hundreds in a single year. He distinguished *Mills v Peak (VO)* (1965) 195 EG 551, which referred to a country mansion, Bisterne D House in Hampshire, of 28 main rooms, which, although occupied by the ratepayer for reasons of family tradition, was found for all practical purposes to be unlettable in the open market. In reviewing previous decisions the Lands Tribunal indicated considerations which may be helpful in determining gross value for mansions lettable in the open market which conform to the ordinary rules of supply and demand but are not helpfully applied to unlettable houses. The appeal was allowed to the extent of comparison being accepted with a smaller house of 18 rooms in the former category.

Disadvantages as well as Advantages

E I have inspected Asgill House, its grounds and locality and I have viewed from the exterior the other houses referred to. I accept the appellant's evidence that the house has its disadvantages as well as attractions. I also accept that the property suffers from a restriction, which may affect rental value, that the house had to be open to the public, but by written appointment only, and not to all-comers; this must, I think, limit the inconvenience even if Mr Hauptfuhrer goes further in the spirit than the letter of his obligation. But I cannot accept that such obligation excludes full private occupation of the very handsome rooms which are for public F view.

I heard no argument whether or not it is correct in law to take account of such restriction in valuing the occupation of the property for rating purposes; be that as it may, I think the valuation court has adequately recognised all the disabilities by its award of a reduction of £200 against the gross value in the list of £1,600. I cannot, however, find that "exploitation by the Richmond Council" without recompense has any bearing on the assessment. I agree with the valuation officer that the house, vacant and to let, would not lack tenants at a competitive market rent.

Thus I have treated the £200 as an end-allowance, and I turn now to a closer examination of the unreduced basic G gross value of £1,475 against the comparables.

A gross value of £1,475 is equivalent to 23¾p per sq ft in a house some 20 to 25 per cent larger than those in Maids of Honour Row, and therefore to be found in a different area/value band. But it is detached while the others are terraced; only 6 per cent is basement as against 18 per cent to 20½ per cent semi-basement at Maids of Honour Row. The Old Palace is semi-detached with no basement and devalues as a 25 per cent smaller house at 23.3p. Asgill House has a better situation and garden than all the comparables above, and in this can be more closely compared with The Wick, of slightly larger area and one which has a lower ground floor, one third H in area of the whole, albeit adapted to a high standard. This devalues at 30p per sq ft.

It is a matter of valuation judgment and I come to the conclusion that the valuation officer's relativities are on the whole sound and that the basic gross value is not excessive to any measurable degree. The fact that the Old Palace does not conform entirely with the pattern of value does not in itself prove Asgill House to be wrong.

Accordingly I uphold the determination of the valuation court and dismiss the appeal. The appellant will contribute £25 towards the valuation officer's costs.

Mr Hauptfuhrer acknowledged the consideration and J assistance he had received from the valuation officer in presenting his case. For my part, I would like to compliment the latter on his very full and helpful documentation.

LIVERPOOL CITY COUNCIL v BRITISH TRANSPORT HOTELS LTD AND GIBBONS (VO)

(LVC/876/1976)

(V G WELLINGS QC) K

December 19 1977

Estates Gazette May 6 1978

(1977) 246 EG 491

Meaning of "assessment"—British Railways Board hotel at Liverpool used as hotel until 1971 and only for storage purposes thereafter—Two proposals by ratepayers—First in 1975 that assessment be "deleted"—Second in 1976, and before LVC hearing on first, that assessment be "substantially reduced"—Agreement by VO, ratepayers and rating authority for reduction—LVC directs alteration L to agreed amount—Tribunal holds that "assessment" refers to totality of line of entries in valuation list, not merely to values—In any event, 1975 proposal could not be interpreted as contending for a reduction—Courses open to LVC were either to strike out line or to reject proposal altogether—Original values restored

A M Jarman (Solicitor to Liverpool City Council) appeared for the appellant rating authority; J A R Grove (instructed by the Solicitor to the British Railways Board) for the first respondent; D M Macdonagh (Solicitor for the Inland Revenue) for the second respondent, the valuation officer.

This is an appeal from the Merseyside local valuation court. M The only issue in the appeal is one of law as to the meaning of a proposal for an alteration in the valuation list. The hereditament is an hotel and premises known as the Exchange Hotel, Tithebarn Street, Liverpool 2. The hereditament was shown in the 1973 valuation list as having gross value £18,000 and rateable value £14,972. The owners of the hotel were the British Railways Board and the lessees were British Transport Hotels Ltd. On March 21 1975 the owners and lessees jointly made a proposal with respect to the hereditament. The proposal was in the usual form. Under the printed words "Name of the person or body by whom or on whose behalf the proposal is made" there were inserted the words "British Railways Board/British Transport Hotels Ltd." Under the printed words "Alteration which that person or body hereby proposes should be made in the valuation list" there were inserted the words "That the assessment should be deleted from the valuation list." Under the printed words "Grounds on which that alteration is proposed" there were inserted "That the assessment is incorrect, excessive and bad in law."

Against the space for "Name of occupier" there was no entry, but against "Name of owner" there were inserted the words "British Railways Board." The form was signed by N Brown "for R C J Bray FRICS, Estate Surveyor and Manager"; it was stated that the form was signed as owner and an address to which communications could be sent was given, namely that of the British Rail Property Board. A telephone number was also provided. The hereditament had been occupied and used by British Transport Hotels Ltd as an hotel until August 20 1971. It then ceased to be used as an hotel and thereafter has been used for storage purposes only.

In due course the valuation officer objected to the above-mentioned proposal (which I will call "the 1975 proposal"). On September 24 1976 the local valuation court heard the appeal of British Transport Hotels Ltd and on September 29 1976 gave its decision, from which the Liverpool City Council now appeals.

In the interval between the making of the 1975 proposal and the hearing in the local valuation court British Transport Hotels Ltd, on March 12 1976, made another proposal (which I will call "the 1976 proposal") with respect to the hereditament. The 1976 proposal was also in the usual form. In the space provided for "Name of the person or body by whom or on whose behalf the proposal is made" the name of British Transport Hotels Ltd was inserted. The "Alteration which that person or body hereby proposes should be made in the valuation list" was stated to be "That the assessment should be substantially reduced." The "Grounds on which that alteration is proposed" were stated to be "That the assessment is incorrect and excessive." The space for name of occupier was not filled in. The name of owner was stated to be British Transport Hotels. The document was signed by one Davies "for R C J Bray FRICS." The capacity in which the document was signed was stated to be agent for owner. The address of British Rail Property Board was provided as was a telephone number.

Although the valuation officer originally objected to the 1976 proposal he, British Transport Hotels and the Liverpool City Council, by an agreement in writing dated September 7 1976, agreed, in relation to the 1976 proposal and in accordance with section 72 of the General Rate Act 1967, that the valuation list should be altered and the appeal hereditament entered therein with the description and address "Premises, Exchange Hotel, Tithebarn Street" at gross value £1,000, rateable value £805.

That agreement, of course, preceded the hearing in the local valuation court of the appeal with respect to the 1975 proposal. The reasons for the decision, which was given on September 29 1976, show that evidence was given before the court supporting a gross value of £1,000 and the decision of the court was that the gross value in the valuation list should be reduced to £1,000 and the rateable value to £805. The question which I have to decide on this appeal is whether the local valuation court had power to decide the appeal in that way, and the answer to that question in turn depends upon the meaning which ought to be ascribed to the 1975 proposal.

The entitlement of a ratepayer to make a proposal for an alteration in the valuation list is contained in section 69 of the Act of 1967:

(1) Subject to subsection (6) of this section, any person (including a rating authority) who is aggrieved:

(a) by the inclusion of any hereditament in the valuation list; or

(b) by any value ascribed in the list to a hereditament or by any other statement made or omitted to be made in the list with respect to a hereditament; or

(c) in the case of a building or portion of a building occupied in parts, by the valuation in the list of that building or portion of a building as a single hereditament,

may at any time make a proposal for the alteration of the list so far as it relates to that hereditament.

Only para (a) and (b) are material to the present case.

Subsection (4) of section 76 describes the persons entitled to appear on the hearing of an appeal to a local valuation court. Subsection (5) provides that

after hearing the persons mentioned in subsection (4) of this section, or such of them as desire to be heard, the local valuation court shall give such directions with respect to the manner in which the hereditament in question is to be treated in the valuation list as appear to them to be necessary to give effect to the contention of the appellant if and so far as that contention appears to the court to be well founded; and the valuation officer shall cause the valuation list to be altered accordingly.

Reasons for LVC's Decisions

The local valuation court's reasons for their decision were given in writing and included the following observations:

After full consideration of the arguments put to us and after consulting the various authorities to which we were referred, we have concluded that we are empowered under section 76(5) of the General Rate Act 1967 and by the terms of the proposal the subject of this appeal, to direct an alteration to the existing assessment other than its total deletion. The upper and lower limits denoting the ambit of the appeal are, in our opinion, the present assessment and nil and we are competent to decide the appeal at any figure between those two extremes.

Standing alone, the stated alteration required is perhaps narrowed in its form by the use of the word "deleted" but taking guidance from the dicta of Scott LJ and Asquith LJ (though admittedly not directly in point) in the *Winchester* case we think we may look elsewhere in the proposal and find that the statement of grounds—that the assessment is incorrect and excessive and bad in law—removes the need for so strict an interpretation as that submitted by Mr Jarman and entitles us to consider an intermediate course.

We agree with the valuation officer that the word "assessment" refers to the valuation expressed by the figures in the valuation list, but this concurrence is not germane to our decision. In our opinion, without this restriction, we are entitled to take the broader view of the appellants' contention even if it applied to the complete entry.

Similarly with the distinguishing of "contention" from "intention." We think we must take the proposal on its face value and give what we consider a proper interpretation to the words used. While the original intention of the appellants might well be relevant in other circumstances we feel that within the bounds of the present dispute it is not. Although we need not decide the issue in this appeal, we are not at all certain that our decision would be different even if section 32 of the General Rate Act 1967 had been mentioned in the proposal. We are however strengthened in our view by its omission.

Section 76(5) requires us to give effect to the contention of the appellant if and so far as that contention appears to us to be well founded. We are of the opinion that a reduction in the assessment gives partial effect to that contention as we have interpreted it and to that extent it is well founded.

Parties' Submissions

All three advocates were agreed that the word "deleted" in the 1975 proposal means "deleted" and does not mean amendment by substitution. Mr Jarman, for the rating authority, submitted that the contention of the appellant derived from the 1975 proposal was that there should be complete erasure of the assessment and that by "assessment" was meant all the entries relating to the hereditament in the valuation list, that is to say, the entire line in the list relating thereto including description of the hereditament, its address, its gross value and rateable value and the serial numbers in the list appertaining thereto. Mr Jarman said that under section 76(5) of the Act of 1967 the local valuation court had a limited choice. They could give effect to the contention of the appellant by deleting the entire line in the list. If they thought that that was not an appropriate course, the only alternative open to them was to reject the proposal. They had, however, and quite wrongly, taken a third course: they had treated the contention of the appellant as though it called for an amendment of the gross and rateable values shown in the list by substituting a nil valuation therefor.

Mr Grove and Mr Macdonagh, for the respectively British Transport Hotels Ltd and the valuation officer, submitted that the primary meaning of the word "assessment" was a quantification of amount. The local valuation court had, they said, jurisdiction to find as they did because a nil valuation was sought in the 1975 proposal. The court did not accept that contention but they found in favour of a gross value of £1,000, which was a course they were entitled to take. Mr Macdonagh, additionally, said that even if the contention required erasure of the entire line in the list the court was nevertheless still competent to decide not to erase the entire line but to leave the description as it was and to alter the values.

I was referred to *Floyd v J Lyons & Co* [1897] 1 Ch 633; *R v Winchester Area Assessment Committee, ex p Wright* [1948] 2 KB 455; *re the Appeal of F Jones (VO)* (1951) 44 R & IT 70; *Ellerby v March (VO)* [1954] 2 QB 357; *Morecambe and Heysham Corporation v Robinson (VO)* [1961] 1 All ER 721. I was also referred to *Ryde on Rating* (13th ed) pp 722 *et seq* and to the meaning ascribed to "assessment" in the *Shorter Oxford English Dictionary*.

The word "assessment" in the 1976 proposal clearly was intended to refer to the values in the valuation list. There was a specific reference to reduction of values and the grounds relied upon were limited to an allegation that the assessment was incorrect and excessive, the words "bad in law" being omitted. Of course the 1976 proposal is, and can be, no guide to the meaning of the 1975 proposal; indeed the actual signatories are different. There is therefore no certainty that the word "assessment" in the 1975 proposal has the same meaning as in the 1976 proposal.

The grounds supporting the proposed alteration in the 1975 proposal (namely, "that the assessment is incorrect, excessive and bad in law") are well known in this tribunal. They appear in most of the proposals which come before it. Their ambit is as wide as it could possibly be and the intention behind their use is no doubt to cover all possible eventualities and to avoid the need for particularity. Whether taken together or individually they give no clue whatsoever as to the intention of the proposer. The word "excessive" no doubt refers to the values, presumably both of them. The word "incorrect" is capable of referring to the rateable value or to description. Presumably, "bad in law" does not refer to the values, but rather to the basis on which the valuation was made, to the description of the hereditament or to the fact that an assessment was made at all. It is a matter of extreme difficulty to determine from the 1975 proposal, even in the light of the circumstances surrounding its making (including, more particularly, the fact that the hereditament had ceased to be used as an hotel and was being used for storage purposes), precisely what was the contention of the appellant and whether the word "assessment" referred to the values or to the entire line of entries in the valuation list. It may be that the document ought to be held to be void for ambiguity or uncertainty, but the point was not taken before me. Further, since the proposal was not prepared by some mute, inglorious Pooter but by or on behalf of a Fellow of the Royal Institution of Chartered Surveyors in the employment of the British Railways Board, perhaps I ought to construe the 1975 proposal *contra proferentem*, but that point, too, was not taken before me.

Doing the best I can in the circumstances, I accept the submissions of Mr Jarman. It appears to me that the proposer was in effect saying: "The hereditament should not be the subject of an assessment at all." What his reasons may have been behind such an assertion is a matter of speculation. It is a possibility, but not, I think, a probability, that he thought that the hereditament, since it had ceased to be used as an hotel, was a railway hereditament, required to be excluded from the valuation list by section 32 of the Act of 1967. In truth I really do not know what his thinking was. Nevertheless he plainly intended something to be deleted or scored through. It would be unlikely that he intended the values to be struck out but the description or address to be left intact. I think therefore that the word "assessment" refers to the totality of the line of entries in the valuation list relating to the hereditament.

Interpretation of 1975 Proposal

If I am wrong in that opinion and if the word "assessment" is confined in meaning to the values shown in the list against the hereditament, the proposer nevertheless still intended that the values should be struck out. He was not in so many words inviting the substitution of a nil valuation. However, even if that had been his intention, it would not, in my judgment, justify the course which the local valuation court took of substituting for the figures shown in the list the figures of £1,000 and £805 respectively. In their reasons the court said: "The upper and lower limits denoting the ambit of the appeal are, in our opinion, the present assessment and nil and we are competent to decide the appeal at any figure between those two extremes." It appears to me from those words that the court was reading into the 1975 proposal a contention that the values ought to be reduced, the amount of the reduction not being specified. In my opinion that is an interpretation which cannot be put upon the 1975 proposal. The courses open to the court were either to strike out the line in the list relating to the hereditament or some part of it (if they took the limited view) or, if they took the view that striking out was not appropriate, to reject the proposal altogether.

For these reasons I allow the appeal and direct that the entries in the valuation list in respect of the hereditament of gross value £18,000 and rateable value £14,972 be restored, at all events until the 1976 proposal shall have taken effect.

The first and second respondents will jointly pay the appellant its costs of this appeal, such costs, if not agreed, to be taxed by the Registrar of the Lands Tribunal on the High Court Scale.

WOLVERHAMPTON ASSOCIATION OF RATEPAYERS v STAMP (VO)

(LVC/861/1977)

(SIR DOUGLAS FRANK QC, PRESIDENT)

November 21 1977

Estates Gazette May 27 1978

(1977) 246 EG 751

Jurisdiction of Lands Tribunal—Alleged contravention of s 76(3) of General Rate Act by LVC—One appeal heard before another settled—Not within tribunal's jurisdiction to quash LVC's decision or order rehearing—Matter for Divisional Court—Power of ratepayers' association to make proposal or to appeal to LVC "an open question"

C Knight, Secretary of the Wolverhampton Association of Ratepayers, appeared for the appellants; P L Ridd, Solicitor of Inland Revenue, for the respondent valuation officer, A J Stamp.

Giving his decision, THE PRESIDENT said:

The short facts which I require for deciding this preliminary point of law are that a proposal was made by the Wolverhampton Association of Ratepayers in respect of the appeal hereditament, 236 Stafford Road, Wolverhampton, that the appeal was heard by the local valuation court on May 19 1977 and that the decision of that court was to reduce the gross value from £570 to £366. That £204 reduction was at the instance of the valuation officer, apparently because the occupancy of the hereditament had been changed from that of an apartment house to a house in single occupation.

The association appealed to the Lands Tribunal and I will quote the grounds of the appeal verbatim:

> 1. The above is a test case to prove multiple decisions were bad in law.
>
> 2. Court procedure was in contravention of General Rate Act 1967, section 76(3) in that a further proposal was heard before earlier proposal for same hereditament was settled. We appeal for quashing of LVC decision, order on LVC to rehear.

It appears that on the same day the local valuation court had heard an appeal against an earlier proposal relating to the same hereditament and had given a decision on it on the same day, July 1 1977.

Mr Ridd, for the valuation officer, argued that the only power which the Lands Tribunal has under section 77 is to "give any directions which the local valuation court might have given" and under subsection (5) of section 76 the power of the local valuation court is to:

> "give such directions with respect to the manner in which the hereditament in question is to be treated in the valuation list as appear to them to be necessary to give effect to the contention of the appellant if and so far as that contention appears to the court to be well founded."

So argues Mr Ridd in effect that the Lands Tribunal can only deal with the quantum of the assessment and not interfere with any procedural irregularity there may have been, although he denies that there was any irregularity. The contravention of subsection (3) of section 76 alleged by the association is that one appeal was heard before another. That section, so far as it is material, provides:

> the court may hear and determine the appeal together with any appeal against objections to earlier proposals relating to that hereditament, but except as aforesaid the court shall not hear the first-mentioned appeal until all earlier proposals relating to the hereditament are settled.

As I understand it, from the notice of appeal, it is alleged that that rule is contravened. Whether it is contravened or not, and Mr Ridd denies it was, it seems to me not a matter for the Lands Tribunal. I agree with Mr Ridd that the provisions of section 77 concerning the power of the Lands Tribunal to give directions, should be construed strictly, and the reason is that conventionally it is the function of the Divisional Court to exercise supervision of tribunals and inferior courts in matters of jurisdiction and procedure, and therefore if the local valuation court errs in procedural matters, in other words acts in excess of jurisdiction or otherwise in accordance with its jurisdiction, then the matter is one for the Divisional Court and not for the Lands Tribunal.

That view seems to me to follow from the case of *Hulme v Bucklow Area Assessment Committee* [1940] 2 KB 255 where the powers of quarter sessions were similar to those now held by the Lands Tribunal. There the complaint was that the assessment committee had no power to enter and inspect the hereditament and that the county valuation committee had been allowed to appear and be heard through the county valuation officer, in each case without notice having been given. Quarter sessions held that they had no jurisdiction to hear an appeal on those grounds and the Divisional Court agreed with them. In the course of his judgment, Charles J said at p 258:

> . . . if the appellant wanted to attack the propriety, the regularity, or the conduct of the inquiry before the assessment committee, the proper way for him to do it would be to apply to this court for a writ of mandamus directing the committee to hear and determine in accordance with law.

Humphreys J, said at p 259:

> When one comes to look at the notice of appeal and the contentions on that notice which are stated in the case, it appears that the one thing which the appellant did not contend was that his premises were wrongly rated. There is not a word to be found in his contentions indicating any reason why his premises should not be rated at £40 gross and £30 rateable. There are eight contentions and every one of them has relation to the proceedings and conduct of the assessment committee and none to the subject-matter of the appeal.

Looking at the instant case, the notice of appeal says nothing at all about the valuations being wrong. That disposes of the second ground appeal in this case. The first ground (if it has any meaning) appears to be directly related to the second ground, but if it is not, then certainly it is no function of the Lands Tribunal to try a test case to prove that "multiple decisions were bad in law", whatever that expression may mean.

I considered whether I should invite Mr Knight, who appeared for the association, to amend the notice of appeal. Although he has apparently great experience in these matters, he did not make any application, so really that question does not arise. But, having gone into the facts it is unlikely that I would have granted leave, because at the local valuation court the association asked for a reduction of £57 in gross value but the local valuation court gave a reduction of £204, and as what the association were doing was contending for a lower assessment, they can hardly be heard to complain that they got even more than they asked for. Mr Knight says that the main complaint is against an inconsistency on the disability allowance, but it seems clear to me that the same percentage of disability allowance has been given in both cases. However, I am not concerned with that but with the preliminary point of law which is before me.

I find that the case for the valuation officer is well founded and that the grounds of appeal do not raise any matter which is within the jurisdiction of the Lands Tribunal to decide or direct, and that it is not within the jurisdiction of the Lands Tribunal to quash the decision of the local valuation court or to order a rehearing by that court. On that last point, the procedural error of the local valuation court alleged by the appellant can only be put right by a rehearing by that court and the Lands Tribunal has no power to order a rehearing by a local valuation court. I would add this: I have not dealt with the standing of the ratepayers' association in this matter and in particular the question whether they had any power to make any proposal or to appeal to the local valuation court, and that remains an open question which can be dealt with, if it falls to be dealt with, on another occasion. From what Mr Knight said, his main purpose in pursuing this appeal is to get that question determined, but as it does not arise, certainly on these preliminary hearings, it is unnecessary for me to determine.

A

MIDLAND BANK LTD v LANHAM (VO)

(LVC/28-29/1975)

(J H EMLYN JONES FRICS)

October 17 1977

Estates Gazette June 17 1978 and June 24 1978

(1977) 246 EG 1017 and 1117

Question fundamental to rating law—Rebus sic stantibus rule—Bank premises in Holborn—Certain upper floors used as bank training school—Whether one such floor and part of another, originally used as offices, to be valued as part of training school or as offices—Held that rebus sic B **stantibus rule restricts consideration of value not only to physical state but to uses within same mode or category as existing use—Further, that all alternative uses to which hereditament in its existing state could be put in real world, and which would be in minds of competing bidders, to be taken as being within same mode or category where existence of such competition can be established by evidence—Interpretation of planning factors "a matter of evidence and of expert valuation evidence"—S68(4) list of changes of circumstances neither comprehensive nor conclusive—Bank's appeal allowed at agreed value on basis of use as part of training school**

C Lord Silsoe QC and G R G Roots (instructed by Lovell White & King) appeared for the appellants; A P Fletcher (instructed by the Solicitor of Inland Revenue) for the respondent valuation officer.

Giving his decision, MR EMLYN JONES said: The question raised in this appeal is fundamental to the law of rating and concerns the assumptions to be made as to the character and use of a hereditament for the purpose of arriving at its gross value. In other words it involves a consideration of the *rebus sic stantibus* rule.

The hereditament in question is situated in part of a building known as Buchanan House, which stands at the D corner of Holborn and Fetter Lane in the City of London. It has a frontage of some 120 ft to Holborn and of about 96 ft to Fetter Lane. On the opposite corner of Fetter Lane bounded by Holborn and New Fetter Lane is the large Daily Mirror building and printing works comprising a 169-ft-high, modern-style, 11-storey office tower with lower blocks on the flanks housing the printing works. The western flank of this building, which fronts on to Fetter Lane, comprises a four-storey office block reducing to a 20-ft-high, single-storey building with entrances therein to the garage and printing works. On the opposite side of Holborn, at the corner of E Hatton Garden, there is a large vacant building site upon which stood the former Gamages store. Adjoining the vacant site to the west is the large Prudential building.

In an Office Zone

Buchanan House was built in 1957 in pursuance of planning permission authorising "the rebuilding of numbers 24-31 Holborn and numbers 75-81 Fetter Lane for use as offices with a bank on the ground floor." The appeal hereditament comprises part of the basement, the ground, first, second and fourth floors and part of the seventh floor. The remainder of the building is occupied and used as offices by a number F of different occupiers and I am not concerned with them. The site of Buchanan House is shown on the approved development plan as being within an office zone with shops to be permitted on the frontage to Holborn.

On May 29 1957 the Midland Bank (whom I shall refer to as the bank) was granted an underlease of the basement, ground, first and second floors for 99 years less 10 days at a rent of £30,000 per annum exclusive on full-repairing convenants plus an annual service charge in respect of cleaning and lighting common parts and central heating, hot water and maintenance of lifts. The bank proceeded to use most of the ground floor and basement as a branch bank. The rest of the accommodation on the ground floor extending G behind the bank with a separate entrance in Fetter Lane was used as a bank training school together with part of the basement and the whole of the first and second floors.

In August 1967 the bank took a lease of part of the seventh floor for the period from August 1 1966 to April 29 1992 at a rent of £2,600 per annum exclusive with no rent review until 1985. Under the terms of the lease the user was to be as offices, canteen and rest room, but there was a proviso that so long as the bank remained the lessee of the premises "the lessee may use the premises for the purpose of or ancillary to the lessee's trade or business."

In September 1968 the bank took an assignment of a lease H on the fourth floor, together with two car-parking spaces in the basement for a term of 13 years from Christmas 1964 at a rent of £17,950 per annum exclusive, without rent review.

These arrangements enabled the training school on the ground floor to be transferred to these additional floors, so that the bank was extended to cover the whole of the ground floor and basement. The position at the relevant date, therefore, was that the hereditament comprised, on the ground floor and basement, a branch bank with the customary banking hall, private offices, and machine rooms together with safe deposit and strong-room accommodation in the basement and staff rooms of the usual sort. J Access to the banking hall was from the corner of Holborn and Fetter Lane and there was a separate access to the rear of the banking hall from Fetter Lane. The access to the upper floors was separate and was gained from Holborn through a common entrance hall with stairways and a lift serving the seven upper floors of the building. The first and second floors were adapted by the bank to form part of the training school and there were certain minor alterations carried out to the fourth and part seventh floors in the bank's occupation in order to make them more convenient for their use as a training school.

K

Training School Accommodation

There are no unusual physical features about the occupation of the ground floor and basement as a bank and I need say nothing more about them. The training school is run by the bank for its own employees, who attend for courses, lectures, films and discussions. It forms the principal Midland Bank training school in the London area, including the home counties, but for some special courses for senior staff students may come from any part of England. The courses vary in length between four days up to a maximum of four weeks. Students attend at different stages of their career so that L some students may come back to the school on many different occasions. The daily average attendance is some 130 students, but at peak times there may be as many as 220 attending different courses. There is, additionally, a permanent staff of about 40. The seventh floor is used largely for trust company employees specialising in trustee and executor work. There is no residential accommodation, but the students are normally expected to remain on the premises between 9 am and 4.30 pm. There is a canteen on the first floor, a cinema on the fourth floor, and there are various rooms set aside as classrooms, locker rooms, cloakrooms and tutors' common rooms. All the classrooms have blackboards and many have M fixed screens or a white roller which enables slides to be shown on all courses. There are a large number of machines of different types, for example typewriters, calculators and so on, and one room is set out as a model bank with counter so that students may gain experience in counter duties. The fourth and seventh floors had been used as offices prior to their occupation by the bank as part of the training school and it is agreed that, vacant and to let, these parts of the hereditament are capable of being used as offices in their existing physical state. There is no such agreement with regard to the first and second floors but my own impression is that any

A alterations necessary in order to use those floors as offices would be very minor in character.

As I have indicated, the fourth floor was originally let as offices and the lessee at that time covenanted, *inter alia,* "to use the demised premises for the purpose of offices only with ancillary canteen and rest-room accommodation." This covenant was consistent with a clause in the bank's underlease whereby their superiors covenanted not to let any other portion of Buchanan House "or suffer any such portion thereof to be occupied or used as a bank nor for the purpose of carrying on the business of trusteeship and executorship" so long as the bank were in occupation of their portion of the B premises. Following the assignment of the lease of the fourth floor it became necessary to provide that the fourth floor could properly be used for the purposes of a bank or as offices or as a training school for bank employees only.

In June 1968 the bank obtained planning permission for a change of use of the fourth floor from offices to training school "for a period until December 31 1977." In a letter which accompanied the decision of the Common Council the town clerk informed the bank "that at the expiration of the period involved the corporation would be prepared to consider a further extension of time, depending on the circumstances then prevailing."

C No similar planning permission has been applied for in respect of the small part of the seventh floor occupied by the bank, although it is conceded by the valuation officer that planning permission would now be needed for a change of use back to offices.

The hereditament which I have described was entered in the valuation list for the City of London rating area on April 1 1973 with an assessment of gross value £302,000, rateable value £251,638. Following a proposal made on behalf of the appellants on September 7 1973 an appeal against this assessment was heard by the Greater London Central Local D Valuation Court on November 26 1974. By a decision dated December 11 1974 the court determined that the value ascribed to the hereditament should be altered to gross value £196,800, rateable value £163,972. Both the Midland Bank and the valuation officer appealed against this decision. The two appeals were subseqently consolidated, with the bank as appellants and the valuation officer as the respondent. In the event the valuation officer no longer seeks to defend the decision of the local valuation court. In the pleadings he contends for an assessment of gross value £193,500, and the appellants contend for an assessment of gross value £140,000.

E

Agreed Values

Prior to the hearing the parties reached agreement on certain alternative values which I here set out in the form in which they were presented to me:

Agreed Statement of Alternative Values

Definitions

Basis A—Value on the basis of use as part of a training school.
Basis B—Value on the basis of use as offices.

Four alternative values:

		Gross value	*Rateable value*
F 1.	If both the fourth and seventh floors are to be valued on basis A	£156,000	£129,972
2.	If the fourth floor is to be valued on basis A and the seventh floor on basis B	£159,500	£132,888
3.	If the fourth floor is to be valued on basis B and the seventh floor on basis A	£177,500	£147,888
4.	If both the fourth and the seventh floors are to be valued on basis B	£181,000	£150,805

The appellants now contend that the fourth and seventh floors are to be valued on the basis of their use as part of a G training school and that the assessment to be determined should be gross value £156,000. The valuation officer contends that the fourth and seventh floors are to be valued on the basis of their use as offices leading to an assessment of gross value £181,000. The local valuation court, in reaching their decision, had come to the conclusion that it would be incorrect to value the upper floors as if they were let and occupied as offices. The valuation officer, however, does not now challenge this finding in relation to the first and second floors and concedes that in any event the values determined by the local valuation court are excessive.

The respondent valuation officer, E R G Lanham FRICS, H gave evidence in support of the higher figure, which was derived from the level of values which emerged from an analysis of office rents in the locality. He pointed to the fact that the third, fifth, sixth and the remaining part of the seventh floors had at all material times been occupied and used as offices, as had the fourth floor and the remaining part of the seventh floor, before being taken over and used as part of the training school. In his opinion any hypothetical tenant could reasonably have anticipated that an application for a change of use for these two floors back to office user would have been successful and that office rentals accordingly provide him with the best evidence of value. He conceded, J however, that although the level of rents for training schools and offices were originally the same, there must have been some changes because there was evidence of a comparable training school let on a lower basis in a commercial building. For this reason he was prepared to accept the lower values "for use as a training school" contained in the statement of agreed alternative values.

Interpretation of "Rebus Sic Stantibus" Rule

The case for the appellants was put by Lord Silsoe. The K hereditament, he said, is to be valued as it is, *rebus sic stantibus*; and relying on the decision in *Fir Mill Ltd v Royton Urban District Council and Jones (Valuation Officer)* (1960) 175 EG 1029, 1111, he submitted that the *res* to be taken into account in valuations for rating related not only to physical factors but also to the mode or category of occupation of the hereditament. In support of this contention, he referred also to section 20 of the General Rate Act 1967, where the same words "mode or category of occupation" are used in the definition of relevant factors to be taken into account in the valuation of a hereditament "as at the time of valuation" under that section. In the present case, said L Lord Silsoe, the fourth and seventh floors, which are the only parts of the hereditament on which there is any dispute, are clearly occupied and used as part of a training school. The valuation officer conceded that such a use attracts a lower value, by the acceptance of the local valuation court's decision regarding the first and second floors. A training school use and an office use must therefore represent different modes and categories of occupation and it is not open to the valuation officer to contend that the fourth and seventh floors are to be valued as offices. In any event, says Lord Silsoe, even in the absence of the agreement on alternative values there can be no certainty that planning permission would be M obtained for use of the disputed parts of the hereditament as offices.

The issue as defined by Mr Fletcher, for the respondent valuation officer, is whether the assessment of the fourth floor and part seventh floor of the appeal hereditament is to be based on office rents or training school rents. He accepts that under the *rebus sic stantibus* rule, those floors must be assumed to remain in the physical state in which they existed at the date of the proposal and no alternative use may be taken into account unless they are capable of being used for such alternative use in their existing physical state, subject only to the *de minimis* rule.

Mr Fletcher recognises three possible interpretations of the *rebus sic stantibus* rule as it is applied to use. First, any alternative use may be considered for which the premises are physically suitable; secondly, any alternative use within the same mode or category as the actual use; and thirdly, no alternative use, other than the actual use of the actual occupier. This last alternative must be wrong, he submits—and indeed it forms no part of the appellants' case in these proceedings. I would merely say that from the cases to which I shall presently be referring there is ample authority for the view that the valuer is not in law restricted to a consideration of the value arising from the actual use made of the premises by the actual occupier.

In the present case, it is accepted that the disputed parts of the hereditament are physically capable of being used and occupied as offices. Mr Fletcher's first submission is that that is as far as the *rebus sic stantibus* rule can be taken. He says that the tribunal was wrong in *Fir Mill* and also in *London Transport Executive v Croydon and Phillips (VO)* (1974) 230 EG 987 in taking into account also questions of use. It must follow, if he be right, that the correct value to be applied to the disputed parts is a value derived from office rents.

Secondly, and in the alternative, Mr Fletcher submits that if it be correct to say that the *rebus sic stantibus* rule extends to cover mode and category of use, then on the facts of the present case the "office" mode is wide enough to cover the use of these premises as a commercial training school.

As for town planning, and the restrictions on the use of premises which are imposed by the Town and Country Planning Acts, Mr Fletcher submits that alternative uses may be taken into account—even though planning permission may be necessary, if it can be shown that such permission can reasonably be expected. On the question of onus, Mr Fletcher expresses certain reservations as to whether it lies on the party alleging a reasonable expectation of planning permission or the reverse.

Changes in Circumstances

In support of his main submission Mr Fletcher refers to the provisions of section 68(4) and section 79(1) and (2) of the General Rate Act 1967. Section 68(4) provides:

(4) In subsection (3) of this section, the expression "material change of circumstances" means a change of circumstances which consists of:

(a) the coming into occupation of a newly-erected or newly-constructed hereditament or of a hereditament which has been out of occupation on account of structural alterations; or

(b) a change in the value of a hereditament caused by the making of structural alterations or by the total or partial destruction of any building or other erection by fire or any other physical cause; or

(c) the happening of any event whereby a hereditament or part of a hereditament becomes, or ceases to be, not liable to be rated; or

(d) a change in the extent to which any railway or canal premises within the meaning of section 32 of this Act are occupied for non-rateable purposes within the meaning of that section; or

(e) property previously rated as a single hereditament becoming liable to be rated in parts; or

(f) property previously rated in parts becoming liable to be rated as a single hereditament; or

(g) a hereditament becoming or ceasing to be:

(i) a dwelling-house; or

(ii) a private garage or private storage premises within the meaning of Schedule 11 to this Act; or

(h) a hereditament being, in accordance with Schedule 13 to this Act, used to a greater or lesser extent for the purposes of a private dwelling or private dwellings,

and the expression "the time of valuation," in relation to a change of circumstances, means the time by reference to which the valuation officer prepared so much of the list as is affected by that change of circumstances.

The relevant part of subsection (1) of section 79 provides:

(1) . . . where an alteration is made in a valuation list . . . then, in relation to any rate current at the date when the proposal . . . was served on the valuation officer, or, . . . on the occupier of the hereditament in question, that alteration shall be deemed to have had effect as from the commencement of the period in respect of which the rate was made.

Subsection (2) provides:

(2) Notwithstanding anything in subsection (1) of this section, where an alteration in the valuation list:

(a) consists of the inclusion in the valuation list of a newly-erected or newly-constructed hereditament or an altered hereditament which has been out of occupation on account of structural alterations; or

(b) is made by reason of any of the events, specified in section 68(4)(b) to (h) of this Act,

the alterations shall have effect only as from the date when the new or altered hereditament comes into occupation or as from the happening of the event by reason of which the alteration is made as the case may be.

These provisions indicate, submits Mr Fletcher, that where there are changes in circumstances leading to a change in value, then following the making of a proposal, the alteration in the valuation list should take effect not from the beginning of the rate period but from the date when the change in value occurred. Since the alterations specified are—with the exception of the reference to dwelling-houses in section 68(4)(g) and (h)—largely attributable to physical changes, the inference is that changes in use *per se* do not give rise to a change in value. Because a change in use can, as with a physical change, take effect from an identifiable date it would be illogical for a change in value arising therefrom to take effect from the beginning of the rate period during which the proposal was made. This inference is all the stronger, says Mr Fletcher, when one looks at section 37(10) of the Rating and Valuation Act 1925 from which section 79(1) and (2) are derived; where the words used are:

. . . . Provided that, in the case of an amendment consisting of the inclusion in the valuation list either of a newly-erected or newly-constructed hereditament or an altered hereditament which has been out of occupation on account of structural alterations, or of the alteration in the valuation list of the value of any hereditament, where the value thereof has been affected by the making of structural alterations or by the total or partial destruction of any building or other erection by fire or any other physical cause, the amendment shall have effect only as from the date when the new or altered hereditament comes into occupation, or as from the happening of the event giving rise to the alteration of the value of the hereditament, as the case may be. . . .

Mr Fletcher relied on the following authorities: *R v Metropolitan Board of Works* (1868) LR 4 QB 15; *Metropolitan Board of Works v West Ham Overseers* (1870) LR 6 QB 193; *Chorlton-upon-Medlock Overseers v Chorlton-upon-Medlock Union* (1882) 47 LT 96; *London County Council v Erith* [1893] AC 562; *Metropolitan Water Board v Chertsey Union Assessment Committee* [1916] 1 AC 337; *Robinson Brothers (Brewers) Ltd v Houghton and Chester-le-Street Assessment Committee* (In the Court of Appeal) [1937] 2 KB 445.

Mr Fletcher also referred to a number of decisions of the Lands Tribunal, some of which he submitted supported his submissions, but he accepted that there were other decisions, including the *Fir Mill* case and *London Transport Executive v Croydon*, which were against him.

If I were to find against him on his main submission he then relied on his alternative submission, by which he would accept that the *rebus sic stantibus* rule restricted consideration of value not only to physical circumstances but to occupations within the same mode or category. At this point, his case thus follows very much the arguments advanced by Lord Silsoe except for the fact that the appellants claim that a training school use could not be said to be within the same mode or category as office use, whereas in Mr Fletcher's submission the training school in the present case clearly was. He suggested that there would be considerable difficulty in accepting a " training school " use as a mode or category of its own and sought to draw a distinction between the training school in the present case used to train bank employees in commercial duties very similar to those carried on in offices, and other forms of training school, for example, for engineering apprentices or catering staff, where different types of premises might be necessary.

Back to First Principles

In a consideration of this appeal it is necessary to go back to the first principles of rating law. I am tempted to echo the words of Lord Chelmsford in *Jones v Mersey Docks and Harbour Board* [1861-73] All ER Rep 78, in expressing surprise that after more than three-and-a-half centuries there still remains a necessity for interpreting the Poor Relief Act 1601 in order to determine the basis upon which " such competent sums of money " shall be raised from occupiers of lands within the parish.

The Poor Relief Act was not finally repealed until 1967, but its basic principle has been preserved, namely, that occupiers of land within the parish should contribute to the needs of the parish according to their ability. Thus the Parochial Assessment Act 1836 provided that the rate for the relief of the poor shall be made:

> upon an estimate of the net annual value of the several hereditaments rated thereunto; that is to say, of the rent at which the same might reasonably be expected to be let from year to year, free of all usual tenants rates and taxes, and tithe commutation rent-charge, if any, and deducting therefrom the probable average annual cost of repairs, insurance, and other expenses, if any, necessary to maintain them in a state to command such rent: provided always, that nothing herein contained shall be construed to alter or affect the principles or different relative liabilities (if any) according to which different kinds of hereditaments are now by law rateable.

In *R v Mast* (1795) 6 TR 154, Lord Kenyon ChJ said at p 155:

> The assessment for the relief of the poor ought to be so contrived, that each inhabitant should contribute in proportion to his ability, which is to be ascertained by his possessions in the parish. Every inhabitant ought to be rated according to the present value of his estate.

In *R v Liverpool Exchange* (1834) 1 Ad & El 465, after a reference to a number of earlier authorities, Littledale J said at p 474:

> These cases establish the principle, that the advantages attendant upon a building, either in respect of its situation or the mode of its occupation, are to be taken into the account in estimating its rateable annual value, wherever those advantages would enable the owner of the building to let it at a higher rent than it would otherwise fetch.

These then are the principles which by the proviso to section 1 of the Parochial Assessment Act 1836 remain unaltered by the passing of that Act.

In *R v Grand Junction Railway Company* (1843) 4 Ad & El 18 Lord Denman CJ said at p 36:

> To this state of facts we applied the established principles of rating—that the rate is to be on the occupier in respect of the beneficial nature of his occupation, in estimating which as to amount, or, to put it in other words, in ascertaining how much net rent such and such an occupation may be expected to command parish officers are to consider, not drily and only what would legally pass by a demise of it, but all the existing circumstances, whether permanent or temporary, wherever situated, however arising or secured, which would reasonably influence the parties to a negotiation for a tenancy as to the amount of rent to be asked or given.

and later at p 43:

> This supposition of a tenancy is only a mode of ascertaining the existing value of the occupation to the existing occupier.

Economic Inquiry

In the light of these authorities it is clearly established that the rate is properly to be levied on the occupiers of hereditaments and that their contribution is to be measured by the annual value of the hereditaments expressed in the form of the rent at which the hereditaments might reasonably be expected to let. Thus the hereditament is real, although in order to determine the rent it becomes necessary to envisage the parties to the negotiations as hypothetical. The answer to the question is therefore one of fact, to be determined on evidence and largely, of course, upon expert evidence. As Scott LJ said in the *Robinson Brothers* case in the Court of Appeal at p 470:

> The objective being the real value of the actual hereditament, the inquiry is primarily economic and not legal; it is only legal in so far as logical relevance is the measure of legal admissibility. That the value is to be expressed in the terms of the statutory definition of gross value is " nothing to the matter ", as Lord Holt would have said; that definition does not prevent the objective being the real value; it merely states certain data in regard to the terms of the hypothetical tenancy which are to be assumed in order to get a fixed instead of variable figure. Some terms must be assumed by every valuer in any calculation: the statute merely chooses the terms, in order that all resultant figures may be strictly comparable; but the statutory definition contains no provision which prevents real value being the one and only objective of the inquiry.
>
> On such an inquiry every factor, intrinsic or extrinsic, which tends to increase or decrease either demand or supply is economically relevant and is, therefore, admissible evidence for the assessment committee or its valuer or the quarter sessions on appeal to consider.

and later at p 477:

> The chain of reasoning, to which I have just given expression, is, in my view, justified by the authorities, particularly that no element of demand should be ignored.

The question to be answered, therefore, is what factors are to be taken into account in carrying out this economic inquiry to which Scott LJ refers? First of all it is appropriate to consider further passages in the judgment of Scott LJ. At p 468 he said:

> The hereditament to be valued under section 22 of the Act is always the actual house or other property for the occupation of which the occupier is to be rated, and that hereditament is to be valued as it in fact is—*rebus sic stantibus*.

and at p 469:

> In weighing up the evidence bearing upon value, it is the duty of the valuer to take into consideration every intrinsic quality and every intrinsic circumstance which tends to push the rental value either up or down, just because it is relevant to the valuation and ought therefore to be cast into the scales of the balance before he looks to see the resultant figure on the dial at which the pointer finally rests.

and a little later:

> Wherever the direct evidence test is not available, no fact which would in all the actual circumstances of the case tend to raise or lower the amount of rent likely to be given by probable competitors can be either irrelevant or inadmissible.
>
> The rent to be ascertained is the figure at which the hypothetical landlord and tenant would, in the opinion of the valuer or the tribunal, come to terms as a result of bargaining for that hereditament, in the light of competition or its absence in both demand and supply, as a result of " the higgling of the market." I call this the true rent because it corresponds to real value.

In *Metropolitan Water Board v Chertsey,* Lord Shaw of Dunfermline said at p 357:

> My Lords, land and buildings should be taken account of just as they are, to the eye of a hypothetical tenant—that is to say, with all their advantages, fitnesses, facilities, and capacities. That is the one cardinal rule of all valuation.

and at p 361 Lord Parker of Waddington said:

> No doubt in ascertaining the annual value of a hereditament, as indeed in ascertaining its capital value, it is necessary and proper to take into account its fitness or adaptability for every purpose for which it can legitimately be used, including the purpose for which it is actually used.

In *Port of London Authority v Orsett* [1920] AC 273 Lord Buckmaster said at p 305:

> The actual hereditament of which the hypothetical tenant is to be determined must be the particular hereditament as it stands, with all its privileges, opportunities, and disabilities created or imposed either by its natural position or by the artificial conditions of an Act of Parliament.

It appears, therefore, that we are faced with a paradox. On the one hand there is the requirement that the value of the occupation to the existing occupier is to be taken as the measure of his liability for a rate; on the other hand is the need to take into account all the characteristics of the hereditament.

The first limb seems to support the view that only the actual use is to be considered and there is some support for this in *dicta* of Lord Halsbury in *North and South West Junction Railway Co v Brentford Union* (1888) 13 Ap Cas 592 and of Lord Parmoor in *Great Western and Metropolitan Railway Companies v Kensington Assessment Committee* [1916] 1 AC 23. It is observed, however, that in both these cases the hereditament consisted of part of a railway line and no suggestion was made that any other use than the actual use was possible. As I have indicated, it is not suggested by counsel on either side in the present case that as a matter of law the inquiry as to value is limited to a consideration of the actual use by the actual occupier. Such a proposition would in my opinion be quite untenable. Quite apart from the authorities, the provisions of section 26(2) of the General Rate Act 1967 would be quite otiose if it were true that at common law a dwelling-house occupied by a person employed in agricultural operations as there expressed could be valued only on the assumption that it was so occupied.

The other limb at first sight appears to support Mr Fletcher's first submission, or at any rate would appear to go part of the way. It is to be observed, however, that in a reference to other uses Lord Parker qualified the alternatives by using the word "legitimately" and Lord Buckmaster referred to the artificial conditions created or imposed by an Act of Parliament.

Competition in the Real World

A closer consideration of the two propositions, however, suggests to me that the paradox is a false one, and that there is no inconsistency in adopting both of them at the same time. The value of the occupation to the occupier is measured by the rent which he is prepared to pay; but although the tenant and the landlord are hypothetical, the hereditament is real and the negotiations take place in the real world. One factor in the real world is that landlords seek to obtain the highest rent possible when they agree to let their properties. The hypothetical landlord, I apprehend, is motivated in the same way subject only, perhaps, to the restraints voluntarily accepted in pursuing a policy of enlightened estate management. In the real world the actual occupier finds that in making his rental bid he may well be in competition with other bidders. It is at this stage that the question of the mode or category of use arises, in the sense that potential tenants of premises within the same mode or category of use are competitors in the real world whose bids will influence the rent which the occupier can reasonably be expected to pay.

In this way it becomes necessary to carry out the economic inquiry to which Scott LJ referred in the *Robinson Brothers* case. Such an approach seems to be consistent with the remarks of Blackburn J in *Mersey Docks and Harbour Board v Liverpool Overseers* (1873) LR9 QB 84 where he said at p 96:

> Where the hereditament, or hereditaments of a similar kind, are in practice actually let at a rent, the amount of which is ascertained by what has been called "the higgling of the market," the application of this definition is easy and simple. Where the hereditaments are not in practice let the problem becomes more difficult. The facts and circumstances, which would be taken into consideration by those who in the case of a real tenancy do in the higgling of the market fix the rent, are to be taken into consideration, and on a view of all those the net annual value of the occupation is to be determined; . . .

In the *Robinson Brothers case,* Greer LJ said at p 461:

> It is quite clear that there are many cases where, if a hereditament were in the market, its annual value would necessarily be decided by the offer that could reasonably be expected from a special class of tenants who would desire to occupy the premises for the purpose of carrying on a successful business. Houses in Harley Street fetch a higher rent by reason of the fact that, if they were in the market, physicians and surgeons who desired to have the higher standing of an address in Harley Street would compete with one another, and the tenancy would go to the highest bidder in this class. If a valuer were asked to value a house in Harley Street, he would obviously have to take into consideration the price that could reasonably be expected to be given in the market in which the principal offerers were medical men who for reasons of their own were prepared to give a larger rent than any other class of tenant. It is pointed out by Channell J in his judgment in the *Bradford* case that the value of any property for the purposes of rating is to be determined by the "higgling" of the market. If it is to be so determined, it seems to me quite illogical to say that you are entitled to consider that some of the individuals so higgling in the market would give a higher price than any one else, and at the same time dismiss from your consideration what would be the highest rental established in the market. You are surely not entitled to say that the brewers should be allowed to go into the market and there compete with one another, and at the same time to exclude the price that they would be willing to pay as the result of such competition, and confine the value of the public-house to the point at which the ordinary tenant would refuse to go further in the competition. The value of any property may reasonably be expected to depend on the rent that would be produced by allowing all the contestants to bid against one another in the market. You cannot find the value due to the higgling of the market by excluding those who are nowadays the principal higglers—namely, the brewers.

Similarly in *Staley v Castleton Overseers* (1864) 10 LT 606, Blackburn J said at p 608:

> So here we have a mill with machinery which would be valuable to one with a capacity to use it. It is obvious, therefore, that it is property for which rent would be given by a millowner. But, if it be regarded further as a mere warehouse, in which machinery might be fitly and safely stored, it is still property for which a rental might and would be paid. And as the sum which would be paid for such a purpose has been agreed upon, I think that is the amount at which the mill should be rated.

At this point I must turn back to a consideration of Mr Fletcher's two submissions; first, that any alternative might be considered for which the premises are physically suited. This as a proposition I think is far too wide, unless it can be shown that any alternative uses have an effect on value. Quite clearly there must be many examples where premises are physically capable of being used for some other purposes but for which purpose there is no demand. For example, it is probably true to say that most dwelling-houses are structurally capable of being used as offices, or as consulting rooms, but it is only where there is evidence showing the existence of actual competition, as in the example of Harley Street suggested by Greer LJ, that there can be any effect on value to be taken into account.

Structural Alterations

There remains the question as to how far it is permissible to go in arriving at the annual value of a hereditament in its existing physical state having regard to possible adaptability for alternative use. In order to illustrate the point I take the example of a hereditament which might be the subject of structural alterations which would lead to an enhancement in value. That enhancement might in a particular case be substantially greater than the actual cost of carrying out the alterations. Quite clearly the full amount of the increased value can be taken into account after the structural alterations have been completed and the cost of the structural alterations cannot properly be taken into account until they are carried out. I should observe in passing that any such increase will take effect in accordance with the provisions of section 79(2) of the General Rate Act 1967 from the date when the structural alterations were carried out.

But what is the position where a tenant, in the knowledge and in anticipation of the increased value after the structural alterations have been carried out, is prepared to pay by way of rent under a tenancy from year to year some part of that enhanced value not covered by the cost of the structural alterations required to bring it about? This seems to me to be again entirely a question of evidence. Such evidence is, however, clearly admissible having regard to the *ratio decidendi* in *Consett Iron Co Ltd v North-West Durham Assessment Committee* [1931] AC 396. This approach also seems to be consistent with a number of other dicta. For example, in *R v Grand Junction Railway* (1843) 4 Ad & El 18 Lord Denman CJ said at p 36:

> What under the Act was possible by law, what in point of fact might be in future, however near, we thought immaterial as to the principle, *though very fit to be taken into account when making the calculation as to the quantum;* but in principle the parish officers were to look to the actual state and value of the occupation.

and in *R v Fletton* (1861) 3 E & E 450, Lord Cockburn CJ said at p 465:

> The true principle, according to which the value of the occupation to the hypothetical tenant contemplated by the Parochial Assessments Act is to be estimated, is to assume the continuance of those circumstances which constitute the value to the existing occupier, *unless it be made to appear that those circumstances are about to undergo a change.*

and in *Townley Mill Co Ltd v Oldham Assessment Committee* [1937] AC 419, Lord Maugham said at p 436:

> There was thus a hypothetical tenant and a hypothetical rent, but I think a real and concrete hereditament. . . . The hypothetical rent which the tenant could give was estimated with reference to the hereditament in its actual physical condition (rebus sic stantibus), and a continuance of the existing state of things was *prima facie* to be presumed. In my opinion it is possible to make very wide assumptions as to the hypothetical tenant if these propositions are borne in mind.

The italics in these three passages are mine.

Finally, in *Dawkins (VO) v Ash Brothers & Heaton Ltd* [1969] 2 AC 366, Lord Wilberforce at p 385 said:

> Let us start from the actual. The principle that the property must be valued as it exists at the relevant date is an old one, certainly older than the Parochial Assessments Act 1836. It has been spelt out in modern terminology in *Poplar Metropolitan Borough Assessment Committee v Roberts* [1922] 2 AC 93, 120, and in *Robinson Bros (Brewers) Ltd v Houghton and Chester-le-Street Assessment Committee* [1937] 2 KB 469 in passages which have been cited. The principle was mainly devised to meet, and it does deal with, an obvious type of case where the character or condition of the property either has undergone a change or is about to do so: thus, a house in course of construction cannot be rated: nor can a building be rated by reference to changes which might be made in it either as to its structure or its use. But it would surely be unreasonable to suppose that the hypothetical tenant is so inescapably imprisoned in the present that no anticipation is permitted of what is to come. Whether the test is what would influence his judgment, or what intrinsic qualities the hereditament possesses, any occupier in real life has to ascertain and to consider whatever may make his tenancy more or less advantageous over the period for which he takes it.

Earlier Decisions of Lands Tribunal

From a study of these authorities I can see nothing to cause me to reject, and much to support, the proposition that in determining the annual value of a hereditament, although one is required to arrive at the value of the occupation to the occupier one is also required to take into account such competition as is likely to arise in the real world from other possible tenants who must, nevertheless, be assumed to take the premises in their existing physical state. I have looked again at the many cases decided in the Lands Tribunal to which I was referred:

Fir Mill Ltd v Royton Urban District Council and Jones (VO) (1960) 175 EG 1029, 1111; *Grand Lodge of Mark Master Masons v Cane (VO)* (1957) 169 EG 63; *Sussex Club (Hove) Ltd v Burton (VO)* (1958) 171 EG 635; *Westbury (VO) v Wiggins & Co Ltd* (1959) 174 EG 235; *Hudson & Son Ltd v Rennick (VO)* (1959) 174 EG 327; *Tomlinson (VO) v Mills* (1959) 6 RRC 146; *Thomas Ware & Sons Ltd v Toovey (VO)* (1959) 175 EG 521; *Hope (VO) v Wellcome Foundation Ltd* (1960) 175 EG 991; *Dawkins (VO) v Warwickshire County Council* (1961) 178 EG 293, 365, 461; *Liverpool Corn Trade Association v Pritchard (VO)* (1962) 182 EG 893; *Spencer v Hampsher (VO)* (1968) 206 EG 1274; *Burling v Escott (VO)* (1968) 16 RRC 319; *Robinson v Le Grys (VO)* (1968) 208 EG 1322; *Duckworth (VO) v Manchester City Council* (1970) 16 RRC 235; *Prince v Baker (VO) and Widnes Borough Council* (1973) 228 EG 2297; *London Transport Executive v Croydon London Borough Council and Phillips (VO)* (1974) 230 EG 987.

In all these cases, with the exception of *Duckworth v Manchester Corporation, London Transport Executive v Croydon* and *Ware v Toovey*, it seems to me that the tribunal in effect rejected a lower value which could be attributable to the actual occupation taken by itself, in favour of a higher value influenced by the competition which in reality attached to the hereditament and which the actual occupier was taken as being prepared to pay.

I need refer in particular to only three of these cases. In *Fir Mill* the tribunal said at p 1033:

> We cannot believe the section contemplates any such restriction as this upon the hypothetical user. In our opinion only two assumptions are permitted. The first assumption is that the hereditament is vacant and to let—vacant in the physical sense and in the sense that the existing business has ended and any process machinery has been removed. The second assumption—and here we accept [counsel for the respondents] second proposition—is that the mode or category of occupation by the hypothetical tenant must be conceived, as the same mode or category as that of the actual occupier. A dwelling-house must be assessed as a dwelling-house; a shop as a shop, but not as any particular kind of shop; a factory as a factory, but not as any particular kind of factory. Some alteration to a hereditament may be, and often is, effected on a change of tenancy. Provided it is not so substantial as to change the mode or category of use, the possibility of making a minor alteration of a non-structural character, which the hypothetical tenant may be assumed to have in mind when making his rental bid, is a factor which may properly be taken into account without doing violence to the statute or to the inference we draw from the authorities.
>
> For all the foregoing reasons, subject to the qualifications we have indicated, we are of opinion that the doctrine *rebus sic stantibus* relates only to the physical characteristics of the hereditament.

Taking this passage as a whole it is quite clear that in the opinion of the tribunal the doctrine of *rebus sic stantibus* does not relate only to the physical characteristics of the hereditament because that expression of opinion is made subject to the qualifications contained in the previous paragraph where two basic assumptions are made; first, that the hereditament is vacant and to let, and secondly, that the mode or category must be "conceived as the same mode or category as that of the actual occupier."

At p 1115 of the report the tribunal went on to say:

> From the appellants, therefore, we have no evidence that we can accept that if at the material date a cotton mill was vacant and to let, a cotton spinner or a cotton weaver would have paid less rent for it than any other class of tenant. As we have said, there is no rental evidence for spinning mills at all. But when we examine the rental evidence for weaving mills, we are led inevitably to the opposite conclusion; it points to the fact that cotton weavers pay as much rent for their premises as other industrialists. Cotton spinning and cotton weaving are integral parts of the same industry. In the absence of any factual evidence to the contrary—and there is no such evidence—we draw the inference that what applies to cotton weavers applies equally to cotton spinners. In short, had we found in the appellants' favour on their basic point of law, our decisions upon values would have been the same.

As I understand this finding, what the tribunal are there saying is that, taking into account the evidence, such as it was, they were satisfied that cotton spinners and weavers would be prepared to match the competing bids of other industrialists.

On the other hand, in the *Croydon* case the tribunal came to the conclusion that the bus garage was probably *sui generis*, which could be interpreted as indicating that there was no evidence of any competing bid. In fact, of course, no evidence of any sort was put forward by the respondents in that case, nor were any submissions made on their behalf.

I would refer thirdly to the approach adopted by the valuation officer in *Dawkins v Warwickshire County Council*. This approach was approved by the tribunal and is set out on p 367 of the report as follows:

> Mr Dawkins was asked by the Solicitor-General what led him to value the land at £1,000 per acre. He replied that as the vendor had been the Leamington Borough Council, he had valued it in accordance with section 82 of the Town and Country Planning Act 1947, which concerns transactions in land between one local authority and another. The price, under that provision, must be the "prevailing use value of land in the locality in which the school is situated"; and he said he had taken into account sales of residential land in the neighbourhood. It was suggested to Mr Dawkins in cross-examination, having regard to the definition of effective capital value as the open-market value for the use to which the hereditament is put, that he ought to have valued the land for use as a school and nothing else. He replied to the contrary. Envisage, he said, a piece of land subject to a covenant restricting its uses to the site of a school. Assume that both vendor and purchaser recognise that the prevailing use of similar land nearby is residential, and they both know the price which the land in question will fetch for that use. The vendor, said Mr Dawkins, would be unwilling to sell below the residential price and would hold on rather than do so. The purchaser would be unwilling to pay a high residential price. In the outcome, they would strike a bargain and the purchase price would be the value of the land for residential purposes slightly deflated: that, he said, was the basis he had adopted.

This last example seems to provide a good illustration of what happens in practice and of the correct approach to be adopted.

In two other cases, the tribunal followed the same approach. In *McBroome v Cox (VO)* (1959) 4 RRC 220—and I quote from the headnote—the appeal hereditament was a retail shop with living accommodation. It was one of a terrace of 16 similar hereditaments all of which had at one time been retail shops, but of which only eight were still shops; of the remainder, two were empty, five had become factories, and one had become a tally shop. The tribunal (John Watson) said at p 223:

> I am sorry for the appellant and I am prepared to believe that his business has declined in recent years. But I must have regard to the rent which his premises would command in the open market if they were vacant and to let—not necessarily as a retail shop.

In *Vinall and Vinall v Covell (VO)* [1974] RA 177 the tribunal (V G Wellings QC) decided that a hairdressing salon should be valued as a shop—"not as any particular kind of shop." It is true that he expressly excluded the possibility of its use as an extension of a motor-car showroom on planning grounds. But in his decision he referred to the practical realities of the situation, and the reduction in assessment which he determined was closely linked to those actual circumstances.

In a recent case, *James v Clee (VO)* (1977) 241 EG 620 the tribunal (R C Walmsley) was concerned with a hereditament in a parade of shops used as an insurance office with living accommodation over. The rent paid was lower than that paid for shops in the parade, but the appeal hereditament had remained unlet for some time. The tribunal based the assessment on the rent instead of on the assessments of the shops; but as he pertinently observed, the case raised the question of the correctness of the shop assessments. In other words, the rent obtainable for the shop occupied as offices was apparently the highest rent which the landlord could obtain.

Finally, I should say that I have had an opportunity of reading the decision in *Henriques v Garland (VO)* (LVC/63 and 769/1976) which is being given today by V G Wellings. That decision appears to be entirely consistent with my findings in the present case.

I thus accept the submission of Lord Silsoe and Mr Fletcher's alternative submission so far as it goes—that in valuing for rating purposes it is necessary to take the hereditament as it stands *rebus sic stantibus:* that the value as thus restricted must relate to the hereditament in its existing physical state; and that the use of the hereditament must be taken to be within the same mode or category as the existing use. So much, I think, is established by authority, as I have interpreted it; and certainly now confirmed by statute law in section 20 of the General Rate Act 1967. Finally, all alternative uses to which the hereditament in its existing state could be put in the real world, and which would be in the minds of competing bidders in the market, are to be taken as being within the same mode or category, where the existence of such competition can be established by evidence.

No Help from Scottish Cases

I would briefly add three further observations. In the *Croydon* case there was a reference by the tribunal to the practice in Scotland and submissions were put to me by counsel on both sides on the relevance of Scottish authority. For my part I do not derive any help from the Scottish cases, mainly because most of them date from the period prior to 1956, when the law of rating in Scotland differed in material ways from the law in England and Wales. In the first place, a separate rate was levied on the proprietor in respect of his ownership, and secondly, where the subjects were rented the rents passing were broadly speaking taken as conclusive evidence of value.

The second observation is with regard to planning law in England and Wales. As the tribunal has said in earlier cases, it does not seem to be part of the role of the tribunal to determine issues of planning. Planning legislation can be regarded as a modern innovation when compared with the ancient institution of the poor rate, and quite clearly it cannot be taken to have amended any of the principles upon which valuations for rating are to be made. Its significance in my opinion is that the existence of planning restrictions and the necessity of obtaining planning permission for certain changes of use are all factors which affect the minds of potential tenants in the real world, and to that extent they must have an influence on value. But the interpretation of these factors remains a matter of evidence and of expert valuation evidence.

The third observation concerns Mr Fletcher's reliance on sections 68(4) and 79(1) and (2) of the General Rate Act 1967. In my judgment no such conclusion as he suggests can be drawn from these sections. As Lord Silsoe points out, the list of changes of circumstances in section 68(4) is by no means exhaustive of those events which are identifiable from a particular date. In general terms, physical changes taking place outside the hereditament which have an effect on value will lead to an alteration in value taking effect from the beginning of the rate period during which the proposal is made, in accordance with the provisions of section 79(1). The list contained in section 68(4) is therefore neither comprehensive nor conclusive.

"Fresh and Further Obstacles"

Up to this point I have dealt entirely in general terms with the basic propositions. But the issue before me is to determine in the light of the evidence presented to me the gross value of the hereditament at Buchanan House occupied by the appellants. If the route which I have followed hitherto has been dark and difficult I can only say that the final stages of the journey are fraught with fresh and further obstacles. In the first place, the hereditament itself is a most peculiar one. No one suggests that it is anything but a single hereditament, although it could be said that it consists of several structurally separate parts. The ground floor and basement are used as a conventional branch bank which in the Town and Country Planning (Use Classes) Order 1972 is comprised in the "office" class. As one hereditament the whole is to be assumed to be occupied for one purpose, although on one view of the facts the functional link between the bank and the training school is a tenuous one. Be that as it may, the hypothetical tenant is someone who requires for his occupation a bank extending to the ground floor and basement and, as is admitted, a training school on the first, second, fourth and part seventh floors.

This brings me to the second difficulty which I face in applying the general propositions of law to the facts in the present case. I have suggested that the inquiry to be made in order to determine the gross value of the appeal hereditament is primarily an economic one. The answer is to be derived from the evidence, but the only evidence as to value before me, apart from certain answers given to me by Mr Lanham, the valuation officer, is to be found in the agreed statement of alternative values.

No argument was developed on what distinction, if any, was to be made between the fourth and seventh floors. It is true that there were differences in tenure, and the use of the fourth floor as part of the training school was covered by planning permission. There was a certain amount of disagreement as to whether an office development permit would be required for a reversion to office use in these upper floors. But I am not persuaded that in the rating context there is anything to distinguish these two floors, the one from the other.

In effect, therefore, I am faced with two alternative values—a gross value of £156,000 if both the disputed floors are to be valued "on the basis of use as part of a training school" and a gross value of £181,000 "on the basis of use as offices." It is common ground that the four upper floors in the occupation of the bank are in fact used as part of a training school. It is also agreed that the first and second floors are to be valued on the basis of their use as part of a training school and thereby at a lower level of value. On the other hand the rents originally paid by the bank in respect of all these upper parts of the building, whether under the original underlease in 1957, or on assignment of the fourth floor in 1968, or on the grant of a lease of the seventh floor part in 1967, were clearly negotiated in a market where office tenants would be in competition. The inference to be drawn from that is that the rents payable by the bank were at those dates at least as high as the rents which office tenants would be prepared to pay. If that were not so, it is difficult to see why the landlords should agree to let those portions at rents lower than they could otherwise obtain.

Parties' Agreement "Not Entirely Clear"

The final difficulty which I face, therefore, is in the interpretation to be put on the formula used by the parties in their "agreed statement of alternative values." I commend the attempt to agree as much as possible of non-contentious items prior to the hearing; but unfortunately the meaning of the present agreement is not entirely clear. At one stage I was inclined to think that it would be necessary for me to give the parties an opportunity of calling further evidence in the light of my findings. But at the end of the day I think the matter is finally determined by the evidence of Mr Lanham, who told me that although the rents for training schools were formerly the same as those for offices, there must have been changes, because there was evidence of a comparable training school let in a commercial building at a rent lower than that which would have applied to offices.

It is of course difficult, as Mr Fletcher points out, to envisage the rents of training schools as being in a class of their own having regard to the different character of premises likely to be required for different training disciplines. But I must assume that Mr Lanham, in forming his opinion, had in mind a training school possessing similar characteristics to the bank training school in the present case; particularly since he accepts the lower values in respect of the first and second floors. The evidence of Mr Lanham can only be taken to mean that on the facts of this case there would be no competing bid from prospective office tenants which would at the relevant date influence the rent which a tenant of this hereditament, taken as a whole, would be expected to pay in respect of the upper floors, used as part of a training school. This rent, in respect of the whole hereditament and in terms of gross value is agreed at £156,000.

The appeal is accordingly allowed, and the assessment is to be determined at gross value £156,000, rateable value £129,972. The respondent valuation officer will pay the costs of the appellants, such costs failing agreement to be taxed by the Registrar of the Lands Tribunal on the High Court Scale.

For Product Safety Concerns and Information please contact our EU representative GPSR@taylorandfrancis.com Taylor & Francis Verlag GmbH, Kaufingerstraße 24, 80331 München, Germany

Batch number: 10732708

Printed by Printforce, the Netherlands